STUDY GUIDE AND WORKBOOK
AN INTERACTIVE APPROACH

for
Starr and Taggart's
BIOLOGY
The Unity and Diversity of Life
SEVENTH EDITION

JANE B. TAYLOR

Northern Virginia Community College

JOHN D. JACKSON

North Hennepin Community College

Wadsworth Publishing Company
I(T)P™ An International Thomson Publishing Company

Belmont • Albany • Bonn • Boston • Cincinnati • Detroit • London • Madrid • Melbourne
Mexico City • New York • Paris • San Francisco • Singapore • Tokyo • Toronto • Washington

Biology Publisher: Jack Carey
Assistant Editor: Kristen Milotich
Editorial Assistant: Kerri Abdinoor
Production Editor: Sheryl Gilbert
Print Buyer: Karen Hunt
Permissions Editor: Peggy Meehan
Copy Editor: Cathy Lundmark
Proofreader: Meg Korones
Compositor: Joan Olson, Wadsworth Digital Productions
Printer: Courier Companies, Inc./Kendallville
Cover Photo: © Thomas D. Mangelsen

COPYRIGHT © 1995 by Wadsworth Publishing Company
A Division of International Thomson Publishing Inc.
I(T)P The ITP logo is a trademark under license.

Printed in the United States of America
1 2 3 4 5 6 7 8 9 10—01 00 99 98 97 96 95

For more information, contact Wadsworth Publishing Company.

Wadsworth Publishing Company
10 Davis Drive
Belmont, California 94002, USA

International Thomson Editores
Campos Eliseos 385, Piso 7
Col. Polanco
11560 México D.F. México

International Thomson Publishing Europe
Berkshire House 168-173
High Holborn
London, WC1V 7AA, England

International Thomson Publishing GmbH
Königswinterer Strasse 418
53227 Bonn, Germany

Thomas Nelson Australia
102 Dodds Street
South Melbourne 3205
Victoria, Australia

International Thomson Publishing Asia
221 Henderson Road
#05-10 Henderson Building
Singapore 0315

Nelson Canada
1120 Birchmount Road
Scarborough, Ontario
Canada M1K 5G4

International Thomson Publishing Japan
Hirakawacho Kyowa Building, 3F
2-2-1 Hirakawacho
Chiyoda-ku, Tokyo 102, Japan

ISBN 0–534–21070–8

CONTENTS

PREFACE

Tell me and I will forget; show me and I might remember; involve me and I will understand.
—Chinese Proverb

The proverb outlines three levels of learning, each successively more effective than the method preceding it. The writer of the proverb understood that humans learn most efficiently when they *involve* themselves in the material to be learned. This study guide is like a tutor; when properly used it will increase the efficiency of your studies. The interactive exercises actively involve you in the most important terms and central ideas of your text. Specific tasks ask you to recall key concepts and terms and apply them to life; they test your understanding of the facts and indicate items to reexamine or clarify. They also estimate your next test score based on specific material. Most important, though, the biology study guide and text together help you make informed decisions about matters that affect your own well-being and that of your environment. In the years to come our survival will require administrative and managerial decisions based on an informed biological background.

HOW TO USE THIS STUDY GUIDE

Following this preface, you will find an outline that will show you how the study guide is organized and will help you use it efficiently. Each chapter begins with the title and a list of the 1- and 2-level headings in that chapter. The interactive exercises follow, wherein each main heading is labeled 1-I, 1-II, and so on, each with page references. Interactive exercises often begin with a convenient list of selected italicized words. This is followed by a list of page-referenced chapter terms under each main heading; space is provided next to each term for you to formulate a definition in your own words. Interactive exercises in each section variously include completion/short answer, true-false, fill-in-the-blanks, matching, choice, dichotomous choice, label-match, crossword puzzles, problems, labeling, sequencing, multiple-choice, and completion of tables. A self-quiz immediately follows the interactive exercises. This quiz is composed primarily of multiple choice questions, although sometimes we may present another examination device. Any wrong answers in the self-quiz indicate which portions of the text you need to reexamine. A series of chapter objectives and review questions follows each self-quiz section. These are tasks that you should be able to accomplish if you have understood the assigned reading in the text. Some objectives require you to compose a short answer or long essay, while others may require a sketch or supplying correct words. The sections called *Integrating and Applying Key Concepts* invite you to try your hand at applying major concepts to situations in which there is not necessarily a single pat answer; therefore, answers are not provided in the chapter answer section (except for a problem in Chapter 11). Your text generally will provide enough clues to get you started on an answer, but these sections are intended to stimulate your thoughts and provoke group discussions. At the end of each chapter you'll find answers to the interactive exercises and self-quizzes.

A person's mind, once stretched by a new idea, can never return to its original dimension.
—Oliver Wendell Holmes

STRUCTURE OF THE STUDY GUIDE

The following outline indicates how each chapter in this study guide is organized:

Chapter Number ————————————————————▶ **10**

Chapter Title ————————————————————▶ **MEIOSIS**

Chapter Outline ————————————————————▶ **ON ASEXUAL AND SEXUAL REPRODUCTION**

OVERVIEW OF MEIOSIS
  Think "Homologues"
  Overview of the Two Divisions

KEY EVENTS DURING MEIOSISI
  Prophase I Activities
  Metaphase I Alignments

FORMATION OF GAMETES
  Gamete Formation in Plants
  Gamete Formation in Animals

MORE GENE SHUFFLINGS AT FERTILIZATION

MEIOSIS AND MITOSIS COMPARED

Interactive Exercises ————————————————————▶ Where appropriate, selected italicized words from each chapter section are listed for student reference. The exercises are divided into numbered sections by main headings with page references; each section begins with page-referenced, boldfaced terms with space for student-composed definitions. Each section ends with groups of interactive exercises that vary in type and require constant interaction with the important chapter information.

Self-Quiz ————————————————————▶ Usually includes multiple-choice questions that sample important blocks of text information.

Chapter Objectives/Review Questions ————————▶ Tasks consisting of combinations of relevant objectives and questions to be answered.

Integrating and Applying Key Concepts ————————▶ Applications of text material to questions for which there may be more than one correct answer.

Answers to Interactive Exercises ————————————▶ Answers for all interactive exercises found
and Self-Quiz under main headings at the end of each chapter (with page references); answers to self-quiz.

1

METHODS AND CONCEPTS IN BIOLOGY

Interactive Exercises

Note: In the answer sections of this book, a specific molecule is most often indicated by its abbreviation. For example, adenosine triphosphate is ATP.

1-I. SHARED CHARACTERISTICS OF LIFE (pp. 2–8)

Selected Italicized Words

energy transfers, variations, hemophilia A

Boldfaced, Page-Referenced Terms

The page-referenced terms are important; they were in boldface type in the chapter. Compose a written definition for each term in your own words without looking at the text. Next, compare your definition with that given in the chapter or in the text glossary. If your definition seems to lack accuracy, allow some time to pass and repeat this procedure until you can define each term rather quickly (rapidity of answering is a gauge of the effectiveness of your learning).

(3) energy _____

(3) DNA _____

_____

(4) cell _____

(4) multicelled organism _____

(4) population _____

_____

(4) community _____

_____

(4) ecosystem _____

_____

(4) biosphere _____

_____

(4) metabolism _____

_____

(4) photosynthesis _____

_____

(4) ATP _____

_____

(5) aerobic respiration _____

_____

(5) producers _____

_____

(5) consumers _____

_____

_____

(5) decomposers _____

_____

(6) receptors _____

_____

(6) homeostasis _____

_____

(6) reproduction _____
(7) inheritance _____

_____

(7) mutations _____
(7) adaptive trait _____

_____

Fill-in-the-Blanks

(1) _____ is the capacity to make things happen, to do work. The special molecule, (2) _____, sets living things apart from the nonliving world; it contains instructions for assembling new organisms from "lifeless" molecules that contain carbon and a few other kinds of atoms. (3) _____ refers to the cell's capacity to extract and convert energy from its surroundings and use energy to maintain itself, grow, and reproduce. Many of the cells in a plant are capable of the process of (4) _____, in which they trap sun-

light energy and convert it into chemical energy that is used to build sugars, starch, and other substances. As a part of this process, molecules of the energy carrier (5) _____ are constructed. In most animals and plants, the stored energy in food molecules is released and transferred to (5) by (6) aerobic _____.

There is an interdependency among organisms in terms of energy flow through them; plants are the food (7) _____ and animals are the (8) _____. At some time, all organisms die; bacteria and fungi are the (9) _____ when they feed on tissues or remains and break down biological molecules to simple raw materials to be cycled back to producers.

With the help of (10) _____, organisms can sense changes in the environment and make controlled responses to them. Life's journey is able to continue because of (11) _____, the production of offspring.

Matching

Choose the most appropriate answer to match with each of the following terms.

12. ___ organ system
13. ___ cell
14. ___ community
15. ___ ecosystem
16. ___ molecule
17. ___ organelle
18. ___ population
19. ___ subatomic particle
20. ___ tissue
21. ___ biosphere
22. ___ multicellular organism
23. ___ organ
24. ___ atom

A. One or more tissues interacting as a unit
B. A proton, neutron, or electron
C. A well-defined structure within a cell, performing a particular function
D. Regions of the Earth where organisms can live
E. The smallest unit of life
F. Two or more organs whose separate functions are integrated to perform a specific task
G. Two or more atoms bonded
H. All of the populations interacting in a given area
I. The smallest unit of a pure substance that has the properties of that substance
J. A community interacting with its nonliving environment
K. An individual composed of cells arranged in tissues, organs, and often organ systems
L. A group of individuals of the same species in a particular place at a particular time
M. A group of cells that work together to carry out a particular function

1-II. LIFE'S DIVERSITY (pp. 8–10)

Selected Italicized Words

unity, diversity; family, order, class, phylum, division, kingdoms; artificial selection, natural selection, evolves

Boldfaced, Page-Referenced Terms

(8) species _____

_____

(8) genus _____

_____

(8) monerans _____

_____

(9) protistans _____

_____

(9) fungi _____

_____

(9) plants _____

_____

(9) animals _____

_____

(10) evolution _____

_____

(10) natural selection _____

_____

Fill-in-the-Blanks

Different kinds of organisms are referred to as (1) _____. A(n) (2) _____ is the first of a two-part name of each organism and encompasses all the species having perceived similarities to one another. The pronghorn antelope is known by the two-part name *Antilocapra americana*; *Antilocapra* is the (3) _____ name, and *americana* is the (4) _____ name.

Complete the Table

5.   Fill in the table below by entering the correct name of each kingdom of life described.

Kingdom	Description
a.	Multicelled consumers
b.	Complex single cells; producers or consumers
c.	Mostly multicelled decomposers
d.	Internally simple, single cells; either producers or consumers
e.	Mostly multicelled producers

Sequence

Arrange in correct hierarchical order with the largest, most inclusive category first and the smallest, most exclusive category last. This exercise classifies a plant with the common name of "False Solomon's Seal." Refer to p. 8 in the text and Appendix I, pp. A1–A3.

6. ___
7. ___
8. ___
9. ___
10. ___
11. ___
12. ___

A. Class: Monocotyledonae
B. Family: Liliaceae
C. Genus: *Smilacina*
D. Kingdom: Plantae
E. Order: Liliales
F. Division: Anthophyta
G. Species: *racemosa*

Short Answer

13. Compose a short definition for evolution (p. 10). _____

_____

_____

True-False

If the statement is true, place a T in the blank. If the statement is false, make it correct by changing the underlined word(s) and writing the correct word(s) in the answer blank.

_____ 14. The most inclusive (largest) taxonomic category is the <u>phylum</u>.

_____ 15. There is a <u>larger</u> number of different species in a class than in an order.

_____ 16. If some organisms in a population inherit traits that lend them a survival advantage, they would be <u>less</u> likely to produce offspring.

_____ 17. Darwin described the selection occurring in pigeons as <u>natural</u> selection.

_____ 18. All bacteria are single-celled, <u>possess</u> nuclei, and are therefore monerans.

1-III. THE NATURE OF BIOLOGICAL INQUIRY (pp. 10–13)
THE LIMITS OF SCIENCE (pp. 12–13)

Selected Italicized Words

subjective

Boldfaced, Page-Referenced Terms

(11) hypotheses _____

_____

(11) prediction _____

_____

(11) theory _____

_____

Complete the Table

1. Complete the following table of concepts important to understanding the scientific method of problem solving. Choose from scientific experiment, variable, prediction, control group, hypothesis, and theory.

Concept	Definition
a.	an educated guess about what the answer (or solution) to a scientific problem might be
b.	a statement of what one should be able to observe in nature if one looks; the "if-then" process
c.	a related set of hypotheses that, taken together, form a broad explanation of a fundamental aspect of the natural world
d.	a carefully designed test that manipulates nature into revealing one of its secrets
e.	used in scientific experiments to evaluate possible side effects of a test being performed on an experimental group
f.	the control group is identical to the experimental group except for the key factor under study

Sequence

Arrange the following steps of the scientific method in a possible chronological sequence. Find the letter of the first process and write it next to 2. The letter of the final process is written next to 8.

2. ___
3. ___
4. ___
5. ___
6. ___
7. ___
8. ___

A. Develop one or more hypotheses about what the solution or answer to a problem might be.
B. Devise ways to test the accuracy of predictions drawn from the hypothesis (use of observations, models, and experiments).
C. Repeat or devise new tests (different tests might support the same hypothesis).
D. Make a prediction using the hypothesis as a guide.
E. If the tests do not provide the expected results, check to see what might have gone wrong.
F. Objectively report the results from tests and the conclusions drawn.
G. Identify a problem or ask a question about nature.

Labeling

Assume that you have to identify what object is hidden inside a sealed, opaque box. Your only tools to test the contents are a bar magnet and a triple-beam balance. Label each of the following with an O (for observation) or a C (for conclusion).

9. ___ The object has two flat surfaces.

10. ___ The object is composed of nonmagnetic metal.

11. ___ The object is not a quarter, a half-dollar, or a silver dollar.

12. ___ The object weighs x grams.

13. ___ The object is a penny.

Completion

14. Questions that are _____ in nature do not readily lend themselves to scientific analysis.

15. Scientists often stir up controversy when they explain a part of the world that was considered beyond natural explanation—that is, belonging to the _____.

Self-Quiz

___ 1. About 12 to 24 hours after the last meal, a person's blood sugar level normally varies from about 60 to 90 milligrams per 100 milliliters of blood, though it may attain 130 mg/100 ml after meals high in carbohydrates. That the blood sugar level is maintained within a fairly narrow range despite uneven intake of sugar is due to the body's ability to carry out _____.

    a. prediction
    b. inheritance
    c. metabolism
    d. homeostasis

___ 2. As an eel migrates from saltwater to freshwater, the salt concentration in its environment decreases from as much as 35 parts of salt per 1,000 parts of seawater to less than 1 part of salt per 1,000 parts of freshwater. The eel stays in the freshwater environment for many weeks because of its body's ability to carry out _____.

    a. adaptation
    b. inheritance
    c. puberty
    d. homeostasis

___ 3. A boy is color-blind just as his grandfather was, even though his mother had normal vision. This situation is the result of _____.

    a. adaptation
    b. inheritance
    c. metabolism
    d. homeostasis

___ 4. The digestion of food, the production of ATP by respiration, the construction of the body's proteins, cellular reproduction by cell division, and the contraction of a muscle are all part of _____.

    a. adaptation
    b. inheritance
    c. metabolism
    d. homeostasis

___ 5. Which of the following does *not* involve using energy to do work?

    a. Atoms being bound together to form molecules
    b. The division of one cell into two cells
    c. The digestion of food
    d. None of these

_____ 6. The experimental group and control group are identical except for _____.

    a. the number of variables studied
    b. the variable under study
    c. two variables under study
    d. the number of experiments performed on each group

_____ 7. A hypothesis should _not_ be accepted as valid if _____.

    a. the sample studied is determined to be representative of the entire group
    b. a variety of different tools and experimental designs yield similar observations and results
    c. other investigators can obtain similar results when they conduct the experiment under similar conditions
    d. several different experiments, each without a control group, systematically eliminate each of the variables except one

_____ 8. The principal point of evolution by natural selection was that _____.

    a. it measures the difference in survival and reproduction that has occurred among individuals that differ from one another in one or more traits

    b. even bad mutations can improve survival and reproduction of organisms in a population
    c. evolution does not occur when some forms of traits increase in frequency and others decrease or disappear with time
    d. individuals lacking adaptive traits make up more of the reproductive base for each new generation

_____ 9. The purpose of experiments is to _____.

    a. identify a problem
    b. develop a hypothesis
    c. test predictions
    d. provide conclusions
    e. reject as many hypotheses as possible

_____ 10. The least inclusive of the taxonomic categories listed is _____.

    a. family
    b. phylum
    c. class
    d. order
    e. genus

Chapter Objectives/Review Questions

This section lists general and detailed chapter objectives that can be used as review questions. You can make maximum use of these items by writing answers on a separate sheet of paper. Fill in answers where blanks are provided. To check for accuracy, compare your answers with information given in the chapter or glossary.

Page	Objectives/Questions
(3)	1. _____ is a capacity to make things happen, to do work.
(3)	2. Name the special molecule that sets living things apart from the nonliving world; generally explain the functions of this molecule.
(4)	3. A _____ is the basic living unit.
(4)	4. Distinguish between single-celled organisms and multicelled organisms.
(4)	5. Arrange in order, from smallest to largest, the levels of organization that occur in nature. Define each as it is listed.
(4)	6. _____ means "energy transfers" within the cell.
(4)	7. Organisms use a molecule known as _____ to transfer chemical energy from one molecule to another.

(4–5)   8. Contrast the general functions of the processes known as photosynthesis and aerobic respiration.

(5)   9. Explain how the basic processes of energy capture and release create interdependencies among organisms.

(6)   10. Fully describe the role of receptors in the life of organisms.

(6)   11. _____ is defined as a state in which the conditions of an organism's internal environment are maintained within tolerable limits.

(6)   12. _____ is the means by which each new organism arises.

(7)   13. Explain the origin of trait variations that function in inheritance.

(7)   14. A trait that assists an organism in survival and reproduction in a certain environment is said to be _____.

(8)   15. Explain the use of genus and species names by consideration of your Latin name, *Homo sapiens*.

(8–9)   16. List the five kingdoms of life that are currently recognized by most scientists; tell generally what kinds of organisms are classified in each kingdom.

(8)   17. Arrange, in order from greater to fewer organisms included, the following categories of classification: class, family, genus, kingdom, order, phylum, and species.

(10)   18. As organisms move through time by successive generations, the character of populations changes; this is called _____.

(10)   19. Evolution by _____ accounts for the great diversity of organisms.

(10)   20. Darwin used _____ selection as a model for natural selection.

(10)   21. Define natural selection and briefly describe what is occurring when a population is said to evolve.

(11)   22. Outline a set of six steps that might be used in the scientific method of investigating a problem.

(11)   23. Describe what is meant by a theory; cite an actual example.

(12)   24. Tests performed to reveal nature's secrets are called _____.

(12)   25. Explain why a control group is used in an experiment.

(12)   26. Generally, members of a control group should be identical to those of the experimental group except for the key factor under study, the _____.

(12–13) 27. Explain how the methods of science differ from answering questions by subjective thinking and systems of belief.

(13)   28. The external _____, not internal _____, must be the testing ground for scientific beliefs.

Integrating and Applying Key Concepts

1. Humans have the ability to maintain body temperature very close to 37°C.
   a. What conditions would tend to make the body temperature drop?
   b. What measures do you think your body takes to raise body temperature when it drops?
   c. What conditions would cause body temperature to rise?
   d. What measures do you think your body takes to lower body temperature when it rises?
2. Do you think that all humans on Earth today should be grouped in the same species?
3. What sorts of topics are usually regarded by scientists as untestable by the kinds of methods that scientists generally use?

Answers

Interactive Exercises

1-I. SHARED CHARACTERISTICS OF LIFE (pp. 2–8)
1. Energy; 2. DNA; 3. Metabolism; 4. photosynthesis; 5. ATP; 6. respiration; 7. producers; 8. consumers; 9. decomposers; 10. receptors; 11. reproduction; 12. F; 13. E; 14. H; 15. J; 16. G; 17. C; 18. L; 19. B; 20. M; 21. D; 22. K; 23. A; 24. I.

1-II. LIFE'S DIVERSITY (pp. 8–10)
1. species; 2. genus; 3. genus; 4. species; 5. a. Animalia; b. Protista; c. Fungi; d. Monera; e. Plantae; 6. D; 7. F; 8. A; 9. E; 10. B; 11. C; 12. G; 13. Evolution occurs when the features that characterize populations of organisms change through successive generations; 14. kingdom; 15. T; 16. more; 17. artificial; 18. lack.

1-III. THE NATURE OF BIOLOGICAL INQUIRY (pp. 10–13)
THE LIMITS OF SCIENCE (pp. 12–13)
1. a. Hypothesis; b. Prediction; c. Theory; d. Scientific experiment; e. Control group; f. Variable; 2. G; 3. A; 4. D; 5. B; 6. E; 7. C; 8. F; 9. O; 10. O; 11. C; 12. O; 13. C; 14. subjective; 15. supernatural.

Self-Quiz

1. d; 2. a; 3. b; 4. c; 5. d; 6. b; 7. d; 8. a; 9. c; 10. e.

2

CHEMICAL FOUNDATIONS FOR CELLS

INTERACTIVE EXERCISES

2-I. ORGANIZATION OF MATTER (pp. 16–21)

Selected Italicized Words

trace elements, net charge

Boldfaced, Page-Referenced Terms

(19) molecule _____

_____

(19) compound _____

_____

(19) protons _____

(19) electrons _____

(19) neutrons _____

(19) atomic number _____

(19) mass number _____

(19) isotopes _____

_____

(19) radioisotopes _____

_____

Fill-in-the-Blanks

$^{12}_{6}C$ is a shorthand method for recording the symbol of the element, the atomic mass, and the atomic number. C is the symbol for the element (1) _____; 12 is the atomic (2) _____; there are (number) (3) _____ neutrons and (number) (4) _____ protons; there are (number) (5) _____ electrons in the atom.

6. Only four kinds of elements compose most of the human body. List them in any order: _____, _____, _____, and _____.

Complete the Table

7. Complete the following table (refer to Tables 2.1 and 2.2 in the text) by entering the name of the element and its symbol in the appropriate spaces.

Element	Symbol	Atomic Number	Mass Number	Electron Distribution First Shell	Second Shell	Third Shell
a.		1	1	1		
b.		6	12	2	4	
c.		7	14	2	5	
d.		8	16	2	6	
e.		15	31	2	8	5
f.		16	32	2	8	6

Identification

8. Following the model below (number of protons and neutrons shown in the nucleus), construct the indicated atoms of the elements listed in question 7. Show electrons in this form: (2e⁻).

He

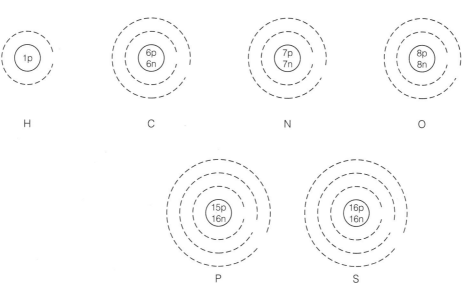

| H | C | N | O |

P   S

True/False

If the statement is true, write a T in the blank. If the statement is false, make it correct by changing the underlined word(s) and writing the correct word(s) in the answer blank.

9. _____ The number of <u>protons</u> in the nucleus of an atom is equal to the number of electrons outside the nucleus.

10. _____ <u>Protons</u> move rapidly around the nucleus and occupy most of the atom's volume.

11. _____ If two particles bear the same electric charge, they <u>attract</u> each other.

12. _____ Carbon 12 and carbon 14 are identical except for the number of <u>neutrons</u> in each atom.

13. _____ <u>Proton</u> activity is the basis for the organization of materials and the flow of energy through the living world.

Matching

Choose the one most appropriate answer for each.

14. ___ atom
15. ___ atomic number
16. ___ electron
17. ___ element
18. ___ electric charge
19. ___ mass number
20. ___ compound
21. ___ isotope
22. ___ trace elements
23. ___ matter
24. ___ molecule
25. ___ neutron
26. ___ proton
27. ___ radioisotopes

A. An uncharged subatomic particle
B. The number of protons in the nucleus of one atom of an element
C. The smallest neutral unit of an element that shows the chemical and physical properties of that element
D. That which occupies space and has mass; solids, liquids, or gases
E. The number of protons and neutrons in the nucleus
F. Two or more atoms linked together by one or more chemical bonds
G. Necessary elements present in insignificant amounts in living things
H. A substance in which the relative proportions of two or more elements never vary
I. A positively charged subatomic particle
J. A term applied to unstable isotopes
K. A negatively charged subatomic particle
L. Ninety-two different types occur in nature
M. A form of an element, the atoms of which contain a different number of neutrons from other forms of the same element
N. Pushes away similar particles but attracts oppositely charged particles

2-II. THE NATURE OF CHEMICAL BONDS (pp. 22–23)

Selected Italicized Words

energy relationship, lowest energy level, higher energy levels, shells

Boldfaced, Page-Referenced Terms

(22) chemical bond _____

_____

(22) orbitals _____

_____

Short Answer

1. State how many electrons are required to fill the one (and only) shell of the hydrogen atom and how many electrons are required to fill the outermost shell of all other atoms. _____

_____

_____

2. State the conditions of atomic structure existing when atoms tend to react with other atoms to form chemical bonds. _____

_____

_____

Matching

Choose the one best answer for each.

3. ___ shells
4. ___ lowest energy level
5. ___ orbitals
6. ___ chemical bond
7. ___ higher energy levels

A. Regions of space around an atom's nucleus where electrons are likely to be at any one instant
B. An energy relationship
C. A series of orbitals arranged around the nucleus
D. Electrons farther from the nucleus than the first orbital
E. Any electron in the orbital closest to the nucleus

2-III. IMPORTANT BONDS IN BIOLOGICAL MOLECULES (pp. 24–25)

Selected Italicized Words

nonpolar covalent bond, polar covalent bond

Boldfaced, Page-Referenced Terms

(24) ion _____

(24) ionic bond _____

(24) covalent bond _____

(25) hydrogen bond _____

Identification

1. Following the model below, sketch the transfer of electron(s) (by arrows) to show how positive and negative ions form ionic bonds to create a molecule of $MgCl_2$ (magnesium chloride).

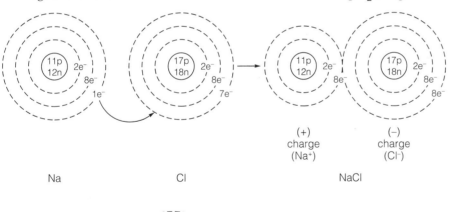

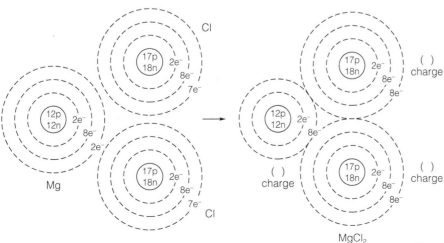

Short Answer

2. Distinguish between a nonpolar covalent bond and a polar covalent bond.

_____

_____

3. Cite one example of a large molecule where hydrogen bonds exist within the molecule.

_____

_____

Identification

4. Following the model of hydrogen gas below, sketch electrons (as dots) in the outer shells to illustrate the nonpolar covalent bonding to form oxygen gas; similarly illustrate polar covalent bonds by completing electron structures to form a water molecule.

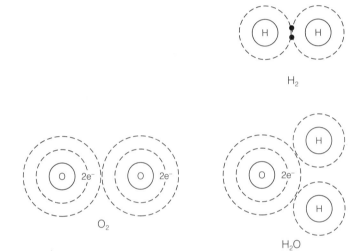

2-IV. PROPERTIES OF WATER (pp. 26–27)

Selected Italicized Words

polarity, temperature-stabilizing, liquid water, cohesive properties, solvent properties

Boldfaced, Page-Referenced Terms

(26) hydrophilic _____

_____

(26) hydrophobic _____

_____

(26) temperature _____

_____

(26) evaporation _____

_____

Fill-in-the-Blanks

The (1) _____ of water molecules allows them to hydrogen-bond with each other. Water molecules hydrogen-bond with polar substances that are (2) _____ (water loving). Polarity causes water to repel oil and other nonpolar substances that are (3) _____ (water dreading). Water changes its temperature more slowly than air because of the great amount of heat required to break the (4) _____ bonds between water molecules. The escape of gaseous water molecules from a fluid surface to the surrounding air is known as (5) _____. Below 0°C, water molecules become locked in the less dense bonding pattern of (6) _____. (7) _____ is the property of water molecules that explains how insects walk on water and how long, narrow water columns rise to the tops of tall trees. Water is an excellent (8) _____ in which ions and polar molecules readily dissolve. Dissolved substances in water are known as (9) _____. A substance is (10) _____ in water when spheres of (11) _____ form around its individual ions or molecules.

2-V. WATER, DISSOLVED IONS, AND PH (pp. 28–29)
CHEMICAL INTERACTIONS AND THE WORLD OF CELLS (p. 30)

Selected Italicized Words

neutrality, H⁺ concentration, pH value, acidic solutions, basic solutions, acidosis

Boldfaced, Page-Referenced Terms

(28) hydrogen ions _____

_____

(28) hydroxide ions _____

_____

(28) pH scale _____

_____

(28) acids _____

_____

(28) bases _____

_____

(29) salts _____

_____

(29) buffer _____

_____

Fill-in-the-Blanks

Hydrogen ions (H^+) are the same thing as free or unbound (1) _____. The ionization of water into H^+ and OH^- is the basis of the (2) _____ scale. At $25°C$, pure water always has just as many H^+ as OH^-; this condition represents (3) _____ on the pH scale and is assigned a pH value of (4) _____. The greater the H^+ concentration, the (5) _____ the pH value. Most living cells maintain an H^+ concentration close to pH (6) _____. A pH of 8 has an H^+ concentration one hundred times higher than a pH of (7) _____.

When acids dissolve in water, they release (8) _____ ions; (9) _____ are substances that combine with H^+ ions. Acids commonly combine with bases to produce ionic compounds called (10) _____. A(n) (11) _____ is any molecule that can combine with hydrogen ions or release them, or both, and so help stabilize pH. HCO_3^-, or (12) _____, helps restore pH when blood becomes too acidic. It then combines with excess H^+ to form (13) _____ _____. In addition, bicarbonate may release H^+ when blood is not acid enough. When a lung disease interferes with carbon dioxide elimination, the blood level of (14) _____ _____ rises (also H^+ level rises). This abnormal condition is known as (15) _____ and makes breathing difficult and weakens the body.

The charged regions and polar groups on the surface of a protein attract water molecules and ions, which in turn attract more (16) _____ molecules. In this way, an electrically charged "cushion" of ions and water forms around the protein surfaces. This keeps the (17) _____ dispersed in cellular fluid rather than settling against cellular structures and rendering their surfaces unavailable for critical cellular chemical reactions.

Complete the Table

17. Complete the following table by consulting Figure 2.13 in the text.

Fluid	pH Value	Acid/Base
a. Blood		
b. Saliva		
c. Urine		
d. Stomach acid		

Self-Quiz

___ 1. A molecule is _____.
   a. a combination of two or more atoms
   b. less stable than its constituent atoms separated
   c. electrically charged
   d. a carrier of one or more extra neutrons

___ 2. If lithium has an atomic number of 3 and an atomic mass of 7, it has _____ neutron(s) in its nucleus.
   a. one
   b. two
   c. three
   d. four
   e. seven

___ 3. An ionic bond is one in which _____.
   a. electrons are shared equally
   b. electrically neutral atoms have a mutual attraction
   c. two charged atoms have a mutual attraction due to electron transfer
   d. electrons are shared unequally

___ 4. A hydrogen bond is _____.
   a. a sharing of a pair of electrons between a hydrogen nucleus and an oxygen nucleus
   b. a sharing of a pair of electrons between a hydrogen nucleus and either an oxygen or a nitrogen nucleus
   c. an attractive force that involves a hydrogen atom and an oxygen or a nitrogen atom that are either in two different molecules or within the same molecule
   d. none of the above

___ 5. A mixture of sugar and water is an example of a(n)_____.
   a. compound
   b. solution
   c. suspension
   d. colloid
   e. ion

___ 6. Radioactive isotopes have _____.
   a. excess electrons
   b. excess protons
   c. excess neutrons
   d. insufficient neutrons
   e. insufficient protons

___ 7. The shapes of large molecules are controlled by _____ bonds.
   a. hydrogen
   b. ionic
   c. covalent
   d. inert
   e. single

___ 8. A pH solution of 10 is _____ times as basic as a pH of 7.
   a. 2
   b. 3
   c. 10
   d. 100
   e. 1000

___ 9. Substances that are nonpolar and repelled by water are _____.
   a. hydrolyzed
   b. nonpolar
   c. hydrophilic
   d. hydrophobic

___ 10. Any molecule that combines with hydrogen ions or releases them, or both, and helps stabilize pH is known as a _____.
   a. neutral molecule
   b. salt
   c. base
   d. acid
   e. buffer

Chapter Objectives/Review Questions

This section lists general and detailed chapter objectives that can be used as review questions. You can make maximum use of these items by writing answers on a separate sheet of paper. Fill in answers where blanks are provided. To check for accuracy, compare your answers with information given in the chapter or glossary.

Page   *Objectives/Questions*

(18)   1. Solids, liquids, and gases are forms of _____.
(18)   2. _____ elements represent less than 0.01 percent of all the atoms in any organism.
(19)   3. Know the symbols for the elements listed in Table 2.1 in the text.
(19)   4. Define and relate the terms *atom*, *molecule*, and *compound*.
(19)   5. Explain how protons, electrons, and neutrons are arranged into atoms and ions.
(19)   6. If the most common mass number of phosphorus is 31 and the atomic number is 15, there are _____ neutrons in the nucleus.
(19)   7. Atoms having the same atomic number but a different mass number are _____.
(19–21)   8. Define *radioisotopes* and list three ways they are useful.
(22)   9. The union between the electron structures of atoms is known as a chemical _____.
(22)   10. Describe the distribution of electrons in the space around the nucleus of an atom.
(22)   11. Electrons farthest away from the nucleus of an atom are said to be at (choose one) ❏ lower ❏ higher energy levels.
(22)   12. An atom tends to react with other atoms when its outermost shell is only partly filled with _____.
(24)   13. Define *ion* and describe the conditions under which ionic bonds form between positive and negative ions.
(24)   14. In a _____ bond, atoms share electrons.
(24–25)   15. Distinguish between a nonpolar covalent bond and a polar covalent bond; be able to give an example of each.
(25)   16. Define *hydrogen bond*; describe conditions under which hydrogen bonds form and cite one example.
(26)   17. Explain what is meant by the polarity of the water molecule.
(26)   18. Describe how the polarity of water molecules allows them to interact with one another.
(26)   19. Define *hydrophilic* and *hydrophobic*; relate these terms to different types of substances that contact water.
(26)   20. Explain why water cools and warms more slowly than air.
(26)   21. The escape of water molecules from fluid surfaces to enter surrounding air is called _____.
(27)   22. The surface tension of water and the movement of long water columns to the tops of trees is explained by a property of water known as _____.
(27)   23. Dissolved substances are called _____; a fluid in which one or more substances can dissolve is called a _____.
(27)   24. Substances are dissolved in water when spheres of _____ form around their individual ions or molecules.
(28)   25. Define *acid* and *base*; cite an example of each.
(28)   26. The concentration of free hydrogen ions in solutions is measured by the _____ scale.
(28)   27. The pH of hair remover is 13; it is a(n) (choose one) ❏ base ❏ acid. The pH of vinegar is 3; it is a(n) (choose one) ❏ base ❏ acid.
(28)   28. Describe the pH existing within the chemistry of living systems.
(29)   29. Define *buffer*; cite an example.
(29)   30. A chemical reaction between an acid and a base produces a _____.
(30)   31. Organization of nearly all large biological molecules is influenced by their interactions with water; describe this interaction as it exists with protein molecules.

Integrating and Applying Key Concepts

1. Explain what would happen if water were a nonpolar molecule instead of a polar molecule. Would water be a good solvent for the same kinds of substances? Would the nonpolar molecule's specific heat likely be higher or lower than that of water? What about its surface tension? Cohesive nature? Ability to form hydrogen bonds? Is it likely that the nonpolar molecules could form unbroken columns of liquid? What implications would that hold for trees?

2. If the ways that atoms bond affect molecular shapes, do the ways that molecules behave toward one another influence the shapes of cellular organelles?

Answers

Interactive Exercises

2-I. ORGANIZATION OF MATTER (pp. 16–21)
1. carbon; 2. mass; 3. six; 4. six; 5. six; 6. carbon, oxygen, hydrogen, nitrogen; 7. a. Hydrogen, H; b. Carbon, C; c. Nitrogen, N; d. Oxygen, O; e. Phosphorus, P; f. Sulfur, S; 8.

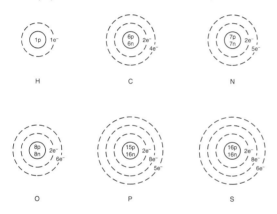

9. T; 10. Electrons; 11. repel; 12. T; 13. Electron; 14. C; 15. B; 16. K; 17. L; 18. N; 19. E; 20. H; 21. M; 22. G; 23. D; 24. F; 25. A; 26. I; 27. J.

2-II. THE NATURE OF CHEMICAL BONDS (pp. 22–23)
1. The only shell of H is filled with two electrons; all other atoms require eight electrons to fill their outermost shell; 2. An atom tends to react with other atoms when its outermost shell is only partly filled with electrons; 3. C; 4. E; 5. A; 6. B; 7. D.

2-III. IMPORTANT BONDS IN BIOLOGICAL MOLECULES (pp. 24–25)
1.

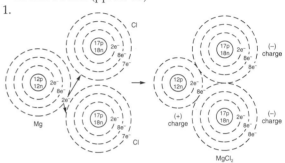

2. In a covalent bond, atoms share electrons to fill their outermost shells. In a nonpolar covalent bond, atoms attract shared electrons equally. In a polar covalent bond, atoms do not share electrons equally, and the bond is positive at one end, negative at the other (for example, the water molecule).

3. In the DNA molecule, the two nucleotide chains are held together by hydrogen bonds.

4.

2-IV. PROPERTIES OF WATER (pp. 26–27)
1. polarity; 2. hydrophilic; 3. hydrophobic; 4. hydrogen; 5. evaporation; 6. ice; 7. Cohesion; 8. solvent; 9. solutes; 10. dissolved; 11. hydration.

2-V. WATER, DISSOLVED IONS, AND PH (pp. 28–29)
CHEMICAL INTERACTIONS AND THE WORLD OF CELLS (p. 30)

1. protons; 2. pH; 3. neutrality; 4. 7 (neutral); 5. lower; 6. 7 (neutral); 7. 10; 8. hydrogen; 9. bases; 10. salts; 11. buffer; 12. bicarbonate; 13. carbonic acid; 14. carbonic acid; 15. acidosis; 16. water; 17. protein;

18. a. blood: 7.3–7.5, slightly basic; b. saliva: 6.2–7.4, slightly acidic or slightly basic; c. urine: 5.0–7.0, slightly acidic or neutral; d. stomach acid: 1.0–3.0, acid.

Self Quiz

1. a; 2. d; 3. c; 4. c; 5. b; 6. c; 7. a; 8. e; 9. d; 10. e.

3

CARBON COMPOUNDS IN CELLS

Interactive Exercises

3-I. PROPERTIES OF ORGANIC COMPOUNDS (pp. 32–35)
3-II. HOW CELLS USE ORGANIC COMPOUNDS (p. 36)

Selected Italicized Words

double covalent bond

Boldfaced, Page-Referenced Terms

(34) organic compound _____

_____

(35) hydrocarbons _____

_____

(35) functional groups _____

_____

(35) alcohols _____

_____

(36) enzymes _____

_____

(36) functional-group transfer _____

_____

(36) electron transfer _____

_____

(36) rearrangement _____

_____

(36) condensation _____

_____

(36) cleavage _____

_____

(36) hydrolysis_____

_____

Short Answer

1. Describe the bonding properties that are responsible for carbon's central role in the molecules of life.

_____

_____

Labeling

Study the above structural formulas of organic compounds; by reference to Table 3.1 in the text, identify the circled functional groups (sometimes repeated) by entering the correct name in the blanks with matching numbers below the sketches; complete the exercise by circling the compounds (in the parentheses following each blank) in which the functional group might appear.

2. _____ (fats - waxes - oils - sugars - amino acids - proteins - phosphate compounds, e.g. ATP)

3. _____ (fats - waxes - oils - sugars - amino acids - proteins - phosphate compounds, e.g. ATP)

4. _____ (fats - waxes - oils - sugars - amino acids - proteins - phosphate compounds, e.g. ATP)

5. _____ (fats - waxes - oils - sugars - amino acids - proteins - phosphate compounds, e.g. ATP)

6. _____ (fats - waxes - oils - sugars - amino acids - proteins - phosphate compounds, e.g. ATP)

7. _____ (fats - waxes - oils - sugars - amino acids - proteins - phosphate compounds, e.g. ATP)

8. _____ (fats - waxes - oils - sugars - amino acids - proteins - phosphate compounds, e.g. ATP)

Fill-in-the-Blanks

Complete each sentence by supplying the described functional group.

A(n) (9) _____ group contains an oxygen atom and a hydrogen atom. A(n) (10) _____ group contains two oxygen atoms, a carbon atom, and a hydrogen atom. A(n) (11) _____ group can contain four oxygen atoms, two hydrogen atoms, and a phosphorus atom. A(n) (12) _____ group may contain one nitrogen atom and two hydrogen atoms. A(n) (13) _____ group contains one carbon atom and three hydrogen atoms.

True-False

If the statement is true, write a T in the blank. If the statement is false, make it correct by changing the underlined word(s) and writing the correct word(s) in the answer blank.

_____ 14. Molecules are assembled into polymers by means of <u>hydrolysis</u>.

_____ 15. Carbon atoms are part of so many different substances because, in one of their electron configurations, there are <u>four</u> unpaired electrons in the outermost occupied energy level. Thus, each unpaired electron can form a covalent bond with a variety of other atoms, including other carbon atoms.

_____ 16. All organic molecules contain <u>oxygen</u>.

_____ 17. When a ring like ⬡ is shown, it is understood that a <u>carbon</u> atom occurs at every corner and that no other atoms are attached. (Think!)

_____ 18. Enzymes are a special class of <u>carbohydrates</u> that speed up specific metabolic reactions.

Identification

19. The structural formulas of two adjacent amino acids are shown below. Identify how enzyme action causes formation of a covalent bond and a water molecule by circling an H atom from one amino acid and an -OH group from the other amino acid. Circle the covalent bond that formed the dipeptide.

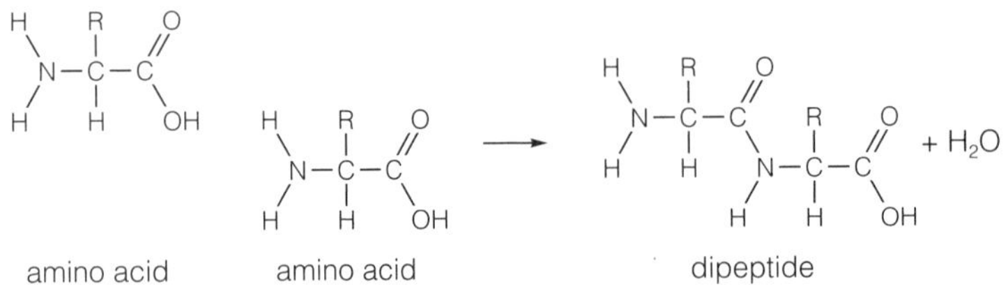

amino acid   amino acid   dipeptide

Short Answer

20. Describe hydrolysis through enzyme action for the molecules in question 19. _____

_____

_____

21. List the four families of small organic molecules that serve as subunits for synthesis of larger carbohydrates, lipids, proteins, and nucleic acids. _____

_____

3-III. THE SMALL CARBOHYDRATES (p. 37)
3-IV. COMPLEX CARBOHYDRATES (pp. 38–39)

Selected Italicized Words

disaccharide

Boldfaced, Page-Referenced Terms

(37) carbohydrate _____

_____

(37) monosaccharides _____

_____

(37) oligosaccharides _____

_____

(37) polysaccharides _____

_____

Identification

1. In the diagram below, identify condensation reaction sites between the two glucose molecules by circling the components of the water removed that allow a covalent bond to form between the glucose molecules. Note that the reverse reaction is hydrolysis and that both condensation and hydrolysis reactions require enzymes to proceed.

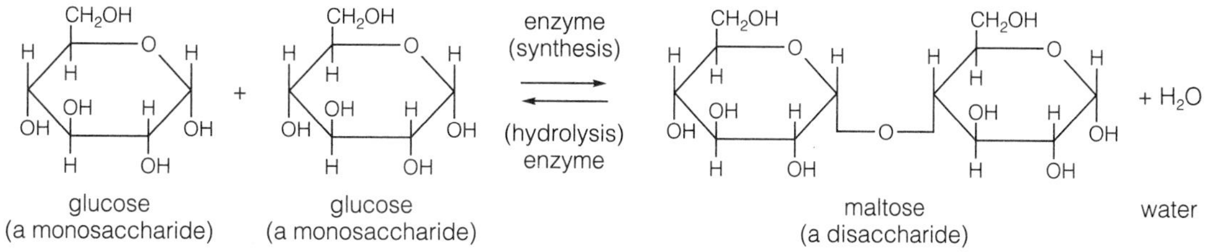

Complete the Table

2. In the table below, enter the name of the carbohydrate described by its carbohydrate class and functions.

Carbohydrate	Carbohydrate	Class Function
a.	Oligosaccharide (disaccharide)	Most plentiful sugar in nature; transport form of carbohydrates
b.	Monosaccharide	Five-carbon sugar occurring in RNA
c.	Monosaccharide	Main energy source for most organisms; precursor of many organic organisms
d.	Polysaccharide	Structural material of plant cell walls
e.	Monosaccharide	Five-carbon sugar occurring in DNA
f.	Oligosaccharide (disaccharide)	Sugar present in milk
g.	Polysaccharide	Main structural material in some external skeletons and other hard body parts of some animals and fungi
h.	Branched polysaccharide	Animal starch
i.	Polysaccharide	Sugar storage form in plants

3-V. LIPIDS (pp. 40–41)

Selected Italicized Words

unsaturated, saturated

Boldfaced, Page-Referenced Terms

(40) lipids _____

_____

(40) fatty acids _____

_____

(40) triglycerides _____

_____

(41) phospholipid _____

_____

(41) waxes _____

_____

(41) sterols _____

_____

Labeling

1. In the appropriate blanks, label the fatty acid molecules shown below as saturated or unsaturated.

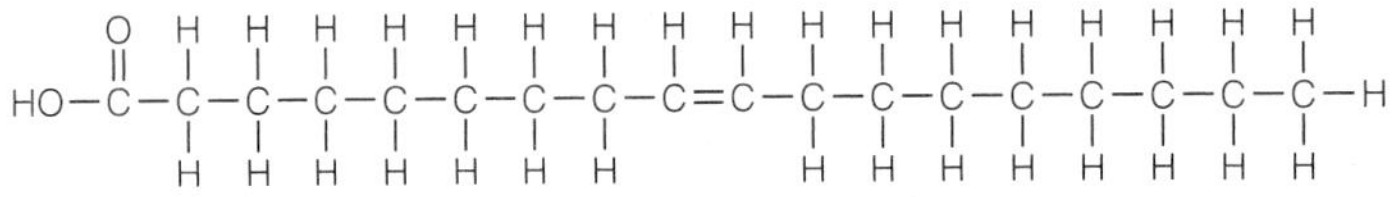

a. oleic acid

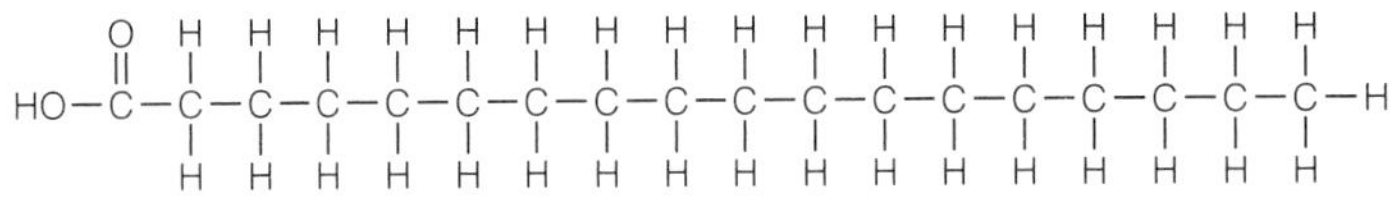

b. stearic acid

a. _____   b. _____

Identification

2. Combine glycerol with three fatty acids below to form a triglyceride by circling the participating atoms that will form three covalent bonds; circle the resulting covalent bonds in the triglyceride.

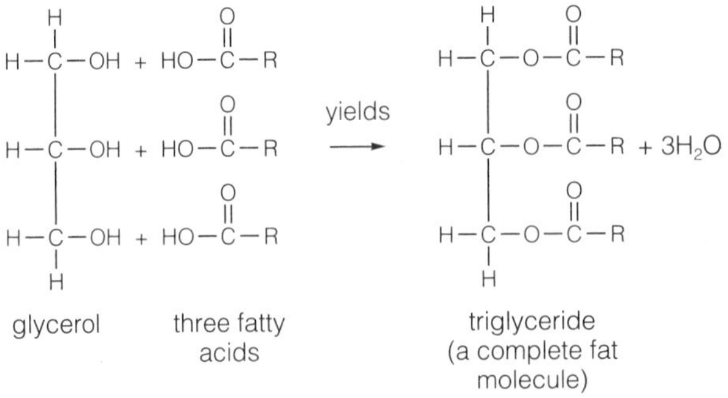

glycerol   three fatty   triglyceride
                  acids   (a complete fat
                                       molecule)

Short Answer

3. Describe the structure and biological functions of phospholipid molecules (p. 41, text). _____

_____

_____

_____

Matching

Choose the appropriate answer for each. Some letter choices may be used more than once.

4. ___ richest source of body energy

5. ___ honeycomb material

6. ___ cholesterol

7. ___ saturated tails

8. ___ butter, lard, oils

9. ___ main cell membrane component

10. ___ cutin

11. ___ triglycerides

12. ___ precursors of testosterone, estrogen, and bile salts

13. ___ unsaturated tails

14. ___ vegetable oil

A. Fatty acids
B. Neutral fats
C. Phospholipids
D. Waxes
E. Sterols

3-VI. PROTEINS (pp. 42–46)

Selected Italicized Words

peptide bonds, primary structure, secondary structure, tertiary structure, quaternary structure

Boldfaced, Page-Referenced Terms

(42) proteins _____

_____

(42) amino acid _____

_____

(46) lipoproteins _____

_____

(46) glycoproteins _____

_____

(46) denaturation _____

_____

Matching

For questions 1, 2, and 3, match the major parts of every amino acid by entering the letter of the part in the blank corresponding to the number on the molecule.

1. ___

2. ___

3. ___

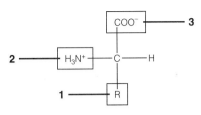

A. R group (a symbol for a characteristic group of atoms that differ in number and arrangement from one amino acid to another)
B. Carboxyl group (ionized)
C. Amino group (ionized)

Identification

4. In the illustration of four amino acids in cellular solution (ionized state) below, circle the atoms and ions that form water when covalent (peptide) bonds form between adjacent amino acids to make a polypeptide. On the completed polypeptide below, circle the newly formed peptide bonds.

Matching

5. ___ primary protein structure

6. ___ secondary protein structure

7. ___ tertiary protein structure

8. ___ quaternary protein structure

A. A coiled or extended pattern, based on hydrogen bonding at regular intervals
B. Incorporates two or more polypeptide chains yielding a protein that is globular, fiberlike, or both
C. The unique sequence of amino acids for each kind of protein
D. Bending and twisting of the protein chain due to R-group interactions

3-VII. NUCLEOTIDES AND NUCLEIC ACIDS (pp. 46–48)

Boldfaced, Page-Referenced Terms

(46) nucleotides _____

_____

(46) ATP _____

_____

(46) coenzymes _____

_____

(46) chemical messengers _____

_____

(47) nucleic acids _____

_____

(47) RNA _____

_____

(47) DNA _____

_____

Matching

For questions 1, 2, and 3, match the following answers to the parts of a nucleotide shown in the diagram at the right.

1. ___   A. A five-carbon sugar (ribose or deoxyribose)
2. ___   B. Phosphate group
          C. A nitrogen-containing base that has either a
3. ___   single-ring or double-ring structure

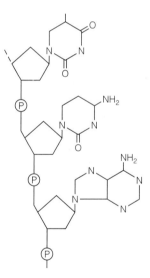

Identification

4. In the diagram of a single-stranded nucleic acid molecule at the right, encircle as many complete nucleotides as possible. How many complete nucleotides are presented?

Matching

Choose the appropriate answer for each.

5. ___ adenosine triphosphate

6. ___ RNA

7. ___ DNA

8. ___ NAD⁺ and FAD

9. ___ cAMP

A. Single nucleotide strand; functions in processes by which genetic instructions are used to build proteins
B. ATP, a cellular energy carrier
C. Nucleotide chemical messenger
D. Single nucleotide units; coenzymes; transport hydrogen ions and their associated electrons from one cell reaction site to another
E. Double nucleotide strand; encodes genetic instructions with base sequences

Complete the Table

10. Complete the table below by entering the correct name of the major cellular organic compounds suggested in the "types" column (choose from carbohydrates, proteins, nucleic acids, and lipids).

Cellular Organic Compounds

Cellular Organic Compounds	Types
a.	Phospholipids
b.	Antibodies
c.	Enzymes
d.	Genes
e.	Glycogen, starch, cellulose, and chitin
f.	Glycerides
g.	Saturated and unsaturated fats
h.	Coenzymes
i.	Steroids, oils, and waxes
j.	Glucose and sucrose

Self-Quiz

___ 1. Carbon is part of so many different substances because _____.
   a. carbon generally forms two bonds with a variety of other atoms
   b. carbon generally forms four bonds with a variety of atoms
   c. carbon ionizes easily
   d. carbon is a polar compound

___ 2. _____ are compounds used by cells as transportable packets of quick energy, storage forms of energy, and structural materials.
   a. Lipids
   b. Nucleic acids
   c. Carbohydrates
   d. Proteins

___ 3. Proteins _____.
   a. include all hormones
   b. are composed of nucleotide subunits
   c. are not very diverse in structure and function
   d. include all enzymes
   e. translates protein-building instructions into actual protein structures

___ 4. Glucose dissolves in water because it _____.
   a. ionizes
   b. is a polysaccharide
   c. forms many hydrogen bonds with the water molecules
   d. has a very reactive primary structure

___ 5. Hydrolysis could be correctly described as the _____.
   a. heating of a compound to drive off its excess water and concentrate its volume
   b. breaking of a long-chain compound into its subunits by adding water molecules to its structure between the subunits
   c. linking of two or more molecules by the removal of one or more water molecules
   d. constant removal of hydrogen atoms from the surface of a carbohydrate

___ 6. DNA _____.
   a. is one of the adenosine phosphates
   b. is one of the nucleotide coenzymes
   c. contains protein-building instructions
   d. are composed of monosaccharides

___ 7. Amino acids are linked by _____ bonds to form the primary structure of a protein.

    a. disulfide

    b. hydrogen

    c. ionic

    d. peptide

___ 8. Lipids _____.

    a. serve as food reserves in many organisms

    b. include cartilage and chitin

    c. include fats that are broken down into one fatty acid molecule and three glycerol molecules

    d. are composed of monosaccharides

___ 9. Most of the chemical reactions in cells must have _____ present before they proceed.

    a. RNA

    b. salt

    c. enzymes

    d. fats

___10. Genetic instructions are encoded in the bases of _____; molecules of _____ function in processes using genetic instructions to construct proteins.

    a. DNA; DNA

    b. DNA; RNA

    c. RNA; DNA

    d. RNA; RNA

Chapter Objectives/Review Questions

This section lists general and detailed chapter objectives that can be used as review questions. You can make maximum use of these items by writing answers on a separate sheet of paper. Fill in answers where blanks are provided. To check for accuracy, compare your answers with information given in the chapter or glossary.

Page	Objectives/Questions
(34)	1. A(n) _____ compound consists of carbon and one or more additional elements, covalently bonded to one another.
(34)	2. Each carbon atom can form as many as _____ covalent bonds with other carbon atoms as well as with other elements.
(35)	3. Define the term *hydrocarbon*.
(35)	4. Define *functional group*; be able to recognize the major functional groups of organic compounds as shown in Table 3.1 of the text.
(35)	5. Sugars and other organic compounds that have one or more hydroxyl groups attached to carbon atoms are classified as _____.
(36)	6. _____ are a special class of proteins that speed specific metabolic reactions in cells.
(36)	7. Define the following as they apply to the chemistry of organic compounds: *functional group-transfer*, *electron transfer*, *rearrangement*, *condensation*, and *cleavage*.
(36)	8. A chain of bonded monomers is called a _____.
(36)	9. _____ is one type of cleavage reaction, like condensation in reverse.
(36)	10. List the four main families of small organic molecules.
(37)	11. Define *carbohydrates* and list their major functions.
(37)	12. The simplest carbohydrates are sugar monomers, the _____; be able to give examples and their functions.
(37)	13. An _____ is a short chain of two or more sugar monomers; be able to give examples of well-known disaccharides and their functions.
(38–39)	14. A _____ is a straight or branched chain of hundreds or thousands of sugar monomers, of the same or different kinds; be able to give common examples and their functions; list common examples.
(40)	15. Define *lipids* and list their major functions.
(40)	16. Describe a fatty acid.
(40)	17. A _____ molecule has one to three fatty acid tails attached to a backbone of glycerol.
(40)	18. Distinguish a saturated fat from an unsaturated fat.

Integrating and Applying Key Concepts

1. Humans can obtain energy from many different food sources. Do you think this ability is an advantage or a disadvantage in terms of long-term survival? Why?

2. If the ways that atoms bond affect molecular shapes, do the ways that molecules behave toward one another influence the shapes of organelles? Do the ways that organelles behave toward one another influence the structure and function of cells?

3. If proteins have the most diverse shapes and the most complex structure of all molecules, why do you suppose that proteins are not the code molecules used to construct new proteins?

Answers

Interactive Exercises

3-I. PROPERTIES OF ORGANIC COMPOUNDS
(pp. 32–36)
3-II. HOW CELLS USE ORGANIC COMPOUNDS
(p. 36)
1. One carbon atom is able to form as many as four covalent bonds with other carbon atoms as well as with atoms of other elements; carbon atoms may bond to form organic compounds in chains, branched chains, and rings; 2. methyl (fats - waxes - oils);
3. hydroxyl (sugars); 4. ketone (sugars);
5. amino (amino acids, proteins);
6. phosphate (phosphate compounds);
7. carboxyl (fats, sugars, amino acids);
8. aldehyde (sugars); 9. hydroxyl;
10. carboxyl; 11. phosphate; 12. amino;
13. methyl; 14. condensation; 15. T;
16. carbon; 17. F; 18. proteins.
19.

amino acid   amino acid   dipeptide

20. Hydrolysis reactions reverse the chemistry of condensation reactions; in the presence of water, large molecules are split into their component smaller molecules. Both condensation and hydrolysis require the presence of enzymes specific to the particular molecules involved.
21. Simple sugars, fatty acids, amino acids, and nucleotides.

3-III. THE SMALL CARBOHYDRATES (p. 37)
3-IV. COMPLEX CARBOHYDRATES (pp. 38–39)
1.

glucose   glucose   maltose   water
(a monosaccharide) (a monosaccharide)   (a disaccharide)

2. a. Sucrose; b. Ribose; c. Glucose; d. Cellulose;
e. Deoxyribose; f. Lactose; g. Chitin; h. Glycogen;
i. Starch.

3-V. LIPIDS (pp. 40–41)

1. a. unsaturated; b. saturated.

2.

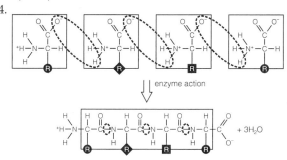

glycerol   three fatty   triglyceride
            acids   (a complete fat
                           molecule)

3. Phospholipids have two fatty acid tails attached to a glycerol backbone; they have hydrophilic heads that dissolve in water. Phospholipids are the main structural materials of cell membranes.

4. B; 5. D; 6. E; 7. A; 8. A; 9. C; 10. D; 11. B; 12. E; 13. A; 14. A.

3-VI. PROTEINS (pp. 42–46)

1. A; 2. C; 3. B.

4.

enzyme action

5. C; 6. A; 7. D; 8. B.

3-VII. NUCLEOTIDES AND NUCLEIC ACIDS (pp. 46–47)

1. B; 2. A; 3. C; 4. Three as shown:

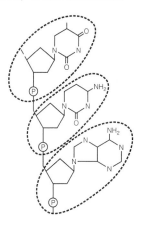

5. B; 6. A; 7. E; 8. D; 9. C; 10. a. Lipids; b. Proteins; c. Proteins; d. Nucleic acids; e. Carbohydrates; f. Lipids; g. Lipids; h. Nucleic acids; i. Lipids; j. Carbohydrates.

Self-Quiz

1. b; 2. c; 3. d; 4. c; 5. b; 6. c; 7. d; 8. a; 9. c; 10. b.

4

CELL STRUCTURE AND FUNCTION

Interactive Exercises

4-I. THE CELL THEORY (pp. 50–51)
4-II. THE NATURE OF CELLS (pp. 52–55)

Selected Italicized Words

compound light microscope, phase-contrast microscope, transmission electron microscope, high-voltage electron microscope, scanning electron microscope

Boldfaced, Page-Referenced Terms

(51) cell _____

---

(51) cell theory (3 major generalizations) _____

_____

_____

(52) plasma membrane _____

_____

(52) DNA-containing region _____

_____

(52) cytoplasm _____

_____

(52) eukaryotic cells _____

_____

(52) nucleus _____

_____

(52) prokaryotic cells _____

_____

(53) surface-to-volume ratio _____

_____

(54) micrograph _____

_____

Complete the Table

1. Summarize important contributions to the emergence of the cell theory by completing the following table.

Contributor	Years	Contribution
a.	(1564–1642)	First to record a microscopic biological observation
b.	(1635–1703)	Looked at cork cells; originated the word "cell"
c.	(1632–1723)	Skilled in lens construction; observed a bacterium
d.	(1773–1858)	First observed and named the nucleus of a cell
e.	(1804–1881) (1810–1882)	Concluded that all plant and animal tissues are composed of cells
f.	(1821–1902)	Established that all cells arise from preexisting cells

Label-Match

Although eukaryotic cells vary in many specific ways, they are all alike in a few basic respects. Identify each part of the illustration below. Complete the exercise by matching and entering the letter of the proper description in the parentheses following each label.

2. _____ _____ ( )

3. _____ ( )

4. _____ ( )

A. Everything in the cell that is enclosed by the plasma membrane, except the nucleus (in all but bacterial cells, specific compartments exist where different metabolic reactions occur)
B. A membrane-bound compartment that contains the hereditary instructions (DNA)
C. Outermost membrane of the cell; separates the internal events from the external environment

Matching

Select the single best answer.

5. ___ transmission electron microscope
6. ___ compound light microscope
7. ___ high-voltage electron microscope
8. ___ phase-contrast microscope
9. ___ micrograph
10. ___ scanning electron microscope

A. A photograph of an image formed with a microscope
B. The beam moves back and forth across a specimen's surface which has been coated with a thin metal layer
C. Converts small differences in the way different structures bend light to large variations in brightness
D. To observe fine details at very high magnification, one must slice cells extremely thin
E. Light emanates from the specimen to form an enlarged image
F. Highly energized electrons used to see structures of intact cells

4-III. PROKARYOTIC CELLS—THE BACTERIA (pp. 56–57)

Selected Italicized Words

prokaryotic, *Escherichia coli*

Boldfaced, Page-Referenced Terms

(56) bacterial flagella _____

_____

(56) cell wall _____

_____

(57) ribosome _____

_____

Fill-in-the-Blanks

(1) _____ are the smallest and most structurally simple cells. The word *prokaryotic* means "before the (2) _____," which implies that bacteria evolved before cells possessing nuclei existed. Some bacteria have one or more long, threadlike motile structures, the (3) _____, that extend from the cell surface; they permit rapid movements through fluid environments. Most bacteria have a cell (4) _____ that surrounds the plasma membrane; this membrane controls movement of substances into and out of the (5) _____. In all cells, (6) _____ are structures composed of two molecular subunits; proteins are synthesized there. Most of the DNA in bacterial cells is found in an irregularly shaped cytoplasmic region, the nucleoid, that lacks an enclosing membrane. Within the nucleoid area the bacterial DNA is a single, (7) _____ molecule.

4-IV. EUKARYOTIC CELLS (pp. 58–61)

Selected Italicized Words

eukaryotic

Boldfaced, Page-Referenced Terms

(58) organelle _____

_____

_____

Matching

Match functions to the list of features typical of most eukaryotic cells. Choose the one best answer for each.

1. ___ diverse vesicles
2. ___ mitochondria
3. ___ Golgi bodies
4. ___ nucleus
5. ___ endoplasmic reticulum
6. ___ ribosomes
7. ___ cytoskeleton

A. Efficient ATP formation
B. Physical isolation and organization of DNA
C. Cell movement, shape, internal organization
D. Further modifies polypeptide chains into mature proteins; sorting, shipping of proteins and lipids for secretion or use in the cell
E. Transport or storage of substances; digestion inside cell; other functions
F. Synthesis of polypeptide chains
G. Initial modification of new polypeptide chains; lipid synthesis

4-V. THE NUCLEUS (pp. 62–63)

Boldfaced, Page-Referenced Terms

(62) nucleolus _____

_____

(63) nuclear envelope _____

_____

(63) chromatin _____

_____

(63) chromosome _____

_____

Complete the Table

1. Complete this table about the eukaryotic nucleus. Enter the name of each nuclear component described.

Nuclear Component	Description
a.	An individual DNA molecule and its associated proteins in the thread-like or condensed form
b.	Total collection of DNA molecules and associated proteins in the nucleus
c.	Two lipid bilayer membranes thick; proteins and protein pores span the bilayers
d.	One or more masses in the nucleus; sites where the ribosome subunits are assembled
e.	Total fluid portion of the interior of the nucleus

4-VI. CYTOMEMBRANE SYSTEM (pp. 64–66)

Selected Italicized Words

rough ER, smooth ER

Boldfaced, Page-Referenced Terms

(64) cytomembrane system _____

_____

(65) ER _____

_____

(65) peroxisomes _____

_____

(66) Golgi body _____

_____

(66) lysosomes _____

_____

Complete the Table

1. Complete this table summarizing components of the cytomembrane system.

Cytomembrane Component	Function
a.	Sacs of enzymes; break down fatty acids and amino acids; convert harmful H_2O_2 to H_2O and O_2
b.	Resemble pancake stacks; modify lipids and proteins as well as sort, package, and ship them out
c.	Bud from Golgi membranes; organelles of intracellular digestion
d.	Many ribosomes attached; polypeptide chains assembled here
e.	Free of ribosomes, curves through the cytoplasm; main site of lipid synthesis
f.	A vesicle in plant cells; contains enzymes that help convert stored fats and oils to sugars

Matching

Study the illustration below and match each component of the cytomembrane system with the most correct description of function. Some components may be used more than once.

a. smooth ER (SER)
b. nucleus
c. Golgi body
d. vesicles from Golgi
e. vesicles from rough ER (RER)
f. endocytosis
g. exocytosis
h. rough ER (RER)
i. ribosomes
j. vesicles budding
k. lysosomes

2. ___ Assembly of polypeptide chains

3. ___ Lipid assembly

4. ___ DNA instructions for building polypeptide chains

5. ___ Transport substances into the cytoplasm from outside cell

6. ___ Initiate protein modification

7. ___ Proteins and lipids take on final form

8. ___ Sort and package lipids and proteins for transport

9. ___ Vesicles formed at plasma membrane transport substances into cytoplasm

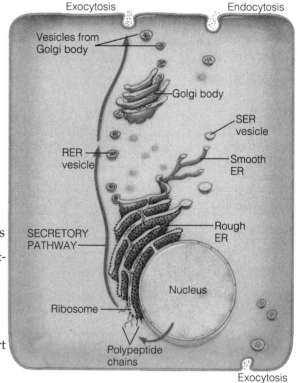

___ 10. Special vesicle budding from Golgi body; major digestion organelle

___ 11. Transport unfinished proteins and lipids to a Golgi body

___ 12. Transport finished Golgi products to the plasma membrane

___ 13. Release Golgi products at the plasma membrane

___ 14. May transport proteins and lipids to other organelles or to the plasma membrane

4-VII. MITOCHONDRIA (p. 67)

Boldfaced, Page-Referenced Terms

(67) mitochondria _____

_____

True-False

If the statement is true, write a T in the blank. If the statement is false, make it correct by changing the underlined word(s) and writing the correct word(s) in the answer blank.

_____ 1. Mitochondria <u>lack</u> DNA and can divide.

_____ 2. Mitochondria have <u>two</u> membranes, which creates two compartments inside each mitochondrion.

_____ 3. Within a mitochondrion, breakdown products of glucose are used to form a <u>small</u> yield of ATP.

_____ 4. The membrane folds within the mitochondrion are known as <u>cristae</u>.

_____ 5. All <u>prokaryotic</u> cells have one or more mitochondria.

4-VIII. SPECIALIZED PLANT ORGANELLES (p. 68)

Boldfaced, Page-Referenced Terms

(68) chloroplasts _____

_____

(68) chromoplasts _____

_____

(68) amyloplasts _____

_____

(68) central vacuole _____

_____

Choice

For questions 1–10, choose from the following:

    a. chloroplasts   b. amyloplasts   c. central vacuole   d. chromoplasts

____ 1. A plastid found in living plant cells; fluid-filled.

____ 2. Plastids that lack pigments.

____ 3. Plastids that have an abundance of carotenoids but no chlorophylls.

____ 4. Organelles that absorb sunlight energy and produce ATP.

____ 5. The source of the yellow-to-red colors of many flowers, leaves, and fruits.

____ 6. Plastid with internal areas known as *grana* and *stroma*.

____ 7. Plastids that resemble certain photosynthetic bacteria.

____ 8. Store starch grains and are abundant in cells of stems, tubers, and seeds.

____ 9. The site of photosynthesis in plant cells.

____ 10. Stores amino acids, sugars, ions, and toxic wastes.

4-IX. THE CYTOSKELETON (pp. 69–72)

Boldfaced, Page-Referenced Terms

(69) cytoskeleton _____

_____

(69) microtubules _____

_____

(69) microfilaments _____

_____

(69) intermediate filaments _____

_____

(70) pseudopods _____

_____

(70) contraction _____

_____

(70) amoeboid motion _____

_____

(70) cytoplasmic streaming _____

_____

(70) flagellum _____

_____

(70) cilium _____

_____

(71) centrioles _____

_____

(71) MTOCs _____

_____

Dichotomous Choice

Circle one of two possible answers given between parentheses in each statement.

1. The cytoskeleton gives (prokaryotic/eukaryotic) cells their shape, internal organization, and movement.
2. (Protein/Carbohydrate) subunits form the basic components of microtubules, microfilaments, and the intermediate filaments of animal cells.
3. Microtubules consist of (tubulin/actin) subunits arranged in parallel rows.
4. Cell movement through fluid environments by forward bulging and production of temporary surface lobes called pseudopods is (contraction/amoeboid motion).
5. Muscle cell movement through the sliding action of actin and myosin microfilaments is called (cytoplasmic streaming/contraction).
6. Cellular structures are pushed or dragged through the cytoplasm by (binding and releasing of myosin and actin/controlled assembly and disassembly of microtubule or microfilament subunits).
7. In response to the sun's position, chloroplasts (with attached myosin filaments) flow in plant cells due to myosin "walking" over actin; this is known as (amoeboid motion/cytoplasmic streaming).
8. (Flagella/Cilia) are long, taillike motile structures on cells.
9. (Flagella/Cilia) are short but very numerous motile structures on cells.
10. The cross-sectional pattern of 9 + 2 microtubules is found in (cilia/centrioles).
11. The human respiratory tract is lined with beating (flagella/cilia).
12. Microtubules found in cilia and flagella have their origin from (basal bodies/centrosomes).
13. MTOCs are sites of dense material that generate large numbers of (microtubules/microfilaments).
14. The (centrosome/kinetochore) is a type of MTOC found near the cell nucleus; it gives rise to spindle microtubules.
15. (Centrosomes/Kinetochores) are MTOCs found on the surface of chromosomes, where they give rise to microtubules that become attached to those of the spindle.

4-X. CELL SURFACE SPECIALIZATIONS (p. 72)

Selected Italicized Words

primary cell wall, secondary cell wall, tight junctions, adhering junctions, gap junctions

Matching

Choose the most appropriate answer to match with each of the following terms.

1. ___ gap junctions

2. ___ ground substance

3. ___ primary wall

4. ___ plasmodesmata

5. ___ adhering junctions

6. ___ pectin

7. ___ secondary wall

8. ___ tight junctions

A. A pliable plant cell wall composed of cellulose strands
B. Deposits cementing adjacent plant cell walls
C. Animal cells scattered in cellular secretions of polysaccharides and protein fibers (e.g., cartilage)
D. Like spot welds cementing animal cells together in a tissue (e.g., skin and heart)
E. Channels linking adjacent animal cells; for passage of signals and substances
F. Link cells of animal epithelial tissues; form seals to prevent molecules crossing tissues
G. Cytoplasmic strands forming transport channels across adjacent living plant cells
H. A very rigid plant cell wall; formed after the first wall

Self-Quiz

Label-Match

Identify each indicated part of the accompanying illustrations. Complete the exercise by matching and entering the letter of the proper function description in the parentheses following each label. Some letter choices must be used more than once.

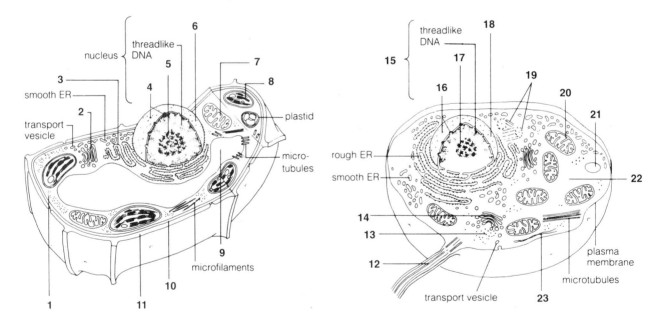

1. _____ ( )

2. _____ _____ ( )

3. _____ _____ _____ ( )

4. _____ _____ ( )

5. _____ ( )

6. _____ _____ ( )

7. _____ ( )

8. _____ ( )

9. _____ _____ ( )

10. _____ _____ ( )

11. _____ _____ ( )

12. _____ ( )

13. _____ ( )

14. _____ _____ ( )

15. _____ ( )

16. _____ _____ ( )

17. _____ ( )

18. _____ _____ ( )

19. _____ ( )

20. _____ ( )

21. _____ ( )

22. _____ ( )

23. _____ ( )

A. Two-membrane structure; outermost part of the nucleus
B. Protection and structural support
C. Increases cell surface area and storage
D. Everything enclosed by the plasma membrane, except for the nucleus
E. Formed as buds from Golgi membranes; contain enzymes for intracellular digestion
F. Small cylinders composed of triplet microtubules; produce microtubules of cilia and flagella; act as basal bodies
G. Microtubular structures for propelling eukaryotic cells; longer than cilia but with similar microtubular structure
H. Site of protein synthesis
I. A membrane-bound compartment that houses DNA in eukaryotic cells
J. Further modification, sorting, and shipping of proteins and lipids for secretion or for use in the cell
K. Sites where the protein and RNA subunits of ribosomes are assembled
L. A major component of the cytoskeleton
M. Photosynthesis and some starch storage
N. Control of material exchanges; mediates cell-environment interactions
O. Site of aerobic respiration
P. Initial modification of protein structure after formation on ribosomes
Q. Protein openings that span both bilayers of the nuclear envelope

Multiple Choice

__ 24. _____ distilled the meaning of the new microscopic observations by himself, _____, and earlier investigators into the first two generalizations of the cell theory.

a. van Leeuwenhoek; Hooke
b. Brown; Schleiden
c. Schwann; Schleiden
d. Schwann; Brown

__ 25. Which of the following is *not* found as a part of prokaryotic cells?

a. Ribosomes
b. DNA
c. Nucleus
d. Cytoplasm
e. Cell wall

__ 26. Which of the following statements most correctly describes the relationship between cell surface area and cell volume?

a. As a cell expands in volume, its diameter increases at a rate faster than its surface area does.
b. Volume increases with the square of the diameter, but surface area increases only with the cube.
c. If a cell were to grow four times in diameter, its volume of cytoplasm increases sixteen times and its surface area increases sixty-four times.
d. Volume increases with the cube of the diameter, but surface area increases only with the square.

___ 27. Animal cells dismantle and dispose of waste materials by _____.

    a. using centrally located vacuoles

    b. several lysosomes fusing with a vesicle that encloses the wastes

    c. microvilli packaging and exporting the wastes

    d. mitochondrial breakdown of the wastes

___ 28. The nucleolus is a dense region of _____ where the subunits that later will be constructed into _____ are made.

    a. nucleoplasm, ribosomes

    b. cytoplasm, vesicles

    c. cytoplasm, chromosomes

    d. nucleoplasm, chromatin

___ 29. The _____ is free of ribosomes, curves through the cytoplasm, and is the main site of lipid synthesis.

    a. lysosome

    b. Golgi body

    c. smooth ER

    d. rough ER

___ 30. Which of the following is *not* present in all cells?

    a. Cell wall

    b. Plasma membrane

    c. Ribosomes

    d. DNA molecules

___ 31. As a part of the cytomembrane system, the _____ modify lipids and proteins to permit sorting and packaging for specific locations.

    a. endoplasmic reticulum

    b. Golgi bodies

    c. peroxisomes

    d. lysosomes

___ 32. Mitochondria convert energy stored in _____ to forms that the cell can use, principally ATP.

    a. water

    b. carbon compounds

    c. $NADPH_2$

    d. carbon dioxide

___ 33. _____ are vesicles that bud from ER; they contain enzymes that use oxygen to break down fatty acids and amino acids.

    a. Lysosomes

    b. Glyoxysomes

    c. Golgi bodies

    d. Peroxisomes

Choice

Cells of the organisms in the five kingdoms of life share the following characteristics: plasma membrane, DNA, RNA, and ribosomes. For questions 34–43, choose the kingdom(s) possessing the following characteristics. Some questions will require more than one kingdom as the answer.

    a. Monera   b. Protista   c. Fungi   d. Plantae   e. Animalia

_____ 34. cytoskeletons

_____ 35. cell wall

_____ 36. nucleus

_____ 37. nucleoli

_____ 38. central vacuoles

_____ 39. photosynthetic pigments

_____ 40. endoplasmic reticulum

_____ 41. Golgi bodies

_____ 42. lysosomes

_____ 43. complex flagella and cilia

Chapter Objectives/Review Questions

This section lists general and detailed chapter objectives that can be used as review questions. You can make maximum use of these items by writing answers on a separate sheet of paper. Fill in answers where blanks are provided. To check for accuracy, compare your answers with information given in the chapter or glossary.

Page	Objectives/Questions
(50)	1. Be able to cite the contributions of the following investigators to the cell theory: Galileo Galilei, Robert Hooke, Antony van Leeuwenhoek, Robert Brown, Theodor Schwann, Matthias Schleiden, and Rudolf Virchow.
(51)	2. List the three generalizations that constitute the cell theory.
(52)	3. Describe these basic eukaryotic cellular features and their functions: plasma membrane, cytoplasm, and nucleus.
(52)	4. Prokaryotic cells lack a _____.
(52)	5. Two thin sheets of _____ molecules serve as the structural framework for cell membranes; proteins situated in the two sheets or at its surface carry out most membrane functions.
(53)	6. Be able to explain the meaning of *surface-to-volume ratio* as applied to the reason that cell size is necessarily limited.
(54–55)	7. Briefly describe the operation principles of light microscopes, phase-contrast microscopes, transmission electron microscopes, high-voltage electron microscopes, and scanning electron microscopes.
(56–57)	8. Describe the basic structure of prokaryotic cells; cite an example of these cells.
(58)	9. In eukaryotic cells, _____ separate different chemical reactions in space and time.
(60)	10. Give the function and cellular location of the following basic eukaryotic organelles and structures: nucleus, ribosomes, endoplasmic reticulum, Golgi bodies, diverse vesicles, mitochondria, and cytoskeleton.
(62)	11. _____ are sites where the protein and RNA subunits of ribosomes are assembled.
(63)	12. Describe the nature of the nuclear envelope and relate its function to its structure.
(63)	13. _____ is the total collection of DNA molecules and associated proteins; a _____ is an individual DNA molecule and associated proteins.
(64–66)	14. Explain how the endoplasmic reticulum, peroxisomes, Golgi bodies, lysosomes, and a variety of vesicles function together as the cytomembrane system.
(67)	15. In _____, energy stored in organic molecules is used to form many ATP molecules.
(68)	16. Give the function of the following plant organelles: chloroplasts, chromoplasts, amyloplasts, and central vacuole.
(68)	17. Describe the details of the structure of the chloroplast, the site of photosynthesis.
(69)	18. Elements of the _____ give eukaryotic cells their internal organization, overall shape, and capacity to move.
(69)	19. List the three major elements of the cytoskeleton.
(70)	20. Briefly define *pseudopods, contraction, amoeboid movement,* and *cytoplasmic movement.*
(70)	21. Both cilia and flagella have an internal microtubule arrangement called the "_____ array."
(71)	22. Microtubules of flagella and cilia arise from _____, which remain at the base of those completed structures as basal bodies.
(71)	23. _____ are MTOCs on the surface of chromosomes.
(72)	24. Define the following cell surface specializations: *primary and secondary plant cell walls, tight junctions, adhering junctions,* and *gap junctions.*

Integrating and Applying Key Concepts

1. Which parts of a cell constitute the minimum necessary for keeping the simplest of living cells alive?
2. How did the existence of a nucleus, compartments, and extensive internal membranes confer selective advantages on cells that developed these features?

ANSWERS

Interactive Exercises

4-I. THE CELL THEORY (pp.50– 51)

4-II. THE NATURE OF CELLS (pp. 52–55)
1. a. Galileo Galilei; b. Robert Hooke; c. Antony van Leeuwenhoek; d. Robert Brown; e. Matthias Schleiden, Theodor Schwann; f. Rudolf Virchow; 2. plasma membrane (C); 3. cytoplasm (A); 4. nucleus (B); 5. D; 6. E; 7. F; 8. C; 9. A; 10. B.

4-III. PROKARYOTIC CELLS—THE BACTERIA (pp. 56–57)
1. Bacteria; 2. nucleus; 3. flagella; 4. wall; 5. cytoplasm; 6. ribosomes; 7. circular.

4-IV. EUKARYOTIC CELLS (pp. 58–61)
1. E; 2. A; 3. D; 4. B; 5. G; 6. F; 7. C.

4-V. THE NUCLEUS (pp. 62–63)
1. a. chromosome; b. chromatin; c. nuclear envelope; d. nucleolus; e. nucleoplasm.

4-VI. THE CYTOMEMBRANE SYSTEM (pp. 64–66)
1. a. peroxisomes; b. Golgi bodies; c. lysosomes; d. rough ER; e. smooth ER; f. glyoxysomes; 2. i; 3. a; 4. b; 5. f; 6. h; 7. c; 8. c; 9. f; 10. k; 11. e; 12. d; 13. d, g; 14. j.

4-VII. MITOCHONDRIA (p. 67)
1. have (contain); 2. T; 3. high (large); 4. T; 5. eukaryotic.

4-VIII. SPECIALIZED PLANT ORGANELLES (p. 68)
1. a; 2. b; 3. d; 4. a; 5. d; 6. a; 7. a; 8. b; 9. a; 10. c.

4-IX. THE CYTOSKELETON (pp. 69–72)
1. eukaryotic; 2. Protein; 3. tubulin; 4. amoeboid motion; 5. contraction; 6. controlled assembly and disassembly of microtubule or microfilament subunits; 7. cytoplasmic streaming; 8. Flagella; 9. Cilia; 10. cilia; 11. cilia; 12. basal bodies; 13. microtubules; 14. centrosome; 15. Kinetochores.

4-X. CELL SURFACE SPECIALIZATIONS (p. 72)
1. E; 2. C; 3. A; 4. G; 5. D; 6. B; 7. H; 8. F.

Self-Quiz
1. ribosomes (H); 2. Golgi complex (body) (J); 3. rough endoplasmic reticulum (P); 4. nuclear pore (Q); 5. nucleolus (K); 6. nuclear envelope (A); 7. mitochondrion (O); 8. chloroplast (M); 9. central vacuole (C); 10. cell (plasma) membrane (N); 11. cell wall (B); 12. flagellum (G); 13. ribosomes (H); 14. Golgi complex (body) (J); 15. nucleus (I); 16. nuclear pore (Q); 17. nucleolus (K); 18. nuclear envelope (A); 19. centrioles (F); 20. mitochondrion (O); 21. lysosome (vacuole) (E); 22. cytoplasm (D); 23. microfilaments (L); 24. c; 25. c; 26. d; 27. b; 28. a; 29. c; 30. a; 31. b; 32. b; 33. d; 34. b, c, d, e; 35. a, b, c, d; 36. b, c, d, e; 37. b, c, d, e; 38. c, d; 39. a, b, d; 40. b, c, d, e; 41. b, c, d, e; 42. b, c, d, e; 43. b, c, d, e.

5

A CLOSER LOOK AT CELL MEMBRANES

Interactive Exercises

5-I. MEMBRANE STRUCTURE AND FUNCTION (pp. 76–79)

Boldfaced, Page-Referenced Terms

(78) phospholipid _____

_____

(78) lipid bilayer _____

_____

(79) fluid mosaic model _____

_____

Matching

Choose the one most appropriate answer for each.

1. ___ embedded proteins
2. ___ phospholipids, glycolipids, sterols
3. ___ "fluid" behavior of cell membranes
4. ___ "mosaic" quality of cell membranes
5. ___ phospholipid
6. ___ Singer and Nicolson, 1972
7. ___ lipid bilayer
8. ___ phytosterol
9. ___ role in formation of budding vesicles from ER or Golgi membranes
10. ___ cholesterol

A. Abundant sterol in animal cell membranes
B. Self-sealing behavior of membrane lipids by H_2O-lipid interactions
C. Proposed the first fluid mosaic membrane model
D. Major structural basis of cell membranes
E. Composite of lipids and proteins
F. Abundant sterol in plant cell membranes
G. Consists of a hydrophilic head and two hydrophobic tails
H. Constant motion of lipid molecules and their interactions
I. Incorporated in lipid bilayers of cell membranes
J. Carry out most membrane functions

5-II. FUNCTIONS OF MEMBRANE PROTEINS (pp. 80–81)

Selected Italicized Words

channel proteins, carrier proteins, freeze-fracturing, observed ratio of proteins to lipids, predicted ratio of proteins to lipids, freeze-etching, protein coat model

Boldfaced, Page-Referenced Terms

(80) transport proteins _____

_____

(80) receptor proteins _____

_____

(80) recognition proteins _____

_____

(80) adhesion proteins _____

_____

(80) centrifuge _____

_____

Labeling

Identify each numbered membrane protein in the illustration below.

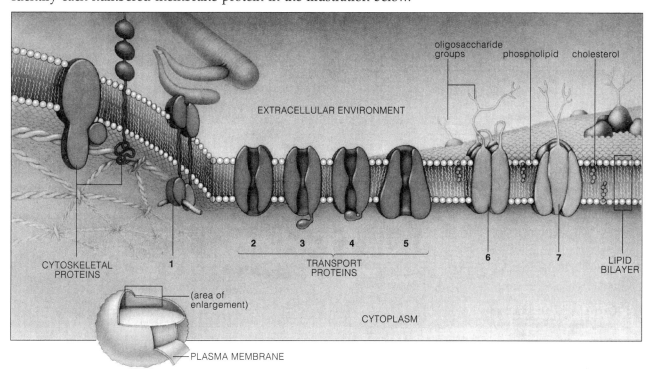

1. _____ protein

2. _____ _____ protein

3. _____ _____ protein (open)

4. _____ _____ protein (closed)

5. _____ protein

6. _____ protein

7. _____ protein

Complete the Table

8. Complete the table below by entering the name of the membrane protein whose function is described. Choose from adhesion proteins, recognition proteins, transport proteins (channel and carrier), and receptor proteins.

Membrane Protein	Function
a.	The interior remains closed off until a shape change shunts bound ions across the membrane; requires an energy boost
b.	Glycoproteins with attached oligosaccharides; helps connect cells within a tissue
c.	Have binding sites for hormones and other extracellular substances that can trigger alterations in cell activities
d.	Like molecular fingerprints; identify a cell as being a specific type; body cells have "self" types
e.	Some gates are always open; others have gates that close or open; helps control directional ion flow across plasma membranes

5-III. DIFFUSION (p. 82)

Selected Italicized Words

net movement, negatively or positively charged membranes

Boldfaced, Page-Referenced Terms

(82) concentration gradients _____

_____

(82) diffusion _____

_____

(82) electric gradient _____

_____

(82) pressure gradient _____

_____

Fill-in-the-Blanks

Cellular life depends on the (1) _____ inherent in molecules or ions that keeps them in constant motion. (2) _____ refers to the number of molecules of a substance in a stated volume of fluid. (3) _____ is a term applied when one region of a fluid contains more molecules than an adjacent region. When like molecules move *down* their concentration gradient, it is called (4) _____; this is a key factor in moving substances across cell membranes and through fluid portions of cytoplasm.

When the concentration gradient is steep, diffusion is (5) (choose one) ❏ slower ❏ faster. As the gradient decreases and the number of molecules moving down the gradient decreases, diffusion is (6) (choose one) ❏ slower ❏ faster. When the net distribution of molecules in a diffusion system is nearly uniform in the two adjoining regions, it is called dynamic (7) _____. Diffusion is (8) (choose one) ❏ slower ❏ faster at higher temperature; this is due to more rapidly moving molecules. If a difference in charge exists between two adjoining regions, the rate and direction of diffusion may be changed by an (9) _____ gradient. Within a cell, the side of the membrane that is negatively charged will exert a(n) (10) (choose one) ❏ attraction ❏ repulsion for positive sodium ions and enhance the flow of sodium across the membrane. A difference in pressure, or the (11) _____ gradient between two adjoining diffusion regions, can also influence the rate and direction of diffusion.

5-IV. OSMOSIS (pp. 82–83)

Selected Italicized Words

isotonic fluids, hypotonic solution, hypertonic solution

Boldfaced, Page-Referenced Terms

(83) tonicity _____

_____

(83) bulk flow _____

_____

(83) osmosis _____

_____

True-False

If the statement is true, write a T in the blank. If the statement is false, make it correct by changing the underlined word(s) and writing the correct word(s) in the answer blank.

_____ 1. Due to the presence of hydrogen bonds, pure water <u>can</u> become more concentrated.

_____ 2. Because membranes exhibit selective permeability, concentrations of dissolved substances can <u>increase</u> on one side of the membrane or the other.

_____ 3. A water concentration gradient is influenced by the number of molecules of all the <u>solvents</u> that are present on both sides of the membrane.

_____ 4. The relative concentrations of solutes in two fluids is called <u>tonicity</u>.

_____ 5. An animal cell placed in a <u>hypertonic</u> solution would swell and perhaps burst.

_____ 6. Water tends to move from <u>hypotonic</u> solutions to areas with more solutes.

_____ 7. Physiological saline is 0.9 percent NaCl; red blood cells placed in such a solution will not gain or lose water; therefore, one could state that the fluid in red blood cells is <u>hypertonic</u>.

_____ 8. A solution of 80 percent solvent, 20 percent solute is <u>more</u> concentrated than a solution of 70 percent solvent, 30 percent solute.

_____ 9. The movement of water molecules and solutes in the same direction in response to a pressure gradient is called <u>bulk flow</u>.

_____ 10. Plant cells placed in a <u>hypotonic</u> solution will swell.

5-V. ROUTES ACROSS CELL MEMBRANES (p. 85)
5-VI. PROTEIN-MEDIATED TRANSPORT (pp. 86–87)

Selected Italicized Words

solute binding site

Boldfaced, Page-Referenced Terms

(85) passive transport _____

_____

(85) active transport _____

_____

(87) calcium pump _____

_____

(87) sodium-potassium pump _____

_____

Choice

For questions 1–12, choose from the following:

    a. passive transport   b. active transport   c. applies to both active and passive transport

___ 1. The calcium pump

___ 2. The transport mechanism sometimes called "facilitated diffusion"

___ 3. Involves a carrier protein that has not been energized

___ 4. A carrier protein has a binding site that attracts a *specific* substance

___ 5. At any given time, the *net* direction of movement depends on how many solute molecules make random contact with vacant binding sites

___ 6. The carrier protein must receive an energy boost, usually from ATP

___ 7. Solute binding to a carrier protein leads to changes in protein shape

___ 8. The sodium-potassium pump

___ 9. Molecules of the solute can be moved both ways across the membrane, depending on which way the carrier's binding site faces

___ 10. A solute is pumped across the cell membrane *against* its concentration gradient

___ 11. The solute binding site becomes altered when ATP donates energy to the carrier protein to allow easier solute binding

___ 12. Two-way transport would continue until solute concentrations became equal on both sides of the membrane; other processes such as cellular solute use may influence the outcome

5-VII. EXOCYTOSIS AND ENDOCYTOSIS (pp. 88–89)

Selected Italicized Words

receptor-mediated endocytosis

Boldfaced, Page-Referenced Terms

(88) exocytosis _____

_____

(88) endocytosis _____

_____

Matching

Choose the one most appropriate answer for each.

1. ___ pinocytosis

2. ___ exocytosis

3. ___ lysosomes

4. ___ receptor-mediated
endocytosis

5. ___ endocytosis

6. ___ phagocytosis

A. Contain digestive enzymes; fuse with incoming vesicles carrying biological molecules

B. Means "cell eater"; such cells may engulf bacterial cells by extending pseudopods around them

C. "Coated pits," lined with membrane receptors specific for lipoproteins

D. A cytoplasmic vesicle moves to the plasma membrane and fuses with it; contents are expelled to the surroundings

E. Part of the plasma membrane sinks inward and balloons around particles, fluid, or tiny prey and seals on itself to form a vesicle

F. Means "cell drinking"; an endocytic vesicle forms around solute-rich extracellular fluid

---

Self-Quiz

___ 1. White blood cells use _____ to devour disease agents invading your body.
   a. diffusion
   b. bulk flow
   c. osmosis
   d. phagocytosis

___ 2. _____ cells depend on the calcium-pump mechanism.
   a. Intestine
   b. Nerve
   c. Muscle
   d. Amoeba

___ 3. Water is such an excellent solvent primarily because _____.
   a. it forms spheres of hydration around substances and can form hydrogen bonds with many nonpolar substances
   b. it has a high heat of fusion
   c. of its cohesive properties
   d. it is a liquid at room temperature

___ 4. In a lipid bilayer, tails point inward and form a _____ region that excludes water.
   a. acidic
   b. basic
   c. hydrophilic
   d. hydrophobic

___ 5. A protistan adapted to life in a freshwater pond is collected in a bottle and transferred to a saltwater bay. Which of the following is likely to happen?
   a. The cell bursts.
   b. Salts flow out of the protistan cell.
   c. The cell shrinks.
   d. Enzymes flow out of the protistan cell.

___ 6. Which of the following is *not* a form of active transport?
   a. Sodium-potassium pump
   b. Endocytosis
   c. Exocytosis
   d. Bulk flow

___ 7. Which of the following is *not* a form of passive transport?
   a. Osmosis
   b. Facilitated diffusion
   c. Bulk flow
   d. Exocytosis

___ 8. O_2, CO_2, H_2O, and other small, electrically neutral molecules move across the cell membrane by _____.
   a. facilitated diffusion
   b. receptor-mediated endocytosis
   c. simple diffusion
   d. active transport

___ 9. Ions such as H$^+$, Na$^+$, K$^+$, and Ca^{++} move across cell membranes by _____.

    a. facilitated diffusion
    b. receptor-mediated endocytosis
    c. simple diffusion
    d. active transport

___ 10. Coated pits, receptors, and transport vesicle formation participate in _____.

    a. facilitated diffusion
    b. receptor-mediated endocytosis
    c. simple diffusion
    d. active transport

Chapter Objectives/Review Questions

This section lists general and detailed chapter objectives that can be used as review questions. You can make maximum use of these items by writing answers on a separate sheet of paper. Fill in answers where blanks are provided. To check for accuracy, compare your answers with information given in the chapter or glossary.

Page Objectives/Questions

(78)   1. _____ molecules are the most abundant component of cell membranes.
(78)   2. Describe the lipid bilayer.
(79)   3. A lipid bilayer has a _____ behavior as a result of constant motion of lipid molecules.
(79)   4. Being a composite of lipids and proteins, the membrane is said to have a _____ quality.
(79)   5. Most membrane functions are carried out by _____ embedded in the bilayer.
(80)   6. Distinguish between the transport proteins known as carrier and channel proteins.
(80)   7. _____ proteins have binding sites for hormones and other extracellular substances that can trigger alterations in cell activities.
(80)   8. State the functions of recognition proteins and adhesion proteins.
(80–81)   9. What is the value of freeze-fracturing to cellular studies?
(82)   10. The net movement of like molecules down a concentration gradient is called _____.
(82)   11. In addition to concentration gradients, diffusion can be influenced by temperature, _____ gradients, and _____ gradients.
(83)   12. Define *tonicity*.
(83)   13. Water tends to move from a _____ solution (less solutes) to a _____ solution (more solutes).
(83)   14. The movement of water across membranes in response to solute concentration gradients, fluid pressure, or both is called _____.
(83)   15. Define and cite an example of *bulk flow*.
(85)   16. Distinguish passive transport from active transport.
(86)   17. Describe the mechanism and result of solute binding to a carrier protein.
(87)   18. When _____ donates energy to the carrier protein, the solute binding site becomes altered to allow easier solute binding.
(87)   19. The calcium and sodium-potassium pump are examples of _____ transport.
(88)   20. Distinguish exocytosis from endocytosis.
(88)   21. _____ refers to "cell drinking"; _____ refers to "cell eating."
(88)   22. Describe mechanisms involved in receptor-mediated endocytosis.

Integrating and Applying Key Concepts

1. If there were no such thing as active transport, how would the lives of organisms be affected?

Answers

Interactive Exercises

5-I. MEMBRANE STRUCTURE AND FUNCTION (pp. 76–79)
1. J; 2. I; 3. H; 4. E; 5. G; 6. C; 7. D; 8. F; 9. B; 10. A.

5-II. FUNCTIONS OF MEMBRANE PROTEINS (pp. 80–81)
1. adhesion; 2. open channel; 3. gated channel; 4. gated channel; 5. carrier; 6. receptor; 7. recognition;
8. a. transport proteins (carrier); b. adhesion protein; c. receptor protein; d. recognition protein; e. transport proteins (channel).

5-III. DIFFUSION (pp. 82–83)
1. energy; 2. Concentration; 3. Gradient; 4. diffusion; 5. faster; 6. slower; 7. equilibrium; 8. faster; 9. electric; 10. attraction; 11. pressure.

5-IV. OSMOSIS (pp. 81–84)
1. cannot; 2. T; 3. solutes; 4. T; 5. hypotonic; 6. T; 7. isotonic; 8. less; 9. T; 10. T.

5-V. ROUTES ACROSS CELL MEMBRANES (p. 85)

5-VI. PROTEIN-MEDIATED TRANSPORT (pp. 86–87)
1. b; 2. a; 3. a; 4. c; 5. a; 6. b; 7. c; 8. b; 9. a; 10. b; 11. b; 12. a.

5-VII. EXOCYTOSIS AND ENDOCYTOSIS (pp. 88–89)
1. F; 2. D; 3. A; 4. C; 5. E; 6. B.

Self-Quiz
1. d; 2. c; 3. a; 4. d; 5. c; 6. d; 7. d; 8. c; 9. d; 10. b.

6

GROUND RULES OF METABOLISM

ENERGY AND LIFE
    How Much Energy Is Available?
    The One-Way Flow of Energy

ENERGY AND THE DIRECTION OF METABOLIC REACTIONS
    Energy Losses and Energy Gains
    Reversible Reactions

METABOLIC PATHWAYS

ENZYMES
    Characteristics of Enzymes
    Enzyme-Substrate Interactions
    Effects of Temperature and pH on Enzyme Activity
    Control of Enzyme Activity

ENZYME HELPERS
    Coenzymes
    Metal Ions

ELECTRON TRANSFERS IN METABOLIC PATHWAYS
    Focus on Health: You Light Up My Life—Visible Effects of Excited Electrons

ATP—THE MAIN ENERGY CARRIER
    Structure and Function of ATP
    The ATP/ADP Cycle

Interactive Exercises

Growing Old with Molecular Mayhem (pp. 92–93)

6-I. ENERGY AND LIFE (pp. 93–95)

Selected Italicized Words

kilocalorie

Boldfaced, Page-Referenced Terms

(93) metabolism _____

_____

(93) energy _____

_____

(94) first law of thermodynamics _____

_____

(95) second law of thermodynamics _____

_____

True-False

If the statement is true, write a T in the blank. If the statement is false, make it correct by changing the underlined word(s) and writing the correct word(s) in the answer blank.

_____ 1. The <u>first</u> law of thermodynamics states that entropy is constantly increasing in the universe.

_____ 2. Your body steadily gives off heat equal to that from a <u>100-watt</u> light bulb.

_____ 3. When you eat a potato, some of the stored chemical energy of the food is converted into <u>mechanical</u> energy that moves your muscles.

_____ 4. <u>Energy</u> is the capacity to accomplish work.

_____ 5. The amount of low-quality energy in the universe is <u>decreasing</u>.

Matching

In the blank preceding each item, indicate if the first law of thermodynamics (I) or the second law of thermodynamics (II) is best described.

6. ___ Cooling of a cup of coffee

7. ___ Evaporation of gasoline into the atmosphere

8. ___ A hydroelectric plant at a waterfall, producing electricity

9. ___ The creation of a snowman by children

10. ___ The glow of an incandescent bulb following the flow of electrons through a wire

11. ___ A typesetter arranging the type for a page of the biology text

12. ___ The movement of a gas-powered automobile

13. ___ The glow of a firefly

14. ___ Humans running the 100-meter dash following usual food intake

15. ___ The death and decay of an organism

6-II. ENERGY AND THE DIRECTION OF METABOLIC REACTIONS (pp. 96–97)

Selected Italicized Words

inputs of energy, exergonic reaction, endergonic reaction, phenylketonuria

Boldfaced, Page-Referenced Terms

(96) chemical equilibrium _____

_____

Fill-in-the-Blanks

The substances present at the end of a reaction, the (1) _____, may have less or more energy than did the starting substances, the (2) _____. Most reactions are (3) _____ in that they can proceed in forward and reverse directions. Such reactions tend to approach chemical (4) _____, a state in which the reactions are proceeding at about the same (5) _____ in both directions.

Labeling

Classify each of the following reactions as *endergonic* or *exergonic*.

_____ 6. Burning wood at a campfire

_____ 7. The products of a chemical reaction have more energy than the reactants.

_____ 8. Glucose + oxygen → carbon dioxide + water + energy

_____ 9. The reactants of a chemical reaction have more energy than the product.

_____ 10. The reaction releases energy.

6-III. METABOLIC PATHWAYS (p. 98)
6-IV. ENZYMES (pp. 98–100)

Selected Italicized Words

allosteric control, biosynthetic pathways, degradative pathways

Boldfaced, Page-Referenced Terms

(98) metabolic pathway _____

_____

(98) substrates _____

_____

(98) intermediates _____

_____

(98) enzymes _____

_____

(98) cofactors _____

_____

(98) energy carriers _____

_____

(98) end products _____

_____

(98) active site _____

_____

(98) induced-fit model _____

_____

(99) activation energy _____

_____

(100) feedback inhibition _____

_____

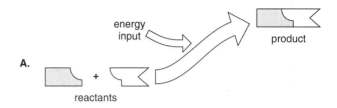
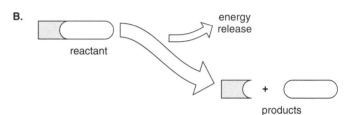

Labeling

Study the diagram below; choose the most appropriate answer for each: A or B.

1. _____ a degradative reaction

2. _____ an endergonic reaction

3. _____ a biosynthetic reaction

4. _____ an exergonic reaction

Matching

Match the most appropriate letter to its number.

5. ___ intermediates

6. ___ degradative pathways

7. ___ dynamic equilibrium

8. ___ cofactors

9. ___ end products

10. ___ metabolic pathway

11. ___ reactants

12. ___ energy carriers

13. ___ biosynthetic pathways

14. ___ enzymes

A. An orderly series of reactions catalyzed by enzymes
B. Small organic molecules are assembled into larger organic molecules
C. Mainly ATP; donates energy to reactions
D. Small molecules and metal ions that assist enzymes or serve as carriers
E. Substances (substrates or precursors) able to enter into a reaction
F. Compounds formed between the beginning and end of a metabolic pathway
G. Organic compounds are broken down in stepwise reactions
H. Proteins that catalyze reactions
I. Rate of forward reaction = rate of reverse reaction
J. Substances present at the end of a metabolic pathway

Matching

Study the sequence of reactions below. Identify the components of the reactions by selecting items from the following list and entering the correct letter in the appropriate blank.

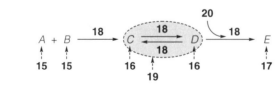

15. _____

16. _____

17. _____

18. _____

19. _____

20. _____

A. cofactor
B. intermediates
C. reactants

D. end product
E. reversible reaction
F. enzymes

Fill-in-the-Blanks

(21) _____ are highly selective proteins that act as catalysts, which means they greatly enhance the rate at which specific reactions approach (22) _____. The specific substance upon which a particular enzyme acts is called its (23) _____; this substance fits into the enzyme's crevice, which is called its (24) _____ _____. The (25) _____-_____ model describes how a substrate contacts the site without a perfect fit. Enzymes increase reaction rates by lowering the required (26) _____ _____.

(27) _____ and (28) _____ are two factors that influence the rates of enzyme activity. Extremely high fevers can destroy the three-dimensional shape of an enzyme, which may adversely affect (29) _____ and cause death. The increased heat energy disrupts weak (30) _____ holding the enzyme in its three-dimensional shape. Most enzymes function best at about pH (31) _____.

One example of a control governing enzyme activity is the (32) _____ pathway in bacteria. The bacterium synthesizes tryptophan and other (33) _____ _____ to construct its proteins. When no more tryptophan is needed, the rate of protein synthesis slows and the cellular concentration of tryptophan (the end-product) (34) (choose one) ❑ rises ❑ falls, and a control mechanism called (35) _____ _____ begins to operate. In this case, excess tryptophan molecules shut down their own production when the unused molecules of tryptophan act to inhibit a key (36) _____ in the biochemical pathway. The inhibited enzyme is governed by (37) _____ control. Such enzymes have an active site and a (38) _____ site where specific substances may bind and alter enzyme activity. When the pathway producing tryptophan is blocked, fewer tryptophan molecules are present to inhibit the key enzyme and production (39) (choose one) ❑ rises ❑ falls. Another type of biochemical control in humans and other multicelled organisms operates through signaling agents called (40) _____.

Matching

Match the items on the sketch below with the list of descriptions. Some answers may require more than one letter.

41. ___

42. ___

43. ___

44. ___

45. ___

46. ___

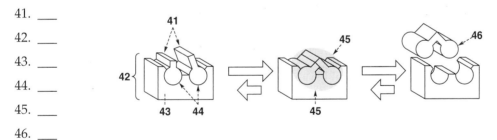

A. Transition state, the time of the most precise fit between enzyme and substrate
B. Complementary active site of the enzyme
C. Enzyme, a protein with catalytic power
D. Product or reactant molecules that an enzyme can specifically recognize
E. Product or reactant molecule
F. Bound enzyme-substrate complex

6-V. ENZYME HELPERS (p. 102)

Boldfaced, Page-Referenced Terms

(101) coenzymes _____

_____

(101) NAD$^+$ _____

_____

(101) FAD _____

_____

(101) NADP$^+$ _____

_____

Matching

Choose the one most appropriate answer for each.

1. ___ metal ions
2. ___ FAD
3. ___ coenzymes
4. ___ NADP$^+$
5. ___ NAD$^+$

A. Large organic molecules; derived in part from vitamins; transfer hydrogens and electrons

B. A coenzyme, nicotinamide adenine dinucleotide phosphate; in reduced form carries H$^+$ and electrons to other reaction sites

C. A coenzyme, nicotinamide adenine transfers hydrogens and electrons to other reaction sites

D. Example: Fe^{++} serves as a cofactor and is a component of enzymatic cytochromes

E. A coenzyme, flavin adenine dinucleotide; transfers hydrogens and electrons to other reaction sites

6-VI. ELECTRON TRANSFERS IN METABOLIC PATHWAYS (pp. 102–103)

Selected Italicized Words

oxidation-reduction reactions, Pyrophorus, Mycobacterium tuberculosis, luciferase

Boldfaced, Page-Referenced Terms

(102) electron transport systems _____

_____

Fill-in-the-Blanks

The release of energy from glucose in cells proceeds in controlled steps of a degradative pathway, so that

(1) _____ molecules form along the route from glucose to carbon dioxide and water. At each step in a

metabolic pathway, a specific (2) _____ lowers the activation energy for the formation of an intermedi-

ate compound. At each step in the pathway, only some of the bond energy is released. In the internal mem-

brane systems of chloroplasts and mitochondria, the liberated electrons released from the breaking of

chemical bonds are sent through (3) _____ _____ systems; these organized systems consist of

enzymes and (4) _____, bound in a cell membrane, that transfer electrons in a highly organized

sequence.

Oxidation-reduction refers to (5) _____ transfers. In terms of oxidation-reduction reactions, a molecule in the sequence that donates electrons is said to be (6) _____, while molecules accepting electrons are said to be (7) _____. Electron transport systems "intercept" excited electrons and make use of the (8) _____ they release. If we think of the electron transport system as a staircase, excited electrons at the top of the staircase have the (9) (choose one) ❑ most ❑ least energy. As the electrons are transferred from one electron carrier to another, some (10) _____ can be harnessed to do biological (11) _____. One type of biological work occurs when energy released during electron transfers (down the steps) is used to move ions in ways that set up ion concentration and electric gradients across a membrane. These gradients are essential for the formation of (12) _____.

Matching

Match the lettered statements to the numbered items on the sketch.

13. _____
14. _____
15. _____
16. _____
17. _____

A. Represent the cytochrome molecules in an electron transport system
B. Electrons at their highest energy level
C. Released energy harnessed and used to produce ATP
D. Electrons at their lowest level
E. The separation of hydrogen atoms into protons and electron

6-VII. ATP—THE MAIN ENERGY CARRIER (p. 104)

Boldfaced, Page-Referenced Terms

(104) ATP _____

_____

(104) ATP/ADP cycle_____

_____

_____

(104) phosphorylation _____

_____

Fill-in-the-Blanks

Photosynthetic cells possess the chemical mechanisms to convert sunlight energy to the chemical energy of molecules known as (1) _____ , a type of nucleotide. It is important to recognize that all cells lack the ability to extract energy *directly* from food molecules such as glucose. The energy in the food molecules must first be harnessed and converted to the energy of (2) _____ _____ . Cells must obtain their energy from these molecules. An ATP molecule consists of (3) _____ (a nitrogen-containing compound), the five-carbon sugar (4)_____ , and a string of three (5) _____ groups. These components are held together by (6) _____ bonds, some of which are rather unstable. Many different enzymes carry out (7) _____ reactions to release the terminal phosphate group. This releases (8) _____ that can be used by cells to carry out many types of biological tasks. ATP molecules serve as the cell's (9) _____ currency.

In the ATP/ADP cycle, an energy input drives the binding of ADP to a phosphate group or to unbound phosphate (P_i), forming (10) _____ . Adding phosphate to a molecule is called (11) _____ ; this increases the energy level of that molecule and primes it to enter a specific reaction in a metabolic pathway.

Labeling

Identify the molecule below and label its parts.

12. _____

13. _____ _____

14. _____

15. The name of this molecule is _____

_____ .

Self-Quiz

___ 1. An important principle of the second law of thermodynamics states that _____ .

    a. energy can be transformed into matter, and because of this, we can get something for nothing

    b. energy can only be destroyed during nuclear reactions, such as those that occur inside the sun

    c. if energy is gained by one region of the universe, another place in the universe also must gain energy in order to maintain the balance of nature

    d. matter tends to become increasingly more disorganized

___ 2. Essentially, the first law of thermodynamics states that _____ .

    a. one form of energy cannot be converted into another

    b. entropy is increasing in the universe

    c. energy cannot be created or destroyed

    d. energy cannot be converted into matter or matter into energy

___ 3. An enzyme is best described as _____ .

    a. an acid

    b. protein

    c. a catalyst

    d. a fat

    e. both b and c

___ 4. Which is *not* true of enzyme behavior?

    a. Enzyme shape may change during catalysis.

    b. The active site of an enzyme orients its substrate molecules, thereby promoting interaction of their reactive parts.

c. All enzymes have an active site where substrates are temporarily bound.

d. An individual enzyme can catalyze a wide variety of different reactions.

_____ 5. When NAD$^+$ combines with hydrogen, the NAD$^+$ is _____.

a. reduced
b. oxidized
c. phosphorylated
d. denatured

_____ 6. A substance that gains electrons is _____.

a. oxidized
b. a catalyst
c. reduced
d. a substrate

_____ 7. In _____ pathways, carbohydrates, lipids, and proteins are broken down in stepwise reactions that lead to products of lower energy.

a. intermediate
b. biosynthetic
c. induced
d. degradative

_____ 8. As to major function, NAD$^+$, FAD, and NADP$^+$ are classified as _____.

a. enzymes

b. phosphate carriers
c. cofactors that function as coenzymes
d. end products of metabolic pathways

_____ 9. When a phosphate bond is linked to ADP, the bond _____.

a. absorbs a large amount of free energy when the phosphate group is attached during hydrolysis
b. is formed when ATP is hydrolyzed to ADP and one phosphate group
c. is usually found in each glucose molecule; that is why glucose is chosen as the starting point for glycolysis
d. formed releases a large amount of usable energy when the phosphate group is split off during hydrolysis

_____ 10. An allosteric enzyme _____.

a. has an active site where substrate molecules bind and another site that binds with intermediate or end-product molecules
b. is an important energy-carrying nucleotide
c. carries out either oxidation reactions or reduction reactions but not both
d. raises the activation energy of the chemical reaction it catalyzes

Chapter Objectives/Review Questions

This section lists general and detailed chapter objectives that can be used as review questions. You can make maximum use of these items by writing answers on a separate sheet of paper. Fill in answers where blanks are provided. To check for accuracy, compare your answers with information given in the chapter or glossary.

Page	Objectives/Questions
(93)	1. _____ is the controlled capacity to acquire and use energy for stockpiling, breaking apart, building, and eliminating substances in ways that contribute to survival and reproduction.
(93–95)	2. Define *energy*; be able to state the first and second laws of thermodynamics.
(94)	3. Explain what is meant by a "system" as related to the laws of thermodynamics.
(95)	4. _____ is a measure of the degree of randomness or disorder of systems.
(95)	5. Explain how the world of life maintains a high degree of organization.
(96–97)	6. What is meant by a reversible reaction?
(96–97)	7. Describe the condition known as chemical equilibrium.
(98)	8. A _____ pathway is an orderly sequence of reactions with specific enzymes acting at each step; the pathways are always linear or circular.
(98)	9. Give the function of each of the following participants in metabolic pathways: substrates, intermediates, enzymes, cofactors, energy carriers, and end products.
(98)	10. What are enzymes? Explain their importance.

(98)   11. _____ are small molecules or metal ions that assist enzymes or carry atoms or electrons from one reaction site to another.

(98)   12. The location on the enzyme where specific reactions are catalyzed is the active _____.

(98)   13. According to Koshland's _____-_____ model, each substrate has a surface region that almost but not quite matches chemical groups in an active site.

(99)   14. When the "energy hill" is made smaller by enzymes so that particular reactions may proceed, it may be said that the enzyme has lowered the _____ energy.

(100)   15. Explain what happens to enzymes in the presence of extreme temperatures and pH.

(100–101)   16. An enzyme control called _____ inhibition operates in the tryptophan pathway when excess tryptophan molecules inhibit a key allosteric enzyme in the pathway.

(101)   17. Cofactors called _____ are organic molecules that may be vitamins or are derived in part from vitamins; _____ ions may also serve as cofactors.

(101)   18. Explain the differences between NAD^+ and NADH, $NADP^+$ and NADPH, and FAD and $FADH_2$.

(102)   19. _____-_____ reactions refer to the electron transfers occurring in an electron transport system.

(103)   20. Fireflies, various beetles, and some other organisms display _____ when luciferases excite the electrons of luciferans.

(104)   21. _____ molecules are the cell's common currency of energy.

(104)   22. Explain the functioning of the ATP/ADP cycle.

(104)   23. Adding a phosphate to a molecule is called _____.

Integrating and Applying Key Concepts

A piece of dry ice left sitting on a table at room temperature vaporizes. As the dry ice vaporizes into CO_2 gas, does its entropy increase or decrease? Tell why you answered as you did.

Answers

Interactive Exercises

6-I. ENERGY AND LIFE (pp. 93–95)
1. second; 2. T; 3. T; 4. T; 5. increasing; 6. II; 7. II; 8. I; 9. I; 10. I; 11. II; 12. I; 13. I; 14. I; 15. II.

6-II. ENERGY AND THE DIRECTION OF METABOLIC REACTIONS (pp. 96–97)
1. products; 2. reactants; 3. reversible; 4. equilibrium; 5. rate; 6. exergonic; 7. endergonic; 8. exergonic; 9. exergonic; 10. exergonic.

6-III. METABOLIC PATHWAYS (p. 98)
6-IV. ENZYMES (pp. 98–101)
1. B; 2. A; 3. A; 4. B; 5. F; 6. G(A); 7. I; 8. D; 9. J; 10. A; 11. E; 12. C; 13. B(A); 14. H; 15. C; 16. B; 17. D; 18. F; 19. E; 20. A; 21. Enzymes; 22. equilibrium; 23. substrate; 24. active site; 25. induced-fit; 26. activation energy; 27. Temperature (Enzymes); 28. enzymes (temperature); 29. metabolism; 30. bonds; 31. 7; 32. tryptophan; 33. amino acids; 34. falls; 35. feedback inhibition; 36. enzyme; 37. allosteric; 38. control; 39. rises; 40. hormones; 41. D; 42. F; 43. C; 44. B; 45. A; 46. E.

6-V. ENZYME HELPERS (p. 101)
1. D; 2. E; 3. A; 4. B; 5. C.

6-VI. ELECTRON TRANSFERS IN METABOLIC PATHWAYS (pp. 102–103)
1. intermediate; 2. enzyme; 3. electron transport; 4. cofactors; 5. electron; 6. oxidized; 7. reduced; 8. energy; 9. most; 10. energy; 11. work; 12. ATP; 13. E; 14. B; 15. C; 16. A; 17. D.

6-VII. ATP—THE MAIN ENERGY CARRIER (p. 104)
1. ATP; 2. adenosine triphosphate; 3. adenine; 4. ribose; 5. phosphate; 6. covalent; 7. hydrolysis; 8. energy; 9. energy; 10. ATP; 11. phosphorylation; 12. triphosphate; 13. ribose sugar; 14. adenine (a nitrogen-containing molecule); 15. adenosine triphosphate.

Self-Quiz
1. d; 2. c; 3. e; 4. d; 5. a; 6. c; 7. d; 8. c; 9. d; 10. a.

7

ENERGY-ACQUIRING PATHWAYS

Interactive Exercises

Sun, Rain, and Survival (pp. 105–107)

7-I. PHOTOSYNTHESIS: AN OVERVIEW (pp. 108–109)

Selected Italicized Words

photoautotrophs, chemoautotrophs, aerobic respiration, light-dependent reactions, light-independent reactions

Boldfaced, Page-Referenced Terms

(107) autotrophs _____

_____

(107) heterotrophs _____

_____

(109) chloroplast _____

_____

(109) stroma _____

_____

(109) thylakoid membrane system _____

_____

Fill-in-the-Blanks

(1) _____ are organisms that obtain carbon and energy directly from the environment and are self-nourishing. (2) _____ is the energy source for photoautotrophs. (3) _____ extract energy from inorganic substances such as sulfur. Animals, fungi, many protistans, and most bacteria feed on autotrophs, each other, and organic wastes; such organisms are referred to as (4) _____, which means "other feeders." The survival of nearly all organisms ultimately depends on (5) _____, the main path by which carbon and energy enter the world of life. The prominent energy-releasing pathway is called aerobic (6) _____.

7. In the space below, supply the missing information to complete the summary equation for photosynthesis:

$$12 \underline{\hspace{1cm}} \text{ plus } \underline{\hspace{1cm}} CO_2 \rightarrow \underline{\hspace{1cm}} O_2 \text{ plus } C_6H_{12}O_6 \text{ plus } 6 \underline{\hspace{1cm}}$$

8. Supply information appropriately to state the equation (above) for photosynthesis in words:

(a) _____ molecules of water plus six molecules of (b) _____ _____ (in the presence of pigments, enzymes, and ultraviolet light) yields six molecules of (c) _____ plus one molecule of (d) _____ plus (e) _____ molecules of water.

Choice

For questions 9–20, choose the area of the chloroplast that correctly relates to the listed structures and events.

a. grana   b. stroma

___ 9. Light-independent reactions

___ 10. The coenzyme NADP⁺ picks up liberated hydrogen and electrons

___ 11. Thylakoid membrane system

___ 12. Sugars are assembled

___ 13. Light-dependent reactions

___ 14. ATP production

___ 15. Carbon dioxide provides the carbon

___ 16. Sunlight energy is absorbed

___ 17. Where the first stage of photosynthesis proceeds

___ 18. Water molecules are split

___ 19. NADPH delivers the hydrogen received from water

___ 20. Oxygen is formed

7-II. LIGHT-TRAPPING PIGMENTS (pp. 110–111)

Boldfaced, Page-Referenced Terms

(110) pigments _____

_____

(110) photons _____

_____

(110) chlorophylls _____

_____

(110) carotenoids _____

_____

(110) phycobilins _____

_____

Matching

Choose the most appropriate answer.

1. ___ chlorophylls
2. ___ chlorophyll *b* and carotenoids
3. ___ carotenoids
4. ___ violet-blue-green-yellow-red
5. ___ photons
6. ___ chlorophyll *a*
7. ___ chloroplast
8. ___ phycobilins
9. ___ grana
10. ___ pigments

A. The main pigment of photosynthesis
B. Packets of energy that have an undulating motion through space
C. The two stages of photosynthesis occur here
D. In general, molecules that can absorb light
E. Absorb violet and blue wavelengths but transmit red, orange, and yellow
F. Visible light portion of the electromagnetic spectrum
G. Pigments that transfer energy to chlorophyll *a*
H. Absorb violet-to-blue and red wavelengths; the reason leaves appear green
I. Red and blue pigments
J. The site of the first stage of photosynthesis

7-III. LIGHT-DEPENDENT REACTIONS (pp. 112–114)

Selected Italicized Words

water molecules, photosystem I

Boldfaced, Page-Referenced Terms

(112) light-dependent reactions _____

_____

_____

(112) photosystems _____

_____

(112) electron transport systems _____

_____ _____

(112) cyclic pathway of ATP formation _____

_____

_____

(112) noncyclic pathway of ATP formation _____

_____

_____

(112) photolysis _____

_____

Complete the Table

1. Complete the following table on elements of the cyclic pathway of the light-dependent reactions.

Component	Function
a. Photosystem I	
b. Electrons	
c. P700	
d. Electron acceptor	
e. Electron transport system	
f. ADP	

Labeling

The diagram below illustrates noncyclic photophosphorylation. Identify each numbered part of the illustration.

2. _____

   _____

3. _____

   _____

   _____

4. _____

   _____

5. _____

   _____

6. _____

7. _____

8. _____

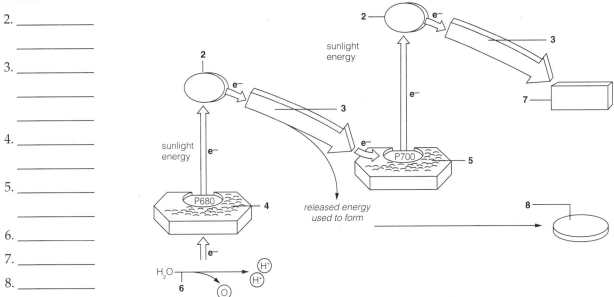

Choice

For questions 9–23, choose the correct light-dependent pathway for the following events and mechanisms; some questions may require both choices as the answer.

        a. cyclic pathway   b. noncyclic pathway

___ 9. ATP formation

___ 10. Photolysis

___ 11. Photosystems, each a cluster of 200 to 300 pigment molecules

___ 12. NADPH

___ 13. Electron transport systems

___ 14. Electrons are not cycled through this pathway

___ 15. ATP is the only product

___ 16. The oldest means of ATP production

___ 17. Chlorophylls P680 and P700 are involved

___ 18. Uses only one photosystem

___ 19. NADPH and ATP are produced

___ 20. Oxygen evolves from split water molecules

___ 21. Uses two different photosystems

___ 22. The dominant form of ATP production today

___ 23. Electrons leave chlorophyll 700, then cycle back to it

Fill-in-the-Blanks

ATP forms in both the cyclic and noncyclic pathways. When (24) _____ flow through the membrane-bound transport systems, they pick up hydrogen ions (H^+) outside the membrane and dump them into the (25) _____ compartment. This sets up H^+ concentration and electric (26) _____ across the membrane. Hydrogen ions that were split away from (27) _____ molecules increase the gradients. The ions respond by flowing out through the interior of (28) _____ proteins that span the membrane. Energy associated with the flow drives the binding of unbound phosphate to ADP, the result being (29) _____. The above description is known as the (30) _____ theory of ATP formation.

7-IV. LIGHT-INDEPENDENT REACTIONS (p. 115)

Boldfaced, Page-Referenced Terms

(115) light-independent reactions _____

_____

_____

(115) RuBP _____

_____

(115) carbon dioxide fixation _____

_____

_____

_____

(115) Calvin-Benson cycle _____

_____

_____

(115) PGA _____

_____

(115) PGAL _____

_____

Label-Match

Identify each part of the accompanying illustration. Complete the exercise by matching and entering the letter of the proper function description in the parentheses following each label.

1. _____ ( )

2. _____ _____ _____ ( )

3. _____ ( )

4. _____ ( )

5. _____ ( )

6. _____ ( )

7. _____ _____ ( )

8. _____-_____ _____ ( )

9. _____ ( )

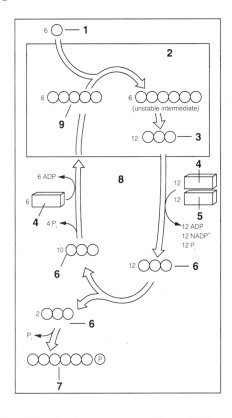

A. A three-carbon sugar, the first sugar produced; goes on to form sugar phosphate and RuBP

B. Typically used at once to form carbohydrate end products of photosynthesis

C. A five-carbon compound produced from PGALs; attaches to incoming CO_2

D. This compound diffuses into leaves; photosynthetic cells have enzymes that attach this molecule to RuBP

E. Includes all the chemistry that "fixes" carbon into an organic compound

F. Two three-carbon molecules of this form from the splitting of the six-carbon intermediate compound

G. A molecule that was reduced in noncyclic photophosphorylation; furnishes hydrogen atoms (from H_2O molecules) to construct sugar molecules

H. A product of the light-dependent reactions; necessary in the light-independent chemistry to energize molecules for reactions

I. Includes all the chemistry that converts PGA to PGAL and PGAL to RuBP and sugar phosphates

Fill-in-the-Blanks

The light-independent reactions can proceed without sunlight as long as (10) _____ and

(11) _____ are available. The reactions begin when an enzyme links (12) _____ to

(13) _____, a five-carbon compound. The resulting six-carbon compound is highly unstable and

breaks apart at once into two molecules of a three-carbon compound, (14) _____. This entire reaction

sequence is called carbon dioxide (15) _____. ATP gives a phosphate group to each (16) _____.

This intermediate compound takes on H^+ and electrons from NADPH to form (17) _____. It takes

(18) _____ carbon dioxide molecules to produce twelve PGALs. Most of the PGAL becomes

rearranged into new (19) _____ molecules, which can be used to fix more (20) _____. Two

(21) _____ are joined together to form a (22) _____ _____, primed for further reactions. The

Calvin-Benson cycle yields enough RuBP to replace those used in carbon dioxide (23) _____. ADP,

NADP$^+$, and phosphate leftovers are sent back to the (24) _____-_____ reaction sites where they

are again converted to (25) _____ and (26) _____. (27) _____ _____ formed in the cycle

serves as a building block for the plant's main carbohydrates.

7-V. THE REACTIONS, START TO FINISH (pp. 116–117)

Complete the Table

Indicate with a check mark which items in the left-hand column apply to each pathway.

Event	Cyclic Pathway	Noncyclic Pathway	Carbon Fixation	Calvin-Benson Cycle
Photolysis of water	(1)	(26)	(51)	(76)
PGAL produced	(2)	(27)	(52)	(77)
P680 and P700 involved	(3)	(28)	(53)	(78)
Uses ATP and NADPH	(4)	(29)	(54)	(79)
ATP via chemiosmosis	(5)	(30)	(55)	(80)
Photosystem I only	(6)	(31)	(56)	(81)
Only ATP produced	(7)	(32)	(57)	(82)
RuBP formed	(8)	(33)	(58)	(83)
Requires CO_2	(9)	(34)	(59)	(84)
Occurs in the stroma	(10)	(35)	(60)	(85)
Uses ADP plus P_i as reactants	(11)	(36)	(61)	(86)
PGA formed	(12)	(37)	(62)	(87)
O_2 is a product	(13)	(38)	(63)	(88)
Reduce NADP to NADPH	(14)	(39)	(64)	(89)
Sugar phosphate produced	(15)	(40)	(65)	(90)
Electrons return to P700	(16)	(41)	(66)	(91)
Photosystems I and II	(17)	(42)	(67)	(92)
Light required	(18)	(43)	(68)	(93)
Six-carbon intermediate	(19)	(44)	(69)	(94)
Glucose formed	(20)	(45)	(70)	(95)
Requires H_2O	(21)	(46)	(71)	(96)
Only P700 involved	(22)	(47)	(72)	(97)
NADPH and ATP produced	(23)	(48)	(73)	(98)
At thylakoid membranes	(24)	(49)	(74)	(99)
Occurs in light or dark	(25)	(50)	(75)	(100)

7-VI. FIXING CARBON—SO NEAR, YET SO FAR (pp. 118–119)

Selected Italicized Words

mesophyll cells, bundle-sheath cells

Boldfaced, Page-Referenced Terms

(118) C3 plants _____

_____

(118) C4 plants _____

_____

(119) succulents _____

_____

(119) CAM plants _____

_____

Choice

For questions 1–10, choose from the following.

a. C3 plants   b. C4 plants

____ 1. Corn, crabgrass, sugarcane

____ 2. When the plants fix carbon, the intermediate compound is oxaloacetate

____ 3. Photorespiration wins out, less PGA forms

____ 4. Eighty percent of the plant species that evolved in Florida

____ 5. The intermediate compound is PGA

____ 6. Fix CO_2 twice

____ 7. Bluegrass

____ 8. Oxaloacetate is transferred to bundle-sheath cells

____ 9. Less sensitive to cold

____ 10. Smaller stomata

7-VII. CHEMOSYNTHESIS (p. 119)

Fill in-the-Blanks

Photosynthesis is not the only energy-acquiring route. A less common route is used by (1) _____ that are classified as (2) _____. These organisms obtain energy by pulling away (3) _____ and (4) _____ from ammonium ions, iron, or sulfur compounds. Many influence the (5) _____ cycling of nitrogen and other vital elements through the biosphere.

Self-Quiz

___ 1. The electrons that are passed to NADPH during noncyclic photophosphorylation were obtained from _____.
   a. water
   b. CO_2
   c. glucose
   d. sunlight

___ 2. Cyclic photophosphorylation functions mainly to _____.
   a. fix CO_2
   b. make ATP
   c. produce PGAL
   d. regenerate ribulose bisphosphate

___ 3. Chemosynthetic autotrophs obtain energy by oxidizing such inorganic substances as _____.
   a. PGA
   b. PGAL
   c. sulfur
   d. water

___ 4. The ultimate electron and hydrogen acceptor in noncyclic photophosphorylation is _____.
   a. $NADP^+$
   b. ADP
   c. O_2
   d. H_2O

___ 5. C4 plants have an advantage in hot, dry conditions because _____.
   a. their leaves are covered with thicker wax layers than those of C3 plants
   b. their stomates open wider than those of C3 plants, thus cooling their surfaces
   c. special leaf cells possess a means of capturing CO_2 even in stress conditions

   d. they also are capable of carrying on photorespiration

___ 6. Chlorophyll is _____.
   a. on the outer chloroplast membrane
   b. inside the mitochondria
   c. in the stroma
   d. in the thylakoids

___ 7. Thylakoid disks are stacked in groups called _____.
   a. grana
   b. stroma
   c. lamellae
   d. cristae

___ 8. Plant cells produce O_2 during photosynthesis by _____.
   a. splitting CO_2
   b. splitting water
   c. degradation of the stroma
   d. breaking up sugar molecules

___ 9. Plants need _____ and _____ to carry on photosynthesis.
   a. oxygen; water
   b. oxygen; CO_2
   c. CO_2; H_2O
   d. sugar; water

___ 10. The two products of the light-dependent reactions that are required for the light-independent chemistry are _____ and _____.
   a. CO_2; H_2O
   b. O_2; NADPH
   c. O_2; ATP
   d. ATP; NADPH

Chapter Objectives/Review Questions

This section lists general and detailed chapter objectives that can be used as review questions. You can make maximum use of these items by writing answers on a separate sheet of paper. Fill in answers where blanks are provided. To check for accuracy, compare your answers with information given in the chapter or glossary.

Page	Objectives/Questions
(107)	1. Distinguish between organisms known as autotrophs and those known as heterotrophs.
(107)	2. _____ can obtain their energy from sunlight; _____ get energy by stripping electrons from sulfur or some other inorganic substance.
(108–109)	3. List the two major stages of photosynthesis as well as where in the cell they occur and what reactions occur there.
(108)	4. The energy-poor molecules that act as raw materials in the photosynthetic equation are _____ and _____.
(108–109)	5. Describe the details of a familiar site of photosynthesis, the green leaf. Begin with the layers of a leaf cross-section and complete your description with the minute structural sites within the chloroplast where the major sets of photosynthetic reactions occur.
(109)	6. The flattened channels and disklike compartments inside the chloroplast are organized into stacks, the _____, which are surrounded by a semifluid interior, the _____; this is the _____ membrane system.
(110)	7. _____ are packets of light energy.
(110)	8. _____ absorb violet-to-blue as well as red wavelengths.
(110)	9. The main pigment of photosynthesis is _____.
(110)	10. Name the wavelengths absorbed and transmitted by the carotenoids.
(110)	11. What pigments are responsible for the red and blue colors of red algae and cyanobacteria?
(111)	12. State what T. Englemann's 1882 experiment with *Cladophora* revealed.
(112)	13. Describe how the pigments found on thylakoid membranes are organized into photosystems and how they relate to photon light energy.
(112)	14. Contrast the components and functioning of the cyclic and noncyclic pathways of the light-dependent reactions.
(112)	15. Explain what the water split during photolysis contributes to the noncyclic pathway of the light-dependent reactions.
(112)	16. Two energy-carrying molecules produced in the noncyclic pathways are _____ and _____; explain why these molecules are necessary for the light-independent reactions.
(113)	17. _____ _____ is the metabolic pathway most efficient in releasing the energy stored in organic compounds.
(113)	18. Following evolution of the noncyclic pathway, _____ accumulated in the atmosphere and made _____ respiration possible.
(114)	19. Explain how the chemiosmotic theory is related to thylakoid compartments and the production of ATP.
(115)	20. Explain why the light-independent reactions are called by that name.
(115)	21. Describe the process of carbon dioxide fixation by stating which reactants are necessary to initiate the process and what stable products result from this process.
(115)	22. When carbon fixation occurs, sunlight energy, which excited electrons during the light-dependent reactions, becomes stored as _____ energy in an organic compound.
(115)	23. Describe the Calvin-Benson cycle in terms of its reactants and products.
(115)	24. State the fate of all the sugar phosphates produced by photosynthetic reactions in autotrophs.
(116)	25. Study the general equation for photosynthesis until you can remember the reactants and products. Reproduce the equation from memory on another piece of paper.
(118–119)	26. Describe the mechanism by which C4 plants thrive under hot, dry conditions; distinguish this CO_2 capturing mechanism from that of C3 plants.

(119)   27. Describe the carbon-fixing adaptation of the CAM plants living in arid environments.

(119)   28. Bacteria that are able to obtain energy from ammonium ions, iron, or sulfur compounds are known as _____.

Integrating and Applying Key Concepts

Suppose that humans acquired all the enzymes needed to carry out photosynthesis. Speculate about the attendant changes in human anatomy, physiology, and behavior that would be necessary for those enzymes actually to carry out photosynthetic reactions.

Answers

Interactive Exercises

7-I. PHOTOSYNTHESIS: AN OVERVIEW
(pp. 105–109)
1. Autotrophs; 2. Sunlight; 3. Chemoautotrophs; 4. heterotrophs; 5. photosynthesis; 6. respiration; 7. 12 H_2O plus 6 $CO_2 \rightarrow$ 6 O_2 plus $C_6H_{12}O_6$ plus 6 H_2O; 8. a. Twelve; b. carbon dioxide; c. oxygen; d. glucose; e. six; 9. b; 10. a; 11. a; 12. b; 13. a; 14. a; 15. b; 16. a; 17. a; 18. a; 19. b; 20. a.

7-II. LIGHT-TRAPPING PIGMENTS (pp. 110–111)
1. H; 2. G; 3. E; 4. F; 5. B; 6. A; 7. C; 8. I; 9. J; 10. D.

7-III. LIGHT-DEPENDENT REACTIONS (pp. 112–114)
1. a. A pigment cluster dominated by P700; b. Electrons represent energy; they are ejected from P700 to an electron acceptor but move over the electron transport system, where some of the energy is used to produce ATP; c. A special chlorophyll molecule that absorbs wavelengths of 700 nanometers and then ejects electrons; d. A molecule that accepts electrons ejected from chlorophyll P700 and then passes electrons down the electron transport system; e. Electrons flow through this system; composed of a series of molecules bound in the thylakoid membrane that drive photophosphorylation; f. ADP undergoes photophosphorylation in cyclic photophosphorylation to become ATP; 2. electron acceptor; 3. electron transport system; 4. photosystem II; 5. photosystem I; 6. photolysis; 7. NADPH; 8. ATP; 9. a, b; 10. b; 11. a, b; 12. b; 13. a, b; 14. b; 15. a; 16. a; 17. b;

18. a; 19. b; 20. b; 21. b; 22. b; 23. a; 24. electrons; 25. thylakoid; 26. gradients; 27. water (H_2O); 28. channel; 29. ATP; 30. chemiosmotic.

7-IV. LIGHT-INDEPENDENT REACTIONS (p. 115)
1. CO_2 (D); 2. carbon dioxide fixation (E); 3. PGA (F); 4. ATP (H); 5. NADPH (G); 6. PGAL (A); 7. sugar phosphates (B); 8. Calvin-Benson cycle (I); 9. RuBP (C); 10. ATP (NADPH); 11. NADPH (ATP); 12. CO_2; 13. RuBP; 14. PGA; 15. fixation; 16. PGA; 17. PGAL; 18. six; 19. RuBP; 20. CO_2; 21. PGALs; 22. sugar phosphate; 23. fixation; 24. light-dependent; 25. ATP (NADPH); 26. NADPH (ATP); 27. Sugar phosphate.

7-V. THE REACTIONS, START TO FINISH
(pp. 116–117)
The following numbers should have a check mark (√):
5, 6, 7, 11, 16, 22, 24, 26, 28, 30, 36, 38, 39, 42, 43, 46, 48, 49, 59, 60, 62, 69, 75, 77, 79, 83, 84, 85, 90, 95, 100.

7-VI. FIXING CARBON—SO NEAR, YET SO FAR
(pp. 118–119)
1. b; 2. b; 3. a; 4. b; 5. a; 6. b; 7. a; 8. b; 9. a; 10. b.

7-VII. CHEMOSYNTHESIS (p. 119)
1. bacteria; 2. chemoautotrophs; 3. hydrogens (electrons); 4. electrons (hydrogens); 5. global.

Self-Quiz
1. a; 2. b; 3. c; 4. a; 5. c; 6. d; 7. a; 8. b; 9. c; 10. d.

8

ENERGY-RELEASING PATHWAYS

Interactive Exercises

8-I. HOW CELLS MAKE ATP (pp. 122–125)

Boldfaced, Page-Referenced Terms

(124) aerobic respiration _____

_____

(124) fermentation pathways _____

_____

(124) anaerobic electron transport _____

_____

(124) glycolysis _____

_____

(124) Krebs cycle _____

_____

(124) electron transport phosphorylation _____

_____

Matching

Choose the one most appropriate answer for each.

1. ___ Krebs cycle
2. ___ oxygen
3. ___ mitochondrion
4. ___ electron transport phosphorylation
5. ___ enzymes
6. ___ ATP
7. ___ glycolysis
8. ___ aerobic respiration
9. ___ cytoplasm
10. ___ fermentation pathways and anaerobic electron transport

A. Starting point for three energy-releasing pathways
B. Main energy-releasing pathway for ATP formation
C. Site of glycolysis
D. Third and final stage of aerobic respiration; high ATP yield
E. Oxygen is not the final electron acceptor
F. Catalyze each reaction step in the energy-releasing pathways
G. Second stage of aerobic respiration; pyruvate is broken down to CO_2 and H_2O
H. Site of the aerobic pathway
I. The final electron acceptor in aerobic pathways
J. The energy form that drives metabolic reactions

Completion

11. Complete the equation below, which summarizes the degradative pathway known as aerobic respiration:

$$\underline{\quad} \text{ plus } \underline{\quad} O_2 \rightarrow 6 \underline{\quad} \text{ plus } 6 \underline{\quad}$$

12. Supply the appropriate information to state the equation (above) for aerobic respiration in words:
One molecule of glucose plus six molecules of (a) _____ (in the presence of appropriate enzymes) yields (b) _____ molecules of carbon dioxide plus (c) _____ molecules of water.

Fill-in-the-Blanks

There are three stages of aerobic respiration. In the first stage, (13) _____, glucose is partially degraded to (14) _____. By the end of the second stage, which includes the (15) _____ cycle, glucose has been completely degraded to carbon dioxide and (16) _____. Neither of the first two stages produces much (17) _____. During both stages, protons and (18) _____ are stripped from intermediate compounds and delivered to a(n) (19) _____ system. That system is used in the third stage of reactions, electron transport (20) _____, which produces a high yield of (21) _____. (22) _____ accepts the "spent" electrons from the transport system and keeps it cleared for repeated operation.

8-II. GLYCOLYSIS: FIRST STAGE OF THE ENERGY-RELEASING PATHWAYS
(pp. 126–127)

Boldfaced, Page-Referenced Terms

(126) pyruvate _____

_____

(126) substrate-level phosphorylation _____

_____

Sequence

Arrange the following events of the glycolysis pathway in correct chronological sequence. Write the letter of the first step next to 1, the letter of the second step next to 2, and so on.

1. ___   A. Two ATPs form by substrate-level phosphorylation; the cell's energy debt is paid off
2. ___   B. Diphosphorylated glucose (fructose 1,6-bisphosphate) molecules split to form two PGALs; this is the first energy-releasing step
3. ___   C. Two three-carbon pyruvate molecules form as the end products of glycolysis
4. ___   D. Glucose is present in the cytoplasm
5. ___   E. Two more ATPs form by substrate-level phosphorylation; the cell gains ATP; net yield of ATP from glycolysis is two ATPs
6. ___   F. The cell invests two ATPs; one phosphate group is attached to each end of the glucose molecule (fructose 1,6-bisphosphate)
7. ___   G. Two PGALs gain two phosphate groups from the cytoplasm
8. ___   H. Hydrogen atoms and electrons from each PGAL are transferred to NAD$^+$, reducing this carrier to NADH

8-III. COMPLETING THE AEROBIC PATHWAY (pp. 128–131)

Labeling

In exercises 1–5, identify the structure or location; in exercises 6–10, identify the chemical substance involved.

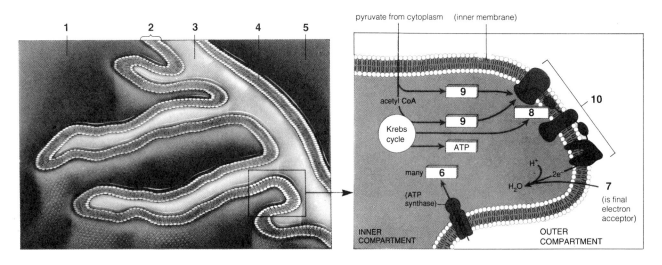

1. _____ _____ of mitochondrion

2. _____ _____ of mitochondrion

3. _____ _____ of mitochondrion

4. _____ _____ of mitochondrion

5. _____

6. _____

7. _____

8. _____

9. _____

10. _____ _____ _____

Choice

For questions 11–32, choose from the following. Some correct answers may require more than one letter.

a. preparatory steps to Krebs cycle   b. Krebs cycle   c. electron transport phosphorylation

___ 11. Chemiosmosis occurs to form ATP molecules

___ 12. Three carbon atoms in pyruvate leave as three CO_2 molecules

___ 13. Chemical reactions occur at transport systems

___ 14. Coenzyme A picks up a two-carbon acetyl group

___ 15. Makes two turns for each glucose molecule entering glycolysis

___ 16. Two NADH molecules form for each glucose entering glycolysis

___ 17. Oxaloacetate forms from intermediate molecules

___ 18. Named for a scientist who worked out its chemical details

___ 19. Occurs within the mitochondrion

___ 20. Two $FADH_2$ and six NADH form from one glucose molecule entering glycolysis

___ 21. Hydrogens collect in the mitochondrion's outer compartment

___ 22. Hydrogens and electrons are transferred to NAD^+ and FAD

___ 23. Two ATP molecules form by substrate-level phosphorylation

___ 24. Free oxygen withdraws electrons from the system and then combines with H^+ to form water molecules

___ 25. No ATP is produced

___ 26. Thirty-two ATPs are produced

___ 27. Delivery point of NADH and $FADH_2$

___ 28. Two pyruvates enter for each glucose molecule entering glycolysis

___ 29. The carbons in the acetyl group leave as CO_2

___ 30. One carbon in pyruvate leaves as CO_2

___ 31. An electron transport system and channel proteins are involved

___ 32. Two $FADH_2$ and ten NADH are sent to this stage

8-IV. ANAEROBIC ROUTES (pp. 132–133)

Selected Italicized Words

Lactobacillus, Saccharomyces cerevisiae, Saccharomyces ellipsoideus

Boldfaced, Page-Referenced Terms

(132) lactate fermentation _____

_____

Fill-in-the-Blanks

(1) _____ pathways do not use oxygen as the final (2) _____ acceptor that ultimately drives the ATP-forming machinery. Anaerobic routes must be used by many bacteria and protistans that live in an oxygen-free environment. (3) _____ is also the first stage of the fermentation pathways. A glucose molecule is split into two (4) _____ molecules, two (5) _____ molecules form, and the net energy yield is two ATP.

In fermentation chemistry, (6) _____ itself accepts hydrogen and electrons from NADH. Pyruvate is then converted to a three-carbon compound, (7) _____, in the fermentation process of a few bacteria and some animal cells. Human muscle cells can carry on lactate fermentation in times of oxygen depletion; this provides a low yield of ATP.

In yeast cells, each pyruvate molecule from glycolysis is rearranged into an intermediate form called (8) _____. This intermediate accepts hydrogen and electrons from NADH and is then converted to (9) _____, the end product of alcoholic fermentation. As yeast cells carry on fermentation, a gas known as (10) _____ _____ is released.

In both types of fermentation pathways, the net energy yield of two ATPs is from (11) _____. The reactions of the fermentation chemistry regenerate the (12) _____ for glycolysis.

Anaerobic electron transport is an energy-releasing pathway occurring among the (13) _____. For example, sulfate-reducing bacteria living in soil or water produce (14) _____ by stripping electrons from a variety of compounds and sending them through membrane transport systems. The inorganic compound (15) _____ ($SO_4^=$) serves as the final electron acceptor and is converted into foul-smelling hydrogen sulfide gas (H_2S).

Complete the Table

Include a check mark (√) in each box that correctly links an occurrence (left-hand column) with a process (or processes).

Occurrence	Glycolysis	Alcoholic Fermentation	Lactate Fermentation	Anaerobic Electron Transport
$6\text{-}C \rightarrow 3C \searrow 3C$	(16)	(29)	(42)	(55)
$NADH \rightarrow NAD^+$	(17)	(30)	(43)	(56)
$NAD^+ \rightarrow NADH$	(18)	(31)	(44)	(57)
$SO_4^= \rightarrow H_2S$	(19)	(32)	(45)	(58)
$NO_3 \rightarrow NO_2$	(20)	(33)	(46)	(59)
CO_2 is a waste product	(21)	(34)	(47)	(60)
$3\text{-}C \rightarrow 3\text{-}C$	(22)	(35)	(48)	(61)
$3\text{-}C \rightarrow 2\text{-}C$	(23)	(36)	(49)	(62)
ATP is used as a reactant	(24)	(37)	(50)	(63)
ATP is produced	(25)	(38)	(51)	(64)
Pyruvate $\rightarrow$ ethanol $\searrow CO_2$	(26)	(39)	(52)	(65)
Occurs in animal cells	(27)	(40)	(53)	(66)
Occurs in yeast cells	(28)	(41)	(54)	(67)

8-V. ALTERNATIVE ENERGY SOURCES IN THE HUMAN BODY (pp. 134–135)

Choice

For questions 1–10, refer to the text and Figure 8.12; choose from the following:

a. glucose   b. glucose-6-phosphate   c. glycogen   d. fatty acids   e. triglycerides

f. PGAL   g. acetyl-CoA   h. amino acids   i. glycerol   j. proteins

____ 1. Fats that are tapped between meals or during exercise as alternatives to glucose

____ 2. Used between meals when free glucose supply dwindles; enters glycolysis after conversion

____ 3. Its breakdown yields much more ATP than glucose

____ 4. Absorbed in large amounts immediately following a meal

____ 5. Represents only 1 percent or so of the total stored energy in the body

___ 6. Following removal of amino groups, the carbon backbones may be converted to fats or carbohydrates or they may enter the Krebs cycle

___ 7. On average, represents 78 percent of the body's stored food

___ 8. Between meals liver cells can convert it back to free glucose and release it

___ 9. Can be stored in cells but not transported across plasma membranes

___ 10. Amino groups undergo conversions that produce urea, a nitrogen-containing waste product excreted in urine

___ 11. Converted to PGAL in the liver; a key intermediate of glycolysis

___ 12. Accumulate inside the fat cells of adipose tissues at strategic points under the skin

___ 13. A storage polysaccharide produced from glucose-6-phosphate following food intake that exceeds cellular energy demand (and increases ATP production to inhibit glycolysis)

___ 14. Building blocks of the compounds that represent 21 percent of the body's stored food

___ 15. A product resulting from enzymes cleaving circulating fatty acids; enters the Krebs cycle

---

Self-Quiz

___ 1. Glycolysis would quickly halt if the process ran out of _____, which serves as the hydrogen and electron acceptor.
   a. $NADP^+$
   b. ADP
   c. NAD^+
   d. H_2O

___ 2. The ultimate electron acceptor in aerobic respiration is _____.
   a. NADH
   b. carbon dioxide (CO_2)
   c. oxygen ($\frac{1}{2} O_2$)
   d. ATP

___ 3. When glucose is used as an energy source, the largest amount of ATP is generated by the _____ portion of the entire respiratory process.
   a. glycolytic pathway
   b. acetyl-CoA formation
   c. Krebs cycle
   d. electron transport phosphorylation

___ 4. The process by which about 10 percent of the energy stored in a sugar molecule is released as it is converted into two small organic-acid molecules is _____.
   a. photolysis
   b. glycolysis
   c. fermentation
   d. the dark reactions

___ 5. During which of the following phases of aerobic respiration is ATP produced directly by substrate-level phosphorylation?
   a. Glucose formation
   b. Ethanol production
   c. Acetyl-CoA formation
   d. The Krebs cycle

___ 6. What is the name of the process by which reduced NADH transfers electrons along a chain of acceptors to oxygen so as to form water and in which the energy released along the way is used to generate ATP?
   a. Glycolysis
   b. Acetyl-CoA formation
   c. The Krebs cycle
   d. Electron transport phosphorylation

___ 7. Pyruvic acid can be regarded as the end product of _____.
   a. glycolysis
   b. acetyl-CoA formation
   c. fermentation
   d. the Krebs cycle

___ 8. Which of the following is *not* ordinarily capable of being reduced at any time?
   a. NAD^+
   b. FAD
   c. Oxygen, O_2
   d. Water

___ 9. ATP production by chemiosmosis involves _____.
   a. H^+ concentration and electric gradients across a membrane
   b. ATP synthases

   c. formation of ATP in the inner mitochondrial compartment
   d. all of the above

___ 10. During the fermentation pathways, a net yield of two ATPs is produced from _____; the NAD^+ necessary for _____ is regenerated during the reactions.
   a. the Krebs cycle; glycolysis
   b. glycolysis; electron transport phosphorylation
   c. the Krebs cycle; electron transport phosphorylation
   d. glycolysis; glycolysis

Matching

Match the following components of respiration to the list of words below. Some components may have more than one answer.

11. ___ lactic acid

12. ___ $NAD^+ \rightarrow NADH$

13. ___ carbon dioxide

14. ___ $NADH \rightarrow NAD^+$

15. ___ pyruvate

16. ___ ATP produced by substrate-level phosphorylation

17. ___ glucose

18. ___ citrate

19. ___ oxygen

20. ___ water

A. Glycolysis
B. Preparatory conversions prior to the Krebs cycle
C. Fermentation
D. Krebs cycle
E. Electron transport

Chapter Objectives/Review Questions

This section lists general and detailed chapter objectives that may be used as review questions. You can make maximum use of these items by writing answers on a separate sheet of paper. Fill in answers where blanks are provided. To check for accuracy, compare your answers with information given in the chapter or glossary.

Page	Objectives/Questions
(124)	1. No matter what the source of energy may be, organisms must convert it to _____, a form of chemical energy that can drive metabolic reactions.
(124)	2. The main energy-releasing pathway is _____ respiration.

(124) 3. List the main anaerobic energy-releasing pathways and some examples of organisms that use them.

(124) 4. Give the overall equation for the aerobic respiratory route.

(124) 5. In the first of the three stages of aerobic respiration, one _____ is partially degraded to two pyruvate molecules.

(124) 6. By the end of the second stage of aerobic respiration, which includes the _____ cycle, _____ has been completely degraded to carbon dioxide and water.

(124) 7. Do the first two stages of aerobic respiration yield a high or low quantity of ATP?

(124) 8. The third stage of aerobic respiration is called electron transport _____, which yields many ATP molecules.

(124) 9. Explain, in general terms, the role of oxygen in aerobic respiration.

(126) 10. Glycolysis occurs in the _____ of the cell.

(126) 11. Explain the purpose served by the cell investing two ATP molecules into the chemistry of glycolysis.

(126–127) 12. Four ATP molecules are produced by _____-_____ phosphorylation for every two used during glycolysis. (Consult Fig. 8.4 in the main text.)

(127) 13. Glycolysis produces _____ (number) NADH, _____ (number) ATP (net), and _____ (number) pyruvate molecules for each glucose molecule entering the reactions.

(129) 14. Consult Figure 8.5 in the main text. Relate the events that happen during acetyl-coenzyme A formation and explain how the process of acetyl-CoA formation relates glycolysis to the Krebs cycle.

(128–129) 15. State the factors responsible for pyruvic acid entering the acetyl-CoA formation pathway.

(128–129) 16. What happens to the CO_2 produced during acetyl-CoA formation and the Krebs cycle?

(130) 17. Explain how chemiosmosis operates in the mitochondrion to account for the production of ATP molecules.

(131) 18. Consult Figure 8.8 in the main text and predict what will happen to the NADH produced during acetyl-CoA formation and the Krebs cycle.

(131) 19. Calculate the number of ATP molecules produced during the Krebs cycle for each glucose molecule that enters glycolysis.

(131) 20. Briefly describe the process of electron transport phosphorylation by stating what reactants are needed and what the products are. State how many ATP molecules are produced through operation of the transport system.

(131) 21. Be able to account for the total *net yield* of 36 ATP molecules produced through aerobic respiration; that is, state how many ATPs are produced in glycolysis, the Krebs cycle, and the electron transport system.

(132) 22. List some environments where there is very little oxygen present and where anaerobic organisms might be found.

(132–133) 23. Describe what happens to pyruvate in anaerobic organisms. Then explain the necessity for pyruvate to be converted to a fermentative product.

(132–133) 24. State which factors determine whether the pyruvate (pyruvic acid) produced at the end of glycolysis will enter into the alcoholic fermentation pathway, the lactate fermentation pathway, or the acetyl-coenzyme A formation pathway.

(132–133) 25. In fermentation chemistry, _____ molecules from glycolysis are accepted to construct either lactate or ethanol; thus, a low yield of _____ molecules continues in the absence of oxygen.

(134–135) 26. List some sources of energy (other than glucose) that can be fed into the respiratory pathways.

(134–135) 27. Predict what your body would do to synthesize its needed carbohydrates and fats if you switched to a diet of 100 percent protein.

(136–137) 28. After reading "Perspective on Life" in the main text, outline the supposed evolutionary sequence of energy-extraction processes.

(136–137) 29. Scrutinize the sketch in the Commentary in the main text closely; then reproduce the carbon cycle from memory.

Integrating and Applying Key Concepts

How is the "oxygen debt" experienced by runners and sprinters related to aerobic and anaerobic respiration in humans?

Answers

Interactive Exercises

8-I. HOW CELLS MAKE ATP (pp. 122–125)

1. G; 2. I; 3. H; 4. D; 5. F; 6. J; 7. A; 8. B; 9. C; 10. E; 11. $C_6H_{12}O_6$ plus 6 O_2 → 6 CO_2 plus 6 H_2O; 12. a. oxygen; b. six; c. six; 13. glycolysis; 14. pyruvate; 15. Krebs; 16. water; 17. ATP; 18. electrons; 19. transport; 20. phosphorylation; 21. ATP; 22. Oxygen.

8-II. GLYCOLYSIS: FIRST STAGE OF THE ENERGY-RELEASING PATHWAYS (pp. 126–127)

1. D; 2. F; 3. B; 4. H; 5. G; 6. A; 7. E; 8. C.

8-III. COMPLETING THE AEROBIC PATHWAY (pp. 128–131)

1. inner compartment; 2. inner membrane; 3. outer compartment; 4. outer membrane; 5. cytoplasm; 6. ATP; 7. oxygen (O_2); 8. $FADH_2$; 9. NADH; 10. electron transport system; 11. c; 12. a, b; 13. c; 14. a; 15. b; 16. a; 17. b; 18. b; 19. a, b, c; 20. b; 21. c; 22. b; 23. b; 24. c; 25. a; 26. c; 27. c; 28. a; 29. b; 30. a; 31. c; 32. c.

8-IV. ANAEROBIC ROUTES (pp. 132–133)

1. Anaerobic; 2. electron; 3. Glycolysis; 4. pyruvate; 5. NADH; 6. pyruvate; 7. lactate; 8. acetaldehyde; 9. ethanol; 10. carbon dioxide; 11. glycolysis; 12. NAD^+; 13. bacteria; 14. ATP; 15. sulfate; Numbers with a check mark (√): 16, 18, 24, 25, 27, 28, 30, 35, 40, 43, 47, 49, 52, 54, 56, 58, 59, 64; all others lack a check.

8-V. ALTERNATIVE ENERGY SOURCES IN THE HUMAN BODY (pp. 134–135)

1. e; 2. b; 3. d; 4. a; 5. c; 6. h; 7. e; 8. b; 9. b; 10. h; 11. i; 12. e; 13. c; 14. h; 15. g.

Self-Quiz

1. c; 2. c; 3. d; 4. b; 5. d; 6. d; 7. a; 8. d; 9. d; 10. d; 11. C; 12. A, B, D; 13. B, (C), D; 14. C, E; 15. A, B; 16. A, D; 17. A; 18. D; 19. E; 20. A, C, D, E.

9

CELL DIVISION AND MITOSIS

Interactive Exercises

9-I. DIVIDING CELLS: THE BRIDGE BETWEEN GENERATIONS (pp. 140–143)

Boldfaced, Page-Referenced Terms

(142) reproduction _____

_____

(142) mitosis _____

_____

(142) meiosis _____

_____

(142) somatic cells _____

_____

(142) germ cells _____

_____

(142) chromosome _____

_____

(142) sister chromatids _____

_____

(142) centromere _____

_____

(143) diploid cell _____

(143) chromosome number _____

Matching

Choose the one most appropriate answer for each.

1. ___ centromere
2. ___ prokaryotic fission
3. ___ diploid cell
4. ___ chromosome
5. ___ germ cells
6. ___ reproduction
7. ___ somatic cells
8. ___ sister chromatids
9. ___ mitosis and meiosis
10. ___ chromosome number

A. Cell lineage set aside for meiosis
B. Producing a new generation of cells or multicelled individuals
C. Any cell having two of each type of chromosome
D. How many of each type of chromosome are present in a cell
E. Each DNA molecule with attached proteins
F. Sort out and package DNA molecules into daughter cell nuclei
G. Constricted chromosome region with attachment sites for microtubules
H. Asexual reproduction, bacterial cell division
I. The two attached DNA molecules of a duplicated chromosome
J. Body cells that reproduce by mitosis and cytoplasmic division

9-II. MITOSIS AND THE CELL CYCLE (pp. 144–146)

Boldfaced, Page-Referenced Terms

(144) cell cycle _____

(144) interphase _____

Labeling

Identify the stage in the cell cycle indicated by each number.

1. _____
2. _____
3. _____
4. _____
5. _____
6. _____
7. _____
8. _____
9. _____
10. _____

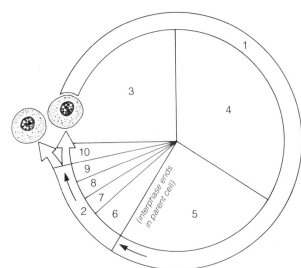

Matching

Link each time span identified below with the most appropriate number in the preceding labeling section.

11. ___ Period after replication of DNA during which the cell prepares for division by further growth and protein synthesis; a second "gap"

12. ___ Period of nuclear division followed by cytoplasmic division, a separate event

13. ___ DNA replication occurs now; a time of "synthesis" of DNA and proteins

14. ___ Period of cell growth before DNA replication; a "gap" of interphase

15. ___ Period when chromosomes are not visible

16. ___ Period of cytoplasmic division

17. ___ Period that includes G_1, S, G_2

9-III. STAGES OF MITOSIS: AN OVERVIEW (pp. 146–147)
9-IV. A CLOSER LOOK AT MITOSIS (pp. 148–149)

Boldfaced, Page-Referenced Terms

(146) prophase _____

_____

(146) metaphase _____

_____

(146) anaphase _____

_____

(146) telophase _____

_____

(146) spindle apparatus _____

_____

(148) centrioles _____

_____

Identify each of the mitotic stages shown below by entering the correct stage in the blank beneath the sketch. Select from late prophase, transition to metaphase, interphase—parent cell, metaphase, early prophase, telophase, interphase—daughter cells, and anaphase. Complete the exercise by matching and entering the letter of the correct phase description in the parentheses following each label.

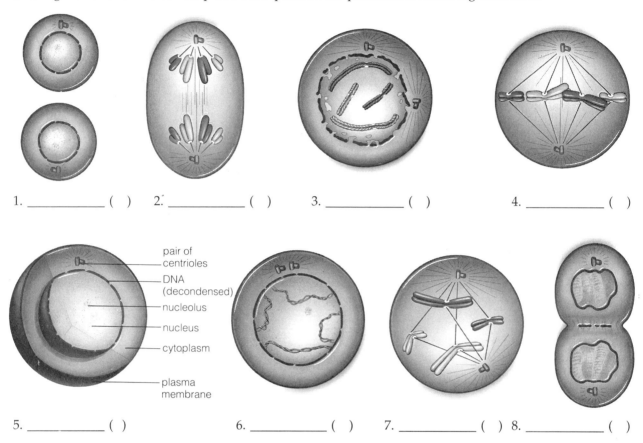

1. _____ ( )   2. _____ ( )   3. _____ ( )   4. _____ ( )

pair of centrioles
DNA (decondensed)
nucleolus
nucleus
cytoplasm
plasma membrane

5. _____ ( )   6. _____ ( )   7. _____ ( ) 8. _____ ( )

A. Attachment between two sister chromatids of each chromosome break; the two are now chromosomes in their own right.
B. Microtubules that form the spindle apparatus penetrate the nuclear region; microtubules become attached to the sister chromatids of each chromosome.
C. The DNA and its associated proteins start to condense into the threadlike chromosome form.
D. Chromosomes are now fully condensed and lined up at the equator of the spindle.
E. DNA is duplicated and the cell prepares for division.
F. Two daughter cells have formed, each diploid.
G. Chromosomes continue to condense. New microtubules are assembled, and they move one of two centrioles toward the opposite end of the cell. The nuclear envelope begins to break up.
H. New patches of membrane join to form nuclear envelopes around the decondensing chromosomes. Cytokinesis begins before this stage ends.

9-V. DIVISION OF THE CYTOPLASM (pp. 150–151)

Boldfaced, Page-Referenced Terms

(150) cell plate formation _____

_____

(150) cleavage _____

_____

Choice

For questions 1–5, choose from the following:

a. plant cells   b. animal cells

___ 1. Formation of a cell plate

___ 2. Microfilaments pull the plasma membrane inward and cut the cell in two

___ 3. Cellulose deposits form a crosswall between the two daughter cells

___ 4. Possess rigid walls that cannot be pinched in two

___ 5. Cleavage furrow

Self-Quiz

___ 1. The replication of DNA occurs _____.
   a. between the growth phases of interphase
   b. immediately before prophase of mitosis
   c. during prophase of mitosis
   d. during prophase of meiosis

___ 2. In the cell life cycle of a particular cell,
   _____.

   a. mitosis occurs immediately prior to S
   b. mitosis occurs immediately prior to G_1
   c. G_2 precedes S
   d. G_1 precedes S
   e. mitosis and S precede G_1

___ 3. In eukaryotic cells, which of the following can occur during mitosis?
   a. Two mitotic divisions to maintain the parental chromosome number
   b. The replication of DNA
   c. A long growth period
   d. The disappearance of the nuclear envelope and nucleolus

___ 4. Diploid refers to _____.
   a. having two chromosomes of each type in somatic cells
   b. twice the parental chromosome number
   c. half the parental chromosome number
   d. having one chromosome of each type in somatic cells

___ 5. Somatic cells are _____ cells; germ cells are _____ cells.
   a. meiotic; body
   b. body; body
   c. meiotic; meiotic
   d. body; meiotic

___ 6. If a parent cell has sixteen chromosomes and undergoes mitosis, the resulting cells will have _____ chromosomes.
   a. sixty-four
   b. thirty-two
   c. sixteen
   d. eight
   e. four

___ 7. The correct order of the stages of mitosis is
   _____.
   a. prophase, metaphase, telophase, anaphase
   b. telophase, anaphase, metaphase, prophase
   c. telophase, prophase, metaphase, anaphase
   d. anaphase, prophase, telophase, metaphase
   e. prophase, metaphase, anaphase, telophase

___ 8. "The nuclear envelope breaks up completely into vesicles. Microtubules are now free to interact with the chromosomes." These sentences describe the _____ of mitosis.
   a. prophase
   b. metaphase
   c. transition to metaphase
   d. anaphase
   e. telophase

___ 9. During _____, sister chromatids of each chromosome are separated from each other, and those former partners, now chromosomes, are moved toward opposite poles.

a. prophase
b. metaphase
c. anaphase
d. telophase

___ 10. In the process of cytokinesis, cleavage furrows are associated with _____ cell division, and cell plate formation is associated with _____ cell division.

a. animal; animal
b. plant; animal
c. plant; plant
d. animal; plant

Chapter Objectives/Review Questions

This section lists general and detailed chapter objectives that may be used as review questions. You can make maximum use of these items by writing answers on a separate sheet of paper. Fill in answers where blanks are provided. To check for accuracy, compare your answers with information given in the chapter or glossary.

Page	Objectives/Questions
(142)	1. Name the substance that contains the instructions for making proteins.
(142)	2. Mitosis and meiosis refer to the division of the cell's _____.
(142)	3. Distinguish between somatic cells and germ cells as to their location and function.
(142)	4. The eukaryotic chromosome is composed of _____ and _____.
(142)	5. The two attached threads of a duplicated chromosome are known as sister _____.
(142)	6. Describe the function of the portion of a chromosome known as a centromere.
(143)	7. Any cell having two of each type of chromosome is a _____ cell.
(144)	8. Be able to list, in order, the various activities occurring in the eukaryotic cell life cycle.
(144)	9. Interphase of the cell cycle consists of G_1, _____, and G_2.
(144)	10. S is the time in the cell cycle when _____ replication occurs.
(145)	11. Explain why cells from Henrietta Lacks continue to benefit humans everywhere more than forty years after her death.
(146–150)	12. Be able to describe the cellular events occurring in the prophase, metaphase, anaphase, and telophase of mitosis.
(148–149)	13. The "_____" is the time when the nuclear envelope breaks up into vesicles prior to metaphase.
(150–151)	14. Compare and contrast cytokinesis as it occurs in plant mitosis and animal mitosis; use the following concepts: cleavage furrow, microfilaments at the cell's midsection, and cell plate formation.

Integrating and Applying Key Concepts

Runaway cell division is characteristic of cancer. Imagine the various points of the mitotic process that might be sabotaged in cancerous cells in order to halt their multiplication. Then try to imagine how one might discriminate between cancerous and normal cells in order to guide those methods of sabotage most effective in combating cancer.

Answers

Interactive Exercises

9-I. DIVIDING CELLS: THE BRIDGE BETWEEN GENERATIONS (pp. 140–143)
1. G; 2. H; 3. C; 4. E; 5. A; 6. B; 7. J; 8. I; 9. F; 10. D.

9-II. MITOSIS AND THE CELL CYCLE (pp. 144–146)
1. interphase; 2. mitosis; 3. G_1; 4. S; 5. G_2; 6. prophase; 7. metaphase; 8. anaphase; 9. telophase; 10. cytokinesis; 11. 5; 12. 2; 13. 4; 14. 3; 15. 1; 16. 10; 17. 1.

9-III. STAGES OF MITOSIS: AN OVERVIEW (pp. 146–147)

9-IV. A CLOSER LOOK AT MITOSIS (pp. 148–149)
1. interphase—daughter cells (F); 2. anaphase (A); 3. late prophase (G); 4. metaphase (D); 5. interphase—parent cell (E); 6. early prophase (C); 7. transition to metaphase (B); 8. telophase (H).

9-V. DIVISION OF THE CYTOPLASM (pp. 150–151)
1. a; 2. b; 3. a; 4. a; 5. b.

Self-Quiz
1. a; 2. d; 3. d; 4. a; 5. d; 6. c; 7. e; 8. c; 9. c; 10. d.

10

MEIOSIS

ON ASEXUAL AND SEXUAL REPRODUCTION

OVERVIEW OF MEIOSIS
   Think "Homologues"
   Overview of the Two Divisions

KEY EVENTS DURING MEIOSIS I
   Prophase I Activities
   Metaphase I Alignments

FORMATION OF GAMETES
   Gamete Formation in Plants
   Gamete Formation in Animals

MORE GENE SHUFFLINGS AT FERTILIZATION

MEIOSIS AND MITOSIS COMPARED

Interactive Exercises

10-I. ON ASEXUAL AND SEXUAL REPRODUCTION (pp. 154–156)

Boldfaced, Page-Referenced Terms

(156) asexual reproduction _____

_____

(156) sexual reproduction _____

_____

(156) allele _____

_____

Choice

For questions 1–10, choose from the following:

                  a. asexual reproduction   b. sexual reproduction

____ 1. Brings together new combinations of alleles in offspring

____ 2. Involves only one parent

____ 3. Both parents pass on one of each gene to their offspring

____ 4. Commonly involves two parents

____ 5. The production of "clones"

___ 6. Involves meiosis and fertilization

___ 7. Offspring are genetically identical copies of the parent

___ 8. Produces the variation in traits that forms the basis of evolutionary change

___ 9. The instructions in every pair of genes are identical in all individuals of a species

___10. New combinations of alleles lead to variations in physical and behavioral traits

10-II. OVERVIEW OF MEIOSIS (pp. 156–160)

Boldfaced, Page-Referenced Terms

(156) meiosis _____

_____

(156) gametes _____

_____

(157) diploid _____

_____

(157) homologous chromosomes _____

_____

(157) haploid _____

_____

(157) sister chromatids _____

_____

True-False

Place a T in the blank if the statement is true. If false, make the statement correct by changing the under-lined word.

_____ 1. A unique molecular form of the same gene is called an <u>allele</u>.

_____ 2. Sperms and eggs are cells known as <u>gonads</u>.

_____ 3. <u>Haploid</u> germ cells produce haploid gametes.

_____ 4. <u>Diploid</u> cells possess pairs of homologous chromosomes.

_____ 5. <u>Meiosis</u> produces cells that have one member of each pair of homologous chromosomes possessed by the species.

Matching

To review the major stages of meiosis, match the following written descriptions with the appropriate sketch. Assume that the cell in this model initially has only one pair of homologous chromosomes (one from a paternal source and one from a maternal source) and crossing over does not occur. Complete the exercise by indicating the diploid ($2n = 2$) or haploid ($n = 1$) chromosome number of the cell chosen in the parentheses following each blank.

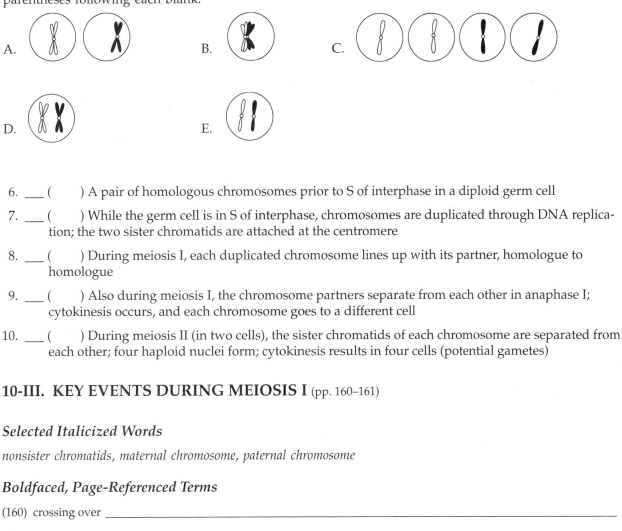

A.   B.   C.

D.   E.

6. ___ (   ) A pair of homologous chromosomes prior to S of interphase in a diploid germ cell

7. ___ (   ) While the germ cell is in S of interphase, chromosomes are duplicated through DNA replication; the two sister chromatids are attached at the centromere

8. ___ (   ) During meiosis I, each duplicated chromosome lines up with its partner, homologue to homologue

9. ___ (   ) Also during meiosis I, the chromosome partners separate from each other in anaphase I; cytokinesis occurs, and each chromosome goes to a different cell

10. ___ (   ) During meiosis II (in two cells), the sister chromatids of each chromosome are separated from each other; four haploid nuclei form; cytokinesis results in four cells (potential gametes)

10-III. KEY EVENTS DURING MEIOSIS I (pp. 160–161)

Selected Italicized Words

nonsister chromatids, maternal chromosome, paternal chromosome

Boldfaced, Page-Referenced Terms

(160) crossing over _____

_____

Label-Match

Identify each of the meiotic stages shown below by entering the correct stage of either meiosis I or meiosis II in the blank beneath the sketch. Choose from prophase I, metaphase I, anaphase I, telophase I, prophase II, metaphase II, anaphase II, and telophase II. Complete the exercise by matching and entering the letter of the correct phase description in the parentheses following each label.

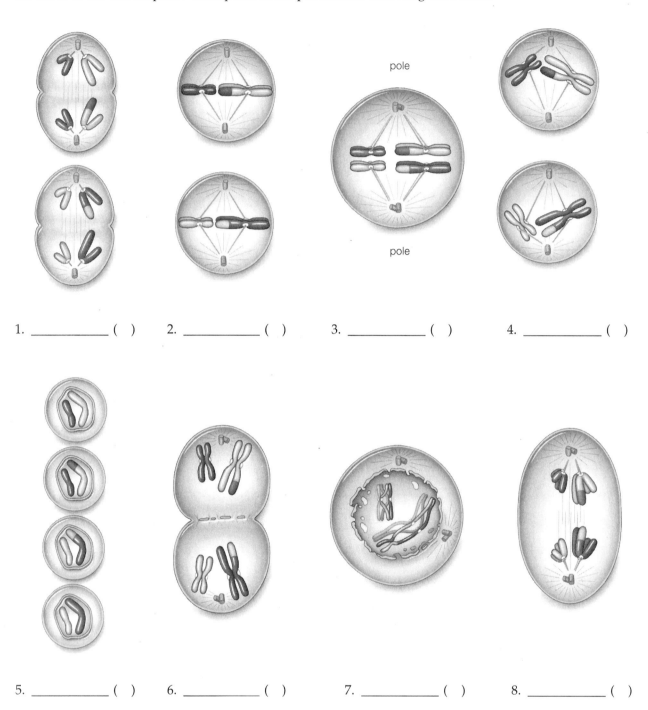

1. _____ ( )   2. _____ ( )   3. _____ ( )   4. _____ ( )

5. _____ ( )   6. _____ ( )   7. _____ ( )   8. _____ ( )

A. The spindle is now fully formed; all chromosomes are positioned midway between the poles of one cell.
B. Centrioles have moved to form the poles in each of two cells; a spindle forms, and microtubules attach the duplicated chromosomes to the spindle and begin moving them toward the equator of each cell.
C. Four daughter nuclei form; when the cytoplasm divides, each new cell has a haploid number of chromosomes, all in the unduplicated state. One or all cells may develop into gametes.
D. In one cell, each duplicated chromosome is pulled away from its homologue; the two are moved to opposite spindle poles.
E. Each chromosome is drawn up close to its homologue; crossing over and genetic recombination occur.
F. All of the chromosomes are now aligned at the spindle's equator; this is occurring in two haploid cells.
G. Two haploid cells form, but chromosomes are still in the duplicated state.
H. Chromatids of each chromosome separate; former "sister chromatids" are now chromosomes in their own right and are moved to opposite poles.

10-IV. FORMATION OF GAMETES (p. 162)

Boldfaced, Page-Referenced Terms

(162) spores _____

_____

(162) sperm _____

_____

(162) egg _____

_____

Choice

For questions 1–10, choose from the following:

      a. animal life cycle   b. plant life cycle   c. both animal and plant life cycles

____ 1. Meiosis results in the production of haploid spores.

____ 2. A zygote divides by mitosis.

____ 3. Meiosis results in the production of haploid gametes.

____ 4. Haploid gametes fuse in fertilization to form a diploid zygote.

____ 5. A zygote divides by mitosis to form a diploid sporophyte.

____ 6. A spore divides by mitosis to produce a haploid gametophyte.

____ 7. A haploid gametophyte divides by mitosis to produce haploid gametes.

____ 8. A haploid spore divides by mitosis to produce a gametophyte.

____ 9. A diploid body forms from mitosis of a zygote.

____ 10. A gamete-producing body and a spore-producing body develop during the life cycle.

Sequence

Arrange the following entities in correct order of development, entering a 1 by the stage that appears first and a 5 by the stage that completes the process of spermatogenesis. Complete the exercise by indicating if each cell is *n* or *2n* in the parentheses following each blank. Refer to Figure 10.9 in the text.

11. ___ (   ) primary spermatocyte
12. ___ (   ) sperm
13. ___ (   ) spermatid
14. ___ (   ) spermatogonium
15. ___ (   ) secondary spermatocyte

Matching

Choose the most appropriate answer to match with each oogenesis concept. Refer to Figure 10.10 in the text.

16. ___ primary oocyte

17. ___ oogonium

18. ___ secondary oocyte

19. ___ ovum and three polar bodies

20. ___ first polar body

A. The cell in which synapsis, crossing over, and recombination occur
B. A cell that is equivalent to a diploid germ cell
C. A haploid cell formed after division of the primary oocyte that does not form an ovum at second division
D. Haploid cells, but only one of which functions as an egg
E. A haploid cell formed after division of the primary oocyte, the division of which forms a functional ovum

10-V. MORE GENE SHUFFLINGS AT FERTILIZATION (pp. 162–163)

Boldfaced, Page-Referenced Terms

(162) fertilization _____

_____

Short Answer

1. List the various mechanisms that contribute to the huge number of new gene combinations that may result from fertilization. _____

_____

_____

10-VI. MEIOSIS AND MITOSIS COMPARED (pp. 164–165)

Complete the Table

1. Complete the table below by entering the word "mitosis" or "meiosis" in the blank adjacent to the statement describing one of these processes.

Description	Mitosis/Meiosis
a. Involves one division cycle	
b. Functions in growth and repair	
c. Daughter cells are haploid	
d. Initiated in germ cells	
e. Involves two division cycles	
f. Daughter cells have one chromosome from each homologous pair	
g. Produces spores in plant life cycles	
h. Daughter cells have the diploid chromosome number	
i. Completed when four daughter cells are formed	

Matching

The cell model used in this exercise has two pairs of homologous chromosomes, one long pair and one short pair. Match the descriptions to the numbers of chromosomes shown in the sketches below.

A

B

C

D

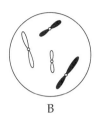
E

F

2. ___ beginning of meiosis II following interkinesis

3. ___ a daughter cell at the end of meiosis II

4. ___ metaphase I of meiosis

5. ___ metaphase of mitosis

6. ___ G₁ in a daughter cell following mitosis

7. ___ prophase of mitosis

The following questions refer to the sketches above; enter answers in the blanks following each question.

8. How many chromosomes are present in cell E? ___

9. How many chromatids are present in cell E? ___

10. How many chromatids are present in cell C? ___

11. How many chromatids are present in cell D? ___

12. How many chromosomes are present in cell F? ___

Self-Quiz

___ 1. Which of the following does not occur in prophase I of meiosis?

a. a cytoplasmic division
b. a cluster of four chromatids
c. homologues pairing tightly
d. crossing over

___ 2. Crossing over is one of the most important events in meiosis because _____.

a. it produces new combinations of alleles on chromosomes
b. homologous chromosomes must be separated into different daughter cells
c. the number of chromosomes allotted to each daughter cell must be halved
d. homologous chromatids must be separated into different daughter cells

___ 3. Crossing over _____.

a. generally results in pairing of homologues and binary fission
b. is accompanied by gene-copying events
c. involves breakages and exchanges between sister chromatids
d. alters the composition of chromosomes and results in new combinations of alleles being channeled into the daughter cells

___ 4. The appearance of chromosome ends lapped over each other in meiotic prophase I provides evidence of _____.

a. meiosis
b. crossing over
c. chromosomal aberration
d. fertilization
e. spindle fiber formation

___ 5. Which of the following does not increase genetic variation?
  a. Crossing over
  b. Random fertilization
  c. Prophase of mitosis
  d. Random homologue alignments at metaphase I

___ 6. Which of the following is the most correct sequence of events in animal life cycles?
  a. meiosis → fertilization → gametes → diploid organism
  b. diploid organism → meiosis → gametes → fertilization
  c. fertilization → gametes → diploid organism → meiosis
  d. diploid organism → fertilization → meiosis → gametes

___ 7. In sexually reproducing organisms, the zygote is _____.
  a. an exact genetic copy of the female parent
  b. an exact genetic copy of the male parent
  c. unlike either parent genetically
  d. a genetic mixture of male parent and female parent

___ 8. Which of the following is the most correct sequence of events in plant life cycles?
  a. fertilization → zygote → sporophyte → meiosis → spores → gametophytes → gametes
  b. fertilization → sporophyte → zygote → meiosis → spores → gametophytes → gametes
  c. fertilization → zygote → sporophyte → meiosis → gametes → gametophyte → spores
  d. fertilization → zygote → gametophyte → meiosis → gametes → sporophyte → spores

___ 9. The cell in the diagram is a diploid cell that has three pairs of chromosomes. From the number and pattern of chromosomes, the cell _____.

  a. could be in the first division of meiosis
  b. could be in the second division of meiosis
  c. could be in mitosis
  d. could be in neither mitosis nor meiosis, because this stage is not possible in a cell with three pairs of chromosomes

___ 10. You are looking at a cell from the same organism as in the previous question. Now the cell _____.

  a. could be in the first division of meiosis
  b. could be in the second division of meiosis
  c. could be in mitosis
  d. could be in neither mitosis nor meiosis, because this stage is not possible in this organism

Chapter Objectives/Review Questions

This section lists general and detailed chapter objectives that may be used as review questions. You can make maximum use of these items by writing answers on a separate sheet of paper. Fill in answers where blanks are provided. To check for accuracy, compare your answers with information given in the chapter or glossary.

Page   *Objectives/Questions*

(156)   1. "One parent always passes on a duplicate of all its genes to offspring" describes _____ reproduction.
(156)   2. Sexual reproduction puts together new combinations of _____ in offspring.
(157)   3. Describe the relationships between the following terms: homologous chromosomes, diploid, and haploid.

(157)   4. During interphase a germ cell duplicates its DNA; a duplicated chromosome consists of two DNA molecules that remain attached to a constriction called the _____.
(157)   5. As long as the two DNA molecules remain attached they are referred to as _____ _____ of the chromosome.
(157–159) 6. During meiosis I, homologous chromosomes pair; each homologue consists of _____ chromatids.
(157–159) 7. During meiosis II, the two sister _____ of each _____ are separated from each other.
(157–159) 8. If the diploid chromosome number for a particular plant species is 18, the haploid number is _____.
(160)   9. _____ _____ breaks up old combinations of alleles and puts new ones together in pairs of homologous chromosomes.
(161)   10. The _____ attachment and subsequent positioning of each pair of maternal and paternal chromosomes at metaphase I leads to different _____ of maternal and paternal traits in offspring.
(162)   11. Using the special terms for the cells at the various stages, describe spermatogenesis in male animals and oogenesis in female animals.
(162)   12. Meiosis in the animal life cycle results in haploid _____; meiosis in the plant life cycle results in haploid _____.
(162)   13. Crossing over, the distribution of random mixes of homologous chromosomes into gametes, and fertilization contribute to _____ in the traits of offspring.
(164)   14. Mitotic cell division produces _____; meiosis and subsequent fertilization promote _____ in traits among offspring.

Integrating and Applying Key Concepts

A few years ago, it was claimed that the actual cloning of a human being had been accomplished. Later, this claim was admitted to be fraudulent. If sometime in the future cloning of humans becomes possible, speculate about the effects of reproduction without sex on human populations.

Answers

Interactive Exercises

10-I. ON ASEXUAL AND SEXUAL REPRODUCTION (pp. 154–156)
1. b; 2. a; 3. b; 4. b; 5. a; 6. b; 7. a; 8. b; 9. a; 10. b.

10-II. OVERVIEW OF MEIOSIS (pp. 156–159)
1. T; 2. gametes; 3. diploid; 4. T; 5. T; 6. E ($2n = 2$); 7. D ($2n = 2$); 8. B ($2n = 2$); 9. A ($n = 1$); 10. C ($n = 1$).

10-III. KEY EVENTS DURING MEIOSIS I (pp. 160–161)
1. anaphase II (H); 2. metaphase II (F); 3. metaphase I (A); 4. prophase II (B); 5. telophase II (C); 6. telophase I (G); 7. prophase I (E); 8. anaphase I (D).

10-IV. FORMATION OF GAMETES (p. 162)
1. b; 2. c; 3. a; 4. c; 5. b; 6. b; 7. b; 8. b; 9. c; 10. b; 11. 2 ($2n$); 12. 5 (n); 13. 4 (n); 14. 1 ($2n$); 15. 3 (n); 16. A; 17. B; 18. E; 19. D; 20. C.

10-V. MORE GENE SHUFFLINGS AT FERTILIZATION (pp. 162–163)
1. During prophase I of meiosis, crossing over and genetic recombination occur. During metaphase I of meiosis, the two members of each homologous chromosome assort independently of the other pairs. Fertilization is a chance mix of different combinations of alleles from two different gametes.

10-VI. MEIOSIS AND MITOSIS COMPARED (pp. 164–165)
1. a. Mitosis; b. Mitosis; c. Meiosis; d. Meiosis; e. Meiosis; f. Meiosis; g. Meiosis; h. Mitosis; i. Meiosis; 2. C; 3. F; 4. D; 5. A; 6. B; 7. E; 8. 4; 9. 8; 10. 4; 11. 8; 12. 2.

Self-Quiz
1. a; 2. a; 3. d; 4. b; 5. c; 6. b; 7. d; 8. a; 9. b; 10. c.

OBSERVABLE PATTERNS
OF INHERITANCE

Interactive Exercises

11-I. MENDEL'S INSIGHT INTO PATTERNS OF INHERITANCE (pp. 168–171)
11-II. MENDEL'S THEORY OF SEGREGATION (pp. 172–173)
11-III. INDEPENDENT ASSORTMENT (pp. 174–175)

Selected Italicized Words

homozygous condition, heterozygous condition, dominant allele, recessive allele, homozygous recessive

Boldfaced, Page-Referenced Terms

(170) true-breeding _____

_____

(171) genes _____

_____

(171) alleles _____

_____

(171) homozygous dominant _____

_____

(171) homozygous recessive _____

_____

(171) heterozygous _____

_____

(171) genotype _____

_____

(171) phenotype _____

_____

(172) monohybrid crosses _____

_____

(173) probability _____

_____

(173) Punnett-square method _____

_____

(173) testcross _____

_____

(174) dihybrid crosses _____

_____

Matching

Choose the one most appropriate answer for each.

1. ___ genotype
2. ___ alleles
3. ___ heterozygous
4. ___ dominant allele
5. ___ phenotype
6. ___ genes
7. ___ recessive allele
8. ___ homozygous
9. ___ diploid cell
10. ___ locus

A. All the different molecular forms of a gene that exist
B. Particular location of a gene on a chromosome
C. Describes an individual having a pair of nonidentical alleles
D. Gene whose effect is masked by its partner
E. Refers to an individual's observable traits
F. Refers to the genes present in an individual organism
G. Gene whose effect "masks" the effect of its partner
H. Describes an individual for which two alleles of a pair are the same
I. Units of information about specific traits; passed from parents to offspring
J. Has a pair of genes for each trait, one on each of two homologous chromosomes

Fill-in-the-Blanks·

A(n) (11) _____ plant is one that is produced from two parents that have bred true for different forms of a single trait. In breeding experiments, (12) _____ is the symbol used for the parental generation; (13) _____ is the symbol used for first-generation offspring; (14) _____ is the symbol used for second-generation offspring. Because fertilization is a chance event, the rules of (15) _____ apply to genetics crosses. The separation of A and a as members of a pair of homologous chromosomes move to different gametes during meiosis is known as Mendel's theory of (16) _____.

Crossing F_1 hybrids (possibly of unknown genotype) back to a plant known to be a true-breeding recessive plant is known as Mendel's (17) _____. From this cross, a ratio of (18) _____ is expected. When F_1 offspring inherit two gene pairs, neither of which consists of identical alleles, the cross is known as a (19) _____ cross. "Gene pairs assorting into gametes independently of other gene pairs located on nonhomologous chromosomes" describes Mendel's theory of (20) _____ _____.

Problems

21. In garden pea plants, Tall (T) is dominant over dwarf (t). In the cross $Tt \times tt$, the Tt parent would produce a gamete carrying T (tall) and a gamete carrying t (short) through segregation; the tt parent could produce only gametes carrying the t (short) gene. Use the Punnett square method (refer to Fig. 11.6 and 11.7 in the text) to determine the genotype and phenotype probabilities of offspring from the above cross, $Tt \times tt$:

Although the Punnett-square (checkerboard) method is a favored method for solving genetics problems, there is a quicker way. Six different outcomes are possible from single factor crosses. Studying the following relationships allows one to obtain the result of any monohybrid cross by inspection.

1. $AA \times AA = $ all AA
(Each of the four blocks of the Punnett square would be AA.)
2. $aa \times aa = $ all aa
3. $AA \times aa = $ all Aa
4. $AA \times Aa$
   or $Aa \times AA$ $= $ ½ AA; ½ Aa
(Two blocks of the Punnett square are AA, and two blocks are Aa.)
5. $aa \times Aa$
   or $Aa \times aa$ $= $ ½ aa; ½ Aa
6. $Aa \times Aa = $ ¼ AA; ½ Aa; ¼ aa
(One block in the Punnett square is AA, two blocks are Aa, and one block is aa.)

Using the gene symbols in exercise 21, apply the six Mendelian ratios listed above to solve the following monohybrid crosses by inspection. State results as genotype ratios.

22. $TT \times TT = $ _____
23. $Tt \times Tt$ $= $ _____
24. $Tt \times tt$ $= $ _____
25. $tt \times tt$ $= $ _____

When working genetics problems dealing with two gene pairs, one can visualize the independent assortment of gene pairs located on nonhomologous chromosomes into gametes by use of a fork-line device. Assume that in humans, pigmented eyes (*B*) are dominant (an eye color other than blue) over blue (*b*), and right-handedness (*R*) is dominant over left-handedness (*r*). To learn to solve a problem, cross the parents *BbRr* × *BbRr*. A sixteen-block Punnett square is required with gametes from each parent arrayed on two sides of the Punnett square (refer to Fig. 11.9 in the text). The gametes receive genes through independent assortment using a fork-line method:

26. Array the gametes at the right on two sides of the Punnett square; combine these haploid gametes to form diploid zygotes within the squares. In the blank spaces below, enter the probability ratios derived within the Punnett square for the phenotypes listed:

    a. _____ pigmented eyes, right-handed
    b. _____ pigmented eyes, left-handed
    c. _____ blue-eyed, right-handed
    d. _____ blue-eyed, left-handed

    B b R r ⤫ × *B b R r* ⤫
    BR, Br, bR, br   BR, Br, bR, br

27. Albinos cannot form the pigments that normally produce skin, hair, and eye color, so albinos have white hair and pink eyes and skin (because the blood shows through). To be an albino, one must be homozygous recessive (*aa*) for the pair of genes that codes for the key enzyme in pigment production. Suppose a woman of normal pigmentation (*A_*) with an albino mother marries an albino man. State the kinds of pigmentation possible for this couple's children, and specify the ratio of each kind of child the couple is likely to have. Show the genotype(s) and state the phenotype(s). _____

    _____

28. In horses, black coat color is influenced by the dominant allele (*B*), and chestnut coat color is influenced by the recessive allele (*b*). Trotting gait is due to a dominant gene (*T*), pacing gait to the recessive allele (*t*). A homozygous black trotter is crossed to a chestnut pacer.

    a. What will be the appearance of the F_1 and F_2 generations?

    _____

    b. Which phenotype will be most common? _____
    c. Which genotype will be most common? _____
    d. Which of the potential offspring will be certain to breed true? _____

11-IV. DOMINANCE RELATIONS (p. 176)
11-V. MULTIPLE EFFECTS OF SINGLE GENES (pp. 176–177)
11-VI. INTERACTIONS BETWEEN GENE PAIRS (pp. 178–179)

Selected Italicized Words

sickle-cell anemia

Boldfaced, Page-Referenced Terms

(176) incomplete dominance _____

_____

(176) codominance _____

_____

(176) pleiotropy _____

_____

(177) ABO blood typing _____

_____

(177) multiple allele system _____

_____

(178) epistasis _____

_____

Complete the Table

1. Complete the following table by supplying the type of inheritance illustrated by each example. Choose from pleiotropy, multiple alleles, incomplete dominance, codominance, and the gene interaction known as epistasis.

Type of Inheritance	Example
a.	Pink-flowered snapdragons produced from red- and white-flowered parents
b.	AB type blood from a gene system of three alleles, A, B, and O
c.	A gene with three or more alleles such as the ABO blood typing alleles
d.	Black, brown, or yellow fur of Labrador retrievers and comb shape in poultry
e.	The multiple phenotypic effects of the gene causing human sickle-cell anemia

Problems

Genes that are not always dominant or recessive may blend to produce a phenotype of a different appearance. This is termed *incomplete dominance*. In four o'clock plants, red flower color is determined by gene R and white flower color by R', while the heterozygous condition, RR', is pink. Determine the phenotypes and genotypes of the offspring from the following crosses:

2. $RR \times R'R'$ = _____
3. $R'R' \times R'R'$ = _____
4. $RR \times RR'$ = _____
5. $RR \times RR$ = _____

Sickle-cell anemia is a genetic disease in which children that are homozygous for a defective gene produce defective hemoglobin ($Hb^S Hb^S$). The genotype of normal persons is $Hb^A Hb^A$. If the level of blood oxygen drops below a certain level in a person with the $Hb^S Hb^S$ genotype, the hemoglobin chains stiffen and cause the red blood cells to form sickle, or crescent, shapes. These cells clog and rupture capillaries, which results in oxygen-deficient tissues where metabolic wastes collect. Several body functions are badly damaged. Severe anemia and other symptoms develop, and death nearly always occurs before adulthood. The sickle-cell gene is considered *pleiotropic*. Persons that are heterozygous ($Hb^A Hb^S$) are said to possess *sickle-cell trait*. They appear normal but their red blood cells will sickle if they encounter oxygen tension (such as at high altitudes).

6. A man whose sister died of sickle-cell anemia married a woman whose blood is found to be normal. What advice would you give this couple about the inheritance of this disease as they plan their family?

_____

_____

7. If a man and a woman, each with sickle-cell trait, planned to marry, what information could you provide for them regarding the genotypes and phenotypes of their future children?

_____

_____

In one example of a *multiple allele system* with *codominance*, the three genes I^A, I^B, and i produce proteins found on the surfaces of red blood cells that determine the four blood types in the ABO system, A, B, AB, and O. Genes I^A and I^B are both dominant over i but not over each other. They are codominant. Recognize that blood types A and B may be heterozygous or homozygous (I^AI^A, I^Ai or I^BI^B, I^Bi), while blood type O is homozygous (ii). Indicate the genotypes and phenotypes of the offspring and their probabilities from the parental combinations in exercises 8–12.

8. $I^Ai \times I^AI^B$ = _____

9. $I^Bi \times I^Ai$ = _____

10. $I^AI^A \times ii$ = _____

11. $ii \times ii$ = _____

12. $I^AI^B \times I^AI^B$ = _____

In one type of gene interaction, two alleles of a gene mask the expression of alleles of another gene, and some expected phenotypes never appear. *Epistasis* is the term given such interactions. Work the following problems on scratch paper to understand epistatic interactions.

In sweet peas, genes C and P are necessary for colored flowers. In the absence of either (_ _ pp or cc _ _), or both (ccpp), the flowers are white. What will be the color of the offspring of the following crosses and in what proportions will they appear?

13. $CcPp \times ccpp$ = _____

14. $CcPP \times Ccpp$ = _____

15. $Ccpp \times ccPp$ = _____

In poultry, an epistatic interaction occurs in two genes to produce a phenotype that neither gene can produce alone. The two interacting genes (R and P) produce comb shape in chickens. The possible genotypes and phenotypes are:

Genotypes	Phenotypes
R_ P_	walnut comb
R_ pp	rose comb
rrP_	pea comb
rrpp	single comb

16. What are the genotype and phenotype ratios of the offspring of a heterozygous walnut-combed male and a single-combed female?

_____

_____

17. Cross a homozygous rose-combed rooster with a homozygous single-combed hen and list the genotype and phenotype ratios of their offspring.

_____

_____

11-VII. LESS PREDICTABLE VARIATIONS IN TRAITS (pp. 180–181)
11-VIII. EXAMPLES OF ENVIRONMENTAL EFFECTS ON PHENOTYPE (p. 182)

Selected Italicized Words

camptodactyly

Boldfaced, Page-Referenced Terms

(180) continuous variation _____

_____

Choice

For questions 1–5, choose from the following primary contributing factors:

    a. environment   b. gene interaction   c. a number of genes affecting a trait

___ 1. Height of human beings

___ 2. Camptodactyly, a human genetic disorder

___ 3. Development of different water buttercup leaf shapes under and above water level

___ 4. The range of eye colors in the human population

___ 5. Heat-sensitive enzyme required for melanin production in Himalayan rabbits

Self-Quiz

___ 1. The best statement of Mendel's principle of independent assortment is that _____.
    a. one allele is always dominant to another
    b. hereditary units from the male and female parents are blended in the off-spring
    c. the two hereditary units that influence a certain trait separate during gamete formation
    d. each hereditary unit is inherited separately from other hereditary units

___ 2. One of two or more alternative forms of a gene for a single trait is a(n) _____.
    a. chiasma
    b. allele
    c. autosome
    d. locus

___ 3. In the F_2 generation of a monohybrid cross involving complete dominance, the expected phenotypic ratio is _____.
    a. 3:1
    b. 1:1:1:1
    c. 1:2:1
    d. 1:1

___ 4. In the F_2 generation of a cross between a red-flowered four o'clock (homozygous) and a white-flowered four o'clock, the expected phenotypic ratio of the offspring is _____.
    a. ¾ red, ¼ white
    b. 100 percent red
    c. ¼ red, ½ pink, ¼ white
    d. 100 percent pink

___ 5. In a testcross, F_1 hybrids are crossed to an individual known to be _____ for the trait.

a. heterozygous
b. homozygous dominant
c. homozygous
d. homozygous recessive

___ 6. A man with type A blood could be the father of _____.

a. a child with type A blood
b. a child with type B blood
c. a child with type O blood
d. a child with type AB blood
e. all of the above

___ 7. A single gene that affects several seemingly unrelated aspects of an individual's phenotype is said to be _____.

a. pleiotropic
b. epistatic
c. mosaic
d. continuous

___ 8. Suppose two individuals, each heterozygous for the same characteristic, are crossed. The characteristic involves complete dominance. The expected genotypic ratio of their progeny is _____.

a. 1:2:1
b. 1:1
c. 100 percent of one genotype
d. 3:1

___ 9. If the two homozygous classes in the F_1 generation of the cross in exercise 8 are allowed to mate, the observed genotypic ratio of the offspring will be _____.

a. 1:1
b. 1:2:1
c. 100 percent of one genotype
d. 3:1

___ 10. The skin color trait in humans exhibits

_____.

a. pleiotropy
b. epistasis
c. mosaic
d. continuous variation

Chapter Objectives/Review Questions

This section lists general and detailed chapter objectives that may be used as review questions. You can make maximum use of these items by writing answers on a separate sheet of paper. Fill in answers where blanks are provided. To check for accuracy, compare your answers with information given in the chapter or glossary.

Page	Objectives/Questions
(169)	1. What was the prevailing method of explaining the inheritance of traits before Mendel's work with pea plants?
(170)	2. Garden pea plants are naturally _____-fertilizing, but Mendel took steps to _____-fertilize them for his experiments.
(171)	3. _____ are units of information about specific traits; they are passed from parents to offspring.
(171)	4. What is the general term applied to the location of a gene on a chromosome?
(171)	5. Define *allele*; how many alleles are present in the genotypes *Tt*? *tt*? *TT*?
(171)	6. When two alleles of a pair are identical, it is a _____ condition; if the two alleles are different, this is a _____ condition.
(171)	7. Distinguish a dominant allele from a recessive allele.
(171)	8. _____ refers to the genes present in an individual; _____ refers to an individual's observable traits.
(172)	9. Offspring of _____ crosses are heterozygous for the one trait being studied.
(172)	10. Explain why probability is useful to genetics.
(173)	11. Be able to use the Punnett-square method of solving genetics problems.
(173)	12. Define *testcross* and cite an example.

(173)　13. Mendel's theory of _____ states that during meiosis, the two genes of each pair separate from each other and end up in different gametes.

(174)　14. Define *dihybrid cross* and distinguish it from a *monohybrid cross*.

(175)　15. Mendel's theory of _____ _____ states that gene pairs on homologous chromosomes tend to be sorted into one gamete or another independently of how gene pairs on other chromosomes are sorted out.

(176)　16. Distinguish between complete dominance, incomplete dominance, and codominance.

(176)　17. Explain why sickle-cell anemia is a good example of pleiotropy.

(176–177)　18. Define *multiple allele system* and cite an example.

(178)　19. Gene interaction involving two alleles of a gene that mask alleles of another gene is called _____.

(180)　20. List possible explanations for less predictable trait variations that are observed.

(182)　21. Himalayan rabbits and water buttercups are good examples of environmental effects on _____.

Integrating and Applying Key Concepts

Solve the following genetics problem: In garden peas, one pair of alleles controls the height of the plant and a second pair of alleles controls flower color. The allele for tall (*D*) is dominant to the allele for dwarf (*d*), and the allele for purple (*P*) is dominant to the allele for white (*p*). A tall plant with purple flowers crossed with a tall plant with white flowers produces ⅜ tall purple, ⅜ tall white, ⅛ dwarf purple, and ⅛ dwarf white. What are the genotypes of the parents?

Answers

Interactive Exercises

11-I. MENDEL'S INSIGHT INTO THE PATTERNS OF INHERITANCE (pp. 168–171)
11-II. MENDEL'S THEORY OF SEGREGATION (pp. 172–173)
11-III. INDEPENDENT ASSORTMENT (pp. 174–175)

1. F; 2. A; 3. C; 4. G; 5. E; 6. I; 7. D; 8. H; 9. J; 10. B; 11. monohybrid; 12. P; 13. F_1; 14. F_2; 15. probability; 16. segregation; 17. testcross; 18. 1:1; 19. dihybrid; 20. independent assortment; 21. genotype: ½ *Tt*; ½ *tt*, phenotype: ½ tall; ½ short; 22. All *TT*; 23. ¼ *TT*; ½ *Tt*; ¼ *tt* or 1 *TT*; 2 *Tt*; 1 *tt*; 24. ½ *Tt*; ½ *tt* or 1 *Tt*:1 *tt*; 25. All *tt*; 26. a. ⁹⁄₁₆ pigmented eyes, right-handed; b. ³⁄₁₆ pigmented eyes, left-handed; c. ³⁄₁₆ blue-eyed, right-handed; d. ¹⁄₁₆ blue-eyed, left-handed (note Punnett square below).

	BR	Br	bR	br
BR	BBRR	BBRr	BbRR	BbRr
Br	BBRr	BBrr	BbRr	Bbrr
bR	BbRR	BbRr	bbRR	bbRr
br	BbRr	Bbrr	bbRr	bbrr

27. Albino = *aa*, normal pigmentation = *AA* or *Aa*. The woman of normal pigmentation with an albino mother is genotype *Aa*; the woman received her recessive gene (*a*) from her mother and her dominant gene (*A*) from her father. It is likely that half of the couple's children will be albinos (*aa*) and half will have normal pigmentation but be heterozygous (*Aa*).

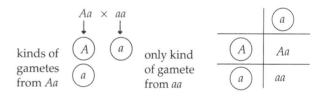

28. a. F_1: black trotter; F_2: nine black trotters, three black pacers, three chestnut trotters, one chestnut pacer; b. black trotter; c. *BbTt*; d. *bbtt* chestnut pacers and *BBTT* black trotters.

11-IV. DOMINANCE RELATIONS (p. 176)
11-V. MULTIPLE EFFECTS OF SINGLE GENES (pp. 176–177)
11-VI. INTERACTIONS BETWEEN GENE PAIRS (pp. 178–179)

1. a. Incomplete dominance; b. Codominance; c. Multiple alleles; d. Epistasis; e. Pleiotropy.

2. phenotype: all pink, genotype: all *RR'*.

3. phenotype: all white, genotype: all *R'R'*.

4. phenotype: ½ red, ½ pink; genotype: ½ *RR*, ½ *RR'*.

5. phenotype: all red, genotype: all *RR*.

6. The man must have sickle-cell trait with the genotype Hb^AHb^S, and the woman he married would have a normal genotype, Hb^AHb^A. The couple could be told that the probability is ½ that any child would have sickle-cell trait and ½ that any child would have the normal genotype.

7. Both the man and the woman have the genotype Hb^AHb^S. The probability of children from this marriage is: ¼ normal, Hb^AHb^A; ½ sickle-cell trait, Hb^AHb^S; ¼ sickle-cell anemia, Hb^SHb^S.

8. genotypes: ¼ I^AI^A; ¼ I^AI^B; ¼ I^Ai; ¼ I^Bi, phenotypes: ½ A; ¼ AB; ¼ B.

9. genotypes: ¼ I^AI^B, ¼ I^Bi, ¼ I^Ai, ¼ ii; phenotypes: ¼ AB, ¼ B, ¼ A, ¼ O.

10. genotypes: all I^Ai, phenotypes: all A.

11. genotypes: all ii, phenotypes: all O.

12. genotypes: ¼ I^AI^A, ½ I^AI^B, ¼ I^BI^B; phenotypes: ¼ A, ½ AB, ¼ B.

13. ¼ color; ¾ white. 14. ¾ color; ¼ white. 15. ¼ color; ¾ white.

16. The genotype of the male parent is *RrPp* and the genotype of the female parent is *rrpp*. The offspring are ¼ walnut comb, *RrPp*; ¼ rose comb, *Rrpp*; ¼ pea comb, *rrPp*; ¼ single comb, *rrpp*.

17. The genotype of the walnut-combed male is *RRpp* and the genotype of the single-combed female is *rrpp*. All offspring are rose comb with the genotype *Rrpp*.

11-VII. LESS PREDICTABLE VARIATIONS IN TRAITS (pp. 180–182)

11-VIII. EXAMPLES OF ENVIRONMENTAL EFFECTS ON PHENOTYPE (p. 182)

1. c; 2. b; 3. a; 4. c; 5. a.

Self-Quiz

1. d; 2. b; 3. a; 4. c; 5. d; 6. e; 7. a; 8. a; 9. c; 10. d.

Integrating and Applying Key Concepts

DdPp × *Ddpp*

12

CHROMOSOMES AND HUMAN GENETICS

Interactive Exercises

12-I. RETURN OF THE PEA PLANT (pp. 186–187)
12-II. THE CHROMOSOMAL BASIS OF INHERITANCE—AN OVERVIEW (pp. 188–189)

Selected Italicized Words

Colchicum autumnale, centrifugation

Boldfaced, Page-Referenced Terms

(188) genes _____

_____

(188) homologous chromosomes _____

_____

(188) alleles _____

_____

(188) crossing over _____

_____

(188) X chromosome _____

_____

(188) Y chromosome _____

_____

(189) sex chromosomes _____

_____

(189) autosomes _____

_____

(189) karyotype _____

_____

Fill-in-the-Blanks

(1) _____ first described the behavior of threadlike bodies (chromosomes) during mitosis.

(2) _____ first proposed that chromosomal number must be reduced by half during gamete formation;

the process became known as (3) _____. (4) _____ chromosomes resemble each other in length,

shape, and gene sequence; one of each type is inherited from the mother and its homologue from the father.

In 1900, researchers came across (5) _____ paper; their results confirmed what he had proposed earlier:

Diploid cells have at least two (6) _____, or units, for each heritable trait, and the units (7) _____

before gametes form.

Slightly different molecular forms of the same gene that arise through mutation are known as

(8) _____. The meiotic event whereby breaks and exchanges accomplish gene recombination between

a chromosome and its homologous partner is known as (9) _____ _____. (10) _____ dia-

grams help answer questions about an individual's chromosomes.

12-III. SEX DETERMINATION IN HUMANS (pp. 190–192)
12-IV. EARLY QUESTIONS ABOUT GENE LOCATIONS (p. 193)
12-V. RECOMBINATION PATTERNS AND CHROMOSOME MAPPING (p. 194)

Selected Italicized Words

Drosophila melanogaster, nonsexual traits, Zea mays

Boldfaced, Page-Referenced Terms

(193) X-linked genes _____

_____

(193) Y-linked genes _____

(193) linkage groups _____

Matching

Choose the single most appropriate letter.

1. ___ crossing over
2. ___ X chromosome
3. ___ Y chromosome
4. ___ sex chromosomes
5. ___ autosomes
6. ___ karyotype
7. ___ X-linked genes
8. ___ Y-linked genes
9. ___ linkage groups
10. ___ SRY gene

A. Examples are the X and Y chromosomes of humans
B. The number of metaphase chromosomes and their defining characteristics
C. Found only on the Y chromosome
D. The master gene for sex determination, on the Y chromosome; regulates reactions necessary for sex determination
E. Any chromosome not concerned with sex determination; identical in males and females
F. Only one is found in a male karyotype; meiotic synapse is with an X chromosome
G. The block of genes located on each type of chromosome; these genes tend to travel together in inheritance
H. Found only on the X chromosome
I. Interrupts gene linkage; brings about the recombination of genes on paternal and maternal homologues
J. Human females possess two; genes are mostly concerned with nonsexual traits

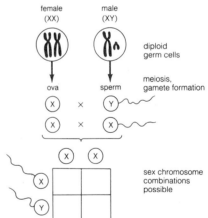

11. Complete the Punnett square below, which brings Y-bearing and X-bearing sperm together randomly in fertilization. Answer the related dichotomous choice questions near the Punnett square.
12. Male humans transmit their Y chromosome only to their (sons/daughters).
13. Male humans receive their X chromosome only from their (mothers/fathers).
14. Human mothers and fathers each provide an X chromosome for their (sons/daughters).
15. Crossing over occurs between (sister/nonsister) chromatids.
16. Morgan's experiments with fruit flies indicated a correlation between sex and eye color; eye color must be carried on the (X/Y) chromosome.

Problems

Early investigators realized that the frequency of crossing over (data from actual genetic crosses) could be used as a tool to map genes on chromosomes. Genetic maps are graphic representations of the relative positions and distances between genes on each chromosome (linkage group). Geneticists arbitrarily equate 1 percent of crossover (recombination) with 1 genetic map unit. In the following questions, a line is used to represent a chromosome with its linked genes.

17. Which of the following represents a chromosome that has undergone crossover and recombination? It is assumed that the organism involved is heterozygous with the genotype: $\begin{array}{c|c} A & a \\ B & b \end{array}$ ————

   a. $\begin{array}{|c} A \\ B \end{array}$ b. $\begin{array}{c|} a \\ b \end{array}$ c. $\begin{array}{c|} B \\ A \end{array}$ d. $\begin{array}{|c} a \\ B \end{array}$

18. Following breeding experiments and a study of crossovers, the following map distances were recorded: gene *C* to gene *B* = 35 genetic map units; *B* to *A* = 10 map units, and *A* to *C* = 45 map units. Which of the following maps is correct?

a. *A B C*
b. *B A C*
c. *B C A*
d. *B A C*

19. Individuals with dominant gene *N* have the nail-patella syndrome and exhibit abnormally developed fingernails and absence of kneecaps; the recessive gene *n* is normal. Geneticists have learned that the genes determining the ABO blood groups and the nail-patella gene are linked, both found on chromosome 9. A woman with blood type A and nail-patella syndrome marries a man who has blood type O and has normal fingernails and kneecaps. This couple has three children. One daughter has blood type A like her mother but has normal fingernails and kneecaps. A second daughter has blood type O with nail-patella syndrome. A son has blood type O and normal fingernails and kneecaps. The chromosomes of the parents are diagrammed below.

a. Complete the exercise by showing the linked genes for each of the children.

♀ ♂

A   n
_____   ×   _____   *(Or An/ON × On/On)*
_____   _____
O   N   *O   n*

Daughter 1   Daughter 2   Son

_____   _____   _____

_____   _____   _____

b. Explain how the son has the genotype *On/On* when only one parent has a chromosome with genes *O* and *n* linked. _____

_____

12-VI. HUMAN GENETIC ANALYSIS (pp. 195–196)
12-VII. PATTERNS OF AUTOSOMAL INHERITANCE (pp. 196–197)
12-VIII. PATTERNS OF X-LINKED INHERITANCE (pp. 198–199)

Selected Italicized Words

galactosemia, familial hypercholesterolemia, achondroplasia, Huntington's disorder, Hemophilia A, Duchenne muscular dystrophy, faulty enamel trait, amyotrophic lateral sclerosis

Boldfaced, Page-Referenced Terms

(195) pedigrees _____

_____

(196) genetic abnormality _____

_____

(196) genetic disorder _____

_____

Choice

For questions 1–18, choose from the following patterns of inheritance; some items may require more than one letter.

a. autosomal recessive   b. autosomal dominant   c. X-linked recessive   d. X-linked dominant

___ 1. The trait is expressed in heterozygous females.

___ 2. Heterozygotes can remain undetected.

___ 3. The trait appears in each generation.

___ 4. The recessive phenotype shows up far more often in males than in females.

___ 5. Both parents may be heterozygous normal.

___ 6. If one parent is heterozygous and the other homozygous recessive, there is a 50 percent chance any child of theirs will be heterozygous.

___ 7. The allele is usually expressed, even in heterozygotes.

___ 8. The trait is expressed in heterozygous females.

___ 9. Heterozygous normal parents can expect that one-fourth of their children will be affected by the disorder.

___ 10. A son cannot inherit the recessive allele from his father, but his daughter can.

___ 11. Females can mask this gene; males cannot.

___ 12. The trait is expressed in heterozygotes of either sex.

___ 13. Heterozygous women transmit the allele to half their offspring, regardless of sex.

___ 14. Individuals displaying this type of disorder will always be homozygous for the trait.

___ 15. The trait is expressed in both the homozygote and the heterozygote.

___ 16. Heterozygous females will transmit the recessive gene to half their sons and half their daughters.

___ 17. If both parents are heterozygous, there is a 50 percent chance that each child will be heterozygous.

___ 18. A son cannot inherit the allele responsible for the trait from an affected father, but all his daughters will.

Problems

19. The autosomal allele that causes albinism (*a*) is recessive to the allele for normal pigmentation (*A*). A normally pigmented woman whose father is an albino marries an albino man whose parents are normal. They have three children, two normal and one albino. Give the genotypes for each person listed.

_____

_____

20. Huntington's disorder is a rare form of autosomal dominant inheritance, *H*; the normal gene is *h*. The disease causes progressive degeneration of the nervous system with onset exhibited near middle age. An apparently normal man in his early twenties learns that his father has recently been diagnosed as having Huntington's disorder. What are the chances that the son will develop Huntington's disorder?

21. A color-blind man and a woman with normal vision whose father was color-blind have a son. Color blindness, in this case, is caused by an X-linked recessive gene. If only the male offspring are considered, what is the probability that their son is color-blind? _____

22. Hemophilia A is caused by an X-linked recessive gene. A woman who is seemingly normal but whose father was a hemophiliac marries a normal man. What proportion of their sons will have hemophilia? What proportion of their daughters will have hemophilia? What proportion of their daughters will be carriers? _____

23. The following pedigree shows the pattern of inheritance of color blindness in a family (persons with the trait are indicated by black circles). What is the chance that the third-generation female indicated by the arrow (below) will have a color-blind son if she marries a normal male? A color-blind male?

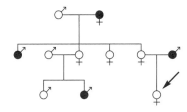

Complete the Table

24. Complete the table below by indicating whether the genetic disorder listed is due to inheritance that is autosomal recessive, autosomal dominant, X-linked recessive, or X-linked dominant.

Genetic Disorder	Inheritance Mode
a. Galactosemia	
b. Familial hypercholesterolemia	
c. Achondroplasia	
d. Huntington's disorder	
e. Hemophilia A	
f. Duchenne muscular dystrophy	
g. Faulty enamel trait	
h. Progeria	
i. Polydactyly	

12-IX. CHANGES IN CHROMOSOME STRUCTURE (p. 200)

Selected Italicized Words

cri-du-chat, fragile X syndrome

Boldfaced, Page-Referenced Terms

(200) deletion _____

_____

(200) inversion _____

_____

(200) translocation _____

_____

(200) duplications _____

_____

Label-Match

On rare occasions, chromosome structure becomes abnormally rearranged. Such changes may have profound effects on the phenotype of an organism. Label the following diagrams of abnormal chromosome structure as a deletion, a duplication, an inversion, or a translocation. Complete the exercise by matching and entering the letter of the proper description in the parentheses following each label.

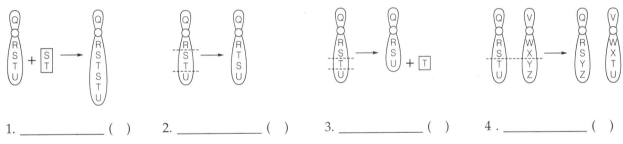

1. _____ ( )   2. _____ ( )   3. _____ ( )   4 . _____ ( )

A. The loss of a chromosome segment
B. A gene sequence in excess of its normal amount in a chromosome
C. A chromosome segment that separated from the chromosome and then was inserted at the same place, but in reverse
D. The transfer of part of one chromosome to a nonhomologous chromosome

12-X. CHANGES IN CHROMOSOME NUMBER (p. 201)
12-XI. WHEN THE NUMBER OF AUTOSOMES CHANGES (p. 202)
12-XII. WHEN THE NUMBER OF SEX CHROMOSOMES CHANGES (pp. 203–205)

Selected Italicized Words

Down syndrome, Alzheimer disease, Turner syndrome, Klinefelter syndrome, phenylketonuria

Boldfaced, Page-Referenced Terms

(201) aneuploidy _____

_____

(201) polyploidy _____

_____

(201) nondisjunction _____

_____

(204) amniocentesis _____

_____

(204) chorionic villi sampling _____

_____

(205) abortion _____

_____

(205) in vitro fertilization _____

_____

Complete the Table

1. Complete the table below to summarize the major mechanisms of chromosome number change in organisms.

Category of Change   *Description*

a. Aneuploidy	
b. Polyploidy	

Short Answer

2. If a nondisjunction occurs at anaphase I of the first meiotic division, what will be the proportion of abnormal gametes (for the chromosome involved in the nondisjunction)? (p. 201) _____

_____

3. If a nondisjunction occurs at anaphase II of the second meiotic division, what will be the proportion of abnormal gametes (for the chromosome involved in the nondisjunction)? (p. 201) _____

_____

4. Contrast the effects of polyploidy in plants and humans. (p. 201) _____

_____

Choice

For questions 5–14, choose from the following:

a. Down syndrome   b. Turner syndrome   c. Klinefelter syndrome   d. XYY condition

___ 5. XXY male

___ 6. Ovaries nonfunctional and secondary sexual traits fail to develop at puberty

___ 7. Testes smaller than normal, sparse body hair, and some breast enlargement

___ 8. Could only be caused by a nondisjunction in males

___ 9. Older children smaller than normal with distinctive facial features; small skin fold over the inner corner of the eyelid

___ 10. X0 female; often abort early; distorted female phenotype

___ 11. Males that tend to be taller than average; some mildly retarded but most are phenotypically normal

___ 12. Injections of testosterone reverse feminized traits but not the mental retardation

___ 13. Trisomy 21; skeleton develops more slowly than normal with slack muscles

___ 14. At one time these males were thought to be genetically predisposed to become criminals

Complete the Table

15. Complete the table below, which summarizes methods of dealing with the problems of human genetics. Choose from phenotypic treatments, genetic screening, genetic counseling, and prenatal diagnosis.

Method	Description
a.	Detects genetic disorders before birth; may use karyotypes, biochemical tests, amniocentesis, and CVS
b.	Parents at risk requesting emotional support and advice from clinical psychologists, geneticists, and social workers
c.	Suppressing or minimizing symptoms of genetic disorders by surgical intervention, controlling diet or environment, or chemically modifying genes
d.	Large-scale programs to detect affected persons or carriers in a population; early detection may allow introduction of preventive measures before symptoms develop

Matching

Choose the one most appropriate answer for each. A letter may be used more than once or not at all.

16. ___ albinism

17. ___ amniocentesis

18. ___ chorionic villi analysis

19. ___ cleft lip

20. ___ galactosemia

21. ___ phenylketonuria

22. ___ karyotype analysis

23. ___ sickle-cell anemia

24. ___ Wilson's disorder

25. ___ pedigree chart

A. A phenotypic defect that can be helped by diet modification
B. A phenotypic defect that can be helped by environmental adjustments
C. A phenotypic defect that can be helped by surgical correction
D. A phenotypic defect that can be helped by chemotherapy
E. Used to arrive at a diagnosis
F. Used to determine the location of a specific gene on a specific chromosome
G. Used to depict genetic relationships among the members of families

---

Self-Quiz

___ 1. All the genes located on a given chromosome compose a _____.
   a. karyotype
   b. bridging cross
   c. wild-type allele
   d. linkage group

___ 2. Chromosomes other than those involved in sex determination are known as _____.
   a. nucleosomes
   b. heterosomes
   c. alleles
   d. autosomes

___ 3. The farther apart two genes are on a chromosome, _____.
   a. the less likely that crossing over and recombination will occur between them
   b. the greater will be the frequency of crossing over and recombination between them
   c. the more likely they are to be in two different linkage groups
   d. the more likely they are to be segregated into different gametes when meiosis occurs

___ 4. Karyotype analysis is _____.
   a. a means of detecting and reducing mutagenic agents
   b. a surgical technique that separates chromosomes that have failed to segregate properly during meiosis II
   c. used in prenatal diagnosis to detect chromosomal mutations and metabolic disorders in embryos
   d. a process that substitutes defective alleles with normal ones

___ 5. Which of the following did Morgan and his research group *not* do?
   a. They isolated and kept under culture fruit flies with the sex-linked recessive white-eyed trait.
   b. They developed the technique of amniocentesis.
   c. They discovered X-linked genes.
   d. Their work reinforced the concept that each gene is located on a specific chromosome.

___ 6. Red-green color blindness is a sex-linked recessive trait in humans. A color-blind woman and a man with normal vision have a son. What are the chances that the son is color-blind? If the parents ever have daughters, what is the chance for each birth that the daughter will be color-blind? (Consider only the female offspring.)

a. 100 percent, 0 percent
b. 50 percent, 0 percent
c. 100 percent, 100 percent
d. 50 percent, 100 percent
e. none of the above

___ 7. Suppose that a hemophilic male (X-linked recessive allele) and a female carrier for the hemophilic trait have a nonhemophilic daughter with Turner syndrome. Nondisjunction could have occurred in _____.

a. both parents
b. neither parent
c. the father only
d. the mother only

___ 8. Nondisjunction involving the X chromosome occurs during oogenesis and produces two kinds of eggs, XX and O (no X chromosome). If normal Y sperm fertilize the two types, which genotypes are possible?

a. XX and XY
b. XXY and YO
c. XYY and XO
d. XYY and YO

___ 9. Of all phenotypically normal males in prisons, the type once thought to be genetically predisposed to becoming criminals was the group with _____.

a. XXY disorder
b. XYY disorder
c. Turner syndrome
d. Down syndrome

___ 10. Amniocentesis is _____.

a. a surgical means of repairing deformities
b. a form of chemotherapy that modifies or inhibits gene expression or the function of gene products
c. used in prenatal diagnosis to detect chromosomal mutations and metabolic disorders in embryos
d. a form of gene-replacement therapy

Chapter Objectives/Review Questions

This section lists general and detailed chapter objectives that may be used as review questions. You can make maximum use of these items by writing answers on a separate sheet of paper. Fill in answers where blanks are provided. To check for accuracy, compare your answers with information given in the chapter or glossary.

Page	Objectives/Questions
(187)	1. Identify the researcher who, in 1882, first described thread-like bodies in the nuclei of dividing cells; state what was so useful about this contribution.
(187)	2. In 1887, _____ proposed that a special division process must reduce the chromosome number by half before gametes form.
(187)	3. Relate the significance of the year 1900 to genetics as a science.
(188)	4. Diploid (2*n*) cells have pairs of _____ chromosomes.
(188)	5. _____ are different molecular forms of the same gene that arise through mutation.
(188)	6. State the circumstances required for crossing over and the results.
(188–189)	7. Define karyotype; briefly describe its preparation and value.
(189)	8. Distinguish between chromosomes and autosomes.
(190)	9. A newly identified region of the Y chromosome called _____ appears to be the master gene for sex determination.
(190)	10. Describe how meiotic segregation of sex chromosomes to gametes and subsequent random fertilization determines sex in many organisms.

(192)   11. In whose laboratory was sex linkage in fruit flies discovered? When?

(193)   12. The tendency of genes located on the same chromosome to end up together in the same gamete is called _____.

(193)   13. State the relationship between crossover frequency and the location of genes on a chromosome.

(195)   14. A _____ chart is used to study genetic connections between individuals.

(196)   15. A genetic _____ is a rare or less common occurrence, whereas a _____ causes mild to severe medical problems.

(196–199) 16. Describe the characteristics of autosomal recessive inheritance, autosomal dominant inheritance, and recessive and dominant X-linked inheritance; cite one example of each.

(200)   17. A(n) _____ is a loss of a chromosome segment; a(n) _____ is a gene sequence separated from a chromosome but then inserted at the same place, but in reverse; a(n) _____ is a repeat of several gene sequences on the same chromosome; a(n) _____ is the transfer of part of one chromosome to a nonhomologous chromosome.

(201)   18. When gametes or cells of an affected individual end up with one extra or one less than the parental number of chromosomes, it is known as _____; relate this to monosomy and trisomy.

(201)   19. Having three or more complete sets of chromosomes is called _____.

(201)   20. _____ is the failure of the chromosomes to separate at either meiosis I or meiosis II.

(202–203) 21. Trisomy 21 is known as _____ syndrome; Turner syndrome has the chromosome constitution _____; XXY chromosome constitution is _____ syndrome; taller than average males with sometimes slightly depressed IQs have the _____ condition.

(204)   22 Define phenotypic treatment and describe one example.

(204)   23. Explain the procedures used in two types of prenatal diagnosis, amniocentesis and chorionic villi analysis; compare the risks.

(204)   24. List some benefits of genetic screening and genetic counseling to society.

(205)   25. Discuss some of the ethical considerations that might be associated with a decision of induced abortion.

(205)   26. _____ fertilization is the fertilizing of eggs in a petri dish.

Integrating and Applying Key Concepts

1. The parents of a young boy bring him to their doctor. They explain that the boy does not seem to be going through the same vocal developmental stages as his older brother. The doctor orders a common cytogenetics test to be done, and it reveals that the young boy's cells contain two X chromosomes and one Y chromosome. Describe the test that the doctor ordered and explain how and when such a genetic result, XXY, most logically occurred.

2. Solve the following genetics problem. Show rationale, genotypes, and phenotypes. A husband sues his wife for divorce, arguing that she has been unfaithful. His wife gave birth to a girl with a fissure in the iris of her eye, a sex-linked recessive trait. Both parents have normal eye structure. Can the genetic facts be used to argue for the husband's suit? Explain your answer.

Answers

12-I. RETURN OF THE PEA PLANT (pp. 186–187)

12-II. THE CHROMOSOMAL BASIS OF INHERITANCE—AN OVERVIEW (pp. 188–189)

1. Flemming; 2. Weismann; 3. meiosis; 4. Homologous; 5. Mendel's; 6. genes; 7. segregate; 8. alleles; 9. crossing over; 10. Karyotype.

12-III. SEX DETERMINATION IN HUMANS (pp. 190–192)

12-IV. EARLY QUESTIONS ABOUT GENE LOCATIONS (p. 193)

12-V. RECOMBINATION PATTERNS AND CHROMOSOME MAPPING (p. 194)

1. I; 2. J; 3. F; 4. A; 5. E; 6. B; 7. H; 8. C; 9. G; 10. D; 11. Two blocks of the Punnett square should be XX and two blocks should be XY; 12. sons; 13. mothers; 14. daughters; 15. nonsister; 16. X; 17. d; 18. a; 19. a. Linked genes of the children: daughter 1 is *On/An*, daughter 2 is *ON/On*, son is *On/On*; b. The son has genotype *On/On* because a crossover must have occurred during meiosis in his mother, resulting in the recombination *On*.

12-VI. HUMAN GENETIC ANALYSIS (pp. 195-196)

12-VII. PATTERNS OF AUTOSOMAL INHERITANCE (pp. 196–197)

12-VIII. PATTERNS OF X-LINKED INHERITANCE (pp. 198–199)

1. b and d; 2. a and c; 3. b; 4. c; 5. a; 6. b; 7. b; 8. b and d; 9. a; 10. c; 11. c; 12. b; 13. b; 14. a; 15. b; 16. c; 17. a; 18. d; 19. The woman's mother is heterozygous normal, *Aa*; the woman is also heterozygous normal, *Aa*. The albino man, *aa*, has two heterozygous normal parents, *Aa*. The two normal children are heterozygous normal, *Aa*; the albino child is *aa*; 20. Assuming the father is heterozygous with Huntington's disorder and the mother normal, the chances are ½ that the son will develop the disease; 21. If only male offspring are considered, the probability is ½ that the couple will have a color-blind son; 22. The probability is that ½ of the sons will have hemophilia; the probability is 0 that a daughter will express hemophilia; the probability is that ½ of the daughters will be carriers; 23. If the woman marries a normal male, the chance that her son would be color-blind is ½. If she marries a color-blind male, the chance that her son would be color-blind is also ½;

24. a. Autosomal recessive; b. Autosomal dominant; c. Autosomal dominant; d. Autosomal dominant; e. X-linked recessive; f. X-linked recessive; g. X-linked dominant; h. Autosomal dominant; i. Autosomal dominant.

12-IX. CHANGES IN CHROMOSOME STRUCTURE (p. 200)

1. duplication (B); 2. inversion (C); 3. deletion (A); 4. translocation (D).

12-X. CHANGES IN CHROMOSOME NUMBER (p. 201)

12-XI. WHEN THE NUMBER OF AUTOSOMES CHANGES (p. 202)

12-XII. WHEN THE NUMBER OF SEX CHROMOSOMES CHANGES (pp. 203–205)

1. a. Gametes or cells of an affected individual that have one extra or one less than the parental number of chromosomes; b. Cells or individuals that have three or more complete sets of chromosomes; 2. All gametes will be abnormal; 3. One-half of the gametes will be abnormal; 4. About half of all flowering plant species are polyploids, but this condition is lethal for humans; 5. c; 6. b; 7. c; 8. d; 9. a; 10. b; 11. d; 12. c; 13. a; 14. d; 15. a. Prenatal diagnosis; b. Genetic counseling; c. Phenotypic treatments; d. Genetic screening; 16. B; 17. E; 18. E; 19. C; 20. A; 21. A; 22. E; 23. B; 24. D; 25. G.

Self-Quiz

1. d; 2. d; 3. b; 4. c; 5. b; 6. a; 7. c; 8. b; 9. b; 10. c.

13

DNA STRUCTURE
AND FUNCTION

DISCOVERY OF DNA FUNCTION
    Early Clues
    Confirmation of DNA Function

DNA STRUCTURE
    Components of DNA
    Patterns of Base Pairing

DNA REPLICATION AND DNA REPAIR

ORGANIZATION OF DNA IN CHROMOSOMES

Interactive Exercises

13-I. DISCOVERY OF DNA FUNCTION (pp. 208–211)

Selected Italicized Words

Streptococcus pneumoniae; pathogenic, Escherichia coli

Boldfaced, Page-Referenced Terms

(209) DNA _____

(210) bacteriophages _____

_____

Complete the Table

1.   Complete the table below, which traces the discovery of DNA function.

Investigators	Year(s)	Contribution
Miescher	1868	a.
b.	1928	discovered the transforming principle in *Streptococcus pneumoniae*; live, harmless R cells were mixed with dead S cells, R cells became S cells
Avery (also MacLeod and McCarty)	1944	c.
d.	mid-1940s	studied the infectious cycle of bacteriophages
Hershey and Chase	1952	e.

13-II. DNA STRUCTURE (pp. 212–213)

Boldfaced, Page-Referenced Terms

(212) nucleotide _____

_____

(212) adenine, A _____

_____

(212) guanine, G _____

_____

(212) thymine, T _____

_____

(212) cytosine, C _____

_____

(212) x-ray diffraction images _____

_____

Short Answer

1. List the three parts of a nucleotide. _____

_____

Labeling

Four nucleotides are illustrated below. In the blank, label each nitrogen-containing base correctly as guanine, thymine, cytosine, or adenine.

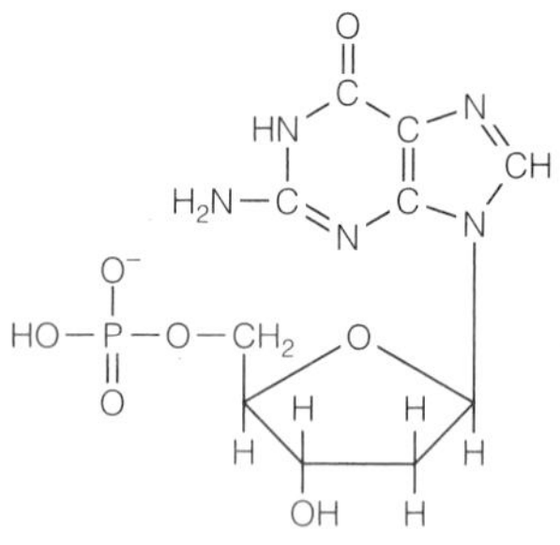

2. 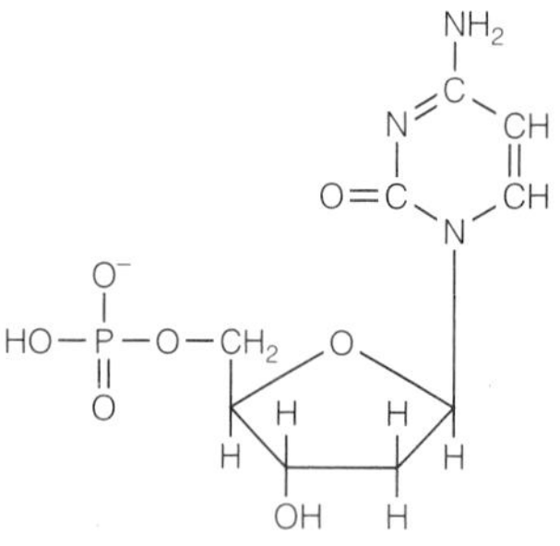 _____   3. 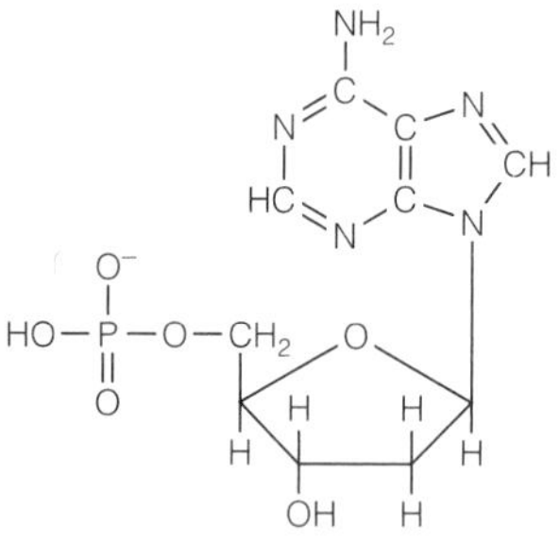 _____   4. 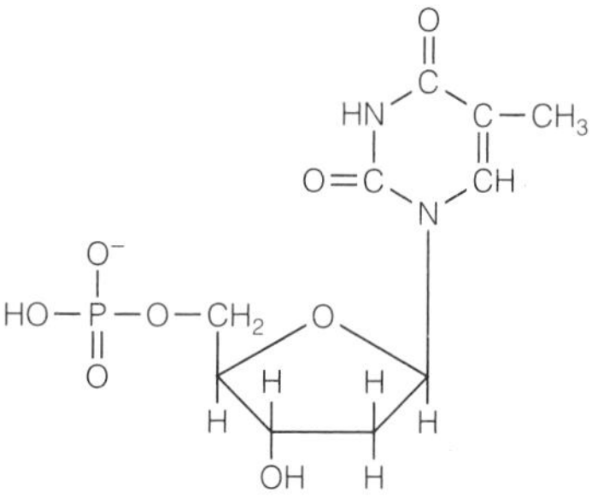 _____   5. _____

True-False

If false, explain why.

_____ 6. DNA is composed of four different types of nucleotides.

_____ 7. In the DNA of every species, the amount of adenine present always equals the amount of thymine, and the amount of cytosine always equals the amount of guanine (A = T and C = G).

_____ 8. In a nucleotide, the phosphate group is attached to the nitrogen-containing base, which is attached to the five-carbon sugar.

_____ 9. Watson and Crick built their model of DNA in the early 1950s.

_____ 10. Guanine pairs with cytosine and adenine pairs with thymine by forming hydrogen bonds between them.

Fill-in-the-Blanks

Base (11) _____ between the two nucleotide strands in DNA is (12) _____ for all species (A=T; G=C). The base (13) _____ (determining which base follows the next in a nucleotide strand) is (14) _____ from species to species.

Label-Match

Identify each indicated part of the DNA illustration at the right. Choose from these answers: phosphate group, purine, pyrimidine, nucleotide, and deoxyribose. Complete the exercise by matching and entering the letter of the proper structure description in the parentheses following each label.

The following DNA memory devices may be helpful: Use _pyr_CUT to remember that the single-ring pyrimidines are cytosine, uracil, and thymine; use _pur_AG to remember that the double-ring purines are adenine and guanine; pyrimidine is a _long_ name for a _narrow_ molecule; purine is a _short_ name for a _wide_ molecule; to recall the number of hydrogen bonds between DNA bases, remember that AT = 2 and CG = 3.

15. _____ ( )

16. _____ _____ ( )

17. _____ ( )

18. _____ ( )

19. _____ ( )

20. _____ ( )

21. _____ ( )

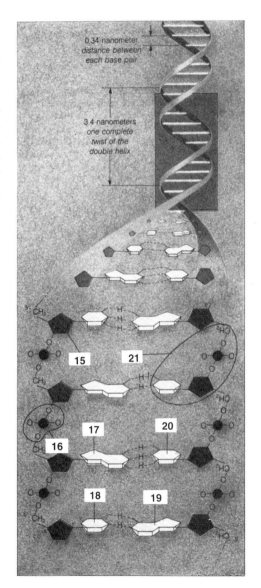

A. The pyrimidine is thymine because it has two hydrogen bonds.
B. A five-carbon sugar joined to two phosphate groups in the upright portion of the DNA ladder.
C. The purine is guanine because it has three hydrogen bonds.
D. The pyrimidine is cytosine because it has three hydrogen bonds.
E. The purine is adenine because it has two hydrogen bonds.
F. Composed of three smaller molecules: a phosphate group, five-carbon deoxyribose sugar, and a nitrogenous base (in this case, a pyrimidine).
G. A chemical group that joins two sugars in the upright portion of the DNA ladder.

22. Explain why understanding the structure of DNA helps scientists understand how living organisms can have so much in common at the molecular level and yet be so diverse at the whole organism level.

_____

_____

13-III. DNA REPLICATION AND DNA REPAIR (pp. 214–215)

Selected Italicized Words

xeroderma pigmentosum, basal cell carcinoma, semiconservative replication

Boldfaced, Page-Referenced Terms

(214) DNA replication _____

(214) DNA polymerases _____

_____

(214) DNA ligases _____

_____

(214) DNA repair _____

_____

Label

1. The term semiconservative replication refers to the fact that each new DNA molecule resulting from the replication process is "half-old, half-new." In the illustration below, complete the replication required in the middle of the molecule by adding the required letters representing the missing nucleotide bases. Recall that ATP energy and the appropriate enzymes are actually required to complete this process.

```
T-  ___      ___  -A
G-  ___      ___  -C
A-  ___      ___  -T
C-  ___      ___  -G
C-  ___      ___  -G
C-  ___      ___  -G
old new     new  old
```

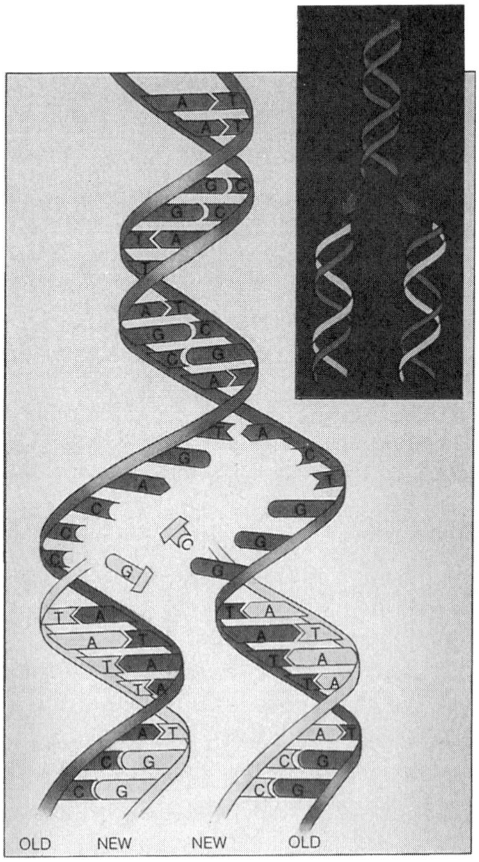

True-False

If false, explain why.

_____ 2. The hydrogen bonding of adenine to guanine is an example of complementary base pairing.

_____ 3. The replication of DNA is considered a conserving process because the same four nucleotides are used again and again during replication.

_____ 4. Each parent strand remains intact during replication, and a new companion strand is assembled on each of those parent strands.

_____ 5. Some of the enzymes associated with DNA assembly repair errors during the replication process.

Matching

Match the numbers on the illustration below to the letters representing structures and functions.

6. ___

7. ___

8. ___

9. ___

10. ___

11. ___

   A. The enzyme (ligase) links short DNA segments together.
   B. One parent DNA strand.
   C. Enzymes unwind DNA double helix.
   D. Enzymes add short "primer" segments to begin chain assembly.
   E. Enzymes (DNA polymerase) assemble new DNA strands.
   F. Newly forming DNA strand.

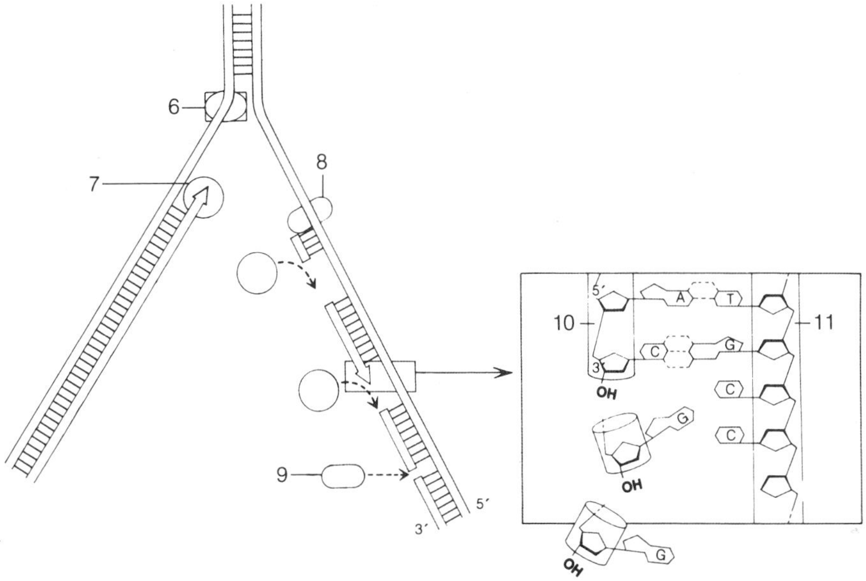

13-IV. ORGANIZATION OF DNA IN CHROMOSOMES (p. 216)

Boldfaced, Page-Referenced Terms

(216) histones _____

_____

(216) nucleosome _____

_____

Fill-in-the-Blanks

Each chromosome has one (1) _____ molecule coursing through it. Eukaryotic DNA is complexed

tightly with many (2) _____. Some (3) _____ proteins act as spools to wind up small pieces of

(4) _____. A (5) _____ is a histone-DNA spool. The way the chromosome is packed is known to

influence the activity of different (6) _____.

Labeling

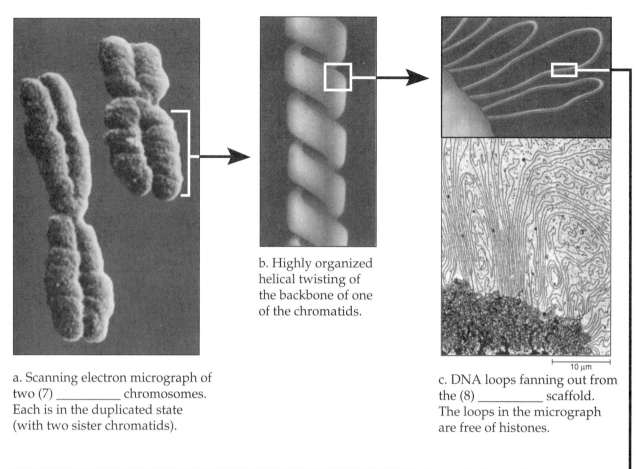

a. Scanning electron micrograph of two (7) _____ chromosomes. Each is in the duplicated state (with two sister chromatids).

b. Highly organized helical twisting of the backbone of one of the chromatids.

c. DNA loops fanning out from the (8) _____ scaffold. The loops in the micrograph are free of histones.

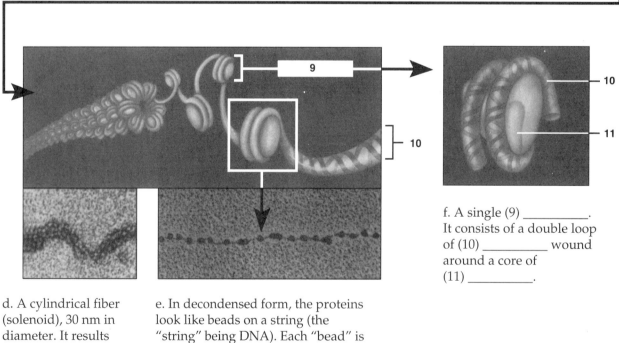

d. A cylindrical fiber (solenoid), 30 nm in diameter. It results from DNA-protein interactions that lead to repeated coiling.

e. In decondensed form, the proteins look like beads on a string (the "string" being DNA). Each "bead" is a (9) _____.

f. A single (9) _____. It consists of a double loop of (10) _____ wound around a core of (11) _____.

Crossword Puzzle: DNA Structure and Function

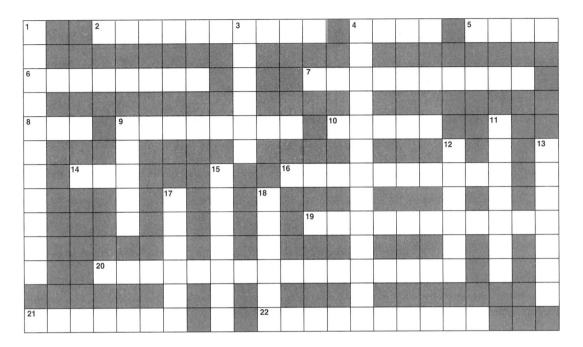

ACROSS

2. DNA _____ assembles DNA nucleotides into a DNA strand.
4. Purine or pyrimidine.
5. _____ diffraction was used to deduce the width and periodicity of DNA.
6. Developed the best x-ray diffraction images of DNA and deduced its basic dimensions.
7. A double loop of DNA wound around a core of histone molecules.
8. Transcription produces this code molecule.
9. Forms three hydrogen bonds with its mate.
10. Coiled structure, like a circular stairway.
14. In most organisms, the molecule that contains the genetic code.
16. Single-ringed nitrogenous base.
19. Duplication of hereditary material during the S phase of the cell cycle.
20. A new DNA strand is assembled on each of the two parent strands of DNA.
21. Forms two hydrogen bonds with its mate.
22. Five carbon sugar + phosphate group + nitrogenous base.

DOWN

1. X-ray _____ is a technique that can yield clues about molecular structure.
3. DNA polymerases and DNA ligases _____ DNA.
4. A class of viruses that infect bacteria.
9. Hershey and _____ demonstrated that DNA contained the hereditary instructions for producing new bacteriophages.
11. Forms two hydrogen bonds with its mate.
12. DNA _____ is an enzyme that joins new short stretches of nucleotides into a continuous strand.
13. Forms three hydrogen bonds with its mate.
15. One of several kinds of proteins that bind tightly to DNA and organize it during its activities.
17. Double-ringed nitrogenous base.
18. Built a model of DNA structure; won Nobel Prize.

Self-Quiz

___1. Each DNA strand has a backbone that consists of alternating _____.

a. purines and pyrimidines
b. nitrogen-containing bases
c. hydrogen bonds
d. sugar and phosphate molecules

___2. In DNA, complementary base pairing occurs between _____.

a. cytosine and uracil
b. adenine and guanine
c. adenine and uracil
d. adenine and thymine

___3. Adenine and guanine are _____.

a. double-ringed purines
b. single-ringed purines
c. double-ringed pyrimidines
d. single-ringed pyrimidines

___4. Franklin used the technique known as _____ to determine many of the physical characteristics of DNA.

a. transformation
b. transmission electron microscopy
c. density-gradient centrifugation
d. x-ray diffraction

___5. The significance of Griffith's experiment that used two strains of pneumonia-causing bacteria is that _____.

a. the conserving nature of DNA replication was finally demonstrated
b. it demonstrated that harmless cells had become permanently transformed through a change in the bacterial hereditary system
c. it established that pure DNA extracted from disease-causing bacteria transformed harmless strains into "pathogenic strains"
d. it demonstrated that radioactively labeled bacteriophages transfer their DNA but not their protein coats to their host bacteria

___6. The significance of the experiments in which ^{32}P and ^{35}S were used is that _____.

a. the semiconservative nature of DNA replication was finally demonstrated

b. it demonstrated that harmless cells had become permanently transformed through a change in the bacterial hereditary system
c. it established that pure DNA extracted from disease-causing bacteria transformed harmless strains into "killer strains"
d. it demonstrated that radioactively labeled bacteriophages transfer their DNA but not their protein coats to their host bacteria

___7. Franklin's research contribution was essential in _____.

a. establishing the double-stranded nature of DNA
b. establishing the principle of base pairing
c. establishing most of the principal structural features of DNA
d. all of the above

___8. When Griffith injected mice with a mixture of dead pathogenic cells—encapsulated S cells and living, unencapsulated R cells of pneumonia bacteria—he discovered that _____.

a. the previously harmless strain had permanently inherited the capacity to build protective capsules
b. the dead mice teemed with living pathogenic (R) cells
c. the killer R strain was encased in a protective capsule
d. all of the above

___9. A single strand of DNA with the base-pairing sequence C-G-A-T-T-G is compatible only with the sequence _____.

a. C-G-A-T-T-G
b. G-C-T-A-A-G
c. T-A-G-C-C-T
d. G-C-T-A-A-C

___10. The nucleosome is a _____.

a. subunit of a nucleolus
b. coiled bead of histone-DNA
c. DNA packing arrangement within a chromosome
d. term synonymous with gene
e. both (b) and (c)

Chapter Objectives/Review Questions

This section lists general and detailed chapter objectives that can be used as review questions. You can make maximum use of these items by writing answers on a separate sheet of paper. Fill in answers where blanks are provided. To check for accuracy, compare your answers with information given in the chapter or glossary.

Page		Objectives/Questions
(208)	1.	Before 1952, _____ molecules and _____ molecules were suspected of housing the genetic code.
(208–211)	2.	Summarize the research carried out by Miescher, Griffith, Avery, and colleagues, and Hershey and Chase; state the specific advances made by each in the understanding of genetics.
(210)	3.	Viruses called _____ were used in early research efforts to discover the genetic material.
(210–211)	4.	Summarize the specific research that demonstrated that DNA, not protein, governed inheritance.
(212)	5.	DNA is composed of double-ring nucleotides known as _____ and single-ring nucleotides known as _____; the two purines are _____ and _____, while the two pyrimidines are _____ and _____.
(213)	6.	Draw the basic shape of a deoxyribose molecule and show how a phosphate group is joined to it when forming a nucleotide.
(213)	7.	Show how each nucleotide base would be joined to the sugar-phosphate combination drawn in objective 6.
(213)	8.	List the pieces of information about DNA structure that Rosalind Franklin discovered through her x-ray diffraction research.
(213–214)	9.	The two scientists who assembled the clues to DNA structure and produced the first model were _____ and _____.
(213–214)	10.	Explain what is meant by the pairing of nitrogen-containing bases (base pairing), and explain the mechanism that causes bases of one DNA strand to join with bases of the other strand.
(214)	11.	Describe how the Watson-Crick model reflects the constancy in DNA observed from species to species yet allows for variations from species to species.
(214)	12.	Assume that the two parent strands of DNA have been separated and that the base sequence on one parent strand is A-T-T-C-G-C; the base sequence that will complement that parent strand is _____.
(214–215)	13.	Describe how double-stranded DNA replicates from stockpiles of nucleotides.
(214)	14.	Explain what is meant by "each parent strand is conserved in each new DNA molecule."
(214)	15.	DNA replication is specifically referred to as _____ replication.
(214–215)	16.	List four functions that enzymes perform during the process of replication.
(214)	17.	During DNA replication, enzymes called DNA _____ assemble new DNA strands.
(217)	18.	The basic histone-DNA packing unit of the chromosome is the _____.
(216)	19.	List possible reasons for the highly organized packing of nucleoprotein into chromosomes.

Integrating and Applying Key Concepts

Review the stages of mitosis and meiosis, as well as the process of fertilization. Include what you have now learned about DNA replication and the relationship of DNA to a chromosome. As you cover the stages, be sure each cell receives the proper number of DNA threads.

Answers

Interactive Exercises

13-I. DISCOVERY OF DNA FUNCTION (pp. 208–211)

1. a. identified "nuclein" from nuclei of pus cells and fish sperm; discovered DNA; b. Griffith; c. reported that the transforming substance in Griffith's bacteria experiments was probably DNA, the substance of heredity; d. Delbrück, Hershey, and Luria; e. worked with radioactive sulfur (protein) and phosphorus (DNA) labels; T4 bacteriophage and *E. coli* demonstrated that labeled phosphorus was in bacteriophage DNA and contained hereditary instructions for new bacteriophages.

13-II. DNA STRUCTURE (pp. 212–213)

1. A five-carbon sugar called deoxyribose, a phosphate group, and one of the four nitrogen-containing bases; 2. guanine in a DNA-containing nucleotide; 3. cytosine in a DNA-containing nucleotide; 4. adenine in a DNA-containing nucleotide; 5. thymine in a DNA-containing nucleotide; 6. T; 7. T; 8. F, phosphate—sugar—base; 9. T; 10. T; 11. pairing; 12. constant; 13. sequence; 14. different; 15. deoxyribose (B); 16. phosphate group (G); 17. purine (C); 18. pyrimidine (A); 19. purine (E); 20. pyrimidine (D); 21. nucleotide (F); 22. Living organisms have so many diverse body structures and behave in different ways because the many different habitats of Earth have selected those genotypes most able to survive in those habitats. The remaining genotypes have perished. The directions that code for the building of those body structures and that enable the specific successful behaviors reside in DNA or, in a few cases, RNA. All living organisms follow the same rules for base pairing between the two nucleotide strands in DNA; adenine always pairs with thymine in undamaged DNA, and cytosine always pairs with guanine. All living organisms must extract energy from food molecules, and the reactions of glycolysis occur in virtually all of Earth's species. That means that similar enzyme sequences enable similar metabolic pathways to occur. While virtually all living organisms on Earth use the same code and the same enzymes during replication, transcription, and translation, the particular array of proteins being formed differs from individual to individual even of the same species according to the sequences of nitrogenous bases that make up an individual's chromosome(s), and therein lies the key to the enormous diversity of life on Earth: No two individuals have the exact same array of proteins in their phenotypes.

13-III. DNA REPLICATION AND DNA REPAIR (pp. 214–215)

1.
T-	A	T -	A
G-	C	G -	C
A-	T	A -	T
C-	G	C -	G
C-	G	C -	G
C-	G	C -	G
old	new	new	old

2. F, adenine bonds to thymine (during replication) or uracil (during transcription); 3. F, it is a semiconservative process because each "new" DNA molecule contains one "old" strand from the parent cell attached to a strand of "new" complementary nucleotides that were assembled from stockpiles in the cell; 4. T; 5. T; 6. C; 7. E; 8. D; 9. A; 10. F; 11. B.

13-IV. ORGANIZATION OF DNA IN CHROMOSOMES (p. 216)

1. DNA; 2. histones; 3. histone; 4. DNA; 5. nucleosome; 6. genes; 7. metaphase; 8. protein; 9. nucleosome; 10. DNA; 11. histones.

Crossword Puzzle: DNA Structure and Function

Self-Quiz

1. d; 2. d; 3. a; 4. d; 5. b; 6. d; 7. c; 8. d; 9. d; 10. e.

14

FROM DNA TO PROTEINS

Interactive Exercises

Focus on Science: Discovering the Connection Between Genes and Proteins
  (pp. 220–221)
14-I. TRANSCRIPTION AND TRANSLATION: AN OVERVIEW (pp. 218–221)

Selected Italicized Words

Neurospora crassa, sickle-cell anemia

Boldfaced, Page-Referenced Terms

(220) gel electrophoresis _____

_____

(221) base sequence _____

_____

(221) transcription _____

_____

(221) translation _____

_____

(221) ribonucleic acid (RNA) _____

_____

(221) messenger RNA (mRNA) _____

_____

(221) ribosomal RNA (rRNA) _____

_____

(221) transfer RNA (tRNA) _____

_____

Fill-in-the-Blanks

Which base follows the next in a strand of DNA is referred to as the base (1) _____. The region of DNA that calls for the assembly of specific amino acids into a polypeptide chain is a(n) (2) _____. The two steps from genes to proteins are called (3) _____ and (4) _____. In eukaryotes, during (5) _____, single-stranded molecules of RNA are assembled on DNA templates in the nucleus. During (6) _____, the RNA molecules are shipped from the nucleus into the cytoplasm, where they are used as templates for assembling (7) _____ chains. Following translation, one or more chains become (8) _____ into the three-dimensional shape of protein molecules. Proteins have (9) _____ and (10) _____ roles in cells, including control of DNA.

Complete the Table

11. Three types of RNA are transcribed from DNA in the nucleus (from genes that code only for RNA). Complete the following table, which summarizes information about these molecules.

RNA Molecule	Abbreviation	Description/Function
a. Ribosomal RNA		
b. Messenger RNA		
c. Transfer RNA		

14-II. TRANSCRIPTION OF DNA INTO RNA (pp. 222–223)

Selected Italicized Words

poly-A tail, pre-mRNA cap

Boldfaced, Page-Referenced Terms

(222) uracil _____

_____

(222) RNA polymerases _____

_____

(222) promoter _____

_____

(223) introns _____

_____

(223) exons _____

_____

Short Answer

1. List three ways in which a molecule of RNA is structurally different from a molecule of DNA.

_____

_____

2. Cite two similarities in DNA replication and transcription.

_____

_____

3. What are the three key ways in which transcription differs from DNA replication? _____

_____

_____

Sequence

Arrange the steps of transcription in hierarchical order with the earliest step first and the latest step last.

4. ___

5. ___

6. ___

7. ___

8. ___

    A. The RNA strand grows along exposed bases until RNA polymerase meets a DNA base sequence that signals "stop."

    B. RNA polymerase binds with the DNA promoter region to open up a local region of the DNA double helix.

    C. An RNA polymerase enzyme locates the DNA bases of the promoter region of one DNA strand by recognizing DNA-associated proteins near a promoter.

    D. RNA is released from the DNA template as a free, single-stranded transcript.

    E. RNA polymerase moves stepwise along exposed nucleotides of one DNA strand; as it moves, the DNA double helix keeps unwinding.

9. Suppose the sequence below represents the DNA strand that will act as a template for the production of mRNA through the process of transcription. Fill in the blanks below the DNA strand with the sequence of complementary bases that will represent the message carried from DNA to the ribosome in the cytoplasm.

Dp Dp Dp Dp Dp Dp Dp Dp Dp Dp Dp Dp Dp Dp Dp Dp Dp Dp Dp Dp Dp Dp Dp Dp Dp Dp Dp Dp Dp

T A C A A G A T A A C A T T A T T T C C T A C C G T C A T C

Rp Rp Rp Rp Rp Rp Rp Rp Rp Rp Rp Rp Rp Rp Rp Rp Rp Rp Rp Rp Rp Rp Rp Rp Rp Rp Rp Rp Rp Rp

(transcribed single strand of mRNA)

Label-Match

Newly transcribed mRNA contains more genetic information than is necessary to code for a chain of amino acids. Before the mRNA leaves the nucleus for its ribosome destination, an editing process occurs as certain portions of nonessential information are snipped out. Identify each indicated part of the illustration below; use abbreviations for the nucleic acids. Complete the exercise by matching and entering the letter of the description in the parentheses following each label.

10. _____ (　)   A. The actual coding portions of mRNA

11. _____ (　)   B. Noncoding portions of the newly transcribed mRNA

12. _____ (　)   C. Presence of cap and tail, introns snipped out, and exons spliced together

13. _____ (　)   D. Acquiring of a poly-A tail by the maturing mRNA transcript

14. _____ (　)   E. The region of the DNA template strand to be copied

15. _____

     F. Reception of a nucleotide cap by the 5′ end of mRNA (the first synthesized)

_____ _____ (　)

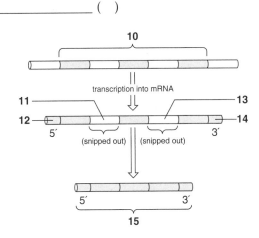

14-III. FROM MRNA TO PROTEINS (pp. 224–225)
         STAGES OF TRANSLATION (pp. 226–227)

Selected Italicized Words

base triplets, molecular "hook," "wobble effect," initiation, elongation, termination

Boldfaced, Page-Referenced Terms

(224) codons _____

_____

(224) genetic code _____

_____

(225) anticodon _____

_____

(225) ribosomes _____

_____

(227) polysomes _____

_____

Matching

1. __ codon
2. __ three at a time
3. __ sixty-one
4. __ the genetic code
5. __ release factors
6. __ ribosome
7. __ anticodon
8. __ the "stop" codons

A. Composed of two subunits, the small subunit with P and A amino acid binding sites as well as a binding site for mRNA
B. Reading frame of the nucleotide bases in mRNA
C. Detach protein and mRNA from the ribosome
D. UAA, UAG, UGA
E. A sequence of three nucleotide bases that can pair with a specific mRNA codon
F. Name for each base triplet in mRNA
G. The number of codons that actually specify amino acids
H. Term for how the nucleotide sequences of DNA and then mRNA correspond to the amino acid sequence of a polypeptide chain

9. Complete the following table, which distinguishes the stages of translation.

Translation Stage	Description
a.	Special initiator tRNA loads onto small ribosomal subunit and recognizes AUG; small subunit binds with mRNA, and large ribosomal subunit joins small one.
b.	Amino acids are strung together in sequence dictated by mRNA codons as the mRNA strand passes through the two ribosomal subunits; two tRNAs interact at P and A sites.
c.	mRNA "stop" codon signals the end of the polypeptide chain; release factors detach the ribosome and polypeptide chain from the mRNA.

10. Given the following DNA sequence, deduce the composition of the mRNA transcript:

TAC   AAG   ATA   ACA   TTA   TTT   CCT   ACC   GTC   ATC

— — —   — — —   — — —   — — —   — — —   — — —   — — —   — — —   — — —   — — —

(mRNA transcript)

11. Deduce the composition of the tRNA anticodons that would pair with the above specific mRNA codons as these tRNAs deliver the amino acids (identified below) to the P and A binding sites of the small ribosomal subunit.

— — —   — — —   — — —   — — —   — — —   — — —   — — —   — — —   — — —   — — —

(tRNA anticodons)

12. From the mRNA transcript in exercise 10, use Figure 14.6 of the text to deduce the composition of the amino acids of the polypeptide sequence.

— — —   — — —   — — —   — — —   — — —   — — —   — — —   — — —   — — —   — — —

(amino acids)

Label-Match

A summary of the flow of genetic information in protein synthesis is useful as an overview. Identify the indicated parts of the accompanying illustration by filling in the blanks with the names of the appropriate structures or functions. Choose from the following: DNA, mRNA, tRNA, polypeptide, rRNA subunits, intron, exon, mature mRNA transcript, new mRNA transcript, anticodon, amino acids, ribosome-mRNA complex. Complete the exercise by matching and entering the letter of the description in the parentheses following each label.

13. _____ ( )

14. _____

_____

_____ ( )

15. _____ ( )

16. _____ ( )

17. _____

_____

_____ ( )

18. _____ ( )

19. _____

_____ ( )

20. _____ ( )

21. _____ ( )

22. _____

_____ ( )

23. _____ ( )

24. _____ -

_____

_____ ( )

25. _____ ( )

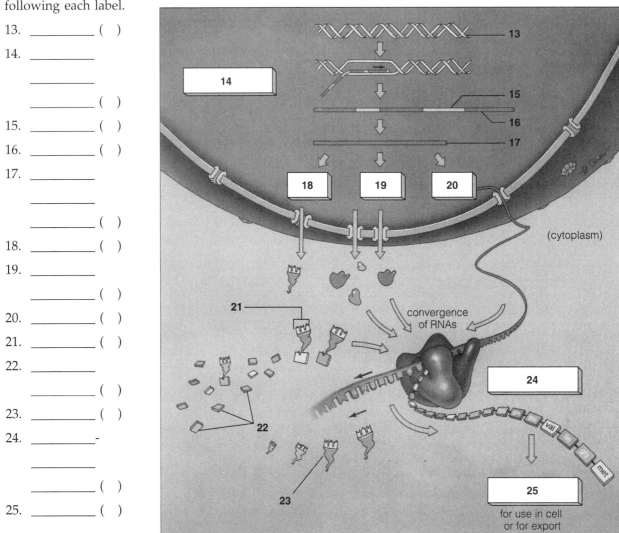

A. Coding portion of mRNA that will translate into a protein
B. Carries the genetic code from DNA in the nucleus to the cytoplasm
C. Transports amino acids to the ribosome and mRNA
D. The building blocks of polypeptides
E. Noncoding portions of newly transcribed mRNA
F. tRNA after delivering its amino acid to the ribosome-mRNA complex
G. Join when translation is initiated
H. Holds the genetic code for protein production
I. Place where translation occurs
J. Includes introns and exons
K. A sequence of three bases that can pair with a specific mRNA codon
L. Snipping out of introns, only exons remaining
M. May serve as a functional protein (enzyme) or a structural protein

Fill-in-the-Blanks

The order of (26) _____ _____ in a protein is specified by a sequence of nucleotide bases. The genetic code is read in units of (27) _____ nucleotides; each unit of three codes for (28) _____ amino acid(s). In the table that showed which triplet specified a particular amino acid, the triplet code was contained in (29) _____ molecules. Each of these triplets is referred to as a(n) (30) _____.

(31) _____ alone carries the instructions for assembling a particular sequence of amino acids from the DNA to the ribosomes in the cytoplasm, where (32) _____ of the polypeptide occurs. (33) _____ RNA acts as a shuttle molecule as each type brings its particular (34) _____ _____ to the ribosome where it is to be incorporated into the growing (35) _____. A(n) (36) _____ is a triplet on mRNA that forms hydrogen bonds with a(n) (37) _____, which is a triplet on tRNA.

14-IV. MUTATION AND PROTEIN SYNTHESIS (pp. 228–229)

Selected Italicized Words

base-pair substitutions, frameshift mutations, jumping genes, mutation rate, lethal gene, beneficial gene

Boldfaced, Page-Referenced Terms

(228) gene mutations _____

_____

(228) mutagens _____

_____

1. Cite several examples of mutagens.

_____

Fill-in-the-Blanks

In addition to changes in chromosomes (crossing over, recombination, deletion, addition, translocation, and inversion), changes can also occur in the structure of DNA; these modifications are referred to as gene mutations. Complete the following exercise on types of spontaneous gene mutations.

Viruses, ultraviolet radiation, and certain chemicals are examples of environmental agents called

(2) _____ that may enter cells and damage strands of DNA. If A becomes paired with C instead of T during DNA replication, this spontaneous mutation is a base-pair (3) _____. Sickle-cell anemia is a genetic disease whose cause has been traced to a single DNA base pair; the result is that one (4) _____ _____ is substituted for another in the beta chain of (5) _____.

A(n) (6) _____ mutation is defined as the insertion or deletion of one to several DNA base pairs; this puts the nucleotide sequence out of phase, and abnormal proteins are produced. Some DNA regions "jump" to new DNA locations and often inactivate the genes in their new environment; such (7) _____ elements may give rise to observable changes in the phenotype of an organism.

(8) _____ is the source of the unity of life; (9) _____ _____ are the original source of life's diversity; they are heritable, small-scale alterations in the (10) _____ sequence of DNA.

Self-Quiz

___ 1. Transcription _____.
   a. occurs on the surface of the ribosome
   b. is the final process in the assembly of a protein
   c. occurs during the synthesis of any type of RNA by use of a DNA template
   d. is catalyzed by DNA polymerase

___ 2. _____ carry(ies) amino acids to ribosomes, where amino acids are linked into the primary structure of a polypeptide.
   a. mRNA
   b. tRNA
   c. Introns
   d. rRNA

___ 3. Transfer RNA differs from other types of RNA because it _____.
   a. transfers genetic instructions from cell nucleus to cytoplasm
   b. specifies the amino acid sequence of a particular protein
   c. carries an amino acid at one end
   d. contains codons

___ 4. _____ dominates the process of transcription.
   a. RNA polymerase
   b. DNA polymerase
   c. Phenylketonuria
   d. Transfer RNA

___ 5. _____ are found in RNA but not in DNA.
   a. Deoxyribose and thymine
   b. Deoxyribose and uracil
   c. Uracil and ribose
   d. Thymine and ribose

___ 6. Each "word" in the mRNA language consists of _____ letters.
   a. three
   b. four
   c. five
   d. more than five

___ 7. If each nucleotide coded for only one amino acid, how many different types of amino acids could be selected?
   a. four
   b. sixteen
   c. twenty
   d. sixty-four

___ 8. The genetic code is composed of _____ codons.
   a. three
   b. twenty
   c. sixteen
   d. sixty-four

___ 9. The cause of sickle-cell anemia has been traced to _____.
   a. a mosquito-transmitted virus
   b. two DNA mutations that result in two incorrect amino acids in a hemoglobin chain
   c. three DNA mutations that result in three incorrect amino acids in a hemoglobin chain
   d. one DNA mutation that results in one incorrect amino acid in a hemoglobin chain

___ 10. An example of a mutagen is _____.
   a. a virus
   b. ultraviolet radiation
   c. certain chemicals
   d. all of the above

For the next five questions, choose from these possibilities:
   a. Beadle and Tatum
   b. Garrod
   c. Khorana, Nirenberg, and others
   d. McClintock
   e. Pauling and Itano

___ 11. In the early 1900s,_____ studied illnesses caused by inheriting one or more defective units of inheritance; concluded that these units function by synthesizing specific enzymes.

___ 12. In the 1940s, _____ discovered transposable elements can move genes from one chromosome to another, or to a different place on the chromosome; finally was awarded the Nobel Prize some fifty years after this news was published.

___ 13. In the late 1940s, _____ developed gel electrophoresis as a way of separating similar proteins from a mixture.

___ 14. In the 1930s, _____ developed the "one gene—one enzyme" hypothesis by studying defective enzymes in the bread mold, *Neurospora crassa*.

___ 15. In the 1950s and 1960s, _____ deduced that the mRNA code language formed base triplet "words" and that each "word" put into position on a ribosome selected a particular amino acid type from an amino acid pool.

Chapter Objectives/Review Questions

This section lists general and detailed chapter objectives that can be used as review questions. You can make maximum use of these items by writing answers on a separate sheet of paper. Fill in answers where blanks are provided. To check for accuracy, compare your answers with information given in the chapter or glossary.

Page	Objectives/Questions
(220–221)	1. Cite an example of a change in one DNA base pair that has profound effects on the human phenotype.
(221)	2. _____ RNA combines with certain proteins to form the ribosome; _____ RNA carries genetic information for protein construction from the nucleus to the cytoplasm; _____ RNA picks up specific amino acids and moves them to the area of mRNA and the ribosome.
(222)	3. Describe the process of transcription and indicate three ways in which it differs from replication.
(222)	4. Transcription starts at a _____, a specific sequence of bases on one of the two DNA strands that signals the start of a gene.
(222)	5. The first end of the mRNA to be synthesized is the _____ end; at the opposite end, the most mature transcripts acquire a _____ tail.
(222)	6. Actual coding portions of a newly transcribed mRNA are called _____; _____ are the noncoding portions.
(222–223)	7. State how RNA differs from DNA in structure and function, and indicate what features RNA has in common with DNA.
(222–224)	8. Explain the nature of the genetic code and describe the kinds of nucleotide sequences that code for particular amino acids.
((223–224)	9. What RNA code would be formed from the following DNA code? TAC-CTC-GTT-CCC-GAA
(223–224)	10. State the relationship between the DNA genetic code and the order of amino acids in a protein chain.
(223–224)	11. Determine which mRNA codons would be formed from the following DNA code. TAC-CTC-GTT-CCC-GAA
(224)	12. Each base triplet in mRNA is called a _____.
224)	13. Scrutinize Figure 14.6 in the text and decide whether the genetic code in this instance applies to DNA, mRNA, or tRNA.
(224)	14. Explain how the DNA message TAC-CTC-GTT-CCC-GAA would be used to code for a segment of protein, and state what its amino acid sequence would be.
(228)	15. List some of the environmental agents, or _____, that can cause mutations.
(228)	16. Briefly describe the spontaneous DNA mutations known as base-pair substitution, frameshift mutation, and transposable element.
(231)	17. Using a diagram, summarize the steps involved in the transformation of genetic messages into proteins (see Fig. 14.14 in the text).

Integrating and Applying Key Concepts

Genes code for specific polypeptide sequences. Not every substance in living cells is a polypeptide. Explain how genes might be involved in the production of a storage starch (such as glycogen) that is constructed from simple sugars.

Answers

Interactive Exercises

14-I. TRANSCRIPTION AND TRANSLATION: AN OVERVIEW (pp. 218–221)

1. sequence; 2. gene; 3. transcription (translation); 4. translation (transcription); 5. transcription; 6. translation; 7. protein; 8. folded; 9. structural (functional); 10. functional (structural); 11. a. rRNA; RNA molecule that associates with certain proteins to form the ribosome, the "workbench" on which polypeptide chains are assembled; b. mRNA; RNA molecule that moves to the cytoplasm, complexes with the ribosome where translation will result in polypeptide chains; c. tRNA; RNA molecule that moves into the cytoplasm, picks up a specific amino acid, and moves it to the ribosome where tRNA pairs with a specific mRNA code word for that amino acid.

14-II. TRANSCRIPTION OF DNA INTO RNA (pp. 222–223)

1. RNA molecules are single-stranded, while DNA has two strands; uracil substitutes in RNA molecules for thymine in DNA molecules; ribose sugar is found in RNA, while DNA has deoxyribose sugar.
2. Both DNA replication and transcription follow base-pairing rules; nucleotides are added to a growing RNA strand one at a time as in DNA replication.
3. Only one region of a DNA strand serves as a template for transcription; transcription requires different enzymes (three types of RNA polymerase); the results of transcription are single-stranded RNA molecules, but replication results in DNA, a double-stranded molecule.
4. C; 5. B; 6. E; 7. A; 8. D; 9. A-U-G-U-U-C-U-A-U-U-G-U-A-A-U-A-A-A-G-G-A-U-G-G-C-A-G-U-A-G; 10. DNA (E); 11. intron (B); 12. cap (F); 13. exons (A); 14. tail (D); 15. mature mRNA transcript (C).

14-III. FROM mRNA TO PROTEINS (pp. 224–225) STAGES OF TRANSLATION (pp. 226–227)

1. F; 2. B; 3. G; 4. H; 5. C; 6. A; 7. E; 8. D; 9. a. Initiation; b. Chain elongation; c. Chain termination; 10. mRNA transcript: AUG UUC UAU UGU AAU AAA GGA UGG CAG UAG; 11. tRNA anticodons: UAC AAG AUA ACA UUA UUU CCU ACC GUC AUC; 12. amino acids: (start) met phe tyr cys asn lys gly try gln (stop); 13. DNA (H); 14. new mRNA transcript (J); 15. intron (E); 16. exon (A); 17. mature mRNA transcript (L); 18. tRNA (C); 19. rRNA subunits (G); 20. mRNA (B); 21. anticodon (K); 22. amino acids (D); 23. tRNA (F); 24. ribosome-mRNA complex (I); 25. polypeptide (M); 26. amino acids; 27. three; 28. one; 29. mRNA; 30. codon; 31. mRNA; 32. assembly (synthesis); 33. Transfer; 34. amino acid; 35. protein (polypeptide); 36. codon; 37. anticodon.

14-IV. MUTATION AND PROTEIN SYNTHESIS (pp. 228–229)

1. Mutagens are environmental agents that attack a DNA molecule and modify its structure. Viruses, ultraviolet radiation, and certain chemicals are examples; 2. mutagens; 3. substitution; 4. amino acid; 5. hemoglobin; 6. frameshift; 7. transposable; 8. DNA; 9. gene mutations; 10. nucleotide (base).

Self-Quiz

1. c; 2. b; 3. c; 4. a; 5. c; 6. a; 7. a; 8. d; 9. d; 10. d; 11. b; 12. d; 13. e; 14. a; 15. c.

15

CONTROL OF GENE EXPRESSION

Interactive Exercises

15-I. THE NATURE OF GENE CONTROL (pp. 234–236)
EXAMPLES OF GENE CONTROL IN PROKARYOTIC CELLS (pp. 236–237)

Selected Italicized Words

benign, malignant, myc gene, Burkitt's lymphoma

Boldfaced, Page-Referenced Terms

(234) tumor _____

_____

(234) metastasis _____

_____

(235) cancer _____

_____

(236) regulatory proteins _____

_____

(236) negative control systems _____

_____

(236) positive control systems _____

_____

Complete the Table

1. All the somatic cells in an organism possess the same genes, and every cell utilizes most of the same genes; yet specialized cells must activate only certain genes. Some agents of gene control have been discovered. Transcriptional controls are the most common. Complete the following table to summarize the agents of gene control.

Agents of Gene Control	Method of Gene Control
a. Repressor protein	
b.	Encourages binding of RNA polymerases to DNA; this is positive control of transcription
c.	Major agents of vertebrate gene control; signaling molecules that move through the bloodstream to affect gene expression in target cells
d. Promoter	
e.	Short DNA base sequences between promoter and the start of a gene; a binding site for control agents

2. When do gene controls come into play? _____

_____

Boldfaced, Page-Referenced Terms

(236) cooperative regulation _____

_____

(236) promoter _____

_____

(236) operator _____

_____

(236) repressor protein _____

_____

(236) operon _____

_____

(237) activator protein _____

_____

Fill-in-the-Blanks

A promoter and operator provide (3) _____ _____. The (4) _____ gene codes for the forma-

tion of mRNA, which assembles a repressor protein. The affinity of the (5) _____ for RNA polymerase

dictates the rate at which a particular operon will be transcribed. A repressor protein allows (6) _____

_____ over the lactose operon. The repressor binds with the operator and overlaps the promoter when

lactose concentrations are (7) _____. This blocks (8) _____ _____ from the genes that will

process lactose. This (9) (choose one) ❐ blocks ❐ promotes production of lactose-processing enzymes.

When lactose is present, lactose molecules bind with the (10) _____ _____. Thus, the repressor

cannot bind to the (11) _____, and RNA polymerase has access to the lactose-processing genes. This

gene control works well because lactose-degrading enzymes are not produced unless they are

(12) _____.

Sequence

Arrange the following events in the positive control of the nitrogen-related operon in the correct order of
occurrence if *Escherichia coli* is first in a low-nitrogen environment and then in a nitrogen-rich environment.

13. ___
14. ___
15. ___
16. ___
17. ___
18. ___
19. ___
20. ___

A. RNA polymerase–to–promoter binding is enhanced.
B. A cascade of molecule-activating reactions brings a large number of activator molecules into service.
C. Enzymes remove phosphates from activator proteins to reverse the earlier response to low nitrogen.
D. Nitrogen again becomes plentiful in the environment of *Escherichia coli*.
E. Enzymes attach phosphates to activator molecules to prepare them for binding with promoter.
F. Activator molecules bind with promoter.
G. Chances for *Escherichia coli* to assimilate any available nitrogen increase.
H. Synthesis of nitrogen-obtaining enzymes increases.

Label-Match

Escherichia coli, a bacterial cell living in mammalian digestive tracts, is able to exert a negative type of gene control over lactose metabolism. Use the numbered blanks to identify each part of the illustration below. Use abbreviations for nucleic acids. Choose from the following: lactose, regulator gene, repressor-operator complex, promoter, mRNA transcript, enzyme genes, lactose operon, lactose enzymes, repressor-lactose complex, repressor protein, RNA polymerase, and operator. Complete the exercise by matching and entering the letter of the proper function description in the parentheses following each label.

21. _____ _____ ( )

22. _____ _____ ( )

23. _____ _____ ( )

24. _____ ( )

25. _____ ( )

26. _____ _____ ( )

27. _____-_____ _____ ( )

28. _____ _____ ( )

29. _____-_____ _____ ( )

30. _____ ( )

31. _____ _____ ( )

32. _____ _____ ( )

A. Includes promoter, operator, and the lactose-metabolizing enzymes

B. Short DNA base sequence between promoter and the beginning of a gene

C. The nutrient molecule that regulates the lactose operon

D. Major enzyme that catalyzes transcription

E. Capable of preventing RNA polymerases from binding with DNA

F. Prevents repressor from binding to the operator

G. Genes that code for lactose-metabolizing enzymes

H. Catalyze the digestion of lactose

I. Binds to operator and overlaps promoter; this prevents RNA polymerase from binding to DNA and initiating transcription

J. Specific base sequence that signals the beginning of a gene

K. Gene that contains coding for production of repressor protein

L. Carries genetic instructions to ribosomes for production of lactose enzymes

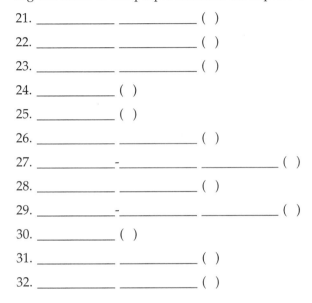

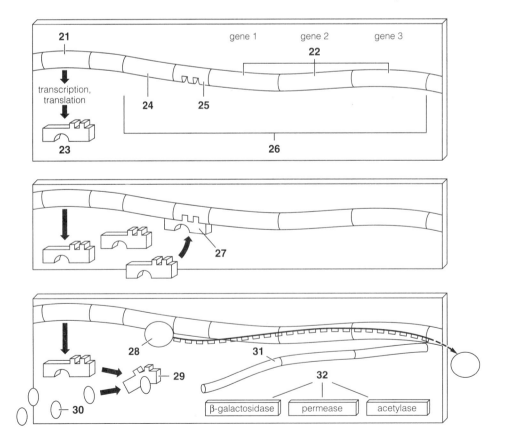

15-II. GENE CONTROL IN EUKARYOTIC CELLS (pp. 238–239)
EVIDENCE OF GENE CONTROL (pp. 240–241)

Selected Emphasized Words

gene amplification, DNA rearrangements, chemical modification, alternative splicing, "masked messengers," "lamp-brush" chromosomes, "mosaic" female, "calico" cat, "spotting gene," anhidrotic ectodermal dysplasia

Boldfaced, Page-Referenced Terms

(238) cell differentiation _____

_____

(241) Barr body _____

_____

(241) Lyonization _____

_____

Short Answer

1. Although a complex organism such as a human being arises from a single cell, the zygote, differentiation occurs in development. Define *differentiation* and relate it to a definition of *selective gene expression*.

_____

_____

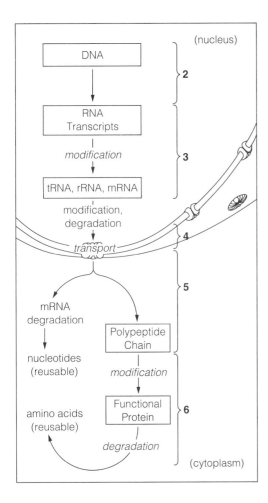

(nucleus)

DNA

RNA Transcripts

modification

tRNA, rRNA, mRNA

modification, degradation

transport

mRNA degradation

nucleotides (reusable)

amino acids (reusable)

Polypeptide Chain

modification

Functional Protein

degradation

(cytoplasm)

Label-Match

Identify each numbered part of the illustration below showing eukaryotic gene control. Choose from translational controls, transport controls, transcriptional controls, transcript processing controls, and post-translational controls. Complete the exercise by correctly matching and entering the letter of the corresponding gene control description in the parentheses following each label.

2. _____ _____ ( )

3. _____ _____ _____ ( )

4. _____ _____ ( )

5. _____ _____ ( )

6. _____-_____ _____ ( )

A. Govern the rates at which mRNA transcripts that reach the cytoplasm will be translated into polypeptide chains at ribosomes

B. Govern modification of the initial mRNA transcripts in the nucleus

C. Govern how the polypeptide chains become modified into functional proteins

D. Dictate which mature mRNA transcripts will be shipped out of the nucleus and into the cytoplasm for translation

E. Influence when and to what degree a particular gene will be transcribed (if at all)

Fill-in-the-Blanks

All the cells in a body contain copies of the same (7) _____. These genetically identical cells become structurally and functionally distinct from one another through a process called (8) _____ _____, which arises through (9) _____ gene expression in different cells. Cells depend on (10) _____, which govern transcription, translation, and enzyme activity. Controls that operate during transcription and transcript processing utilize (11) _____ proteins, especially (12) _____ that are turned on and off by the attachment and detachment of cAMP.

   (13) _____ is also controlled by the way eukaryotic DNA is packed with proteins in chromosomes. A(n) (14) _____ body is a condensed X chromosome; the process by which an XX randomly inactivates an X chromosome is called (15) _____. X chromosome inactivation produces adult human females who are (16) _____ for X-linked traits. This effect is shown in human females affected by (17) _____ _____ _____; the disorder provides evidence that the process of (18) _____ gene expression causes cells to differentiate.

True-False

If false, explain why.

_____ 19. "Lampbrush" chromosomes of meiotic prophase I seen in electron micrographs of amphibian eggs provide evidence of translational control.

_____ 20. When the same primary mRNA transcript is spliced in alternative ways, the result may be different mRNAs, each coding for a slightly different protein.

15-III. EXAMPLES OF SIGNALING MECHANISMS (pp. 242–243)
GENES IMPLICATED IN CANCER (p. 244)

Selected Italicized Words

"polytene chromosome"

Boldfaced, Page-Referenced Terms

(242) hormones _____

_____

(242) enhancers _____

_____

(243) phytochrome _____

_____

(244) oncogene _____

_____

(244) proto-oncogene _____

_____

(244) carcinogens _____

_____

True-False

If the statement is true, write a T in the blank. If false, make it true by changing the underlined word.

_____ 1. <u>Polytene</u> chromosomes of many insect larvae are unusual in that their DNA has been repeatedly replicated; thus, these chromosomes have multiple copies of the same genes. When these genes are undergoing transcription, they "puff."

_____ 2. When cells become cancerous, cell populations <u>decrease</u> to very <u>low</u> densities and <u>stop</u> dividing.

_____ 3. <u>All</u> abnormal growths and massings of new tissue in any region of the body are called tumors.

_____ 4. Malignant tumors have cells that <u>migrate</u> and <u>divide</u> in other organs.

_____ 5. Oncogenes are genes that <u>combat</u> cancerous transformations.

_____ 6. Proto-oncogenes <u>rarely</u> trigger cancer.

_____ 7. The normal expression of proto-oncogenes is vital, even though their <u>normal</u> expression may be lethal.

Self-Quiz

___1. _____ refers to the processes by which cells with identical genotypes become structurally and functionally distinct from one another according to the genetically controlled developmental program of the species.
   a. Metamorphosis
   b. Metastasis
   c. Cleavage
   d. Differentiation

___2. _____ binds to operator whenever lactose concentrations are low.
   a. Operon
   b. Repressor
   c. Promoter
   d. Operator

___3. Any gene or group of genes together with its promoter and operator sequence is a(n) _____.
   a. repressor
   b. operator
   c. promoter
   d. operon

___4. The operon model explains the regulation of _____ in prokaryotes.
   a. replication
   b. transcription
   c. induction
   d. Lyonization

___5. In multicelled eukaryotes, cell differentiation occurs as a result of _____.
   a. growth
   b. selective gene expression
   c. repressor molecules
   d. the death of certain cells

___6. One type of gene control discovered in female mammals is _____.
   a. a conflict in maternal and paternal alleles
   b. slow embryo development
   c. X chromosome inactivation
   d. operon

___7. Due to X inactivation of either the paternal or maternal X chromosome, human females with anhidrotic ectodermal dysplasia _____.
   a. completely lack sweat glands
   b. develop benign growths
   c. have mosaic patches of skin that lack sweat glands
   d. develop malignant growths

___8. Which of the following characteristics seems to be most uniquely correlated with metastasis?
   a. loss of nuclear-cytoplasmic controls governing cell growth and division
   b. changes in recognition proteins on membrane surfaces
   c. "puffing" in the chromosomes
   d. the massive production of benign tumors

___9. Genes with the potential to induce cancerous formations are known as _____.
   a. proto-oncogenes
   b. oncogenes
   c. carcinogens
   d. malignant genes

___10. _____ controls govern the rates at which mRNA transcripts that reach the cytoplasm will be translated into polypeptide chains at the ribosomes.
   a. Transport
   b. Transcript processing
   c. Translational
   d. Transcriptional

Chapter Objectives/Review Questions

This section lists general and detailed chapter objectives that can be used as review questions. You can make maximum use of these items by writing answers on a separate sheet of paper. Fill in answers where blanks are provided. To check for accuracy, compare your answers with information given in the chapter or glossary.

Page		Objectives/Questions
(234)	1.	Define *tumor* and distinguish between benign and malignant tumors.
(234)	2.	_____ is a process in which a cancer cell leaves its proper place and invades other tissues to form new growths.
(236–237)	3.	The negative control of _____ protein prevents the enzymes of transcription from binding to DNA; the positive control of _____ protein enhances the binding of RNA polymerases to DNA.
(236)	4.	A _____ is a specific base sequence that signals the beginning of a gene in a DNA strand.
(236)	5.	Some control agents bind to _____, which are short base sequences between a promoter and the beginning of a gene.
(236–238; 242)	6.	Gene expression is controlled through regulatory _____, hormones, _____ that speed up reaction rates, and _____ proteins that bind to enhancer sites on DNA.
(236)	7.	A gene control arrangement in which the same promoter-operator sequence services more than one gene is called an _____.
(236–237)	8.	Describe the sequence of events that occurs on the chromosome of *E. coli* after you drink a glass of milk.
(236–237)	9.	The cells of *E. coli* manage to produce enzymes to degrade lactose when those molecules are _____ and to stop production of lactose-degrading enzymes when lactose is _____.
(238)	10.	Define *selective gene expression* and explain how this concept relates to cell differentiation in multicelled eukaryotes.
(238)	11.	Cell _____ occurs in multicelled eukaryotes as a result of _____ gene expression.
(238–239)	12.	Be able to list and define the levels of gene control in eukaryotes.
(240)	13.	The "lampbrush" chromosomes seen in amphibian eggs and larvae represent a change in chromosome structure that has been correlated directly with _____ of the genes necessary for growth.
(240–241)	14.	Explain how X chromosome inactivation provides evidence for selective gene expression; use the example of anhidrotic ectodermal dysplasia.
(241)	15.	The condensed X chromosome seen on the edge of the nuclei of female mammals is known as the _____ body.
(241)	16.	Describe the tissue effect known as Lyonization.
(242)	17.	Explain how hormones act as a major agent of gene control.
(242)	18.	Describe the gene control mechanism afforded many insect larvae by polytene chromosomes.
(244)	19.	Describe the relationship of proto-oncogenes, environmental irritants, and oncogenes.
(244)	20.	Cigarette smoke, x-rays, gamma rays, and ultraviolet radiation are examples of _____.

Integrating and Applying Key Concepts

Suppose you have been restricting yourself to a completely vegetarian diet for the past six months. Quite unexpectedly, you find yourself in a social situation that requires you to eat a half-pound sirloin steak. Would you expect to digest the steak as easily as you digest soybean burgers? Explain your yes or no answer in terms of transcriptional controls or feedback inhibition.

Answers

15-I. THE NATURE OF GENE CONTROL
(pp. 234–236)
EXAMPLES OF GENE CONTROL IN PROKARYOTIC CELLS (pp. 236–237)

1. a. Prevents transcription enzymes (RNA polymerases) from binding to DNA; this is negative transcription control; b. Activator protein; c. Hormones; d. Specific base sequences on DNA that serve as binding sites for control agents; before RNA assembly can occur on DNA, the enzymes must bind with the promoter site; e. Operators; 2. In any organism, gene controls operate in response to chemical changes within the cell or its surroundings; 3. transcription controls; 4. regulator; 5. promoter; 6. negative control; 7. low; 8. RNA polymerase (mRNA transcription); 9. blocks; 10. repressor protein; 11. operator; 12. needed (required); 13. B; 14. E; 15. F; 16. A; 17. H; 18. G; 19. D; 20. C; 21. regulator gene (K); 22. enzyme genes (G); 23. repressor protein (E); 24. promoter (J); 25. operator (B); 26. lactose operon (A); 27. repressor-operator complex (I); 28. RNA polymerase (D); 29. repressor-lactose complex (F); 30. lactose (C); 31. mRNA transcript (L); 32. lactose enzymes (H).

15-II. GENE CONTROL IN EUKARYOTIC CELLS
(pp. 238–239)
EVIDENCE OF GENE CONTROL (pp. 240–241)

1. All cells in the body descend from the same zygote; as cells divide to form the body, they become specialized in composition, structure, and function—they differentiate through selective gene expression; 2. transcriptional control (E); 3. transcript processing control (B); 4. transport controls (D); 5. translational controls (A); 6. post-translational controls (C); 7. DNA (genes); 8. cell differentiation; 9. selective; 10. controls; 11. regulatory; 12. activators; 13. Transcription; 14. Barr; 15. Lyonization; 16. mosaic; 17. anhidrotic ectodermal dysplasia; 18. selective; 19. F, transcriptional; 20. T.

15-III. EXAMPLES OF SIGNALING MECHANISMS
(pp. 242–243)
GENES IMPLICATED IN CANCER (p. 244)

1. T; 2. increase, high, start; 3. T; 4. T; 5. F, bring about; 6. T; 7. F, abnormal.

Self Quiz

1. d; 2. b; 3. d; 4. b; 5. b; 6. c; 7. c; 8. a; 9. b; 10. c.

16

RECOMBINANT DNA AND GENETIC ENGINEERING

Interactive Exercises

16-I. RECOMBINATION IN NATURE—AND IN THE LABORATORY (pp. 246–248)

Selected Italicized Words

severe combined immune deficiency (SCID)

Boldfaced, Page-Referenced Terms

(248) recombinant DNA technology _____

_____

(248) genetic engineering _____

_____

(248) plasmids _____

_____

1. Describe and distinguish between the bacterial chromosome and plasmids present in a bacterial cell.

_____

_____

True-False

If the statement is true, write T in the blank. If the statement is false, make it correct by changing the underlined word(s) and writing the correct word(s) in the answer blank.

_____ 2. Plasmids are <u>organelles</u> on <u>the</u> <u>surfaces</u> <u>of</u> <u>which</u> <u>amino</u> <u>acids</u> <u>are</u> <u>assembled</u> <u>into</u> <u>polypep-tides</u>.

_____ 3. <u>Gene</u> <u>transfer</u> and <u>recombination</u> are common in nature.

Fill-in-the-Blanks

Genetic experiments have been occurring in nature for billions of years as a result of gene (4) _____, crossing over and recombination, and other events. Humans now are causing genetic change by using (5) _____ _____ technology in which researchers cut and splice together gene regions from different (6) _____, then greatly (7) _____ the number of copies of the genes that interest them. The genes, and in some cases their (8) _____ products, are produced in quantities that are large enough for (9) _____ and practical applications. (10) _____ _____ involves isolating, modifying, and inserting particular genes back into the same organism or into a different one.

Many bacteria can transfer (11) _____ genes to a bacterial neighbor that may integrate them into its (the recipient's) chromosome, forming a recombinant DNA molecule naturally. In nature, (12) _____ as well as bacteria dabble in gene transfers and recombination of genes, and so do eukaryotic organisms.

16-II. PRODUCING RESTRICTION FRAGMENTS (p. 249)
WORKING WITH DNA FRAGMENTS (pp. 250–251)
RFLP ANALYSIS (p. 252)

Selected Italicized Words

"recombinant plasmids," cloned DNA

Boldfaced, Page-Referenced Terms

(249) DNA ligase _____

_____

(249) DNA library _____

_____

(250) polymerase chain reaction (PCR) _____

_____

(250) DNA polymerase _____

_____

(252) restriction fragment length polymorphisms (RFLPs) _____

_____

(252) DNA fingerprint _____

_____

Complete the Table

1. Complete the table below, which summarizes some of the basic tools and procedures used in recombinant DNA technology.

Tool/Procedure	Definition and Role in Recombinant DNA Technology
a. Restriction enzymes	
b. DNA ligase	
c. DNA library	
d. Cloned DNA	
e. PCR	
f. DNA sequencing	
g. RFLPs	
h. DNA fingerprint	

Matching

Match the steps in the formation of a DNA library with the parts of the illustration below.

2. ___
3. ___
4. ___
5. ___
6. ___
7. ___

A. Joining of chromosomal and plasmid DNA using DNA ligase
B. Restriction enzyme cuts chromosomal DNA at specific recognition sites
C. Cut plasmid DNA
D. Recombinant plasmids containing cloned library
E. Fragments of chromosomal DNA
F. Same restriction enzyme is used to cut plasmids

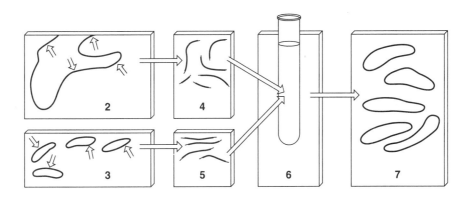

True-False

A genetic engineer used restriction enzymes to prepare fragments of DNA from two different species that were then mixed. Four of these fragments are illustrated below. Fragments (a) and (c) are from one species, (b) and (d) from the other species. Answer exercises 8–12.

_____ TACA	_____ TTCA	_____ CGTA	_____ ATGT
a.	b.	c.	d.

___ 8. Some of the fragments represent sticky ends.

___ 9. The same restriction enzyme was used to cut fragments (b), (c), and (d).

___ 10. Different restriction enzymes were used to cut fragments (a) and (d).

___ 11. Fragment (a) will base-pair with fragment (d) but not with fragment (c).

___ 12. The same restriction enzyme was used to cut the different locations in the DNA of the two species shown.

Matching

Match the most appropriate letter with each numbered partner.

13. ___ polymerase chain reaction (PCR)

14. ___ DNA ligase

15. ___ DNA library

16. ___ cloned DNA

17. ___ DNA fingerprint

18. ___ recombinant DNA technology

19. ___ plasmids

20. ___ restriction enzymes

21. ___ genome

A. All the DNA in a haploid set of chromosomes
B. Process by which a gene is split into two strands and then copied over and over by enzymes; most common type of gene amplification
C. Caused by a unique array of RFLPs inherited from each parent
D. Connects DNA fragments
E. Small circular DNA molecules that carry only a few genes
F. Cuts DNA molecules
G. A collection of DNA fragments produced by restriction enzymes and incorporated into plasmids
H. Multiple, identical copies of DNA fragments from an original chromosome
I. Method of genetic engineering

Fill-in-the-Blanks

Each person has a genetic (22) _____, a unique pattern of RFLPs that can be used to map the human genome, apprehend criminals, and resolve cases of disputed paternity and maternity. Bacterial host cells lack the proper (23) _____ _____ to translate cloned genes unless the (24) _____ have been cut out. Identification of the order and identity of nucleotides in DNA is called (25) _____.

16-III. MODIFIED HOST CELLS (pp. 252–253)

Boldfaced, Page-Referenced Terms

(252) DNA probes _____

_____

(252) nucleic acid hybridization _____

_____

(253) cDNA _____

_____

(253) reverse transcription _____

_____

Complete the Table

1. Complete the table below, which summarizes some of the basic tools and procedures used in recombinant DNA technology.

Tool/Procedure	Definition and Role in Recombinant DNA Technology
a. DNA probe	
b. Nucleic acid hybridization	
c. cDNA	
d. Reverse transcription	

Fill-in-the-Blanks

Host cells that take up modified genes may be identified by procedures involving DNA (2) _____.

Gene (3) _____ does not automatically follow the successful taking-in of a modified gene by a host cell; genes must be suitably (4) _____ first, which involves getting rid of the (5) _____ from the mRNA

transcripts, then using the enzyme reverse (6) _____ to produce cDNA. The process goes like this:

a. An mRNA transcript of a desired gene is used as a(n) (7) _____ for assembling a DNA strand. An enzyme, (6), does the assembling. b. An mRNA–(8) _____ hybrid molecule results. c. (9) _____ action removes the mRNA and assembles a second strand of (10) _____ on the first strand.

.d. The result is double-stranded (11) _____, "copied" from an mRNA (12).

a.

b.

c.

d.

16-IV. GENETIC ENGINEERING OF BACTERIA (p. 254)
GENETIC ENGINEERING OF PLANTS (pp. 254–255)
GENETIC ENGINEERING OF ANIMALS (pp. 256–257)

Selected Italicized Words

"ice-minus bacteria," "fail-safe" genes, "Ti" plasmid, Agrobacterium tumefaciens, "gene gun," "super-mouse," "biotech barnyards," eugenic engineering

Boldfaced, Page-Referenced Terms

(256) human genome project _____

_____

(256) gene therapy _____

_____

Short Answer

1. Explain the goal of the human genome project. _____

_____

In exercises 2–7, summarize the results of the given experimentation dealing with genetic modifications of plants and animals.

2. Cattle may soon be producing human collagen: _____

_____

3. The bacterium, *Agrobacterium tumefaciens*: _____

_____

4. Cotton plants: _____

_____

5. Introduction of the rat and human somatotropin gene into fertilized mouse eggs:_____

_____

6. "Ice-minus bacteria" and strawberry plants: _____

_____

7. Genetically modified lymphoblasts and severe combined immune deficiency (SCID):

_____

Fill-in-the-Blanks

The harmful gene is called the "ice-forming" gene, and the modified bacteria are known as (8) _____-_____ bacteria; genetic engineers were able to remove the harmful gene and test the modified bacterium on strawberry plants with no adverse effects. Inserting one or more genes into the (9) _____ _____ of an organism for the purpose of correcting genetic defects is known as (10) _____ _____. Attempting to modify a human genome by inserting genes into sperm or eggs is called (11) _____ _____.

Self-Quiz

___ 1. Small circular molecules of DNA in bacteria are called _____.
   a. plasmids
   b. desmids
   c. pili
   d. F particles
   e. transferins

___ 2. Base-pairing between nucleotide sequences from different sources is called _____.
   a. nucleic acid hybridization
   b. reverse replication
   c. heterocloning
   d. plasmid formation

___ 3. Enzymes used to cut genes in recombinant DNA research are _____.
   a. ligases
   b. restriction enzymes
   c. transcriptases
   d. DNA polymerases
   e. replicases

___ 4. The total DNA in a haploid set of chromosomes of a species is its _____.
   a. plasmid
   b. enzyme potential
   c. genome
   d. DNA library
   e. none of the above

___ 5. An enzyme that heals random base-pairing of chromosomal fragments and plasmids is _____.
   a. reverse transcriptase
   b. DNA polymerase
   c. cDNA
   d. DNA ligase

___ 6. A DNA library is _____.
   a. a collection of DNA fragments produced by restriction enzymes and incorporated into plasmids
   b. cDNA plus the required restriction enzymes
   c. mRNA-cDNA
   d. composed of mature mRNA transcripts

___ 7. Amplification results in _____.
   a. plasmid integration
   b. bacterial conjugation
   c. cloned DNA
   d. production of DNA ligase

___ 8. Any DNA molecule that is copied from mRNA is known as _____.
   a. cloned DNA
   b. cDNA
   c. DNA ligase
   d. hybrid DNA

___ 9. The most commonly used method of DNA amplification is _____.
   a. polymerase chain reaction
   b. gene expression
   c. genome mapping
   d. RFLPs

___ 10. Restriction fragment length polymorphisms are valuable because _____.
   a. they reduce the risks of genetic engineering
   b. they provide an easy way to sequence the human genome
   c. they allow fragmenting DNA without enzymes
   d. they provide DNA fragment sizes unique to each person

Crossword Puzzle—Recombinant DNA and Genetic Engineering

ACROSS

1. A DNA _____ is a unique array of RFLPs inherited from each parent in a Mendelian pattern.
5. The process of transferring DNA code to any of three forms of RNA.
6. _____ transcription assembles a DNA strand on mRNA and produces a "hybrid" DNA-RNA molecule.
7. _____ has been "copied" from a mature mRNA transcript for a gene from which its introns have been removed.
8. Restriction Fragment Length _____.
12. A collection of DNA fragments, produced by (15) enzymes, that have been incorporated into plasmids.
15. _____ enzymes cut apart foreign DNA that has been injected into the bacterial cell; the short, single-stranded ends of a DNA fragment are "sticky."
17. A small circlet of DNA or RNA that contains only a few genes; in some bacteria.
18. DNA _____ uses short nucleotide sequences as "primers" to begin replication.
20. _____ DNA technology splices together DNA cut from different species and inserts the product into bacteria or other rapidly dividing cells.
21. Molecule that contains the genetic code of almost all organisms.
22. Genetic _____ inserts modified genes back into the same (or a different) organism.

DOWN

2. Prefix that means four (Greek).
3. _____ DNA (multiple, identical copies of DNA fragments) is produced by repeated replications and cell divisions.
4. Gene _____ transfers one or more normal genes into the body cells of an organism to correct a genetic defect.
9. DNA _____ seals base-pairings during replication.
10. Base-pairing between nucleotide sequences from different sources is called nucleic acid _____.
11. The common method of amplifying DNA.
13. DNA _____ uses short DNA sequences assembled from radioactively labeled nucleotides to identify a gene of interest.
14. _____ a gene determines the order of nucleotides contained therein.
16. The human genome _____.
19. Severe combined immune deficiency.

Chapter Objectives/Review Questions

This section lists general and detailed chapter objectives that can be used as review questions. You can make maximum use of these items by writing answers on a separate sheet of paper. Fill in answers where blanks are provided. To check for accuracy, compare your answers with information given in the chapter or glossary.

Page		Objectives/Questions
(247)	1.	List the means by which natural genetic recombination occurs.
(247)	2.	Define *recombinant DNA technology*.
(248)	3.	_____ are small, circular, self-replicating molecules of DNA or RNA within a bacterial cell.
(249)	4.	Some bacteria produce _____ enzymes that cut apart DNA molecules injected into the cell by viruses; such DNA fragments, or "_____ ends," often have staggered cuts capable of base-pairing with other DNA molecules cut by the same _____ enzymes.
(249)	5.	Base-pairing between chromosomal fragments and cut plasmids is made permanent by DNA _____.
(249)	6.	Be able to explain what a DNA library is; review the steps used in creating such a library.
(250)	7.	List and define the two major methods of DNA amplification.
(250)	8.	Polymerase chain reaction is the most commonly used method of DNA _____.
(250)	9.	Multiple, identical copies of DNA fragments produced by restriction enzymes are known as _____ DNA.
(250–251)	10.	Explain how gel electrophoresis is used to sequence DNA.
(252)	11.	List some practical genetic uses of RFLPs.
(252–253)	12.	How is a cDNA probe used to identify a desired gene carried by a modified host cell?
(253)	13.	A special viral enzyme, _____ _____, presides over the process by which mRNA is transcribed into DNA.
(253)	14.	Define *cDNA*.
(253)	15.	Why do researchers prefer to work with cDNA when working with human genes?
(256–257)	16.	Tell about the human genome project and its implications.
(256–257)	17.	Define *gene therapy* and *eugenic engineering*.

Integrating and Applying Key Concepts

How could scientists guarantee that *Escherichia coli*, the human intestinal bacterium, will not be transformed into a severely pathogenic form and released into the environment if researchers use the bacterium in recombinant DNA experiments?

Answers

Interactive Exercises

16-I. RECOMBINATION IN NATURE—AND IN THE LABORATORY (pp. 246–248)
1. The bacterial chromosome, a circular DNA molecule, contains all the genes necessary for normal growth and development. Plasmids, small, circular molecules of "extra" DNA, carry only a few genes and are self-replicating; 2. F, small circles of DNA in bacteria; 3. T; 4. mutations; 5. recombinant DNA; 6. species; 7. amplify; 8. protein; 9. research; 10. Genetic engineering; 11. plasmid; 12. viruses.

16-II. PRODUCING RESTRICTION FRAGMENTS (pp. 249–250)
WORKING WITH DNA FRAGMENTS (pp. 250–251)
RFLP ANALYSIS (p. 252)
1. a. Bacterial enzymes that cut apart DNA molecules injected into the cell by viruses; several hundred have been identified; b. A replication enzyme that joins the short fragments of DNA; c. A collection of DNA fragments produced by restriction enzymes and incorporated into plasmids; d. After a DNA fragment is inserted into a host cell's cloning vector (often a plasmid), repeated replications and divisions of the host cells pro-

duce multiple, identical copies of DNA fragments, or cloned DNA; e. A method for amplifying DNA fragments in a test tube (see Fig. 16.5); f. Several procedures allow researchers to determine the nucleotide sequence of a DNA fragment (see Fig. 16.6); g. Variations in the banding patterns of DNA fragments from different individuals; the variations occur because no two individuals (except identical twins) have identical base sequences in their entire DNA set. Restriction enzymes cut different DNA molecules at different sites and into different numbers of DNA fragments; h. A unique array of RFLPs; 2. B; 3. F; 4. E; 5. C; 6. A; 7. D; 8. T; 9. F; 10. F; 11. T; 12. F; 13. B; 14. D; 15. G; 16. H; 17. C; 18. I; 19. E; 20. F; 21. A; 22. fingerprint; 23. splicing enzymes; 24. introns; 25. sequencing.

16-III. MODIFIED HOST CELLS (pp. 252–253)

1. a. A nucleic acid hybridization technique used to identify bacterial colonies harboring the DNA (gene) of interest; a short nucleotide sequence is assembled from radioactively labeled subunits (part of the sequence must be complementary to that of the desired gene); b. Base-pairing between nucleotide sequences from different sources can indicate that the two different sources carry the same gene(s); c. Any DNA molecule "copied" from mRNA; d. A process that uses a viral enzyme to transcribe mRNA into DNA that can then be inserted into a plasmid for amplification; the same cDNA (in plasmids) can be inserted into bacteria, which may use it to make the specific protein; 2. probes (hybridization); 3. expression (translation); 4. engineered (processed, modified, spliced, altered); 5. introns; 6. transcriptase; 7. template; 8. cDNA; 9. Enzyme; 10. cDNA; 11. cDNA.

16-IV. GENETIC ENGINEERING OF BACTERIA
(p. 254)
GENETIC ENGINEERING OF PLANTS
(pp. 254–255)
GENETIC ENGINEERING OF ANIMALS
(pp. 256–257)

1. Researchers are working to sequence the estimated 3 billion nucleotides present in human chromosomes; 2. If human collagen can be produced, it may be used to correct various skin, cartilage, and bone disorders; 3. The plasmid in *Agrobacterium* can be used as a vector to introduce desired genes into cultured plant cells; *A. tumefaciens* was used to deliver a firefly gene into cultured tobacco plant cells; 4. Certain cotton plants have been genetically engineered for resistance to worm attacks; 5. In separate experiments the rat and human somatotropin genes were integrated into the mouse DNA. The mice grew much larger than their normal littermates; 6. Bacteria that have specific proteins on their cell surfaces facilitate ice crystal formation on whatever substrate the bacteria are located; bacteria without the ability to synthesize those proteins ("ice-minus") have been genetically engineered and were sprayed on strawberry plants. Nothing bad happened; 7. Researchers have been able to introduce the ADA gene into some of the SCID sufferers' lymphoblasts and stimulate them to divide. About a billion copies of the genetically modified cells have been injected into the bloodstream of a girl with SCID, and she is doing well; 8. ice-minus; 9. body cells (genetic material); 10. gene therapy; 11. eugenic engineering.

Self-Quiz
1. a; 2. a; 3. b; 4. c; 5. d; 6. a; 7. c; 8. b; 9. a; 10. d.

17

EMERGENCE OF EVOLUTIONARY THOUGHT

Interactive Exercises

17-I. EARLY BELIEFS AND CONFOUNDING DISCOVERIES (pp. 260–263)

Boldfaced, Page-Referenced Terms

(262) biogeography _____

_____

(263) comparative anatomy _____

_____

(263) sedimentary beds _____

_____

(263) fossils _____

_____

(263) evolution _____

_____

Matching

Select the single best answer.

1. ___ comparative anatomy
2. ___ biogeography
3. ___ fossils
4. ___ school of Hippocrates
5. ___ evolution
6. ___ great Chain of Being
7. ___ Buffon
8. ___ species
9. ___ Aristotle
10. ___ sedimentary beds

A. Rock layers deposited at different times; may contain fossils
B. Came to view nature as continuum of organization, from lifeless matter through complex forms of plant and animal life
C. Each kind of being; represent links in the great Chain of Being
D. Modification of species over time
E. Extended from the lowest forms of life to humans and on to spiritual beings
F. Suggested that perhaps species originated in more than one place and perhaps had been modified over time
G. Studies of body structure comparisons and patterning
H. Studies of the world distribution of plants and animals
I. Suggested that the gods were not the cause of the sacred disease
J. One type of direct evidence that organisms lived in the past

17-II. A FLURRY OF NEW THEORIES (p. 264)

Boldfaced, Page-Referenced Terms

(264) catastrophism _____

_____

(264) theory of inheritance of acquired characteristics _____

_____

Choice

For questions 1–12, choose from the following:

a. Georges Cuvier   b. Jean-Baptiste Lamarck

___ 1. Stretching directed "fluida" to the necks of giraffes, which lengthened permanently

___ 2. Acknowledged there were abrupt changes in the fossil record that corresponded to discontinuities between certain layers of sedimentary beds; thought this was evidence of change in populations of ancient organisms

___ 3. The force for change in organisms is the drive for perfection

___ 4. There was only one time of creation that populated the world with all species

___ 5. Catastrophism

___ 6. Permanently stretched giraffe necks were bestowed on offspring

___ 7. When a global catastrophe destroyed many organisms, a few survivors repopulated the world

___ 8. Theory of inheritance of acquired characteristics

___ 9. The survivors of a global catastrophe were not new species (may have differed from related fossils), but naturalists had not discovered them yet

___ 10. During a lifetime, environmental pressures and internal "desires" bring about permanent changes

___ 11. The force for change is a drive for perfection, up the Chain of Being

___ 12. The drive to change is centered in nerves that direct an unknown "fluida" to body parts in need of change

17-III. VOYAGE OF THE *BEAGLE* (p. 264)
17-IV. DARWIN'S THEORY TAKES FORM (pp. 266–267)
17-V. REGARDING WALLACE (p. 268)
17-VI. REGARDING THE "MISSING LINKS" (p. 268)

Selected Italicized Words

H.M.S. Beagle, observation, inference, descent with modification, Archaeopteryx

Boldfaced, Page-Referenced Terms

(266) theory of uniformity _____

_____

(267) natural selection _____

_____

Complete the Table

1. Several key players and events in the life of Charles Darwin led him to his conclusions about natural selection and evolution. Summarize these influences by completing the table below.

Event/Person	Importance to Synthesis of Evolutionary Theory
a.	Botanist at Cambridge University who perceived Darwin's real interests and arranged for Darwin to become a ship's naturalist
b.	Where Darwin earned a degree in theology but also developed his love for natural history
c.	British ship that carried Darwin (as a naturalist) on a five-year voyage around the world
d.	Wrote *Principles of Geology*; advanced the theory of uniformity; suggested the Earth was much older than 6,000 years
e.	Wrote an influential essay (read by Darwin) on human populations asserting that people tend to produce children faster than food supplies, living space, and other resources can be sustained
f.	Volcanic islands 900 kilometers from the South American coast where Darwin correlated differences in various species of finches with their environmental challenges
g.	The key point in Darwin's theory of evolution; involves reproductive capacity, heritable variations, and adaptive traits
h.	English naturalist contemporary with Darwin; independently developed Darwin's theory of evolution before Darwin published
i.	Unearthed in 1861; the first transitional fossil (between reptiles and birds); provided evidence for Darwin's theory

Self-Quiz

___ 1. An acceptable definition of evolution is _____.

   a. changes in organisms that are extinct
   b. changes in organisms since the flood
   c. changes in organisms over time
   d. changes in organisms in only one place

___ 2. The two scientists most closely associated with the concept of evolution are _____.

   a. Lyell and Malthus
   b. Henslow and Cuvier
   c. Henslow and Malthus
   d. Darwin and Wallace

___ 3. Studies of the distribution of organisms on the Earth is known as _____.

   a. a great Chain of Being
   b. stratification
   c. geological evolution
   d. biogeography

___ 4. Buffon suggested that _____.

   a. tailbones in a human have no place in a perfectly designed body
   b. perhaps species originated in more than one place and may have been modified over time

c. the force for change in organisms was a built-in drive for perfection, up the Chain of Being

d. gradual processes now molding the Earth's surface had also been at work in the past

___ 5. In Lyell's book, *Principles of Geology*, he suggested that _____.

a. tailbones in a human have no place in a perfectly designed body

b. perhaps species originated in more than one place and may have been modified over time

c. the force for change in organisms was a built-in drive for perfection, up the Chain of Being

d. gradual processes now molding the Earth's surface had also been at work in the past

___ 6. One of the central ideas of Lamarck's theory of desired evolution was that _____.

a. tailbones in a human have no place in a perfectly designed body

b. perhaps species originated in more than one place and may have been modified over time

c. the force for change in organisms was a built-in drive for perfection, up the Chain of Being

d. gradual processes now molding the Earth's surface had also been at work in the past

___ 7. The theory of catastrophism is associated with _____.

a. Darwin

b. Cuvier

c. Buffon

d. Lamarck

___ 8. The idea that any population tends to outgrow its resources, and its members must compete for what is available, belonged to _____.

a. Malthus

b. Darwin

c. Lyell

d. Henslow

___ 9. The ideas that all natural populations have the reproductive capacity to exceed the resources required to sustain them and members of a natural population show great variation in their traits are associated with _____.

a. catastrophism

b. desired evolution

c. natural selection

d. Malthus's idea of survival

___ 10. *Archaeopteryx* is evidence for _____.

a. birds descending from reptiles

b. evolution

c. an evolutionary "link" between two major groups of organisms

d. the existence of fossils

e. all of the above

Chapter Objectives/Review Questions

Page   *Objectives/Questions*

(262)   1. Relate Aristotle's view of nature and describe what is meant by the great Chain of Being.

(262)   2. It was the study of _____ that raised this question: How did so many species get from the center of creation to islands and other isolated places?

(263)   3. Scientists working in a field known as _____ _____ wondered why animals as different as humans, whales, and bats had so many similarities.

(263)   4. Give two examples of the type of evidence found by comparative anatomists that suggested living things may have changed with time.

(263)   5. _____ _____ consist of distinct, multiple layers of sedimentary rock that were originally deposited at different time periods.

(263)   6. A _____ is defined as the recognizable remains or body impressions of organisms that lived in the past.

(263)   7. The modification of species over time is called _____.

(264)   8. Be able to discuss Cuvier's theory of catastrophism.

(264)   9. What was the main force for change in organisms as described in Lamarck's theory of desired evolution?

(264–268) 10. Be able to state the significance of the following to the theory of evolution: Henslow, H.M.S. *Beagle*, Lyell, Darwin, Malthus, Wallace, and *Archaeopteryx*.

(267)   11. To what conclusion was Darwin led when he considered the various species of finches living on the separate islands of the Galápagos?

(267)   12. Outline the key observations and inferences of Darwin's theory of evolution by natural selection.

Integrating and Applying Key Concepts

An Austrian biologist named Paul Kammerer (1880–1926) once reported that he had found solid evidence that the nuptial pads on the forelimbs of the male midwife toad (used to grip the female during mating) were *acquired* by the "energy and diligence" of the parent toad and were then transmitted to offspring for the benefit of future generations. When he presented his evidence, several observing scientists suspected his specimens were forgeries. Scientists attempted to repeat Kammerer's experiments, but they were unrepeatable. Kammerer incurred the wrath of the international scientific community and, in disgrace, he eventually ended his life by his own hand. Considering what you have learned in this chapter, what is your assessment of Kammerer's claim?

Answers

Interactive Exercises

17-I. EARLY BELIEFS AND CONFOUNDING DISCOVERIES (pp. 262–263)
1. G; 2. H; 3. J; 4. I; 5. D; 6. E; 7. F; 8. C; 9. B; 10. A.

17-II. A FLURRY OF NEW THEORIES (p. 264)
1. b; 2. a; 3. b; 4. a; 5. a; 6. b; 7. a; 8. b; 9. a; 10. b; 11. b; 12. b.

17-III. VOYAGE OF THE *BEAGLE* (p. 264)
17-IV. DARWIN'S THEORY TAKES FORM (pp. 266–267)
17-V. REGARDING WALLACE (p. 268)
17-VI. REGARDING THE "MISSING LINKS" (p. 268)
1. a. John Henslow; b. Cambridge University; c. H.M.S. *Beagle*; d. Charles Lyell; e. Thomas Malthus; f. Galápagos Islands; g. Natural selection; h. Alfred Wallace; i. *Archaeopteryx*.

Self-Quiz
1. c; 2. d; 3. d; 4. b; 5. d; 6. c; 7. b; 8. a; 9. c; 10. e.

18

MICROEVOLUTION

Interactive Exercises

18-I. INDIVIDUALS DON'T EVOLVE; POPULATIONS DO (pp. 270–273)

Selected Italicized Words

morphological traits, physiological traits, behavioral traits

Boldfaced, Page-Referenced Terms

(272) population _____

_____

(272) gene pool _____

_____

(272) alleles _____

_____

(272) phenotype _____

_____

Labeling

The traits of individuals in a population are often classified as being morphological, physiological, or behavioral traits. In the list of traits below, enter "M" if the trait seems to be morphological, "P" if the trait is physiological, and "B" if the trait is behavioral.

1. ___ Frogs have a three-chambered heart.

2. ___ The active transport of Na$^+$ and K$^+$ ions is unequal.

3. ___ Humans and orangutans possess an opposable thumb.

4. ___ An organism periodically seeking food.

5. ___ Some animals have a body temperature that fluctuates with the environmental temperature.

6. ___ The platypus is a strange mammal: It has a bill like a duck and lays eggs.

7. ___ Some vertebrates exhibit greater parental protection for offspring than others.

8. ___ During short photoperiods, the pituitary gland releases small quantities of gonadotropins.

9. ___ Red grouse defend large multipurpose territories where they forage, mate, nest, and rear young.

10. ___ Lampreys have an elongated cylindrical body without scales.

Matching

Select the one most appropriate answer to match the sources of genetic variation.

11. ___ fertilization

12. ___ changes in chromosome structure or number

13. ___ crossing over at meiosis

14. ___ gene mutation

15. ___ independent assortment at meiosis

A. Leads to mixes of paternal and maternal chromosomes in gametes
B. Produces new alleles
C. Leads to the loss, duplication, or alteration of alleles
D. Brings together combinations of alleles from two parents
E. Leads to new combinations of alleles in chromosomes

Short Answer

16. Briefly discuss the role of the environment and genetics in the production of an organism's phenotype.

_____

_____

18-II. STABILITY VERSUS CHANGE IN ALLELE FREQUENCIES (p. 274)
18-III. MUTATION (p. 275)

Boldfaced, Page-Referenced Terms

(274) allele frequencies _____

_____

(274) genetic equilibrium _____

_____

(274) microevolution _____

_____

(275) mutation _____

_____

Short Answer

1. List the conditions (in any order) that must be met before genetic equilibrium (or nonevolution) will occur._____

_____

_____

Problems

2. For the following situation, assume that the conditions listed in question 1 do exist; therefore, there should be no change in gene frequency, generation after generation. Consider a population of hamsters in which dominant gene B produces black coat color and recessive gene b produces gray coat color (two alleles are responsible for color). The dominant gene has a frequency of 80 percent (or 0.80). It would follow that the frequency of the recessive gene is 20 percent (or 0.20). From this, the assumption is made that 0.80 percent of all sperm have gene B and 80 percent of all eggs have gene B. Also, 20 percent of all sperm carry gene b and 20 percent of all eggs carry gene b. (See page 274 in the text.)

a. Calculate the probabilities of all possible matings in the Punnett square.

b. Summarize the genotype and phenotype frequencies of the F_1 generation.

Sperm	
0.80 B	0.20 b

Eggs		
0.80 B	BB	Bb
0.20 b	Bb	bb

Genotypes	Phenotypes
___BB	
___Bb	___% black
___bb	___% gray

c. Further assume that the individuals of the F_1 generation produce another generation and the assumptions of the Hardy-Weinberg rule still hold. What are the frequencies of the sperm produced?

Parents (F_1)	B sperm	b sperm
___BB	___	___
___Bb	___	___
___bb	___	___
Totals =	___	___

The egg frequencies may be similarly calculated. Note that the gamete frequencies of the F_2 are the same as the gamete frequencies of the last generation. Phenotype percentage also remains the same. Thus, the gene frequencies did not change between the F_1 and the F_2 generations. Again, given the assumptions of the Hardy-Weinberg equilibrium, gene frequencies do not change generation after generation.

3. In a population, 81 percent of the organisms are homozygous dominant, and 1 percent are homozygous recessive. Find (a) the percentage of heterozygotes, (b) the frequency of the dominant allele, and (c) the frequency of the recessive allele. _____

4. In a population of 200 individuals, determine how many individuals are (a) homozygous dominant, (b) homozygous recessive, and (c) heterozygous for a particular locus if $p = 0.8$. _____

5. If the percentage of gene D is 70 percent in a gene pool, find the percentage of gene d. _____

6. If the frequency of gene R in a population is 0.60, what percentage of the individuals are heterozygous Rr? _____

Matching

Select the most appropriate answer for each.

7. ___ allele frequencies

8. ___ neutral mutations

9. ___ microevolution

10. ___ lethal mutation

11. ___ alleles

12. ___ mutation

13. ___ genetic equilibrium

A. A heritable change in DNA
B. Zero evolution
C. Different molecular forms of a gene
D. Small-scale changes in allele frequencies as brought about by mutation, genetic drift, gene flow, and natural selection
E. Expression always leads to the death of the individual
F. The abundance of each kind of allele in the entire population
G. Neither harmful nor helpful to the individual

18-IV. GENETIC DRIFT (p. 276)
18-V. GENE FLOW (p. 277)

Selected Italicized Words

feline infectious peritonitis

Boldfaced, Page-Referenced Terms

(276) genetic drift _____

(276) founder effect _____

(276) bottlenecks _____

(276) endangered species _____

Choice

For questions 1–10, choose from the following microevolutionary forces that can change gene frequency:

a. genetic drift   b. gene flow

___ 1. A random change in allele frequencies over the generations, brought about by chance alone

___ 2. Emigration

___ 3. Prior to the turn of the century hunters killed all but twenty of a large population of northern elephant seals

___ 4. When allele frequencies change due to individuals leaving a population or new individuals entering it

___ 5. Long ago, seabirds, winds, or ocean currents carried a few seeds from the Pacific Northwest to the Hawaiian Islands to establish plant populations

___ 6. Sometimes result in endangered species

___ 7. Founder effects and bottlenecks are two extreme cases

___ 8. Immigration

___ 9. Only 20,000 cheetahs have survived to the present

___ 10. Scrubjays make hundreds of round trips carrying acorns from oak trees as much as a mile away to soil in their home territories for winter storage; this introduces new alleles to oak stands

18-VI. NATURAL SELECTION (p. 278)

Selected Italicized Words

Biston betularia

Boldfaced, Page-Referenced Terms

(278) natural selection _____

(278) directional selection _____

(279) mark-release-recapture method _____

_____

(279) pest resurgence _____

_____

(280) stabilizing selection _____

_____

(280) disruptive selection _____

_____

(282) balanced polymorphism _____

_____

(282) sexual dimorphism _____

_____

(283) sexual selection _____

_____

Complete the Table

1. Complete the following table, which defines three types of natural selection effects.

Effect	Characteristics	Example
a. Stabilizing selection		
b. Directional selection		
c. Disruptive selection		

Labeling

2. Identify the three curves below as stabilizing selection, directional selection, or disruptive selection.

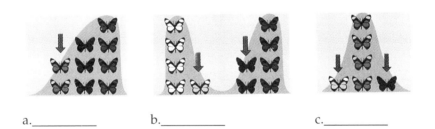

a._____   b._____   c._____

Choice

For questions 3–16 , choose from the following categories of natural selection; in some cases, two letters may be correct.

> a. directional selection   b. stabilizing selection   c. disruptive selection
> d. balanced polymorphism   e. sexual selection

___ 3. May result even when heterozygotes are selected against; unstable *Rh* markers in the human population provide an example

___ 4. Sexual dimorphism

___ 5. Phenotypic forms at both ends of the variation range are favored and intermediate forms are selected against

___ 6. May account for the persistence of certain phenotypes over time; horsetail plants still bear strong resemblance to their ancient relatives

___ 7. Selection maintains two or more alleles for the same trait in steady fashion, generation after generation

___ 8. The most common forms of a trait in a population are favored; over time, alleles for uncommon forms are eliminated

___ 9. Allele frequencies shift in a steady, consistent direction; this may be due to a change in the environment

___ 10. Often results when environmental conditions favor homozygotes over heterozygotes

___ 11. Counters the effects of mutation, genetic drift, and gene flow

___ 12. Human newborns weighing an average of 7 pounds are favored

___ 13. In West Africa, finches known as black-bellied seedcrackers have either large or small bills

___ 14. When a trait gives an individual an advantage in reproductive success

___ 15. The most frequent wing color of peppered moths shifted from a light form to a dark form as tree trunks became soot darkened due to coal used for fuel during the industrial revolution

___ 16. Females of a species choosing mates to directly affect reproductive success

Self-Quiz

For questions 1–3, choose from these answers:

    a. gene pool
    b. genetic variation
    c. gene flow
    d. allele frequency

___ 1. Differences in the combinations of alleles carried by the individuals of a population would be its _____.

___ 2. For sexually reproducing species, the _____ is a source of potentially enormous variation in traits.

___ 3. The relative abundance of each type of allele in a population is the _____.

___ 4. $p + q = 1$ expresses the _____ of a population; $p^2 + 2pq + q^2$ expresses the _____ of a population.

    a. genotype frequency; allele frequency
    b. genotype frequency; genotype frequency
    c. allele frequency; genotype frequency
    d. allele frequency; allele frequency

___ 5. The unit of evolution is the _____.

    a. population
    b. individual
    c. fossil
    d. missing link

___ 6. Changes in allele frequencies brought about by mutation, genetic drift, gene flow, and natural selection are called _____ processes.

    a. genetic equilibrium
    b. microevolution
    c. founder effects
    d. independent assortment

For questions 7–9, choose from these answers:

    a. disruptive selection
    b. genetic drift
    c. directional selection
    d. stabilizing selection

___ 7. Cockroaches becoming increasingly resistant to pesticides is an example of _____.

___ 8. The founder effect is a special case of _____.

___ 9. Different alleles that code for different forms of hemoglobin have led to _____ in the midst of the malarial belt that extends through West and Central Africa.

For questions 10–12, choose from these answers:

    a. balanced polymorphism
    b. sexual dimorphism
    c. genetic equilibrium
    d. sexual selection

___ 10. Hardy and Weinberg invented an ideal population that was in _____ and used it as a baseline against which to measure evolution in real populations.

___ 11. The term _____ is applied when a population has nonidentical alleles for a trait that are maintained at frequencies greater than 1 percent.

___ 12. Phenotypic differences between the males and females of a species are known as _____.

Chapter Objectives/Review Questions

Page	Objectives/Questions
(272)	1. A _____ is a group of individuals occupying a given area and belonging to the same species.
(272)	2. Distinguish between morphological, physiological, and behavioral traits.
(272)	3. All of the genes of an entire population belong to a _____.
(272)	4. Each kind of gene usually exists in one or more molecular forms, called _____.
(273)	5. Review the five categories through which genetic variation occurs among individuals.
(274)	6. The abundance of each kind of allele in the whole population is referred to as allele _____.
(274–275)	7. Be able to calculate allele and genotype frequencies when provided with the genotype frequency of the recessive allele.
(274)	8. A point at which allele frequencies for a trait remain stable through the generations is called genetic _____.
(274)	9. Be able to list the five conditions that must be met before conditions for the Hardy-Weinberg rule are met.
(274)	10. Changes in allele frequencies brought about by mutation, genetic drift, gene flow, and natural selection are called _____.
(275)	11. A _____ is a random, heritable change in DNA.
(275)	12. Distinguish a lethal mutation from a neutral mutation.
(276)	13. Random fluctuations in allele frequencies over time, due to chance, are called _____.
(276)	14. Distinguish the founder effect from a bottleneck.
(276)	15. Relate bottlenecks to endangered species.
(277)	16. Allele frequencies change as individuals leave or enter a population; this is gene _____.
(278)	17. _____ _____ is defined as the change or stabilization of allele frequencies due to differences in survival and reproduction among variant members of a population.
(278–281)	18. Be able to define and provide an example of directional selection, stabilizing selection, and disruptive selection.
(282)	19. Define *balanced polymorphism*.
(282)	20. The occurrence of phenotypic differences between males and females of a species is called _____ _____.
(282–283)	21. _____ selection is based on any trait that gives an individual a competitive edge in mating and producing offspring.

Integrating and Applying Key Concepts

Can you imagine any way in which directional selection may have occurred or may be occurring in humans? Which factor(s) do you suppose are the driving force(s) that sustain(s) the trend? Do you think the trend could be reversed? If so, by what factor(s)?

Answers

Interactive Exercises

18-I. INDIVIDUALS DON'T EVOLVE; POPULATIONS DO (pp. 272–273)

1. M; 2. P; 3. M; 4. B; 5. P; 6. M; 7. B; 8. P; 9. B; 10. M; 11. D; 12. C; 13. E; 14. B; 15. A; 16. Phenotypic variation is due to the effects of genes and the environment. The environment can mediate how the genes governing traits are expressed in an individual. For example, ivy plants grown from cuttings of the same parent all have the same genes, but the amount of sunlight affects the genes governing leaf growth. Ivy plants growing in full sun have smaller leaves than those growing in full shade. Offspring inherit genes, not phenotype.

18-II. STABILITY VERSUS CHANGE IN ALLELE FREQUENCIES (p. 274)
18-III. MUTATION (p. 275)

1. There are five conditions: no genes are undergoing mutation; a very, very large population; the population is isolated from other populations of the species; all members of the population survive and reproduce; and mating is random; 2. a. 0.64 *BB*, 0.16 *Bb*, 0.16 *Bb*, and 0.04 *bb*; b. genotypes: 0.64 *BB*, 0.32 *Bb*, and 0.04 *bb*; phenotypes: 96% black, 4% gray; c.

Parents	B sperm	b sperm
0.64 *BB*	0.64	0
0.32 *Bb*	0.16	0.16
0.04 *bb*	0	0.04
Totals =	0.80	0.20

3. Find (b) first, then (c), and finally (a). a. $2pq = 2 \times (0.9) \times (0.1) = 2 \times (0.09) = 0.18 = 18\%$, which is the percentage of heterozygotes; b. $p^2 = 0.81$, $p = \sqrt{0.81} = 0.9 =$ the frequency of the dominant allele; c. $p + q = 1$, $q = 1 - 0.9 = 0.1 =$ the frequency of the recessive allele; 4. a. homozygous dominant $= p^2 \times 200 = (0.8)^2 \times 200 = 0.64 \times 200 = 128$ individuals; b. homozygous recessive $= q^2 \times 200$ $q = (1.00 - p)^2 = (0.2)^2 \times 200 = (0.04) \times (200) = 8$ individuals; c. heterozygotes $= 2pq \times 200 = 2 \times 0.8 \times 0.2 \times 200 = 0.32 \times 200 = 64$ individuals. Check: $128 + 8 + 64 = 200$; 5. If $p = 0.70$, since $p + q = 1$, $0.70 + q = 1$; then $q = 0.30$, or 30 percent; 6. If $p = 0.60$, since $p + q = 1$, $0.60 + q = 1$; then $q = 0.40$; thus, $2pq = 0.48$, or 48 percent; 7. F; 8. G; 9. D; 10. E; 11. C; 12. A; 13. B.

18-IV. GENETIC DRIFT (p. 276)
18-V. GENE FLOW (p. 277)

1. a; 2. b; 3. a; 4. b; 5. a; 6. a; 7. a; 8. b; 9. a; 10. b.

18-VI. NATURAL SELECTION (p. 278)

1. a. The most common phenotypes are favored; average human birth rate of 7 pounds; b. Allele frequencies shift in a steady, consistent direction in response to a new environment or a directional change in an old one; light to dark forms of peppered moths; c. Forms at both ends of the phenotypic range are favored and intermediate forms are selected against; finch beaks in West Africa; 2. a. directional; b. disruptive; c. stabilizing; 3. b, d; 4. e; 5. c; 6. b; 7. b, d; 8. b; 9. a; 10. b, d; 11. b; 12. b; 13. c; 14. e; 15. a; 16. e.

Self-Quiz

1. b; 2. a; 3. d; 4. c; 5. a; 6. b; 7. c; 8. b; 9. d; 10. c; 11. a; 12. b.

19

SPECIATION

Interactive Exercises

19-I. WHAT IS A SPECIES? (pp. 286–288)
19-II. ISOLATING MECHANISMS (p. 289)

Selected Italicized Words

gene flow, hybrid, Helix aspersa, Agrostis tenuis

Boldfaced, Page-Referenced Terms

(287) speciation _____

_____

(288) species _____

_____

(288) biological species concept _____

_____

(289) genetic divergence _____

_____

(289) isolating mechanisms _____

_____

Choice

For questions 1–10, choose from the following:

        a. Mayr's biological species   b. genetic divergence   c. isolating mechanisms

___ 1. Heritable aspect of body form, physiology, or behavior that prevents gene flow between genetically divergent populations

___ 2. Groups of interbreeding natural populations

___ 3. May occur as a result of stopping gene flow

___ 4. Phenotypically very different but can interbreed and produce fertile offspring

___ 5. Defined as the buildup of differences in the separated pools of alleles of populations

___ 6. A "kind" of living thing

___ 7. Factors that prevent horses and zebras from mating in the wild

___ 8. Prezygotic

___ 9. Does not work well with organisms that reproduce solely by asexual means or those known only from the fossil record

___ 10. Postzygotic

Matching

Select the most appropriate answer to match the isolating mechanisms; complete the exercise by entering "pre" in the parentheses if the mechanism is prezygotic and "post" if the mechanism is postzygotic.

11. ___ ecological (   )

12. ___ temporal (   )

13. ___ hybrid inviability (   )

14. ___ mechanical (   )

15. ___ zygote mortality (   )

16. ___ gametic mortality (   )

17. ___ behavioral (   )

18. ___ hybrid offspring (   )

A. Potential mates occupy overlapping ranges but reproduce at different times.

B. The first-generation hybrid forms but shows very low fitness.

C. Potential mates occupy different local habitats within the same area.

D. Potential mates meet but cannot figure out what to do about it.

E. Sperm is transferred but the egg not fertilized (gametes die or gametes are incompatible).

F. The hybrid is sterile or partially so.

G. Potential mates attempt engagement, but sperm cannot be successfully transferred.

H. The egg is fertilized, but the zygote or embryo dies.

Choice

For questions 19–25, choose from the following isolating mechanisms:

        a. temporal   b. behavioral   c. mechanical   d. gametic   e. ecological   f. postzygotic

___ 19. Sterile zebroids

___ 20. Two sage species; each has its flower petals arranged as a "landing platform" for a different pollinator

___ 21. Two species of cicada; one matures, emerges, and reproduces every thirteen years, the other every seventeen years

___ 22. Populations of the grass *Agrostis tenuis* demonstrate differing tolerances to copper at varying distances from copper mines; speciation may be under way

___ 23. Molecularly mismatched gametes of different sea urchin species

___ 24. Prior to copulation, male and female birds engage in complex courtship rituals recognized only by birds of their own species

___ 25. Sterile mules

Identification

26. The illustration below depicts the divergence of one species as time passes. Each horizontal line represents a population of that species and each vertical line a point in time. Answer the questions below the illustration.

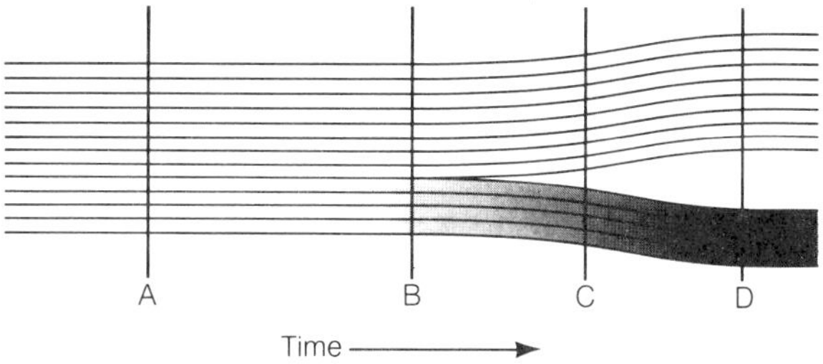

a. How many species are represented at time A? _____

b. How many species exist at time D? _____

c. Between what times (letters) does divergence begin? _____

d. What letter represents the time that complete divergence is reached? _____

e. Between what two times (letters) is divergence clearly under way but not complete? _____

19-III. MODELS OF SPECIATION (p. 292)
19-IV. ALLOPATRIC SPECIATION (pp. 292–293)
19-V. LESS PREVALENT SPECIATION ROUTES (pp. 294–295)
19-VI. PARAPATRIC SPECIATION (p. 295)

Selected Italicized Words

Thalassoma fisciatum, T. lucasanum

Boldfaced, Page-Referenced Terms

(292) allopatric speciation _____

_____

(292) sympatric speciation _____

_____

(292) parapatric speciation _____

_____

(294) polyploidy _____

_____

(295) hybrid zone _____

_____

Complete the Table

1. Complete the following table, which defines three speciation models.

Speciation Model	Description
a.	Suggests that species can form within the range of an existing species, in the absence of physical or ecological barriers; a controversial model; evidence comes from crater lake fishes
b.	Emphasizes disruption of gene flow; perhaps the main speciation route; through divergence and evolution of isolating mechanisms, separated populations become reproductively incompatible and cannot interbreed
c.	Occurs where populations share a common border; the borders may be permeable to gene flow; in some, gene exchange is confined to a common border, the hybrid zone

Dichotomous Choice

Circle one of two possible answers given between parentheses in each statement.

2. Instant speciation by polyploidy in many flowering plant species represents (parapatric/sympatric) speciation.
3. In the past, hybrid orioles existed in a hybrid zone between eastern Baltimore orioles and western Bullock orioles; today, hybrid orioles are becoming less frequent in this example of (allopatric/parapatric) speciation.
4. In the absence of gene flow between geographically separate populations, genetic divergence may become sufficient to bring about (allopatric/parapatric) speciation.
5. The uplifted ridge of land we now call the Isthmus of Panama divides the Atlantic Ocean from the Pacific Ocean and formed a natural laboratory for biologists to study (allopatric/sympatric) speciation.
6. In the 1800s, a major earthquake changed the course of the Mississippi River; this isolated some populations of insects that could not swim or fly and set the stage for (parapatric/allopatric) speciation.

Short Answer

7. Study the illustration below, which shows the possible course of wheat evolution; similar genomes (a genome is a complete set of chromosomes) are designated by the same letter. Answer the questions below the diagram.

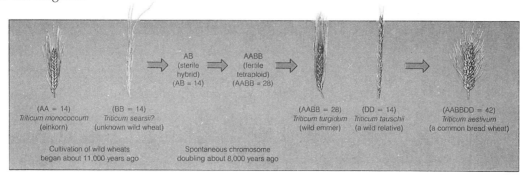

a. How many pairs of chromosomes are found in *Triticum monococcum*?

_____

b. How are the *T. monococcum* chromosomes designated?

_____

c. How many pairs of chromosomes are found in *T. searsii*?

_____

d. How are the *T. searsii* chromosomes designated?

_____

e. Why is the hybrid designated AB sterile? _____

f. What cellular event must have occurred to create the *T. turgidum* genome?

_____

g. Describe the genome of the plant arising from the cross of *T. turgidum* and *T. tauschii* (not shown on the diagram).

_____

h. What cellular event must have occurred to create the *T. aestivum* genome?

_____

i. What is the source of the A genome in *T. aestivum*? The source of the B genome in *T. aestivum*? The source of the D genome in *T. aestivum*?

_____

19-VII. PATTERNS OF SPECIATION (pp. 296–297)
19-VIII. EXTINCTIONS—END OF THE LINE (p. 297)

Selected Italicized Words

physical access, evolutionary access, ecological access

Boldfaced, Page-Referenced Terms

(296) cladogenesis _____

_____

(296) anagenesis _____

_____

(296) evolutionary tree diagrams _____

_____

(296) gradual model of speciation _____

_____

(296) punctuation model of speciation _____

_____

(297) adaptive radiation _____

_____

(297) adaptive zones _____

_____

(297) background extinction _____

_____

(297) mass extinction _____

_____

Complete the Table

1. Complete the table below to summarize information used to interpret evolutionary tree diagrams. Refer to the text, p. 296.

Branch Form	Interpretation
a.	Traits changed rapidly around the time of speciation.
b.	An adaptive radiation occurred.
c.	Extinction
d.	Speciation occurred through gradual changes in traits over geologic time.
e.	Traits of the new species did not change much thereafter.
f.	Evidence of this presumed evolutionary relationship is only sketchy.

Matching

Select the most appropriate answer for each.

2. ___ cladogenesis

3. ___ evolutionary tree diagrams

4. ___ gradual model of speciation

5. ___ punctuation model of speciation

6. ___ adaptive radiation

7. ___ adaptive zones

8. ___ background extinction

9. ___ mass extinction

10. ___ physical, evolutionary, and ecological access

11. ___ anagenesis

A. Expected rate of species disappearance as local conditions change

B. Conditions allowing a lineage to radiate into an adaptive zone

C. Speciation in which most morphological changes are compressed into a brief period of hundreds or thousands of years when populations first start to diverge; evolutionary tree branches make abrupt ninety-degree turns

D. Ways of life such as "burrowing in the seafloor"

E. Speciation with slight angles on the evolutionary tree, which conveys many small changes in form over long time spans

F. An abrupt rise in the rates of species disappearance above background level

G. Speciation occurring within a single, unbranched line of descent

H. A method of summarizing information about the continuities of relationships among species

I. A branching speciation pattern; isolated populations diverge

J. A burst of microevolutionary activity within a lineage; results in the formation of new species in a wide range of habitats

Self-Quiz

___ 1. Of the following, _____ is (are) not a major criterion of the biological species concept.

    a. groups of populations
    b. reproductive isolation
    c. asexual reproduction
    d. interbreeding natural populations
    e. production of fertile offspring

___ 2. When something prevents gene flow between two populations or subpopulations, _____ may occur.

    a. genetic drift, natural selection, and mutation
    b. genetic divergence
    c. a buildup of differences in the separated gene pools of alleles
    d. all of the above

___ 3. When potential mates occupy different local habitats within the same area, it is termed _____ isolation.

    a. temporal
    b. behavioral
    c. mechanical
    d. ecological

___ 4. "Potential mates occupy overlapping ranges but reproduce at different times" is a description of _____ isolation.

    a. temporal
    b. behavioral
    c. mechanical
    d. ecological

___ 5. Of the following, _____ is a postzygotic isolating mechanism.

    a. behavioral isolation
    b. gametic mortality
    c. hybrid inviability
    d. mechanical isolation

For questions 6–8, choose from the following answers:

    a. parapatric speciation
    b. sympatric speciation
    c. allopatric speciation

___ 6. When the Isthmus of Panama was formed, it provided contemporary scientists with an ideal natural laboratory to study _____.

___ 7. Polyploidy and cichlid fishes of crater lakes provide evidence for _____.

___ 8. The identification of a hybrid zone between the ranges of eastern Baltimore orioles and western Bullock orioles is an example of _____.

___ 9. The branching speciation pattern revealed by the fossil record is called _____.

    a. anagenesis
    b. the gradual model
    c. cladogenesis
    d. the punctuation model

___ 10. Species formation by many small changes over long time spans fits the _____ model of speciation; rapid speciation with morphological changes compressed into a brief period of population divergence describes the _____ model of speciation.

    a. anagenesis; punctuation
    b. punctuation; gradual
    c. gradual; cladogenesis
    d. gradual; punctuation

___ 11. Adaptive radiation is _____.

    a. a burst of microevolutionary activity within a lineage
    b. the formation of new species in a wide range of habitats
    c. the spreading of lineages into unfilled adaptive zones
    d. a lineage radiating into an adaptive zone when it has physical, ecological, or evolutionary access to it
    e. all of the above

___ 12. The expected rate of species disappearance over time is called _____.

    a. mass extinction
    b. reverse background radiation
    c. background extinction
    d. speciation extinction

Chapter Objectives/Review Questions

Page		Objectives/Questions
(288)	1.	Be able to give the biological species concept as stated by Ernst Mayr.
(289)	2.	_____ _____ is a buildup of differences in separated pools of alleles.
(289)	3.	Be able to briefly define the major categories of prezygotic and postzygotic isolating mechanisms (see text, Table 19.1).
(291)	4.	The process by which species are formed is known as _____.
(291)	5.	The evolution of reproductive _____ _____ paves the way for genetic divergence and speciation.
(292)	6.	_____ speciation occurs when daughter species form gradually by divergence in the absence of gene flow between geographically separate populations.
(292)	7.	In _____ speciation, daughter species arise, sometimes rapidly, from a small proportion of individuals within an existing population.
(292)	8.	When daughter species form from a small proportion of individuals along a common border between two populations, it is called _____ speciation.
(294–295)	9.	Explain why sympatric speciation by polyploidy is a rapid method of speciation.
(295)	10.	In some cases of parapatric speciation, gene exchange between two species is confined to a _____ zone.
(296)	11.	Distinguish cladogenesis from anagenesis.
(296)	12.	Be able to explain the use of evolutionary tree diagrams and the symbolism used.
(296)	13.	Explain why branches on an evolutionary tree that have slight angles indicate the _____ model of speciation.
(296)	14.	Branches on an evolutionary tree that turn abruptly with ninety-degree turns are consistent with the _____ model of speciation.
(297)	15.	An adaptive radiation is a burst of _____ activity within a lineage that results in the formation of new species in a variety of habitats.
(297)	16.	Cite examples of adaptive zones.
(297)	17.	How does background extinction differ from mass extinction?

Integrating and Applying Key Concepts

Systematics is defined as the practice of describing, naming, and classifying living things; this includes the comparative study of organisms and all relationships among them. Plant systematists who work with flowering plants readily accept Ernst Mayr's definition of a "biological species" as described in this chapter. However, in real practice, systematists often must also work with a concept known as the "morphological species." For example, the statement is sometimes made that "two plant specimens belong to the same morphological species but not to the same biological species." Explain the meaning of that statement. What type of experimental evidence would be necessary as a basis for this statement? From your study of this chapter, can you suggest a reason that application of the term "morphological species" is sometimes necessary? What difficulties and inaccuracies, in terms of identifying a species, might the use of both species concepts present?

Answers

Interactive Exercises

19-I. WHAT IS A SPECIES? (p. 288)
19-II. ISOLATING MECHANISMS (p. 289)
1. c; 2. a; 3. b; 4. a; 5. b; 6. a; 7. c; 8. c; 9. a; 10. c; 11. C
(pre); 12. A (pre); 13. B (post); 14. G (pre); 15. H (post);
16. E (pre); 17. D (pre); 18. F (post); 19. f; 20. c; 21. a;
22. e; 23. d; 24. b; 25. f; 26. a. one; b. two; c. barely
between A and B, but considerable divergence begins
between B and C; d. D; e. B and C.

19-III. MODELS OF SPECIATION (p. 292)
19-IV. ALLOPATRIC SPECIATION (pp. 292–293)
19-V. LESS PREVALENT SPECIATION ROUTES
  (pp. 294–295)
19-VI. PARAPATRIC SPECIATION (p. 295)
1. a. Sympatric; b. Allopatric; c. Parapatric; 2. sympatric; 3. parapatric; 4. allopatric; 5. allopatric; 6. allopatric; 7. a. Seven pairs; b. AA; c. Seven pairs; d. BB;
e. The chromosomes fail to pair in meiosis; f. Fertilization of nonreduced gametes produced a fertile
tetraploid; g. Sterile hybrid, ABD; h. Fertilization of
nonreduced gametes resulted in a fertile hexaploid;
i. *Triticum monococcum*; *T. searsii*; *T. tauschii*.

19-VII. PATTERNS OF SPECIATION (pp. 296–297)
19-VIII. EXTINCTIONS—END OF THE LINE (p. 297)
1. a. Horizontal branching; b. Many branchings of the
same lineage at or near the same point in geologic time;
c. A branch that ends before the present; d. Softly angled
branching; e. Vertical continuation of a branch; f. A
dashed line; 2. I; 3. H; 4. E; 5. C; 6. J; 7. D; 8. A; 9. F;
10. B; 11. G.

Self-Quiz

1. c; 2. d; 3. d; 4. a; 5. c; 6. c; 7. b; 8. a; 9. c; 10. d;
11. e; 12. c.

20

THE MACROEVOLUTIONARY PUZZLE

FOSSILS—EVIDENCE OF ANCIENT LIFE
    Fossilization
    Interpreting the Geologic Tombs
    Interpreting the Fossil Record
EVIDENCE FROM COMPARATIVE EMBRYOLOGY
    Developmental Programs Among the Vertebrates
    Developmental Programs Among the Larkspurs
    Homologous Structures
    Potential Confusion from Analogous Structures
EVIDENCE FROM COMPARATIVE BIOCHEMISTRY
    Molecular Clocks
    Protein Comparisons
    Nucleic Acid Comparisons

ORGANIZING THE EVIDENCE—CLASSIFICATION SCHEMES
    The Linnean Scheme
    Pitfalls Awaiting the Taxonomist
    If It's So Difficult, Why Bother?
RECONSTRUCTING EVOLUTIONARY HISTORY
    Systematics
    Portraying Relative Relationships Among Organisms
    Focus on Science: Constructing a Cladogram
A FIVE-KINGDOM SCHEME

Interactive Exercises

20-I. FOSSILS—EVIDENCE OF ANCIENT LIFE (pp. 300–303)

Selected Italicized Words

trace fossils

Boldfaced, Page-Referenced Terms

(301) macroevolution _____

_____

(302) fossils _____

_____

(303) fossilization _____

_____

(303) stratification _____

_____

Matching

Select the most appropriate answer.

1. ___ fossilization
2. ___ stratification
3. ___ trace fossils
4. ___ continuity of relationship
5. ___ macroevolution
6. ___ fossils
7. ___ biased fossil record
8. ___ formation of sedimentary deposits
9. ___ oldest fossils
10. ___ rupture or tilt of sedimentary layers

A. Evolutionary connectedness between organisms, past to present
B. Found in the deepest rock layers
C. Affected by movements of the Earth's crust, hardness of organisms, population sizes, and certain environments
D. Produced by gradual deposition of silt, volcanic ash, and other materials
E. Refers to large-scale patterns, trends, and rates of change among groups of species
F. A process favored by rapid burial in the absence of oxygen
G. Evidence of later geologic disturbance
H. Recognizable, physical evidence of ancient life; the most common are bones, teeth, shells, spore capsules, seeds, and other hard parts
I. Includes the indirect evidence of coprolites and imprints of leaves, stems, tracks, trails, and burrows
J. A layering of sedimentary deposits

20-II. EVIDENCE FROM COMPARATIVE EMBRYOLOGY (pp. 304–305)
20-III. EVIDENCE FROM COMPARATIVE BIOCHEMISTRY (pp. 308–309)

Selected Italicized Words

Delphinium decorum, D. nudicaule

Boldfaced, Page-Referenced Terms

(304) comparative morphology _____

_____

(306) morphological divergence _____

_____

(306) homology _____

_____

(307) morphological convergence _____

_____

(307) analogy _____

_____

(308) neutral mutations _____

_____

(308) molecular clock _____

_____

(308) DNA-DNA hybridization _____

_____

Labeling

Evidence for macroevolution comes from comparative morphology and comparative biochemistry. For each of the following listed items, place an "M" in the blank if the evidence comes from comparative morphology or a "B" if the evidence comes from comparative biochemistry.

1. ___ Early in the vertebrate developmental program, embryos of different lineages proceed through strikingly similar stages.

2. ___ Comparison of the developmental rates of changes in a chimpanzee skull and a human skull.

3. ___ The evolution of different rates of flower development in two larkspur species also led to the coevolution of two different pollinators, honeybees and hummingbirds.

4. ___ Using accumulated neutral mutations to date the divergence of two species from a common ancestor.

5. ___ The amino acid sequence in the cytochrome *c* of humans precisely matches the sequence in chimpanzees.

6. ___ Establishing a rough measure of evolutionary distance between two organisms by use of DNA hybridization studies that demonstrate the extent to which the DNA from one species base-pairs with another.

7. ___ Regulatory genes control the rate of growth of different body parts and can produce large differences in the development of two very similar embryos.

8. ___ Results from DNA-DNA hybridization studies supported an earlier view that giant pandas, bears, and red pandas are related by common descent.

9. ___ Distantly related vertebrates, such as sharks, penguins, and porpoises, show a similarity to one another in their proportion, position, and function of body parts.

10. ___ Finding similarities in vertebrate forelimbs when comparing the wings of pterosaurs, birds, bats, and porpoise flippers.

20-IV. ORGANIZING THE EVIDENCE—CLASSIFICATION SCHEMES (pp. 310–313)

Selected Italicized Words

relative degrees of relationship

Boldfaced, Page-Referenced Terms

(310) higher taxa _____

_____

(310) classification schemes _____

_____

(310) binomial system _____

_____

(310) genus _____

_____

(311) phylogeny _____

_____

Matching

Choose the single most appropriate letter.

1. ___ black bear
2. ___ phylogeny
3. ___ binomial system
4. ___ *Pinus strobus* and *Pinus banksiana*
5. ___ species
6. ___ genus
7. ___ family
8. ___ taxa
9. ___ *Ursus americanus* (black bear) and *Homarus americanus* (Atlantic lobster)
10. ___ Linnaeus

A. A group of similar species
B. The taxon most often studied
C. Evolutionary relationships among species, reflected in classification schemes
D. The organizing units of classification schemes
E. Classification level that includes similar genera
F. Originated the modern practice of providing two scientific names for organisms
G. An example of a common name
H. Two species belonging to one genus
I. An example of two very different organisms having the same specific epithet (species name)
J. System of assigning a two-part Latin name to a species

Short Answer

11. Arrange the following jumbled taxa in proper order, with the most inclusive first: family, class, species, kingdom, genus, phylum (or division), and order. _____

_____

12. Arrange the following jumbled taxa in proper order, with the most inclusive first: genus: *Archibaccharis*; kingdom: Plantae; order: Asterales; species: *lineariloba*; class: Dicotyledonae; division: Anthophyta; family: Asteraceae. _____

_____

_____

20-V. RECONSTRUCTING EVOLUTIONARY HISTORY (pp. 314–317)

Selected Italicized Words

patterns of diversity, outgroup, derived trait, shared derived traits, monophyletic group

Boldfaced, Page-Referenced Terms

(314) systematics _____

_____

(314) cladogram _____

_____

Short Answer

1. List three methods employed in systematics to elucidate the patterns of diversity in an evolutionary context. _____

_____

_____

_____

Complete the Table

2. Complete the table below to define three different approaches to the problem of reconstructing the history of life.

School of Systematics	Definition
a. Evolutionary systematics	
b. Phenetics	
c. Cladistics	

Short Answer

3. What are monophyletic lineages? _____

_____

4. Describe the construction of a cladogram. _____

_____

5. What is an outgroup? _____

_____

6. What is meant by a derived trait? _____

_____

7. What is the essential information portrayed by a cladogram? _____

_____

Cladogram Interpretation

8. Study the cladogram of seven taxa below. The vertical bars on the stem of the cladogram represent shared derived traits. Various taxa are indicated by numbers. Answer the questions following the cladogram.

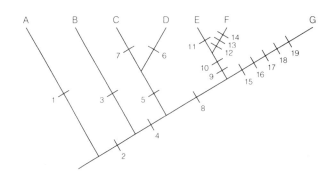

a. On the cladogram above, what types of information could be represented by the numbered traits?

_____

_____

b. Which derived trait is shared by taxa EFG? _____

c. Which derived trait is shared by taxa CDEFG? _____

d. Which is the unique derived trait shared by taxa C and D?_____

_____

e. What does it mean if some taxa are closer together on the cladogram than others? _____

_____

f. Which taxon on the cladogram represents the outgroup condition? _____

_____

g. What is a monophyletic taxon?_____

_____

h. For this cladogram, list the monophyletic taxa shown as well as their shared derived traits.

_____

20-VI. A FIVE-KINGDOM SCHEME (p. 318)

Boldfaced, Page-Referenced Terms

(318) Monera _____

_____

(318) Protista _____

_____

(318) Fungi _____

_____

(318) Plantae _____

_____

(318) Animalia _____

_____

Matching

Choose the one best answer for each.

1. ___ Monera
2. ___ Protista
3. ___ Fungi
4. ___ Plantae
5. ___ Animalia

A. Multicelled heterotrophs that feed by extracellular digestion and absorption
B. Single-celled prokaryotes, some autotrophs, others heterotrophs
C. Diverse multicelled heterotrophs, including predators and parasites
D. Multicelled photosynthetic autotrophs
E. Diverse single-celled eukaryotes, some photosynthetic autotrophs, many heterotrophs

Self-Quiz

___ 1. Phylogeny is _____.
   a. the identification of organisms and assigning names to them
   b. producing a "retrieval system" of several levels
   c. an evolutionary history of organism(s), both living and extinct
   d. a study of the adaptive responses of various organisms

___ 2. The significant contribution by Linnaeus was to develop _____.
   a. the idea that organisms should be placed in two kingdoms
   b. the binomial system
   c. cladograms
   d. the idea that a species could have several common names

___ 3. The process of grouping organisms according to similarities derived from a common ancestor is known as _____.
   a. taxonomy
   b. classification
   c. a binomial system
   d. cladistics

___ 4. Lineages sharing a common evolutionary heritage are termed _____.
   a. phenetic
   b. polyphyletic
   c. monophyletic
   d. homologous

___ 5. All living organisms have eukaryotic cell structure except the _____.
   a. Animalia
   b. Plantae
   c. Fungi
   d. Monera
   e. Protista

For questions 6 and 7, choose from the following answers:
   a. convergence
   b. divergence

___ 6. The wings of pterosaurs, birds, and bats serve as examples of _____.

___ 7. Penguins and porpoises serve as examples of _____.

For questions 8–12, choose from the following answers:
   a. DNA-DNA hybridization
   b. homologous
   c. macroevolution
   d. analogous
   e. regulatory genes

___ 8. _____ structures resemble one another due to common descent.

___ 9. A rough measure of the evolutionary distance between two species is provided by _____.

___ 10. _____ control the rate of growth in different body parts.

___ 11. Examples of _____ structures are the shark fin and the penguin wing.

___ 12. _____ refers to changes in groups of species.

Chapter Objectives/Review Questions

(308)   13. Define *neutral mutations* and discuss their use as molecular clocks.

(308)   14. Some nucleic acid comparisons are based on _____ hybridization.

(310)   15. Describe the binomial system as developed by Linnaeus; why is Latin used as the language of binomials?

(311)   16. Be able to arrange the classification groupings correctly, from most to least inclusive.

(311)   17. Modern classification schemes reflect _____, the evolutionary relationships among species, starting with the most ancestral and including all the branches leading to their descendants.

(312)   18. Explain the difficulties taxonomists may encounter in identifying species.

(314)   19. _____ is a branch of biology that deals with patterns of diversity in an evolutionary context.

(314)   20. Be able to describe the construction, appearance, and use of cladograms.

(318)   21. List the five kingdoms of life in the Whittaker system and cite examples of the organisms in each.

Integrating and Applying Key Concepts

Water crowfoot plants (family Ranunculaceae) often have two or more distinct leaf types within the same species and on the same plant. One type is capillary (hair-like), found submerged in the water or growing well above the water. The other type is laminate (flat and expanded) and is found floating on the water or submerged. In some *Ranunculus* species, three different leaf shapes form on a plant in a sequence. Suppose two taxonomists describe the same crowfoot plant as being two different and distinct species due to these two different leaf shapes. Can you think of difficulties that might arise when other researchers are required to refer to these "two species" in the course of their work?

Answers

Interactive Exercises

20-I. FOSSILS—EVIDENCE OF ANCIENT LIFE (pp. 302–303)

1. F; 2. J; 3. I; 4. A; 5. E; 6. H; 7. C; 8. D; 9. B; 10. G.

20-II. EVIDENCE FROM COMPARATIVE EMBRYOLOGY (pp. 304–305)

20-III. EVIDENCE FROM COMPARATIVE BIOCHEMISTRY (pp. 308–309)

1. M; 2. M; 3. M; 4. B; 5. B; 6. B; 7. M; 8. B; 9. M; 10. M.

20-IV. ORGANIZING THE EVIDENCE— CLASSIFICATION SCHEMES (pp. 310–311)

1. G; 2. C; 3. J; 4. H; 5. B; 6. A; 7. E; 8. D; 9. I; 10. F; 11. kingdom, phylum (or division), class, order, family, genus, species; 12. kingdom: Plantae; division: Anthophyta; class: Dicotyledonae; order: Asterales; family: Asteraceae; genus: *Archibbacharis*; species: *lineariloba*.

20-V. RECONSTRUCTING EVOLUTIONARY HISTORY (pp. 314–317)

1. Systematics relies on (a) taxonomy, the naming and identification of organisms; (b) phylogenetic reconstruction, the identification of evolutionary patterns that unite different organisms; and (c) classification, the grouping of organisms using systems that consist of many hierarchical levels, or ranks, which reflect evolutionary relationships; 2. a. Differences and similarities between organisms are compared in a relatively imprecise, subjective manner. Groups of organisms are defined by traditionally accepted morphological features, such as pollen size and morphology, leaf size, leaf shape, and flower structure. Such subjective methods may result in several different interpretations of the same evidence; b. Using phenetics, organisms are grouped according to their similarities. A pitfall is that perceived similarities may be truly the result of morphological divergences, not convergences; c. Using cladistics, organisms are linked or grouped together that have shared characteristics

(homologies); the similarities are derived from a common ancestor; 3. Monophyletic lineages are lineages that share a common evolutionary heritage; 4. The cladogram is a branching diagram (a type of family tree) that represents the patterns of relationships of organisms, based on their derived traits; 5. An outgroup is the cladogram taxon exhibiting the fewest derived characteristics; 6. A derived trait is one that is shared by members of a lineage but not present in their common ancestor; 7. Cladograms portray relative relationships among organisms. Distribution patterns of many homologous structures are examined. Taxa closer together on a cladogram share a more recent common ancestor than those that are farther apart. Cladograms do not convey direct information about ancestors and descendants; 8. a. The numbered traits could be discrete morphological, physiological, and behavioral traits; b. trait 8; c.

trait 4; d. trait 5; e. Taxa that are closer together share a more recent ancestor than those that are farther apart; f. taxon A; g. A monophyletic taxon is one in which all members have the same derived traits and share a single node on the cladogram; h. The monophyletic taxa and their derived traits are BCDEFG (derived trait 2), CDEFG (derived trait 4), CD (derived trait 5), EFG (derived trait 8), EF (derived traits 9 and 10).

20-VI. A FIVE-KINGDOM SCHEME (p. 318)
1. B; 2. E; 3. A; 4. D; 5. C.

Self-Quiz
1. c; 2. b; 3. d; 4. c; 5. d; 6. b; 7. a; 8. b; 9. a; 10. e;
11. d; 12. c.

21

THE ORIGIN AND EVOLUTION OF LIFE

CONDITIONS ON THE EARLY EARTH
  Origin of the Earth
  The First Atmosphere
  The Synthesis of Organic Compounds
EMERGENCE OF THE FIRST LIVING CELLS
  Origin of Agents of Metabolism
  Origin of Self-Replicating Systems
  Origin of the First Plasma Membranes
LIFE ON A CHANGING GEOLOGIC STAGE
  Drifting Continents and Changing Seas

The Geologic Time Scale
LIFE IN THE ARCHEAN AND PROTEROZOIC ERAS
  Focus on Science: The Rise of Eukaryotic Cells
LIFE IN THE PALEOZOIC ERA
LIFE DURING THE MESOZOIC ERA
  Focus on the Environment: Plumes, Global Impacts, and the Dinosaurs
LIFE DURING THE CENOZOIC ERA

Interactive Exercises

21-I. CONDITIONS ON THE EARLY EARTH (pp. 322–325)
21-II. EMERGENCE OF THE FIRST LIVING CELLS (pp. 326–327)

Choice

Answer questions 1–20 by choosing from the following probable stages of the physical and chemical evolution of life:

    a. Formation of the Earth
    c. The synthesis of organic compounds
    e. Origin of self-replicating systems

    b. The early Earth and the first atmosphere
    d. Origin of agents of metabolism
    f. Origin of the first plasma membranes

___ 1. Most likely, the proto-cells were little more than membrane sacs protecting information-storing templates and various metabolic agents from the environment.

___ 2. Simple systems (of this type) of RNA, enzymes, and coenzymes have been created in the laboratory.

___ 3. Four billion years ago, the Earth was a thin-crusted inferno.

___ 4. Sunlight, lightning, or heat escaping from the Earth's crust could have supplied the energy to drive their condensation into complex organic molecules.

___ 5. During the first 600 million years of Earth history, enzymes, ATP, and other molecules could have assembled spontaneously at the same locations.

___ 6. Were gaseous oxygen (O_2) and water also present? Probably not.

___ 7. Sidney Fox heated amino acids under dry conditions to form protein chains, which he placed in hot water. The cooled chains self-assembled into small, stable spheres. The spheres were selectively permeable.

___ 8. Stanley Miller mixed hydrogen, methane, ammonia, and water in a reaction chamber that recirculated the mixture and bombarded it with a spark discharge. Within a week, amino acids and other small organic compounds had been formed.

___ 9. Without an oxygen-free atmosphere, the organic compounds that started the story of life never would have formed on their own. Free oxygen would have attacked them.

___ 10. Much of the Earth's inner rocky material melted.

___ 11. Imagine an ancient sunlit estuary, rich in clay deposits. Countless aggregations of organic molecules stick to the clay.

___ 12. In other experiments, fatty acids and glycerol combined to form long-tail lipids under conditions that simulated evaporating tidepools. The lipids self-assembled into small, water-filled sacs, which in many instances were like cell membranes.

___ 13. We still don't know how DNA entered the picture.

___ 14. Rocks collected from Mars, meteorites, and the moon contain chemical precursors that must have been present on the early Earth.

___ 15. This differentiation resulted in the formation of a crust of basalt, granite, and other types of low-density rock, a rocky region of intermediate density (the mantle), and a high-density, partially molten core of nickel and iron.

___ 16. When the crust finally cooled and solidified, water condensed into clouds and the rains began. For millions of years, runoff from rains stripped mineral salts and other compounds from the Earth's parched rocks.

___ 17. The close association of enzymes, ATP, and other molecules would have promoted chemical interactions. Sunlight energy alone could have driven the spontaneous formation of RNA molecules.

___ 18. Even if amino acids did form in the early seas, they wouldn't have lasted long. In water, the favored direction of most spontaneous reactions is toward hydrolysis, not condensation.

___ 19. Between 4.6 and 4.5 billion years ago, the outer regions of the cloud cooled.

___ 20. At first, it probably consisted of gaseous hydrogen (H_2), nitrogen (N_2), carbon monoxide (CO), and carbon dioxide (CO_2).

21-III. LIFE ON A CHANGING GEOLOGIC STAGE (pp. 328–329)

Boldfaced, Page-Referenced Terms

(328) plate tectonic theory _____

_____

(329) Proterozoic _____

(329) Paleozoic _____

(329) Mesozoic _____

(329) Cenozoic _____

(329) geologic time scale _____

(329) Archean _____

Sequence

Earth history has been divided into four great eras that are based on four abrupt transitions in the fossil record. The oldest era has been subdivided. Arrange the eras in correct chronological sequence from the oldest to the youngest.

1. ___   A. Mesozoic

2. ___   B. Cenozoic

3. ___   C. Proterozoic

4. ___   D. Archean

5. ___   E. Paleozoic

Matching

Choose the one most appropriate answer for each.

6. ___ Archean

7. ___ Mesozoic

8. ___ Pangea

9. ___ Proterozoic

10. ___ plate tectonic theory

11. ___ Paleozoic

12. ___ Gondwana

13. ___ Cenozoic

14. ___ influenced the evolution of life

15. ___ geologic time scale

A. The "modern era" of geologic time
B. Source of the most ancient fossils prior to subdivision of this era
C. An ancient continent that drifted southward from the tropics, across the south polar region, then northward
D. The boundaries mark the times of mass extinctions
E. The first era of the geologic time scale; a recent subdivision of the Proterozoic
F. An era that follows the Paleozoic
G. Plumes of molten rock from the Earth's mantle spread out beneath the crust or break through it
H. An era whose fossils followed those of the Proterozoic
I. Formed from fusion with Gondwana and other land masses; a single world continent that extended from pole to pole
J. Changes in land masses, the oceans, and the atmosphere

21-IV. LIFE IN THE ARCHEAN AND PROTEROZOIC ERAS (pp. 330–333)

Boldfaced, Page-Referenced Terms

(330) archaebacteria _____

_____

(330) eubacteria _____

_____

(330) stromatolites _____

_____

(331) theory of endosymbiosis _____

_____

(333) protistans _____

_____

Choice

For questions 1–15, choose from the following:

a. Archean   b. Proterozoic

___ 1. By 2.5 billion years ago, the noncyclic pathway of photosynthesis had evolved in some eubacterial species.

___ 2. By 1.2 billion years ago, eukaryotes had originated.

___ 3. The first prokaryotic cells emerged.

___ 4. Oxygen, one of the by-products of photosynthesis, began to accumulate.

___ 5. There was an absence of free oxygen.

___ 6. There was a divergence of the original prokaryotic lineage into three evolutionary directions.

___ 7. About 800 million years ago, stromatolites began a dramatic decline.

___ 8. Between 3.5 and 3.2 billion years ago, the cyclic pathway of photosynthesis evolved in some species of eubacteria.

___ 9. Fermentation pathways were the most likely sources of energy.

___ 10. An oxygen-rich atmosphere stopped the further chemical origin of living cells.

___ 11. Food was available, predators were absent, and biological molecules were free from oxygen attacks.

___ 12. Aerobic respiration became the dominant energy-releasing pathway.

___ 13. Archaebacteria and eubacteria arose.

___ 14. The first cells may have originated in tidal flats or on the seafloor, in muddy sediments warmed by heat from volcanic vents.

___ 15. Near the shores of the supercontinent Laurentia, small, soft-bodied animals were leaving tracks and burrows on the seafloor.

Fill-in-the Blanks

(See *Focus on Science*: The Rise of Eukaryotic Cells, pp. 331–333.)

The most important feature of eukaryotic cells is the abundance of membrane-bound (16) _____ in the cytoplasm. The origin of these structures remains a mystery. Some probably evolved gradually, through (17) _____ mutations and natural selection. Some extant prokaryotic cells do possess infoldings of the (18) _____ membrane, a site where enzymes and other metabolic agents are embedded. In prokaryotic cells that were ancestral to eukaryotic cells, similar membranous infoldings may have served as a (19) _____ from the surface to deep within the cell. These membranes may have evolved into (20) _____ channels and into an envelope around the DNA. The advantage of such membranous enclosures may have been to protect (21) _____ and their products from foreign invader cells. Evidence for this is that yeasts, simple nucleated eukaryotic cells, have an abundance of (22) _____ that they transfer from cell to cell much as prokaryotic cells do.

As the evolution of eukaryotes proceeded, accidental partnerships between different (23) _____ species must have formed countless times. Some partnerships resulted in the origin of mitochondria, chloroplasts, and other organelles. This is the theory of (24) _____, developed in greatest detail by Lynn Margulis. According to this theory, (25) _____ arose after the noncyclic pathway of photosynthesis emerged and oxygen had accumulated to significant levels. (26) _____ transport systems had been expanded in some bacteria to include extra cytochromes to donate electrons to (27) _____ in a system of aerobic respiration. By 1.2 billion years ago or earlier, amoebalike ancestors of eukaryotes were engulfing aerobic (28) _____ and perhaps forming endocytic vesicles around the food for delivery to the cytoplasm for digestion. Some aerobic bacteria resisted digestion and thrived in a new, protected, nutrient-rich environment. In time they released extra (29) _____—which the hosts came to depend on for growth, greater activity, and assembly of other structures. The guest aerobic bacteria came to depend on (30) _____ metabolic functions. The aerobic and anaerobic cells were now (31) _____ of independent existence. The guests had become (32) _____, supreme suppliers of (33) _____.

Mitochondria are bacteria-size and replicate their own (34) _____, dividing independently of the host cell's division process. In addition, (35) _____ may be descended from aerobic eubacteria that engaged in oxygen-producing photosynthesis. Such cells may have been engulfed, resisted digestion, and provided their respiring (36) _____ cells with needed oxygen. Chloroplasts resemble some

(37) _____ in metabolism and overall nucleic acid sequence. Chloroplasts have self-replicating DNA and they divide independently of the host cell's division.

New cells appeared on the evolutionary stage that were equipped with a nucleus, ER membranes, and mitochondria or chloroplasts (or both); they were the first eukaryotic cells, the first (38) _____.

21-V. LIFE IN THE PALEOZOIC ERA (pp. 334–335)
21-VI. LIFE DURING THE MESOZOIC ERA (pp. 336–340)
21-VII. LIFE DURING THE CENOZOIC ERA (pp. 341–343)

Selected Italicized Words

Psilophyton, Cooksonia, Archaeopteris, Lystrosaurus, Triceratops, Velociraptor

Sequence

Refer to Figure 21.16 in the text. Study of the geologic record reveals that as the major events in the evolution of the Earth and its organisms occurred, there were periodic major *extinctions* of organisms followed by major *radiations* of organisms. Using the list below, arrange the letters of the extinctions and radiations in the approximate order in which they occurred, from *youngest* to *oldest*.

1. ___
2. ___
3. ___
4. ___
5. ___
6. ___
7. ___
8. ___
9. ___
10. ___
11. ___
12. ___
13. ___
14. ___

A. Pangea, worldwide ocean forms; shallow seas squeezed out. Major radiations of reptiles, gymnosperms.
B. Glaciations as Gondwana crosses South Pole. Mass extinction of many marine organisms.
C. Mass extinction of many marine invertebrates, most fishes.
D. Pangea breakup begins. Rich marine communities. Major radiations of dinosaurs.
E. Asteroid impact? Mass extinction of all dinosaurs and many marine organisms.
F. Recovery, radiations of marine invertebrates, fishes, dinosaurs. Gymnosperms the dominant land plants. Origin of mammals.
G. Major glaciations. Modern humans emerge and begin what may be the greatest mass extinction of all time on land, starting with Ice Age hunters.
H. Unprecedented mountain building as continents rupture, drift, collide. Major climatic shifts; vast grasslands emerge. Major radiations of flowering plants, insects, birds, mammals. Origin of earliest human forms.
I. Gwondana moves south. Major radiations of marine invertebrates, early fishes.
J. Laurasia forms. Gondwana moves north. Vast swamplands, early vascular plants. Radiation of fishes continues. Origin of amphibians.
K. Pangea breakup continues, broad inland seas form. Major radiations of marine invertebrates, fishes, insects, dinosaurs, origin of angiosperms (flowering plants).
L. Asteroid impact? Mass extinction of many organisms in seas, some on land; dinosaurs, mammals survive.
M. Mass extinction. Nearly all species in seas and on land perish.
N. Tethys Sea forms. Recurring glaciations. Major radiations of insects, amphibians. Spore-bearing plants dominant; gymnosperms present; origin of reptiles.

Chronology of Events—Geologic Time

Refer to Figure 21.16 in the text. From the list of evolutionary events above (A–N), select the letters occurring in a particular era of time by circling the appropriate letters.

15. Cenozoic: A - B - C - D - E - F - G - H - I - J - K - L - M - N
16. Border of Cenozoic-Mesozoic: A - B - C - D - E - F - G - H - I - J - K - L - M - N
17. Mesozoic: A - B - C - D - E - F - G - H - I - J - K - L - M - N
18. Border of Mesozoic-Paleozoic: A - B - C - D - E - F - G - H - I - J - K - L - M - N
19. Paleozoic: A - B - C - D - E - F - G - H - I - J - K - L - M - N

Complete the Table

20. Refer to Figure 21.16 in the text. To review some of the events that occurred in the geologic past, complete the table below by entering the geologic era (Archean, Proterozoic, Paleozoic, Mesozoic, or Cenozoic) and the *approximate* time in millions of years since the time of the events.

Era	Time	Events
a.		A few reptile lineages give rise to mammals and the dinosaurs.
b.		Formation of Earth's crust, early atmosphere, oceans; chemical evolution leading to the origin of life.
c.		Origin of amphibians.
d.		Origin of animals with hard parts.
e.		Rocks 3.5 billion years old contain fossils of well-developed prokaryotic cells that probably lived in tidal mud flats.
f.		Flowering plants emerge; gymnosperms begin their decline.
g.		In the Carboniferous, there were major radiations of insects and amphibians; gymnosperms present, origin of reptiles.
h.		Oxygen accumulated in the atmosphere.
i.		Before the close of this era, the first photosynthetic bacteria had evolved.
j.		Insects, amphibians, and early reptiles flourished in the swamp forests of the Permian.
k.		Most of the major animal phyla evolved in rather short order.
l.		Dinosaurs ruled.
m.		Origin of aerobic metabolism; origin of protistans, algae, fungi, animals.
n.		Grasslands emerge and serve as new adaptive zones for plant-eating mammals and their predators.
o.		Humans destroy habitats and many species.
p.		The first ice age initiated the first global mass extinction; reef life everywhere collapsed.
q.		The invasion of land begins; small stalked plants establish themselves along muddy margins, and the lobe-finned fishes ancestral to amphibians move onto land.

Self-Quiz

___ 1. Between 4.6 and 4.5 billion years ago, _____.

    a. the Earth was a thin-crusted inferno
    b. the outer regions of the cloud from which our solar system formed was cooling
    c. life originated
    d. the atmosphere was laden with oxygen

___ 2. More than 3.8 billion years ago, _____.

    a. the Earth was a thin-crusted inferno
    b. the outer regions of the cloud from which our solar system formed was cooling
    c. life originated
    d. the atmosphere was laden with oxygen

For questions 3–10, choose from these answers:

    a. Archean
    b. Cenozoic
    c. Mesozoic
    d. Paleozoic
    e. Proterozoic

___ 3. Dinosaurs and gymnosperms were the dominant forms of life during the _____ era.

___ 4. The Alps, Andes, Himalayas, and Cascade Range were born during major reorganization of land masses early in the _____ era.

___ 5. The composition of Earth's atmosphere changed during the _____ era from one that was anaerobic to one that was aerobic.

___ 6. Invertebrates, primitive plants, and primitive vertebrates were the principal groups of organisms on Earth during the _____ era.

___ 7. Before the close of the _____ era, the first photosynthetic bacteria had evolved.

___ 8. The _____ era ended with the greatest of all extinctions, the Permian extinction.

___ 9. Late in the _____ era, flowering plants arose and underwent a major radiation.

___ 10. The _____ era included adaptive zones into which plant-eating mammals and their predators radiated.

Chapter Objectives/Review Questions

Page	Objectives/Questions
(324)	1. Cloudlike remnants of stars are mostly _____, but they also contain water, iron, silicates, hydrogen cyanide, methane, ammonia, formaldehyde, and many other simple inorganic and organic substances.
(324)	2. Describe the formation of the early Earth prior to the formation of the first atmosphere.
(324)	3. Be able to list the probable chemical constituents of the Earth's first atmosphere.
(324)	4. _____ and water were probably not present in the early Earth's atmosphere.
(324–325)	5. Describe experimental evidence provided by Stanley Miller (and others) that the formation of biological molecules from simple precursor molecules might have occurred on the early Earth.
(325)	6. _____ might have been the medium on which condensation reactions yielding complex organic compounds occurred.
(326)	7. During the first _____ years of Earth history, enzymes, ATP, and other molecules could have assembled at the same location; their close association would have promoted chemical interactions—and the beginning of _____ pathways.
(327)	8. Although simple, self-replicating systems of RNA, enzymes, and coenzymes have been created in the laboratory, the chemical ancestors of _____ and DNA remain unknown.
(327)	9. Describe the experiments performed by Sidney Fox that aided understanding of the origin of plasma membranes.
(328)	10. Describe plate tectonic theory and its application to the evolution of Earth's geology as we understand it today.
(329)	11. Be able to list the four geologic eras in proper order, from oldest to youngest.
(329)	12. The first era of the geologic time scale is now called the _____, "the beginning."
(330)	13. List the three evolutionary directions of the original prokaryotic lineage during the early Archean Era.
(330)	14. Populations of some early anaerobic eubacteria utilized the cyclic photosynthetic pathway; their populations formed huge mats called _____.
(331)	15. Describe how the endosymbiosis theory may help to explain the origin of eukaryotic cells.
(330–343)	16. Be able to generally discuss the important geological and biological events occurring throughout the Archean, Proterozoic, Paleozoic, Mesozoic, and Cenozoic Eras.

Integrating and Applying Key Concepts

As Earth becomes increasingly loaded with carbon dioxide and various industrial waste products, how do you think living forms on Earth will evolve to cope with these changes?

Answers

Interactive Exercises

21-I. CONDITIONS ON THE EARLY EARTH
(pp. 324–325)

21-II. EMERGENCE OF THE FIRST LIVING CELLS
(pp. 326–327)

1. f; 2. e; 3. a; 4. c; 5. d; 6. b; 7. f; 8. c; 9. b; 10. a; 11. d;
12. f; 13. e; 14. c; 15. a; 16. b; 17. e; 18. c; 19. a; 20. b.

21-III. LIFE ON A CHANGING GEOLOGIC STAGE
(pp. 328–329)

1. D; 2. C; 3. E; 4. A; 5. B; 6. E; 7. F; 8. I; 9. B; 10. G;
11. H; 12. C; 13. A; 14. J; 15. D.

21-IV. LIFE IN THE ARCHEAN AND PROTEROZOIC
ERAS (pp. 330–333)

1. b; 2. b; 3. a; 4. b; 5. a; 6. a; 7. b; 8. a; 9. a; 10. b;
11. a; 12. b; 13. a; 14. a; 15. b; 16. organelles; 17. gene;
18. plasma; 19. channel; 20. ER; 21. genes; 22. plasmids;
23. prokaryotic; 24. endosymbiosis; 25. eukaryotes;
26. Electron; 27. oxygen; 28. bacteria; 29. ATP; 30. host;
31. incapable; 32. mitochondria; 33. ATP; 34. DNA;
35. chloroplasts; 36. host; 37. eubacteria; 38. protistans.

21-V. LIFE IN THE PALEOZOIC ERA (pp. 334–335)

21-VI. LIFE DURING THE MESOZOIC ERA
(pp. 336–340)

21-VII. LIFE DURING THE CENOZOIC ERA
(pp. 341–343)

1. G; 2. H; 3. E; 4. K; 5. D; 6. L; 7. F; 8. M; 9. A; 10. N;
11. C; 12. J; 13. B; 14. I; 15. G, H; 16. E; 17. D, F, K, L;
18. M; 19. A, B, C, I, J, N; 20. a. Mesozoic, 240–205;
b. Archean, 4,600–3,800; c. Paleozoic, 435–360; d. Paleozoic, 550–500; e. Archean, 4,600–3,800; f. Mesozoic,
135–65; g. Paleozoic, 360–280; h. Proterozoic, 2,500–570;
i. Archean, 3,800–2,500; j. Paleozoic, 290–240; k. Paleozoic, 550–500; l. Mesozoic, 181–65; m. Proterozoic,
2,500–570; n. Cenozoic, 65–1.65; o. Cenozoic, 1.65–present; p. Paleozoic, 440–435; q. Paleozoic, 435–360.

Self-Quiz

1. b; 2. c; 3. c; 4. b; 5. e; 6. d; 7. a; 8. d; 9. c; 10. b.

22

BACTERIA AND VIRUSES

Interactive Exercises

22-I. CHARACTERISTICS OF BACTERIA (pp. 346–349)
BACTERIAL REPRODUCTION (p. 350)

Selected Italicized Words

Escherichia coli, parasitic, saprobic, Gram-positive, Gram-negative, streptococci, staphylococci

Boldfaced, Page-Referenced Terms

(346) microorganisms _____

_____

(346) pathogens _____

_____

(348) archaebacteria _____

_____

(348) eubacteria _____

_____

(348) photoautotrophs _____

_____

(348) photoheterotrophs _____

_____

(348) chemoautotrophs _____

_____

(348) chemoheterotrophs _____

_____

(348) coccus, cocci _____

_____

(348) bacillus, bacilli _____

_____

(348) spiral _____

_____

(349) prokaryotic _____

_____

(349) cell wall _____

_____

(349) peptidoglycan _____

_____

(349) glycocalyx _____

_____

(349) bacterial flagella (sing., flagellum) _____

_____

(349) pilus, pili _____

_____

(350) binary fission _____

_____

(350) plasmid _____

_____

(350) bacterial conjugation _____

_____

Choice

a. archaebacteria   b. chemoautotrophic eubacteria   c. chemoheterotrophic eubacteria
d. photoautotrophic eubacteria   e. photoheterotrophic eubacteria

1. ___ Use CO_2 from the environment as a source of carbon atoms and use electrons, hydrogen, and energy released from reactions between various inorganic substances to assemble chains of carbon (food storage).

2. ___ Use CO_2 and H_2O from the environment as sources of carbon, hydrogen, and oxygen atoms and use sunlight to power the assembly of food storage molecules.

3. ___ Cannot use CO_2 from the environment to construct own carbon chains; instead obtain nutrients from the products, wastes, or remains of other organisms; can break down glucose to pyruvate and follow it with fermentation of some sort or another.

4. ___ Cannot use CO_2 from the environment to construct own cellular molecules but can absorb sunlight and transfer some of that energy to the bonds of ATP; must obtain food molecules (carbon chains) produced by other organisms to construct their own molecules.

5. ___ Cell walls lack peptidoglycan.

Fill-in-the-Blanks

Bacteria are microscopic, (6) _____ cells having one bacterial chromosome and, often, a number of smaller (7) _____. The cells of nearly all bacterial species have a(n) (8) _____ around the plasma membrane and a(n) (9) _____ or slime layer surrounding the cell wall. Typically, the width or length of these cells falls between 1 and 10 (10) (choose one) ❑ millimeters ❑ nanometers ❑ centimeters ❑ micrometers. Most bacteria reproduce by (11) _____ _____. Spherical bacteria are (12) _____, rod-shaped bacteria are (13) _____, and helical bacteria are (14) _____. (15) Gram-_____ bacteria retain the purple stain when washed with alcohol.

22–II. BACTERIAL CLASSIFICATION (p. 351)
ARCHAEBACTERIA (p. 352)
EUBACTERIA (pp. 352–353)
REGARDING THE PATHOGENIC BACTERIA (pp. 354-355)
REGARDING THE "SIMPLE" BACTERIA (p. 355)

Selected Italicized Words

Anabaena, Lactobacillus, Rhizobium, tetanus, botulism, Lyme disease, Azospirillum

Boldfaced, Page-Referenced Terms

(352) methanogens _____

_____

(352) halophiles _____

_____

(352) thermophiles _____

_____

(352) heterocysts _____

_____

(354) endospore _____

_____

(354) antibiotic _____

_____

(355) fruiting bodies _____

_____

Fill-in-the-Blanks

In many respects, the cell structure, metabolism, and nucleic acid sequences are unique to the

(1) _____; for example, none has a cell wall that contains (2) _____. On the other hand,

(3) _____ are far more common than the three rather unusual types of (1). The most common pho-

toautotrophic bacteria are the (4) _____ (also called the blue-green algae). *Anabaena* and others of the

(4) group produce oxygen during photosynthesis. Heterocysts are cells in *Anabaena* that carry out

(5) _____ _____. Many species of chemoautotrophic eubacteria affect the global cycling of nitro-

gen, sulfur, (6) _____, and other nutrients. Sugarcane and corn plants benefit from a (7) _____-

fixing spirochete, *Azospirillum*. Beans and other legumes benefit from the (5) activities of (8) _____,

which dwells in their roots.

   When environmental conditions become adverse, many bacteria form (9) _____, which resist mois-

ture loss, irradiation, disinfectants, and even acids. Formation of (9) by bacteria can kill humans if they

enter the food supply and are not killed by high temperature and high (10) _____. Anaerobic bacteria

can live in canned food, reproduce, and produce deadly toxins. Two examples of pathogenic (disease-caus-

ing) bacteria that form (9) harmful to humans are (11) _____ _____ and (12) _____

_____. (13) _____ _____ may be the most common tick-borne disease in the United States

by now; tick bites deliver the (14) _____ from one host to another.

   Bacterial behavior depends on (15) _____ _____, which change shape when they absorb or

connect with chemical compounds. Cyanobacteria require (16) _____ as their source of energy that

drives their metabolic activities. Most of the world's bacteria are (17) (choose one) ❑ producers

❑ consumers ❑ decomposers, so we think of them as "good" heterotrophs. Antibiotics such as

(18) _____ block protein synthesis in their target cells; other antibiotics such as (19) _____ disrupt

the formation of cell walls in their target cells. *Escherichia coli*, which dwell in our gut, synthesize vitamin

(20) ____ and substances useful in digesting fats.

Matching

Match each of the items below with a lowercase letter designating its principal bacterial group and an uppercase letter denoting its best descriptor from the right-hand column.

a. Archaebacteria   b. Chemoautotrophic eubacteria   c. Chemoheterotrophic eubacteria
d. Photoautotrophic eubacteria   e. Photoheterotrophic eubacteria

21. ___, ___ *Anabaena*

22. ___, ___ *Bacillus, Clostridium*

23. ___, ___ *Escherichia coli*

24. ___, ___ *Halobacterium*

25. ___, ___ *Lactobacillus*

26. ___, ___ *Methanobacterium*

27. ___, ___ *Nitrobacter, Nitrosomonas*

28. ___, ___ *Rhizobium, Agrobacterium*

29. ___, ___ *Rhodospirillum*

30. ___, ___ *Salmonella*

31. ___, ___ *Spirochaeta, Treponema*

32. ___, ___ *Staphylococcus, Streptococcus*

33. ___, ___ *Streptomyces, Actinomyces*

34. ___, ___ *Thermoplasma, Sulfolobus*

A. Live in anaerobic sediments of lakes and in animal gut; chemosynthetic; used in sewage treatment facilities
B. Purple; generally in anaerobic sediments of lakes or ponds; do not produce oxygen, do not use water as a source of electrons
C. Endospore-forming rods and cocci that live in the soil and in the animal gut; some major pathogens
D. Gram-positive cocci that live in the soil and in the skin and mucous membranes of animals; some major pathogens
E. Gram-positive nonsporulating rods that ferment plant and animal material; some are important in dairy industry; others contaminate milk, cheese
F. In acidic soil, hot springs, hydrothermal vents on seafloor; may use sulfur as a source of electrons for ATP formation
G. Live in extremely salty water; have a unique form of photosynthesis
H. Gram-negative aerobic rods and cocci that live in soil or aquatic habitats or are parasites of animals or plants or both; some fix nitrogen
I. Nitrifying bacteria that live in the soil, fresh water, and marine habitats; play a major role in the nitrogen cycle
J. Gram-negative anaerobic rod that inhabits the human colon, where it produces vitamin K
K. Major gram-negative pathogens of the human gut that cause specific types of food poisoning
L. Mostly in lakes and ponds; cyanobacteria; produce O_2 from water as an electron donor
M. Major producer of antibiotics; an actinomycete that lives in soil and some aquatic habitats
N. Helically coiled, motile parasites of animals; some are major pathogens

22-III. THE VIRUSES (pp. 356–357)
VIRAL MULTIPLICATION CYCLES (pp. 358–359)

Selected Italicized Words

helical, polyhedral, enveloped, and complex viruses; "host"; sporadic, endemic

Boldfaced, Page-Referenced Terms

(356) virus _____

_____

(356) bacteriophages _____

_____

(358) lytic pathway _____

_____

(358) lysogenic pathway _____

_____

(359) infection _____

_____

(359) disease _____

_____

(359) epidemic _____

_____

(359) pandemic _____

_____

Short Answer

1. a. State the principal characteristics of viruses (p. 356). _____

_____

 b. Describe the structure of viruses (pp. 356–357). _____

_____

 c. Distinguish between the ways viruses replicate themselves (pp. 358–359). _____

_____

2. a. List five specific viruses that cause human illness (pp. 356–357). _____

_____

 b. Describe how each virus in (a) does its dirty work (pp. 358–359). _____

_____

_____

_____

_____

_____

Fill-in-the-Blanks

A(n) (3) _____ is a noncellular, nonliving infectious agent, each of which consists of a central (4) _____ _____ core surrounded by a protective (5) _____ _____. (6) _____ contain the blueprints for making more of themselves but cannot carry on metabolic activities. Chickenpox and shingles are two infections caused by DNA viruses from the (7) _____ category. Naked strands or circles of RNA that lack a protein coat are called (8) _____. (9) _____ are RNA viruses that infect animal cells, cause AIDS, and follow (10) _____ pathways of replication. During a period of (11) _____, viral genes remain inactive inside the host cell and any of its descendants.

    (12) _____ are the usual units of measurement with which to measure viruses, while microbiologists measure bacteria and protistans in terms of (13) _____. A bacterium 86 micrometers in length is (14) _____ nanometers long. Pathogenic protein particles are called (15) _____.

Identification

Identify the virus that causes the following illnesses by writing the name of the virus in the blank preceding the disease. Then tell whether it is a DNA virus or an RNA virus.

_____ _____ 16. Common colds

_____ _____ 17. AIDS, leukemia

_____ _____ 18. Cold sores, chickenpox

Matching

Match each item below with the correct lettered description.

19. ___ antibiotic

20. ___ antiviral drug

21. ___ bacteriophage

22. ___ endemic

23. ___ epidemic

24. ___ lysogenic pathway

25. ___ lytic pathway

26. ___ microorganism

27. ___ pathogen

28. ___ sporadic

29. ___ vector

30. ___ virus

A. An agent that transports to other creatures or temporarily houses pathogens coming from an infected organism
B. Disease that breaks out irregularly, affects few organisms
C. Chemical substance that interferes with gene expression or other normal functions of bacteria
D. Disease abruptly spreads through large portions of a population
E. Disease that occurs continuously but is localized to a relatively small portion of the population
F. Acyclovir and AZT, for example
G. Any organism too small to be seen without a microscope
H. A virus that infects a bacterium
I. Damage and destruction to host cells occurs quickly
J. Noncellular infectious agent that must invade a living cell in order to reproduce itself
K. Viral nucleic acid is integrated into the nucleic acid of the host cell and replicated during this time
L. Any disease-causing organism or agent

Self-Quiz

Multiple Choice

___ 1. Which of the following diseases is *not* caused by a virus?

    a. smallpox
    b. polio
    c. influenza
    d. syphilis

___ 2. Bacteriophages are _____.

    a. viruses that parasitize bacteria
    b. bacteria that parasitize viruses
    c. bacteria that phagocytize viruses
    d. composed of a protein core surrounded by a nucleic acid coat

Matching

Match all applicable letters with the appropriate terms. A letter may be used more than once, and a blank may contain more than one letter.

3. _____ *Anabaena, Nostoc*

4. _____ *Clostridium botulinum*

5. _____ *Escherichia coli*

6. _____ *Herpes simplex*

7. _____ HIV

8. _____ *Lactobacillus*

9. _____ *Staphylococcus*

A. Bacteria
B. Virus
C. Cyanobacteria
D. Gram-positive eubacteria
E. Cause cold sores and a type of venereal disease
F. Associated with AIDS, ARC

Matching

Match the pictures on the opposite page with the names below.

10. ___

11. ___

12. ___

13. ___

14. ___

15. ___

A. *Bacillus*
B. bacteriophage
C. *Clostridium tetani*
D. *Cyanobacterium*
E. *Herpes simplex*
F. HIV

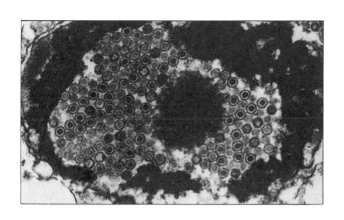

10.

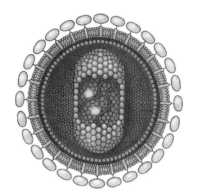

11.

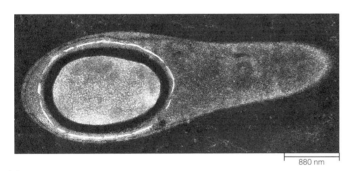

12.

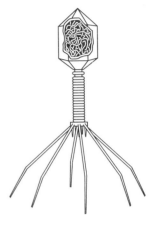

13.

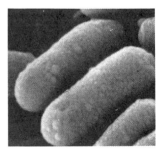

14.

880 nm

15.

Crossword Puzzle—Bacteria and Viruses

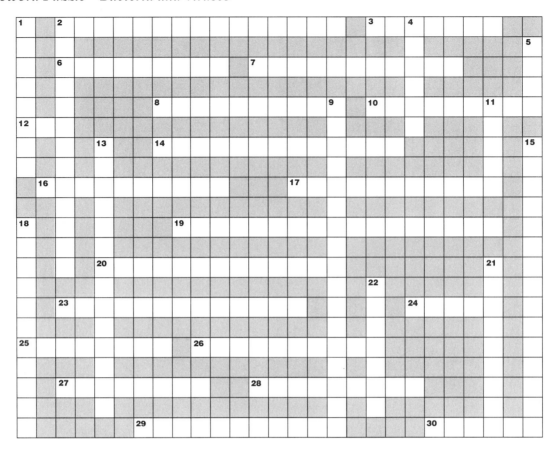

ACROSS

2. _____ can assemble their own food molecules by using light energy to join together carbon, hydrogen, and oxygen atoms.
3. Binary _____ is the process most bacteria use to reproduce.
6. Disease-causing agent.
7. Bacterial _____ transfers a plasmid from a donor to a recipient cell.
8. A chemical substance produced by *actinomycetes* and other microorganisms that kills or inhibits the growth of other microorganisms.
10. A whip-like organelle of locomotion.
12. Boy, young man.
14. Group that includes blue-green algae.
16. Sticky mesh of polysaccharides, polypeptides, or both; alternate name of either a capsule or slime layer.
17. _____ cells existed before the origin of the nucleus and of eukaryotic cells.
19. Cannot synthesize their own food molecules; can use sunlight as an energy source to make ATP, but cannot fix CO_2 from their surroundings; must obtain carbon compounds from other organisms.
20. Virus that infects bacterial host cells.
23. Any organism that can be seen most clearly by using a microscope.
24. A short filamentous protein that projects above a cell wall; helps to tether a bacterium to another bacterium or to a surface.
25. A(n) _____ occurs when, over the entire world, a disease spreads through large portions of many populations for a limited period.
26. _____ are structures that resist heat, drying, boiling, and radiation; can give rise to new bacterial cells.
27. A(n) _____ occurs when a disease abruptly spreads through large portions of a single population for a limited period.
28. Salt-loving archaebacterium.
29. Methane-producing archaebacteria.
30. Spherical bacterium.

DOWN

1. Rod-shaped bacterium.
2. Substance not in archaebacterial cell walls.
4. Helical.
5. _____-positive bacterial cell walls have an affinity for crystal violet stain.

9. _____ include Earth's major decomposers, nitrogen-fixers, and fermenters of milk.
11. The _____ pathway of bacteriophage multiplication assembles new viral particles very soon after infecting the host cell; no integration into host cell's DNA.
13. This group includes the most ancient of Earth's bacteria.

15. Bacteria that can synthesize their own food molecules by using energy released from specific chemical reactions.
18. Heat-tolerant bacterial type.
21. The _____ pathway of bacteriophage multiplication includes integrating viral DNA into its bacterial host's chromosome.
22. Small loop of DNA in addition to the main bacterial "chromosome."

Chapter Objectives/Review Questions

This section lists general and detailed chapter objectives that can be used as review questions. You can make maximum use of these items by writing answers on a separate sheet of paper. To check for accuracy, compare your answers with information given in the chapter or glossary.

Page	*Objectives/Questions*
(348)	1. Distinguish chemoautotrophs from photoautotrophs.
(348–349)	2. Describe the principal body forms of monerans (inside and outside).
(350)	3. Explain how, with no nucleus or few, if any, membrane-bound organelles, bacteria reproduce themselves and obtain energy to carry on metabolism.
(352–355)	4. State the ways in which archaebacteria differ from eubacteria.

Integrating and Applying Key Concepts

The textbook (Figure 50.14) identifies natural gas as a nonrenewable fuel resource, yet there is a group of archaebacteria that produce methane, the burning of which can serve as a fuel for heating and cooking. Recall or imagine how these bacteria could be incorporated into a system that could serve human societies by generating methane in a cycle that is renewable. Why did your text categorize natural gas as a nonrenewable resource? Is methane a constituent of natural gas? Why or why not?

Answers

Interactive Exercises

22-I. CHARACTERISTICS OF BACTERIA
(pp. 348–349)
BACTERIAL REPRODUCTION (p. 350)
1. b; 2. d; 3. c; 4. e; 5. a; 6. prokaryotic; 7. plasmids; 8. wall; 9. capsule; 10. micrometers; 11. binary fission; 12. cocci; 13. bacilli; 14. spiral; 15. positive.

22-II. BACTERIAL CLASSIFICATION (p. 351)
ARCHAEBACTERIA (p. 352)
EUBACTERIA (pp. 352–353)
1. Archaebacteria; 2. peptidoglycan; 3. eubacteria; 4. cyanobacteria; 5. nitrogen fixation; 6. phosphorus; 7. nitrogen; 8. *Rhizobium*; 9. endospores; 10. pressure; 11. *Clostridium botulinum* (*C. tetani*); 12. *Clostridium tetani* (*C. botulinum*); 13. Lyme disease; 14. spirochete; 15. membrane receptors; 16. sunlight; 17. decomposers; 18. streptomycins; 19. penicillins; 20. K; 21. d, L; 22. c,

C; 23. c, J; 24. a, G; 25. c, E; 26. a, A; 27. b, I; 28. c, H; 29. e, B; 30. c, K; 31. c, N; 32. c, D; 33. c, M; 34. a, F.

22-III. THE VIRUSES (pp. 356–357)
VIRAL MULTIPLICATION CYCLES (pp. 358–359)
1. a. Nonliving, infectious agents, smaller than the smallest cells; require living cells to act as hosts for their replication; not acted upon by antibiotics; b. The core can be DNA or RNA; the capsid can be protein or lipid or both; c. Bacteriophage viruses may use the lytic pathway, in which the virus quickly subdues the host cells, replicates itself, and releases descendants as the cell undergoes lysis; or they may use a temperate pathway, in which viral genes remain inactive inside the host cell during a period of latency, which may be a long time, before activation and lysis; 2. There can be multiple answers (see Table 22.3 in text); a. Possible answers include *Herpes simplex* (a herpesvirus), Varicella-zoster (a herpesvirus), rhinovirus (a picornavirus), poliovirus (an enterovirus of

the picorna group), and HIV (a retrovirus); b. *Herpes simplex*: DNA virus. Initial infection is a lytic cycle that causes herpes (sores) on mucous membranes on mouth or genitals. Recurrent infections are temperate. Host cells are in nerves and skin. No immunity. No cure. Varicella-zoster: DNA virus. Initial infection is a lytic cycle that causes sores on skin. Generally, immunity is conferred by one infection, but in some people subsequent infections follow temperate cycles and cause "shingles." Rhinovirus: RNA virus. Causes the common cold. Host cells are generally mucus-producing cells of respiratory tract. Poliovirus: RNA virus. Causes polio. Host cells are in motor nerves that lead to the diaphragm and other important muscles. Destruction of these nerve cells causes paralysis that may be temporary or permanent. Recurrences can occur. Immunize your children! HIV: RNA virus. Host cells are specific white blood cells. Temperate cycle has a latency period that may last longer than a year before host tests positive for HIV. As white blood cells are destroyed, the host's immune system is progressively destroyed (AIDS). No cure exists; 3. virus; 4. nucleic acid; 5. protein coat (viral capsid);

6. Viruses; 7. *Herpesvirus* (or *Varicella*); 8. viroids; 9. Retroviruses (HIV); 10. temperate; 11. latency; 12. Nanometers; 13. micrometers; 14. 86,000; 15. prions; 16. Rhinoviruses, RNA; 17. Retroviruses, RNA; 18. Herpesviruses, DNA; 19. C; 20. F; 21. H; 22. E; 23. D; 24. K; 25. I; 26. G; 27. L; 28. B; 29. A; 30. J.

Self-Quiz
1. d; 2. a; 3. A, C; 4. A; 5. A; 6. B, E; 7. B, F; 8. A, D; 9. A, D; 10. E; 11. F; 12. D; 13. B; 14. C; 15. A.

Integrating and Applying Key Concepts
The text identifies natural gas as a fossil fuel along with coal and petroleum because most supplies of it are produced as a result of drilling and mining rather than through methane generation by bacteria.
$$2\ CH_4 + 4\ O_2 \rightarrow 4\ H_2O + 2\ CO_2 + heat$$ (for cooking and heating)
Plants then photosynthesize CO_2 to produce food to feed carbon-based organismal flesh. Methanogens generate methane (CH_4) from dead organisms or dung.

1 B		2 P	H	O	T	O	A	U	T	O	T	R	O	P	H	S		3 F	I	S	S	I	O	N		
A		E																	P						5 G	
C		6 P	A	T	H	O	G	E	N	S		7 C	O	N	J	U	G	A	T	I	O	N			R	
I		T																R							A	
L		I					8 A	N	T	I	B	I	O	T	I	9 C		10 F	L	A	G	E	L	11 L	U	M
12 L	A	D										H			L						T		15 C			
U		O		13 A		14 C	Y	A	N	O	B	A	C	T	E	R	I	A				I		H		
S		G		R								M														
	16 G	L	Y	C	O	C	A	L	Y	X				17 P	R	O	K	A	R	Y	O	T	I	C	E	
		Y		H										H								M				
18 T		C		A				19 P	H	O	T	O	H	E	T	E	R	O	T	R	O	P	H	S	O	
H		A		E								T											A			
E		N		20 B	A	C	T	E	R	I	O	P	H	A	G	E						21 L		U		
R				A								R		22 P							Y		T			
M		23 M	I	C	R	O	O	R	G	A	N	I	S	M		O		L		24 P	I	L	U	S	O	
O				T								T		A							O		T			
25 P	A	N	D	E	M	I	C		26 E	N	D	O	S	P	O	R	E	S			G		R			
H				R								O		M							E		O			
I		27 E	P	I	D	E	M	I	C		28 H	A	L	O	P	H	I	L	E			N		P		
L				A								H		D							I		H			
E				29 M	E	T	H	A	N	O	G	E	N	S				30 C	O	C	C	U	S			

23

PROTISTANS

Interactive Exercises

23-I. WHAT IS A PROTISTAN? (pp. 362–363)
CHYTRIDS AND WATER MOLDS (p. 364)
SLIME MOLDS (pp. 364–365)

Selected Italicized Words

Saprolegnia, Phytophthora infestans, late blight, Dictyostelium discoideum, cellular slime molds, plasmodial slime molds

Boldfaced, Page-Referenced Terms

(364) chytrids _____

_____

(364) mycelium _____

_____

(364) water molds _____

_____

(364) slime molds _____

_____

Choice

1. For each structure, indicate with a "P" if it is a prokaryotic (bacterial) characteristic, and indicate with an "E" if it is a eukaryotic characteristic.

a. double-membraned nucleus	
b. mitochondria present	
c. reproduce by binary fission	
d. engage in mitosis	
e. circular chromosome present	
f. endoplasmic reticulum present	
g. cilia or flagella with 9 + 2 core	

More Choice

2. For each group below, indicate with a "+" if it has chloroplasts and a "–" if members of that group lack chloroplasts and the ability to photosynthesize.

a. brown algae		f. protozoans	
b. chytrids		g. red algae	
c. chrysophytes		h. slime molds	
d. dinoflagellates		i. sporozoans	
e. green algae		j. water molds	

Fill-in-the-Blanks

(3) _____ and (4) _____ _____ are the only fungi that produce motile spores; this is a primitive trait that may resemble ancestral fungi that lived several hundred million years ago in watery habitats. Water molds are only distantly related to other fungi and are seen to have evolved from (5) _____ algae. The cells of some (6) _____ _____ differentiate and form (7) _____ _____: stalked structures bearing spores at their tips; in this manner, they resemble (8) _____. Some slime-mold spores resemble the spores of many (9) _____. Slime molds also spend part of their life creeping about like (10)_____ and engulfing food.

Many fungi have cells merged lengthwise, forming tubes that have thin transparent walls reinforced with (11) _____; so do some (12) _____. Some fungal species, such as late blight, are (13) (choose one) ❏ parasitic ❏ saprophytic. The vegetative body of most true fungi is a (14) _____, which is a mesh of branched, tubular filaments. Their metabolic activities enable them to act as (15) _____ in ecosystems. Fungi secrete (16) _____ into their surroundings, where large organic molecules are broken down into smaller components that the fungal cells then absorb. Although most fungi are (17) _____ (obtaining their nutrients from nonliving organic matter), some are (18) _____ and get their nutrients directly from their living host's tissues. A common form of asexual reproduction is the growth of a new fungal body from a(n) (19) _____.

23-II. CONCERNING THE ANIMAL-LIKE PROTISTANS (p. 366)
AMOEBOID PROTOZOANS (pp. 366–367)
CILIATED PROTOZOANS (p. 367)
FLAGELLATED PROTOZOANS (p. 368)
SPOROZOANS (pp. 368–369)

Selected Italicized Words

Amoeba proteus, Entamoeba histolytica, amoebic dysentery, Paramecium, African sleeping sickness, Chagas disease, toxoplasmosis, malaria, Trypanosoma brucei, Trichomonas vaginalis, Plasmodium

Boldfaced, Page-Referenced Terms

(366) protozoans _____

_____

(366) cysts _____

_____

(366) amoeboid protozoans _____

_____

(366) amoebas _____

_____

(367) foraminiferans _____

_____

(367) heliozoans _____

_____

(367) radiolarians _____

_____

(367) ciliated protozoans _____

_____

(367) contractile vacuoles _____

_____

(367) protozoan conjugation _____

_____

(368) flagellated protozoans _____

_____

(368) sporozoan _____

_____

Fill-in-the-Blanks

Amoebas move by sending out (1) _____, which surround food and engulf it. (2) _____ secrete a hard exterior covering of calcareous material that is peppered with tiny holes through which sticky, food-trapping pseudopods extend. Needle-like (3) _____ often support the pseudopods. Accumulated shells of (4) _____, which generally have a skeleton of silica (glass), and foraminiferans are key components of many oceanic sediments.

Examples of flagellated protozoans that are parasitic include the (5) _____, two species of which cause African sleeping sickness and Chagas disease.

Paramecium is a ciliate that lives in (6) _____ environments and depends on (7) _____ _____ for eliminating the excess water constantly flowing into the cell. *Paramecium* has a (8) _____, a cavity that opens to the external watery world. Once inside the cavity, food particles become enclosed in (9) _____-_____ _____, where digestion takes place.

(10) _____ is a famous sporozoan that causes malaria. When a particular (11) _____ draws blood from an infected individual, (12) _____ of the parasite fuse to form zygotes, which eventually develop within the mosquito.

Matching

Put as many letters in each blank as are applicable.

13. _____ *Entamoeba histolytica*
14. _____ foraminiferans
15. _____ *Ptychodiscus brevis*
16. _____ *Paramecium*
17. _____ *Plasmodium*
18. _____ *Trichomonas vaginalis*
19. _____ *Trypanosoma brucei*

A. Ciliophora
B. Mastigophora
C. Sarcodina
D. Sporozoa
E. Amoeboid protozoans
F. Flagellated protozoans
G. African sleeping sickness
H. Malaria
I. Amoebic dysentery
J. Red tide
K. Primary component of many ocean sediments

Matching

Match the pictures below with the names below.

20. ___
21. ___
22. ___
23. ___
24. ___

A. *Didinium*, a ciliate
B. *Euglena*
C. Flagellated protozoans
D. Foraminiferans
E. *Paramecium*

20.

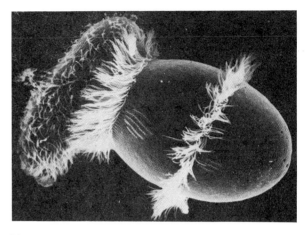

21.

22.

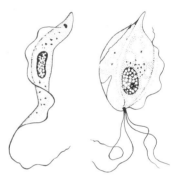

23.

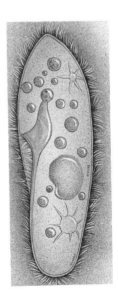

24.

23-III. EUGLENOIDS (p. 370)
CONCERNING "THE ALGAE" (p. 370)
CHRYSOPHYTES AND DINOFLAGELLATES (p. 371)
RED ALGAE (p. 372)
BROWN ALGAE (pp. 372–373)
GREEN ALGAE (pp. 374–375)
ON THE ROAD TO MULTICELLULARITY (p. 376)

Selected Italicized Words

Euglena, algae, Euchema, Postelsia, Macrocystis, Sargassum, Codium, Acetabularia, Volvox, Chlamydomonas, Spirogyra, Trichoplax

Boldfaced, Page-Referenced Terms

(370) euglenoids _____

_____

(370) phytoplankton _____

_____

(370) zooplankton _____

_____

(371) golden algae _____

_____

(371) diatoms _____

_____

(371) yellow-green algae _____

_____

(371) dinoflagellates _____

_____

(371) red tides _____

_____

(372) red algae _____

_____

(372) agar _____

_____

(372) carrageenan _____

_____

(372) brown algae _____

_____

(373) holdfasts _____

_____

(373) stipes _____

(373) blades _____

(373) algin _____

(374) green algae _____

(376) multicellular organism _____

Fill-in-the-Blanks

Euglenoids contain (1) _____,which enable them to carry out photosyntheses. An (2) _____ of
carotenoid pigment granules partly shields a light-sensitive receptor and enables *Euglena* to remain where
light is optimal for its activities. Some strains of *Euglena* can be converted from photosynthetic, chloroplast-
containing forms to strains that are (3) _____.

The term (4) "_____" no longer has formal classification significance, because organisms once
lumped under that term are now assigned to different kingdoms. (5) _____ include 600 species of
"yellow-green algae," about 500 species of "golden algae," and more than 5,600 existing species of golden-
brown (6) _____. Photosynthetic chrysophytes contain xanthophylls and (7) _____; these pigments mask the green color of chlorophyll in golden algae and diatoms. Diatom cells have external thin,
overlapping "shells" of (8) _____ that fit together like a pill box. Each year, 270,000 metric tons of
(9) _____ _____ are extracted from a quarry near Lompoc, California, and are used to make abra-
sives, (10) _____ materials, and insulating materials. Dinoflagellates are photosynthetic members of
marine (11) _____ and freshwater ecosystems; some forms are also heterotrophic. (12) _____
undergo explosive population growth and color the seas red or brown, causing a red tide that may kill hundreds or thousands of fish and, occasionally, people.

Several species of red algae secrete (13) _____ (used in culture media) as part of their cell walls.
Most red algae live in (14) _____ habitats. Some red algae have stonelike cell (15) _____, participate in coral reef building, and are major producers. The (16) _____ algae live offshore or in intertidal
zones and have many representatives with large sporophytes known as kelps; some species produce
(17) _____, a valuable thickening agent. Green algae are thought to be ancestral to more complex
plants because they have the same types and proportions of (18) _____ pigments, have (19) _____
in their cell walls, and store their carbohydrates as (20) _____.

Complete the Table

21. Complete the table below.

Type of alga	Typical pigments	Probably evolved from	Uses by humans	Representatives
Red algae (Rhodophyta)	a.	b.	c.	d.
Brown algae (Phaeophyta)	e.	f.	g.	h.
Green algae (Chlorophyta)	i.	j.	k.	l.

Label-Match

Identify each indicated part of the illustration on the opposite page by entering its name in the appropriate numbered blank. Choose from the following terms: cytoplasmic fusion, gamete production, asexual reproduction, resistant zygote, fertilization, zygote, meiosis and germination, spore mitosis, gametes meet. Complete the exercise by matching each part with a description from the list below, entering the correct letter in the parentheses following each label.

22. _____ (   )

23. _____ _____ (   )

24. _____ and _____ (   )

25. _____ _____ (   )

26. _____ _____ (   )

27. _____ _____ (   )

28. _____ _____ (   )

29. _____ (   )

A. Fusion of two gametes of different mating types
B. A device to survive unfavorable environmental conditions
C. More spore copies are produced
D. Fusion of two haploid nuclei
E. Haploid cells form smaller haploid gametes when nitrogen levels are low
F. Formed after fertilization
G. Two haploid gametes coming together
H. Reduction of the chromosome number

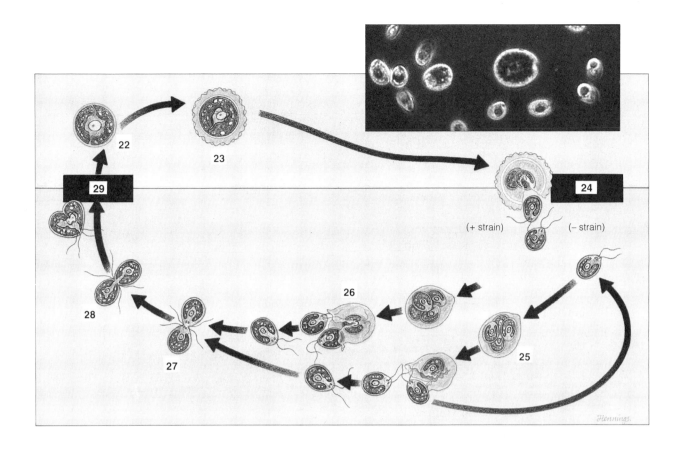

Self-Quiz

Multiple Choice

___ 1. Many biologists believe that chloroplasts are descendants of _____ that were able to survive symbiotically within a predatory host cell.

    a. aerobic bacteria with "extra" cytochromes
    b. endosymbiont prokaryotic autotrophs
    c. dinoflagellates
    d. bacterial heterotrophs

___ 2. Which of the following specialized structures is not correctly paired with a function?

    a. gullet—ingestion
    b. cilia—food gathering
    c. contractile vacuole—digestion
    d. anal pore—waste elimination

___ 3. _____ form a group of related organisms that suggests how lineages of single-celled organisms might have progressed through a colonial stage to multicellularity.

    a. Ciliates such as *Paramecium*, *Didinium*, and *Vorticella*
    b. Volvocales such as *Chlamydomonas* and *Volvox*
    c. Golden algae and diatoms
    d. Sporozoans such as *Plasmodium*, *Neisseria*, and the spirochetes

___ 4. Population "blooms" of _____ cause "red tides" and extensive fish kills.

    a. *Euglena*
    b. specific dinoflagellates
    c. diatoms
    d. *Plasmodium*

___ 5. Exposure to free oxygen is lethal for all _____.

    a. obligate anaerobes
    b. bacterial heterotrophs
    c. chemosynthetic autotrophs
    d. facultative anaerobes

___ 6. Which of the following protists does *not* cause great misery to humans?

a. *Dictyostelium discoideum*
b. *Entamoeba histolytica*
c. *Plasmodium*
d. *Trypanosoma brucei*

___ 7. For an organism to be considered truly multicellular, _____.

a. its cells must be heterotrophic
b. there must be division of labor and cellular specialization
c. the organism cannot be parasitic
d. the organism must at least be motile

___ 8. Red, brown, and green "algae" are found in the kingdom _____.

a. Plantae
b. Monera
c. Protista
d. all of the above

___ 9. Red algae _____.

a. are primarily marine organisms

b. are thought to have developed from green algae
c. contain xanthophyll as their main accessory pigments
d. all of the above

___ 10. Stemlike structure, leaflike blades, and gas-filled floats are found in the species of _____.

a. red algae
b. brown algae
c. bryophytes
d. green algae

___ 11. Because of pigmentation, cellulose walls, and starch storage similarities, the _____ algae are thought to be ancestral to more complex plants.

a. red
b. brown
c. blue-green
d. green

Matching

Match all applicable letters with the appropriate terms. A letter may be used more than once, and a blank may contain more than one letter.

12. _____ *Amoeba proteus*

13. _____ diatoms

14. _____ *Dictyostelium*

15. _____ foraminifera

16. _____ *Ptychodiscus brevis (red tide)*

17. _____ *Paramecium*

18. _____ *Plasmodium*

19. _____ *Volvox*

A. Protista
B. Slime mold
C. Photosynthetic flagellates
D. Dinoflagellates
E. Obtain food by using pseudopodia
F. Causes malaria
G. A sporozoan
H. A ciliate
I. Live in "glass" houses
J. Live in hardened shells that have thousands of tiny holes, through which pseudopods protrude

Matching

20. ___   A. *Amoeba proteus*
21. ___   B. diatoms
22. ___   C. foraminiferans
23. ___   D. *Paramecium*

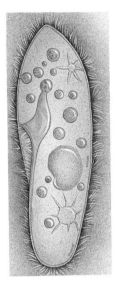

21.

20.

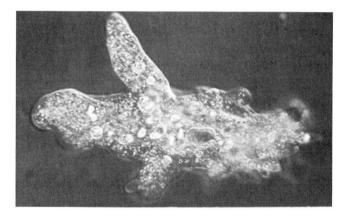

23.

22.

Chapter Objectives/Review Questions

This section lists general and detailed chapter objectives that can be used as review questions. You can make maximum use of these items by writing answers on a separate sheet of paper. Fill in answers where blanks are provided. To check for accuracy, compare your answers with information given in the chapter or glossary.

Page	Objectives/Questions
(362–363)	1. Trace a sequence of events that may have transformed heterotrophic prokaryotes into photosynthetic eukaryotes.
(366–367)	2. State the principal characteristics of the amoebas, radiolarians, and foraminiferans. Indicate how they generally move from one place to another and how they obtain food.

(367)   3. List the features common to most ciliated protozoans.

(368)   4. Two flagellated protozoans that cause human misery are _____ _____ and _____ _____.

(368–369)   5. Characterize the sporozoan group, identify the group's most prominent representative, and describe the life cycle of that organism.

(370)   6. Explain how heterotrophic protistans could have acquired the capacity for photosynthesis, and state the evidence to support your explanation.

(371)   7. How do golden algae resemble diatoms?

(371)   8. Explain what causes red tides.

(372–377)   9. State the outstanding characteristics of organisms of the red, brown, and green algae divisions.

(376)   10. Outline a possible route from unicellularity to a multicellular state with division of labor that could have been traveled by protistan ancestors.

Integrating and Applying Key Concepts

Explain why totally submerged aquatic plants that live in deep water never developed heterosporous life cycles.

Answers

Interactive Exercises

23-I. WHAT IS A PROTISTAN? (p. 363)
  CHYTRIDS AND WATER MOLDS (p. 364)
  SLIME MOLDS (pp. 364–365)
1. a. E; b. E; c. P; d. E; e. P; f. E; g. E; 2. (+): a, c, d, e, g; (–) b, f, h, i, j; 3. Chytrids; 4. water molds (Oomycetes); 5. red; 6. slime molds; 7. fruiting bodies; 8. myxobacteria (monerans); 9. fungi; 10. animals; 11. chitin; 12. chytrids; 13. parasitic; 14. mycelium; 15. decomposers; 16. enzymes; 17. saprobes; 18. parasites; 19. spore.

23-II. CONCERNING THE ANIMAL-LIKE PROTISTANS (p. 366)
  AMOEBOID PROTOZOANS (pp. 366–367)
  CILIATED PROTOZOANS (p. 367)
  FLAGELLATED PROTOZOANS (p. 368)
  SPOROZOANS (pp. 368–369)
1. pseudopods; 2. Foraminiferans; 3. spines; 4. radiolarians; 5. trypanosomes; 6. freshwater; 7. contractile vacuoles; 8. gullet; 9. enzyme-filled vesicles; 10. *Plasmodium*; 11. mosquito; 12. gametes; 13. C, E, I; 14. C, E, K; 15. B, F, J; 16. A; 17. D, H; 18. B, F; 19. B, F, G; 20. B; 21. A; 22. D; 23. C; 24. E.

23-III. EUGLENOIDS (p. 370)
  CONCERNING "THE ALGAE" (p. 370)
  CHRYSOPHYTES AND DINOFLAGELLATES (p. 371)
  RED ALGAE (p. 372)
  BROWN ALGAE (pp. 372–373)
  GREEN ALGAE (pp. 374–375)
  ON THE ROAD TO MULTICELLULARITY (p. 376)
1. chloroplasts; 2. "eyespot"; 3. heterotrophic; 4. algae; 5. Chrysophytes; 6. diatoms; 7. fucoxanthin; 8. silica

(glass); 9. diatoma shells; 10. filtering; 11. phytoplankton; 12. Dinoflagellates; 13. agar; 14. marine; 15. walls; 16. brown; 17. algin; 18. photosynthetic; 19. cellulose; 20. starch; 21. a. chlorophyll *a*, phycobilins; b. cyanobacteria; c. agar, used as a moisture-preserving agent and culture medium; carrageenan is a stabilizer of emulsions; d. *Bonnemaisonia, Euchema*; e. chlorophylls *a* and *b*, various carotenoids such as fucoxanthin; f. chrysophytes; g. algin, used as a thickener, emulsifier, and stabilizer of foods, cosmetics, medicines, paper, and floor polish; also are sources of mineral salts and fertilizer; h. *Postelsia* (sea palm), *Sargassum, Laminaria*; i. chlorophylls *a* and *b*; j. a heterotrophic prokaryote fusing with endosymbiont autotrophic prokaryote; k. chlorophytes form much of the phytoplankton base of many food webs that support humans; l. *Volvox, Ulva, Spirogyra, Chlamydomonas*; 22. zygote, F; 23. resistant zygote, B; 24. meiosis and germination, H; 25. asexual reproduction, C; 26. gamete production, E; 27. gametes meet, G; 28. cytoplasmic fusion, A; 29. fertilization, D.

Self-Quiz

1. b; 2. c; 3. b; 4. b; 5. a; 6. a; 7. b; 8. c; 9. a; 10. b; 11. d; 12. A, E; 13. A, I; 14. A, B; 15. A, E, J; 16. A, C, D; 17. A, H; 18. A, F, G; 19. A, C; 20. B; 21. D; 22. C; 23. A.

24

FUNGI

Interactive Exercises

24-I. MAJOR GROUPS OF FUNGI (pp. 378–379)
24-II. CHARACTERISTICS OF FUNGI (p. 380)

Boldfaced, Page-Referenced Terms

(380) saprobes _____

_____

(380) parasites _____

_____

(380) mycelium _____

_____

(380) hypha _____

_____

(380) spores _____

_____

(380) sporangia _____

_____

(380) gametangia _____

_____

Matching

Choose the single most appropriate answer.

1. ___ saprobes
2. ___ parasites
3. ___ mycelium
4. ___ hypha
5. ___ spores
6. ___ sporangia
7. ___ gametangia

A. Nonmotile reproductive cells or multicelled structures; often walled and germinate following dispersal from the parent body
B. A mesh of branching fungal filaments that grows over and into organic matter, secretes digestive enzymes, and functions in food absorption
C. Gamete-producing structures
D. Fungi that obtain nutrients from nonliving organic matter and so cause its decay
E. Reproductive structures that produce asexual spores
F. Fungi that extract nutrients from tissues of a living host
G. Each filament in a mycelium; consists of tube-shaped cells with chitin-reinforced walls

Labeling

Identify each generalized fungi life cycle stage below by writing its name in the appropriate blank.

8. _____

9. _____

10. _____ _____

11. _____

12. _____

24-III. ZYGOMYCETES (pp. 380–381)

Selected Italicized Words

Rhizopus stolonifer, *Pilobolus*

Boldfaced, Page-Referenced Terms

(380) zygomycetes _____

_____

(380) zygosporangium _____

_____

Fill-in-the-Blanks

The numbered items on the illustration on the next page (*Rhizopus* life cycle) represent missing information; complete the corresponding numbered blanks in the narrative to supply missing information about zygomycetes.

The sexual phase begins when haploid (1) _____ of two different mating strains grow into each other and fuse. Two (2) _____ form between the (1), and several haploid nuclei are produced inside each. Later, their nuclei fuse, forming a zygote with a thick protective wall, called a (3) _____ (the key defining feature of the zygomycetes). Meiosis proceeds and (4) _____ are produced when this structure germinates. Each gives rise to stalked structures that can produce many spores, each of which can be the start of an extensive (5) _____.

Rhizopus spores are very small, dry, and easily dispersed by the winds. They have been carried nearly everywhere, including the North Pole. *Rhizopus stolonifer* is a familiar (6) _____ with a bad reputation as a spoiler of baked goods. Most (7) _____ zygomycetes are at home in soil, decaying plant and animal matter and stored food. Zygomycetes of the genus *Pilobolus* prefer animal (8) _____, from which they forcefully disperse spores. *Rhizopus* is normally harmless to (9) _____ but, as with other saprobic fungi, it is an opportunist in the lungs and skin cuts of individuals with weakened (10) _____ systems.

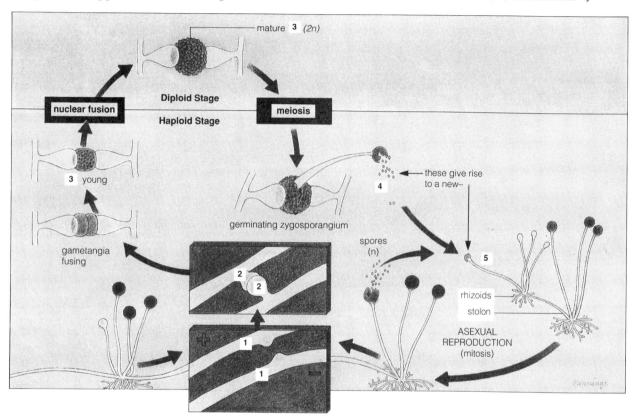

24-IV. SAC FUNGI (pp. 382–383)

Selected Italicized Words

Saccharomyces cerevisiae, Neurospora sitophila, Neurospora crassa, Claviceps purpurea, ergotism

Boldfaced, Page-Referenced Terms

(382) sac fungi _____

_____

(382) conidia _____

_____

(382) ascocarps _____

_____

True-False

If the statement is true, write a T in the blank. If the statement is false, make it correct by changing the underlined word(s) and writing in the correct word(s) in the answer blank.

_____ 1. Most of the sac fungi species are <u>single-celled</u>.

_____ 2. <u>Fermentation</u> carried on by *Saccharomyces cerevisiae* furnishes carbon dioxide for leavening bread and ethanol for alcoholic beverages.

_____ 3. <u>*Neurospora* *sitophila*</u> is a sac fungus important in genetic research.

_____ 4. Trained pigs and dogs are used to snuffle out edible <u>morels</u>.

_____ 5. Asexual reproduction is <u>uncommon</u> among the sac fungi.

_____ 6. Asexual reproduction in yeasts is by <u>budding</u>.

_____ 7. Multicelled sac fungi species form specialized <u>sexual</u> spores called conidia.

_____ 8. <u>Ascospores</u> can be shaped like flasks, globes, and shallow cups.

_____ 9. Spore-producing sacs called <u>asci</u> usually form on the inner surface of ascocarps.

_____ 10. In the sac fungi, <u>mitosis</u> produces haploid spores that are dispersed from a sac; each spore is capable of germination to form a mycelium that grows through soil, decaying wood, and other substrates.

24-V. CLUB FUNGI (pp. 383–385)

Selected Italicized Words

Agaricus brunnescens, *Armillaria bulbosa*, *Amanita muscaria*, *Amanita phalloides*

Boldfaced, Page-Referenced Terms

(384) club fungi _____

_____

(385) basidia _____

_____

(385) basidiocarp _____

_____

(385) dikaryotic mycelium_____

_____

Fill-in-the-Blanks

The numbered items on the illustration of a club fungus life cycle on the next page represent missing information; complete the numbered blanks in the narrative to supply the missing information.

The mature mushroom is actually a short-lived (1) _____; each consists of a cap and a stalk. Club-shaped structures, the spore-producing (2) _____, develop on the gills. Each bears two haploid ($n + n$) nuclei. (3) _____ fusion occurs within the club-shaped structures, which yields a (4) _____ stage. (5) _____ occurs within the club-shaped structures, and four haploid (6) _____ emerge at the tip of each. The spores are released, and each may germinate into a haploid (*n*) mycelium. When hyphae of two compatible mating strains meet, (7) _____ fusion occurs. Following this, a(n) (8) "_____" (*n* + *n*) mycelium gives rise to the spore-bearing mushrooms.

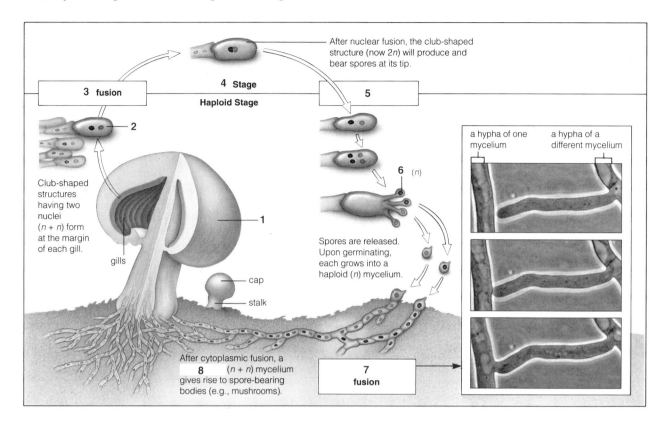

After nuclear fusion, the club-shaped structure (now 2*n*) will produce and bear spores at its tip.

3 fusion

4 Stage
Haploid Stage

5

2

Club-shaped structures having two nuclei (*n* + *n*) form at the margin of each gill.

gills

1

6 (*n*)

Spores are released. Upon germinating, each grows into a haploid (*n*) mycelium.

cap

stalk

After cytoplasmic fusion, a
8 (*n* + *n*) mycelium gives rise to spore-bearing bodies (e.g., mushrooms).

7
fusion

a hypha of one mycelium

a hypha of a different mycelium

Matching

Choose the one most appropriate answer for each.

9. ___ rust and smut fungi

10. ___ *Agaricus brunnescens*

11. ___ *Armillaria bulbosa*

12. ___ *Amanita muscaria* (text, Fig. 24.7*e*)

13. ___ *Amanita phalloides* (text, Fig. 24.7*d*)

A. Fly agaric mushroom, causes hallucinations when eaten; ritualistic use
B. Among the oldest and largest of the club fungi
C. Destroys entire fields of wheat, corn, and other major crops
D. Common cultivated mushroom; multimillion-dollar business
E. Death cap mushroom; kills humans

Complete the Table

14. After reading the *Commentary*, "A Few Fungi We Would Rather Do Without," on text p. 383, complete the following table, which deals with a few pathogenic and toxic fungi.

Fungi - Group Name	Description
a.	Black stem wheat rust, corn smut, severe mushroom poisoning
b.	Plant wilt, various species cause ringworms, including athlete's foot, mucous membrane infections, histoplasmosis
c.	Food spoilage
d.	Dutch elm disease, chestnut blight, apple scab, ergot of rye (ergotism), brown rot of stone fruits

24-VI. BENEFICIAL ASSOCIATIONS BETWEEN FUNGI AND PLANTS (pp. 386–387)

Selected Italicized Words

exomycorrhiza, endomycorrhiza

Boldfaced, Page-Referenced Terms

(386) mutualism _____

_____

(386) symbiosis _____

_____

(386) lichens _____

_____

(386) mycorrhizae _____

_____

Fill-in-the-Blanks

(1) _____ refers to an interaction between species in which positive benefits flow both ways. If such species intimately depend on each other for growth and reproduction, such interactions are a form of mutualism called (2) _____. (3) _____ are mutualistic interactions between a fungus and photosynthetic cells (cyanobacteria, green algae, or both). Thousands of (4) _____ fungi enter into such interactions. Lichens form when a fungal hypha penetrates a (5) _____ cell and begins absorbing carbohydrates from it. If the cell survives, both it and the fungus multiply together into crusty formations. The lichen has a sheltering effect on the (6) _____ species. Lichens grow slowly on bare (7) _____, tree bark, and fence posts. (8) _____ forms slowly through the activities of lichens as they absorb minerals from rock. The death of lichens that live around cities is a signal of (9) _____ deterioration.

The rootlets of many vascular plants associate with fungi in an intimate, mutually beneficial way; such associations are called (10) _____, or "fungus roots." A mycorrhizal fungus (with huge absorptive surfaces) absorbs (11) _____ from the host plant, which absorbs (12) _____ ions from the fungus. Without mycorrhizae, many plants do not grow well, for they cannot readily absorb (13) _____ and other important ions. In an (14) _____, the hyphae form a dense net around living cells in roots but do not penetrate them; they are common in temperate regions. About 5,000 fungal species enter such associations; most are (15) _____ fungi, including truffles. (16) _____ are much more common and form on the roots of about 80 percent of all vascular plants. In such cases, the fungal hyphae penetrate algal cells, as they do in lichens. Fewer than 200 species of (17) _____ serve as the fungal partner. Their (18) _____ branch extensively to form tree-shaped absorptive structures within cells and extend for several centimeters into the surrounding soil. Mycorrhizae are highly susceptible to (19) _____ _____, and this adversely affects the world's forests.

24-VII. "IMPERFECT" FUNGI (pp. 383, 387–388)

Selected Italicized Words

Histoplasma capsulatum, Aspergillus, Penicillium

Short Answer

1. What criterion is used to determine if a particular fungal species is an "imperfect" fungus? Cite two examples of "imperfect" fungi.

_____

_____

_____

_____

_____

_____

Self-Quiz

Label-Match

In the blank beneath each illustration (1–8), identify the organism by common name (or scientific name if a common name is unavailable). Then match each organism with the appropriate item (may be used more than once) from the list after the illustrations by entering the letter in the parentheses.

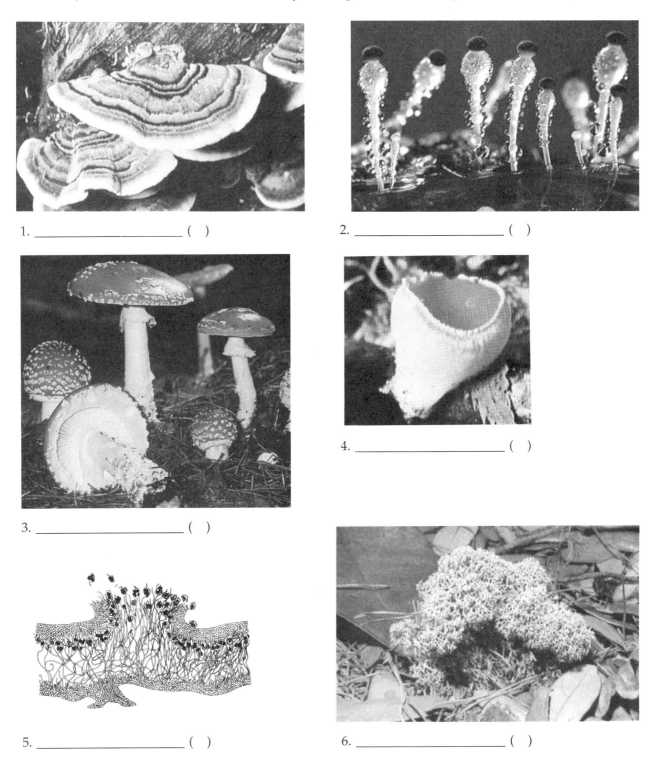

1. _____ ( )

2. _____ ( )

3. _____ ( )

4. _____ ( )

5. _____ ( )

6. _____ ( )

7. _____ ( )

8. _____ ( )

A. sac fungi
B. zygomycetes
C. club fungi
D. imperfect fungi
E. algae and fungi

___ 9. Most true fungi send out cellular filaments called _____.

    a. mycelia
    b. hyphae
    c. mycorrhizae
    d. asci

___ 10. Heterotrophic species of fungi can be _____.

    a. saprobic
    b. parasitic
    c. mutualistic
    d. all of the above

For questions 11–20, choose from the following:

    a. club fungi
    b. imperfect fungi
    c. sac fungi
    d. zygomycetes

___ 11. The group that includes *Rhizopus stolonifer,* the notorious black bread mold, is _____.

___ 12. The group that includes delectable morels and truffles but also includes bakers' and brewers' yeasts is _____.

___ 13. The group that includes shelf fungi, which decompose dead and dying trees, and mycorrhizal symbionts that help trees extract mineral ions from the soil is _____.

___ 14. The group that includes the commercial mushroom *Agaricus brunnescens,* as well as the death cap mushroom, *Amanita phalloides,* is _____.

___ 15. The group that includes *Penicillium,* which has a variety of species that produce penicillin and substances that flavor Camembert and Roquefort cheeses, is _____.

___ 16. The group whose spore-producing structures (asci) are shaped like flasks, globes, and cups is _____.

___ 17. The group that forms a thick wall around the zygote to produce a zygosporangium is _____.

___ 18. The group whose spore-producing structures are called basidia is _____.

___ 19. The groups that are symbiotic with young roots of shrubs and trees in mycorrhizal associations are _____ and _____.

___ 20. A group of fungi whose members were assigned to it because their sexual phase was undetected or absent.

Chapter Objectives/Review Questions

Page	*Objectives/Questions*
(380)	1. Fungi are heterotrophs; most are _____ and obtain nutrients from nonliving organic matter and so cause its decay.
(380)	2. Other fungi are _____; they extract nutrients from tissues of a living host.
(380)	3. Distinguish between the meanings of the following terms: *hypha, hyphae, mycelium,* and *mycelia.*
(380)	4. What is the most used reproductive mode in the fungi?
(380)	5. Sporangia produce _____; gametangia produce _____.
(380)	6. Be able to review the life cycle of *Rhizopus stolonifer.*
(381)	7. The key defining feature of the zygomycetes is the _____.
(382)	8. Only sac fungi produce spores in sacs known as _____ during the sexual phase of their life cycle.
(382)	9. What are some typical shapes of ascocarps?
(382)	10. Multicelled sac fungi form specialized spores called _____.
(383)	11. Give the name of the fungus that causes the disease known as ergotism; list the symptoms of ergotism.
(384)	12. Describe the diverse appearances of the fungi classified as club fungi.
(384)	13. What is the importance of the club fungi in the genus *Amanita*?
(385)	14. Know the general life cycle of the club fungi (mushroom).
(385)	15. Cells that produce club fungi spores are known as _____.
(385)	16. The visible portion of a club fungus is the _____.
(385)	17. A _____ mycelium is one in which the hyphae have undergone cytoplasmic fusion but not nuclear fusion.
(386)	18. Define *mutualism* and explain why a lichen fits that definition.
(386)	19. Describe the fungus–plant root association known as mycorrhizae.
(387)	20. Distinguish exomycorrhizae from endomycorrhiza.
(387)	21. What is the effect of pollution on mycorrhizae?
(388)	22. Explain why a classification group, the "imperfect" fungi, is necessary.

Integrating and Applying Key Concepts

Suppose humans acquired a few well-placed fungal genes that caused them to reproduce in the manner of a "typical" fungus (Fig. 24.8, text). Try to imagine the behavioral changes that humans would likely undergo. Would their food supplies necessarily be different? Table manners? Stages of their life cycle? Courtship patterns? Habitat? Would the natural limits to population increase be the same? Would their body structure change? Would there necessarily have to be separate sexes? Compose a descriptive science-fiction tale about two mutants who find each other and set up "housekeeping" together.

Answers

Interactive Exercises

24-I. MAJOR GROUPS OF FUNGI (p. 379)
24-II. CHARACTERISTICS OF FUNGI (p. 380)
1. D; 2. F; 3. B; 4. G; 5. A; 6. E; 7. C; 8. zygote;
9. spores; 10. absorptive body; 11. spores; 12. gametes.

24-III. ZYGOMYCETES (pp. 380–381)
1. hyphae; 2. gametangia; 3. zygosporangium; 4. spores;
5. mycelium; 6. saprobe; 7. saprobic; 8. feces; 9. humans; 10. immune.

24-IV. SAC FUNGI (pp. 382–383)
1. multicelled; 2. T; 3. *Neurospora crassa*; 4. truffles;
5. common; 6. T; 7. asexual; 8. Ascocarps; 9. T;
10. meiosis.

24-V. CLUB FUNGI (pp. 383–385)
1. basidiocarp; 2. basidia; 3. Nuclear; 4. diploid;
5. Meiosis; 6. spores; 7. cytoplasmic; 8. dikaryotic; 9. C;
10. D; 11. B; 12. A; 13. E; 14. a. Club fungi; b. Imperfect
fungi; c. Zygomycetes; d. Sac fungi.

**24- VI. BENEFICIAL ASSOCIATIONS BETWEEN
  FUNGI AND PLANTS** (pp. 386–387)
1. Mutualism; 2. symbiosis; 3. Lichens; 4. sac; 5. host;
6. photosynthetic; 7. rocks; 8. Soil; 9. environmental;
10. mycorrhizae; 11. carbohydrates; 12. mineral;
13. phosphorus; 14. exomycorrhiza; 15. club; 16. Endomycorrhiza; 17. zygomycetes; 18. hyphae; 19. air pollution.

24-VII. "IMPERFECT" FUNGI (pp. 383, 388)
1. If the sexual phase is unknown for a particular fungus, it is referred to the informal group known as the "imperfect" fungi. If mycologists later detect a sexual phase in the fungus, it is then assigned to (usually) the sac fungi or the club fungi. Examples: *Aspergillus* produces citric acid for candies and soft drinks and ferments soybeans for soy sauce; certain *Penicillium* species are used to "flavor" Camembert and Roquefort cheeses; other species of *Penicillium* produce antibiotics, the penicillins.

Self-Quiz
1. *Polyporus* (C); 2. *Pilobolus* (B); 3. fly agaric mushroom (C); 4. cup fungus (A); 5. lichen (E); 6. lichen (E); 7. mushroom (C); 8. *Penicillium* (D); 9. b; 10. d; 11. d; 12. c; 13. a; 14. a; 15. b; 16. c; 17. d; 18. a; 19. a, c; 20. b.

25

PLANTS

Interactive Exercises

25-I. CLASSIFICATION OF PLANTS (pp. 390–391)
25-II. EVOLUTIONARY TRENDS AMONG PLANTS (pp. 392–393)

Boldfaced, Page-Referenced Terms

(391) vascular plants _____

_____

(391) bryophytes _____

_____

(391) gymnosperms _____

_____

(391) angiosperms _____

_____

(392) root systems _____

_____

(392) shoot systems _____

_____

(392) xylem _____

_____

(392) phloem _____

_____

(392) lignin _____

_____

(392) cuticle _____

_____

(392) stomata _____

_____

(392) sporophyte _____

_____

(393) gametophytes _____

_____

(393) pollen grains _____

_____

Crossword Puzzle—Plant Classification and Evolutionary Trends

1. Complete the crossword puzzle on the opposite page that includes terms and concepts related to plant classification and evolutionary trends.

ACROSS

2. Multicelled, haploid, gamete-producing structures produced from a haploid spore; nourish and protect the forthcoming generation.
4. Plants that possess internal tissues that conduct water and solutes; also have roots, stems, and leaves.
6. An organic compound that strengthens plant cell walls; such reinforced tissues structurally support plant parts that display the leaves to sunlight.
7. A vascular tissue that distributes sugars and other photosynthetic products.
8. Seed-bearing plants that include cycads, ginkgo, gnetophytes, and conifers.

12. A vascular tissue that distributes water and dissolved ions through plant parts.
13. Means "spore-producing body"; forms when a new 2n zygote embarks on a course of mitotic cell divisions, which produces the large multi-celled plant body.

DOWN

1. A waxy surface covering over stems and leaves that reduces water loss on hot, dry days.
3. Include mosses, liverworts, and hornworts that generally grow in moist habitats; small plants lacking true roots, stems, and leaves as well as vascular tissues.

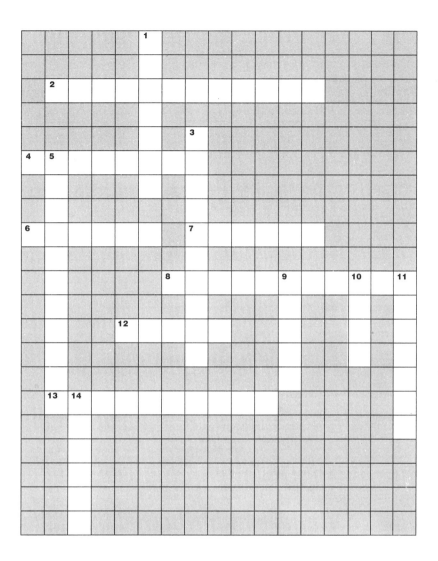

5. Plants that produce seeds by means of flowers; includes two classes: dicots and monocots.
9. System that consists of stems and leaves, which function in the absorption of sunlight energy and carbon dioxide from the air.
10. System that consists of underground, cylindrical absorptive structures with a large surface area for taking up soil and scarce mineral ions.

11. Numerous tiny passageways in surface tissues of leaves and young stems; the main route for absorbing carbon dioxide and controlling evaporative water loss.
14. Develops from one type of haploid plant spore; becomes mature, sperm-bearing male gametophytes.

Complete the Table

2. As plants evolved, several key evolutionary events occurred that solved the problems of living in new land environments. Complete the following table to summarize these events. Choose from phloem, sporophyte (diploid) dominance of the life cycle, well-developed shoot systems, cuticle, heterospory, lignin production, xylem, seeds, well-developed root systems, interaction of young roots and mycorrhizal fungi, stomata, pollen grains.

Evolutionary Trends	Survival Problem Solved
a.	Provides a large surface area for rapidly taking up soil water and scarce mineral ions; often anchors the plant
b.	Consist of stems and leaves that function in the absorption of sunlight energy and CO_2 from the air
c.	Provides pipelines to distribute water and dissolved ions through plant parts
d.	Provides pipelines to distribute sugars and other photosynthetic products
e.	Allows extensive growth of stems and branches; very hard substance that allows support of plant parts to display the leaves
f.	Allows water conservation for plants living on land
g.	Provides the main route for absorbing carbon dioxide while controlling evaporative water loss
h.	A symbiotic relationship that provides many sporophytes with water and scarce nutrients, even in seasonally dry habitats
i.	Allows plants to live and reproduce in higher, drier parts of the world; enables nourishment and protection for their spores and gametes through unfavorable environmental conditions
j.	Allowed the evolution of male gametes specialized for dispersal without liquid water; this led to the evolution of seeds
k.	Allows movement of male gametophytes to the female gametophytes
l.	Embryos became packaged with protective and nutritive tissues for survival on dry land

Label-Match

Match each of the following life cycle sketches with the appropriate description. The order illustrates an evolutionary trend from haploid to diploid dominance during the colonization of land.

3. ___   A. Typical life cycle for some algae (haploid dominance)

4. ___   B. Typical life cycle for bryophytes (haploid dominance)

5. ___   C. Typical life cycle for vascular plants (diploid dominance)

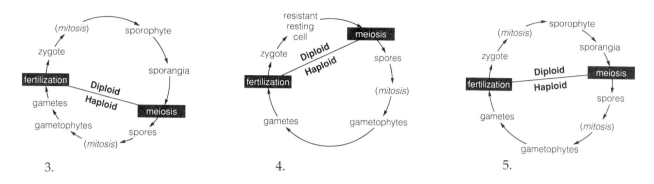

3.   4.   5.

25-III. BRYOPHYTES (pp. 394–395)

Selected Italicized Words

Marchantia

Boldfaced, Page-Referenced Terms

(394) mosses _____

_____

(394) liverworts _____

_____

(394) hornworts _____

_____

(394) rhizoids _____

_____

(395) gametangia _____

_____

(395) sporangium _____

_____

True-False

If the statement is true, write a T in the blank. If the statement is false, make it correct by changing the underlined word(s) and writing the correct word(s) in the answer blank.

_____ 1. Mosses especially are sensitive to <u>water</u> pollution.

_____ 2. Bryophytes <u>have</u> leaflike, stemlike, and rootlike parts, although they do not contain xylem or phloem.

_____ 3. Most bryophytes have <u>rhizomes</u>, elongated cells or threads that attach gametophytes to soil and serve as absorptive structures.

_____ 4. Bryophytes are the simplest plants to exhibit a cuticle, cellular jackets around gamete-producing parts, and large gametophytes that retain nutritionally <u>dependent</u> sporophytes.

_____ 5. <u>Liverworts</u> are the most common bryophytes.

_____ 6. Following fertilization, zygotes give rise to <u>gametophytes</u>.

_____ 7. Each <u>sporophyte</u> consists of a stalk and a sporangium.

_____ 8. <u>Club</u> moss is a bog moss whose large, dead cells in their leaflike parts soak up five times as much water as cotton.

_____ 9. Bryophyte sperm reach eggs by movement through <u>air</u>.

_____ 10. Eggs and sperm of moss plants develop at shoot tips in <u>gametangia</u>.

Fill-in-the-Blanks

The numbered items on the illustration on the next page represent missing information about a typical moss life cycle; complete the numbered blanks in the narrative below to supply the missing information on the illustration.

The gametophytes are the green leafy "moss plants." Sperm develop in jacketed structures at the shoot tip

of the male (11) _____, and eggs develop in jacketed structures at the shoot tip of the female

(12) _____. Raindrops transport (13) _____ to the egg-producing structure. (14) _____

occurs within the egg-producing structure. The (15) _____ grows and develops into a mature

(16) _____ (with sporangium and stalk) while attached to the gametophyte. (17) _____ occurs

within the sporangium of the sporophyte where haploid (18) _____ form, develop, and are released.

The released spores grow and develop into male or female (19) _____.

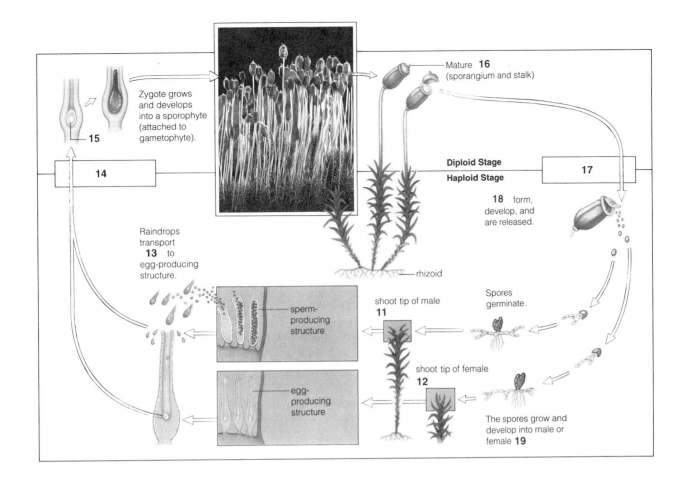

Zygote grows and develops into a sporophyte (attached to gametophyte).

15

14

Mature **16** (sporangium and stalk)

Diploid Stage

Haploid Stage

17

18 form, develop, and are released.

Raindrops transport **13** to egg-producing structure.

rhizoid

sperm-producing structure

shoot tip of male **11**

Spores germinate.

egg-producing structure

shoot tip of female **12**

The spores grow and develop into male or female **19**

25-IV. SEEDLESS VASCULAR PLANTS (pp. 396–399)

Selected Italicized Words

Cooksonia, Psilophyton, Psilotum, Lycopodium, Selaginella, Equisetum

Boldfaced, Page-Referenced Terms

(396) whisk ferns _____

_____

(396) lycophytes _____

_____

(396) horsetails _____

_____

(396) ferns _____

_____

Choice

For questions 1–21 , choose from the following:

    a. whisk ferns   b. lycophytes   c. horsetails   d. ferns   e. applies to a, b, c, and d

___ 1. Only one genus survives, *Equisetum.*

___ 2. Familiar club mosses growing on forest floors

___ 3. Seedless vascular plants

___ 4. *Psilotum*

___ 5. Rust-colored patches, the sori, are on the lower surface of their fronds.

___ 6. Some tropical species are the size of trees.

___ 7. Ancestral plants composed "ancient carbon treasures" of the Carboniferous; became peat and coal.

___ 8. Stems were used by pioneers of the American West to scrub cooking pots.

___ 9. The sporophytes have no roots or leaves.

___ 10. Mature leaves are usually divided into leaflets.

___ 11. The sporophyte has vascular tissues.

___ 12. When the sporangium snaps open, spores catapult through the air.

___ 13. Grow in mud soil of streambanks and in disturbed habitats, such as roadsides and railroad beds.

___ 14. Mycorrhizal fungi assist the rhizomes in absorption.

___ 15. Sporophytes have rhizomes and hollow photosynthetic aboveground stems with scalelike leaves.

___ 16. The sporophyte is the larger, longer-lived phase of the life cycle.

___ 17. *Lycopodium*

___ 18. The young leaves are coiled into the shape of a fiddlehead.

___ 19. A spore develops into a small green, heart-shaped gametophyte.

___ 20. *Selaginella*, a heterosporous genus

___ 21. A group with a genus that is unique among vascular plants

Fill-in-the-Blanks

The numbered items on the illustration of a generalized fern life cycle below represent missing information; complete the numbered blanks in the narrative below to supply the missing information on the illustration.

Fern leaves (fronds) of the sporophyte are usually divided into leaflets. The underground stem of the sporophyte is termed a (22) _____. On the undersides of many fern fronds, rust-colored patches of sporangia (the sori) occur. (23) _____ of diploid cells within each sporangium produces haploid (24) _____. The (25) _____ are catapulted into the air when each sporangium snaps open. A spore may germinate and grow into a (26) _____ that is small, green, and heart-shaped. Jacketed structures develop on the underside of the mature (27) _____. Each male jacketed structure produces many (28) _____, while each female jacketed structure produces a single (29) _____. These gametes meet in (30) _____. The diploid (31) _____ is first formed inside the female jacketed structure; it divides to form the developing (32) _____, still attached to the gametophyte.

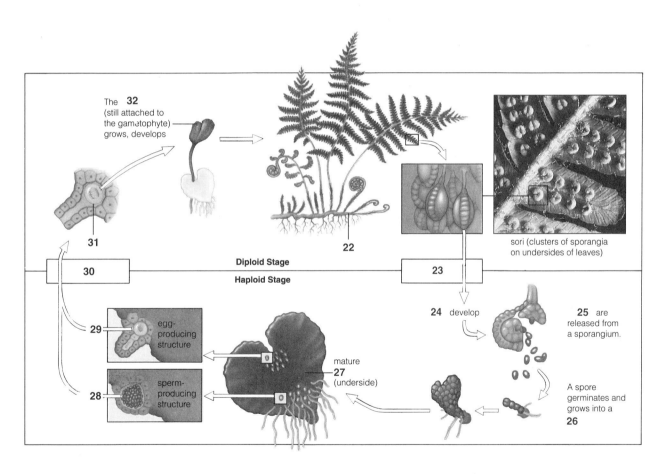

The **32** (still attached to the gametophyte) grows, develops

31

30

Diploid Stage

Haploid Stage

22

23

sori (clusters of sporangia on undersides of leaves)

24 develop

25 are released from a sporangium.

29 egg-producing structure

28 sperm-producing structure

mature **27** (underside)

A spore germinates and grows into a **26**

25-V. THE SEED-BEARING PLANTS (p. 400)
25-VI. GYMNOSPERMS (pp. 401–402)

Selected Italicized Words

gymnosperm, Ginkgo biloba, Gnetum, Ephedra, Welwitschia

Boldfaced, Page-Referenced Terms

(400) ovule _____

_____

(401) conifers _____

_____

(401) cones _____

_____

(401) microspores _____

_____

(401) megaspores _____

_____

(401) pollination _____

_____

(401) deforestation _____

_____

(402) cycads _____

_____

(402) ginkgos _____

_____

(402) gnetophytes _____

_____

Choice

For questions 1–12, choose from the following:

a. cycads   b. ginkgos   c. gnetophytes   d. conifers   e. gymnosperms (applies to a, b, c, and d)

___ 1. Fleshy-coated seeds of female trees produce an awful stench when stepped on.

___ 2. Includes pines, spruces, firs, hemlocks, junipers, cypresses, and redwoods.

___ 3. In parts of Asia, people eat the seeds and a starchy flour is made from the trunks after poisonous alkaloids are rinsed out.

___ 4. Only a single species survives, the maidenhair tree.

___ 5. Includes *Welwitschia* of hot deserts of South and West Africa, *Gnetum* of humid tropical regions, and *Ephedra* of deserts and other arid regions.

___ 6. Have massive, cone-shaped structures that bear either pollen or ovules; superficially resemble palm trees.

___ 7. Seeds are mature ovules.

___ 8. The male trees are now planted in cities because of their attractive, fan-shaped leaves and their resistance to insects, disease, and air pollutants.

___ 9. Their ovules and seeds are not covered; they are borne on surfaces of spore-producing reproductive structures.

___ 10. Some plants in this group are mostly a deep taproot; the exposed part is a woody disk-shaped stem bearing cone-shaped strobili and two strap-shaped leaves that split lengthwise repeatedly as the plant ages.

___ 11. Most species are "evergreen" trees and shrubs with needlelike or scalelike leaves.

___ 12. Includes conifers, cycads, ginkgos, and gnetophytes.

Fill-in-the-Blanks

The numbered items on the illustration below represent missing information; complete the numbered blanks in the narrative below to supply the missing information on the illustration.

The familiar pine tree, a conifer, represents the mature (13) _____. Pine trees produce two kinds of spores in two kinds of cones. Pollen grains are produced in male (14) _____. Ovules are produced in young female (15) _____. Inside each (16) _____, (17) _____ occurs to produce haploid megaspores; one develops into a many-celled female (18) _____ that contains haploid (19) _____. Diploid cells within pollen sacs of male cones undergo (20) _____ to produce haploid microspores. Microspores develop into (21) _____ grains. (22) _____ occurs when spring air currents deposit pollen grains near ovules of female cones. A pollen (23) _____ representing a male gametophyte

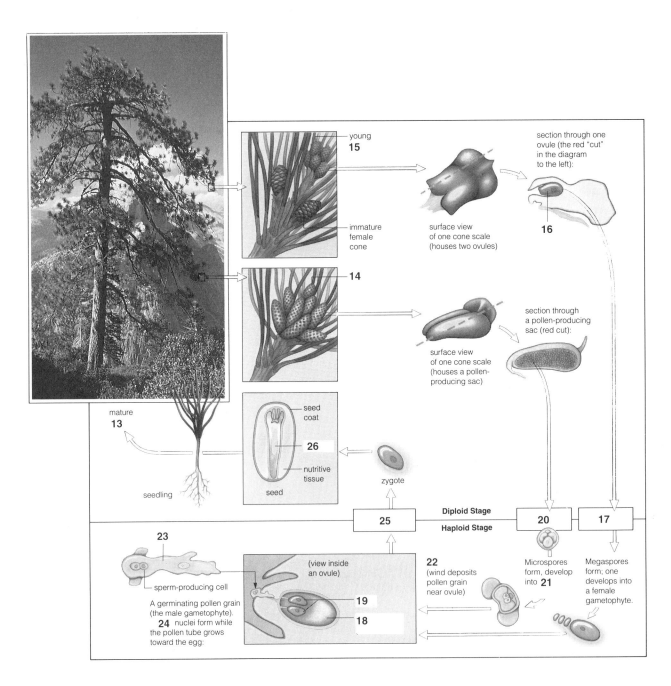

grows toward the female gametophyte. (24) _____ nuclei form within the pollen tube as it grows toward the egg. (25) _____ follows, and the ovule becomes a seed that is composed of an outer seed coat, the diploid sporophyte plant (26) _____, and nutritive tissue.

25-VII. ANGIOSPERMS (pp. 403–405)

Selected Italicized Words

angiosperm, Eucalyptus

Boldfaced, Page-Referenced Terms

(405) fruits _____

_____

(405) pollinators _____

_____

Matching

1. ___ monocot examples
2. ___ flower
3. ___ pollinators
4. ___ endosperm
5. ___ dicot examples
6. ___ seeds

  A. Nutritive seed tissue
  B. Palms, lilies, orchids, wheat, corn, rice, rye, sugarcane, and barley
  C. Unique angiosperm reproductive structure
  D. Insects, bats, birds, and other animals that coevolved with flowers of plant life cycles
  E. Packaged in fruits
  F. Most shrubs and trees, most nonwoody plants, cacti, and water lilies

Complete the Table

7. Complete the table below to compare the plant groups studied in this chapter.

Plant Group	Dominant Generation	Vascular Tissue	Seeds
a. Bryophytes			
b. Lycophytes			
c. Horsetails			
d. Ferns			
e. Gymnosperms			
f. Angiosperms			

Self-Quiz

___ 1. Members of the plant kingdom probably evolved from _____ more than 400 million years ago.

a. unicellular brown algae
b. multicellular green algae
c. unicellular green algae
d. multicellular red algae

___ 2. The _____ is *not* a trend in the evolution of plants.

a. evolution of complex sporophytes
b. shift from homospory to heterospory
c. shift from diploid to haploid dominance
d. development of xylem and phloem

___ 3. Existing nonvascular plants do *not* include _____.

a. horsetails
b. mosses
c. liverworts
d. hornworts

___ 4. Plants possessing xylem and phloem are called _____ plants.

a. gametophyte
b. nonvascular
c. vascular
d. seedless

___ 5. Bryophytes _____.

a. have vascular systems that enable them to live on land
b. include lycopods, horsetails, and ferns
c. have true roots but not stems
d. include mosses, liverworts, and hornworts

___ 6. _____ are not seedless vascular plants.

a. Lycophytes
b. Gymnosperms

c. Horsetails
d. Whisk Ferns
e. Ferns

___ 7. In horsetails, lycopods, and ferns, _____.

a. spores give rise to gametophytes
b. the main plant body is a gametophyte
c. the sporophyte bears sperm- and egg-producing structures
d. all of the above

___ 8. _____ are seed plants.

a. Cycads and ginkgos
b. Conifers
c. Angiosperms
d. all of the above

___ 9. In complex land plants, the diploid stage is resistant to adverse environmental conditions such as dwindling water supplies and cold weather. The diploid stage progresses through this sequence: _____.

a. gametophyte → male and female gametes
b. spores → sporophyte
c. zygote → sporophyte
d. zygote → gametophyte

___ 10. Monocots and dicots are groups of _____.

a. gymnosperms
b. club mosses
c. angiosperms
d. horsetails

Chapter Objectives/Review Questions

Integrating and Applying Key Concepts

Explain why totally submerged aquatic plants that live in deep water never developed heterosporous life cycles.

Answers

Interactive Exercises

25-I. CLASSIFICATION OF PLANTS (p. 391)
25-II. EVOLUTIONARY TRENDS AMONG PLANTS (pp. 392–393)

1.

```
 .  .  .  .  .  C  .  .  .  .  .  .  .  .  .  .  .
 .  .  .  .  .  U  .  .  .  .  .  .  .  .  .  .  .
 .  G  A  M  E  T  O  P  H  Y  T  E  S  .  .  .  .
 .  .  .  .  .  I  .  .  .  .  .  .  .  .  .  .  .
 .  .  .  .  .  C  .  B  .  .  .  .  .  .  .  .  .
 V  A  S  C  U  L  A  R  .  .  .  .  .  .  .  .  .
 .  N  .  .  .  E  .  Y  .  .  .  .  .  .  .  .  .
 .  G  .  .  .  .  .  O  .  .  .  .  .  .  .  .  .
 L  I  G  N  I  N  .  P  H  L  O  E  M  .  .  .  .
 .  O  .  .  .  .  .  H  .  .  .  .  .  .  .  .  .
 .  S  .  .  .  .  G  Y  M  N  O  S  P  E  R  M  S
 .  P  .  .  .  .  .  T  .  .  .  H  .  .  O  .  T
 .  E  .  .  X  Y  L  E  M  .  .  O  .  .  O  .  O
 .  R  .  .  .  .  .  S  .  .  .  O  .  .  T  .  M
 .  M  .  .  .  .  .  .  .  .  .  T  .  .  .  .  A
 .  S  P  O  R  O  P  H  Y  T  E  .  .  .  .  .  T
 .  .  O  .  .  .  .  .  .  .  .  .  .  .  .  .  A
 .  .  L  .  .  .  .  .  .  .  .  .  .  .  .  .  .
 .  .  L  .  .  .  .  .  .  .  .  .  .  .  .  .  .
 .  .  E  .  .  .  .  .  .  .  .  .  .  .  .  .  .
 .  .  N  .  .  .  .  .  .  .  .  .  .  .  .  .  .
```

2. a. Well-developed root systems; b. Well-developed shoot systems; c. Xylem; d. Phloem; e. Lignin production; f. Cuticle; g. Stomata; h. Interaction of young roots and mycorrhizal fungi; i. Sporophyte (diploid) dominance of the life cycle; j. Heterospory; k. Pollen grains; l. Seeds; 3. C; 4. A; 5. B.

25-III. BRYOPHYTES (pp. 394–395)
1. air; 2. T; 3. rhizoids; 4. T; 5. Mosses; 6. sporophytes; 7. T; 8. Peat; 9. water; 10. T; 11. gametophyte; 12. gametophyte; 13. sperm; 14. Fertilization; 15. zygote; 16. sporophyte; 17. Meiosis; 18. spores; 19. gametophytes.

25-IV. SEEDLESS VASCULAR PLANTS (pp. 396–399)
1. c; 2. b; 3. e; 4. a; 5. d; 6. d; 7. e; 8. c; 9. a; 10. d; 11. e; 12. d; 13. c; 14. a; 15. c; 16. e; 17. b; 18. d; 19. d; 20. b; 21. a; 22. rhizome; 23. Meiosis; 24. spores; 25. spores; 26. gametophyte; 27. gametophyte; 28. sperm; 29. egg; 30. fertilization; 31. zygote; 32. embryo.

25-V. THE SEED-BEARING PLANTS (p. 400)
25-VI. GYMNOSPERMS (pp. 401–402)
1. b; 2. d; 3. a; 4. b; 5. c; 6. a; 7. e; 8. b; 9. e; 10. c; 11. d; 12. e; 13. sporophyte; 14. cones; 15. cones; 16. ovule; 17. meiosis; 18. gametophyte; 19. eggs; 20. meiosis; 21. pollen; 22. Pollination; 23. tube; 24. Sperm; 25. Fertilization; 26. embryo.

25-VII. ANGIOSPERMS (pp. 403–405)
1. B; 2. C; 3. D; 4. A; 5. F; 6. E; 7. a. Gametophyte, None or simple vascular tissue, No; b. Sporophyte, Yes, No; c. Sporophyte, Yes, No; d. Sporophyte, Yes, No; e. Sporophyte, Yes, Yes; f. Sporophyte, Yes, Yes.

Self-Quiz
1. b; 2. c; 3. a; 4. c; 5. d; 6. b; 7. a; 8. d; 9. c; 10. c.

26

ANIMALS: THE INVERTEBRATES

Interactive Exercises

Selected Italicized Words

anterior, posterior, dorsal, ventral, "complete" digestive system, thoracic, abdominal, "false coelom," "segmented," "primitive"

Boldfaced, Page-Referenced Terms

(410) animal _____

_____

(410) ectoderm _____

_____

(410) endoderm _____

_____

(410) mesoderm _____

_____

(410) vertebrates _____

_____

(410) invertebrates _____

_____

(410) radial symmetry _____

_____

(410) bilateral symmetry _____

_____

(411) cephalization _____

_____

(411) coelom _____

_____

(412) placozoan _____

_____

(412) sponges _____

_____

(413) collar cells _____

_____

(413) larva _____

_____

(414) cnidarians _____

_____

(414) nematocysts _____

_____

(414) epithelial tissues _____

_____

(414) nerve cells _____

_____

(414) sensory cells _____

_____

(414) contractile cells _____

_____

(414) hydrostatic skeleton _____

_____

(417) gonads _____

_____

(417) planula _____

_____

(417) comb jellies _____

_____

Complete the Table

1. Complete the table below by filling in the appropriate phylum or representative group name.

Phylum	Some Representatives	Number of Known Species
a.	Sponges	8,000
b.	i. _____ , jellyfishes, corals, sea anemones	11,000
c.	Turbellarians, flukes, tapeworms	15,000
d.	Pinworms, hookworms	20,000
Rotifera	Species with crown of cilia	1,800
e.	j. _____ , slugs, clams, squids, octopuses	110,000
f.	k. _____ , leeches, polychaetes	15,000
g.	Crabs, lobsters, spiders, insects	1,000,000+
h.	Sea stars, sea urchins, sea cucumbers	6,000

Fill-in-the-Blanks

More than 95 percent of all animals on Earth are (2) _____ . The simplest multicellular animal known is a (3) _____ . (4) _____ form the most primitive major group of multicellular animals. They are nourished by microscopic organisms extracted from the water that flows in through pores in the body wall by sticky (5) _____ _____ . (6) _____ between sponge cells and integration of activities are poorly developed. Sponges lack (7) _____ cells, muscles, and a gut. Cnidarians are (8) _____ symmetrical and have stinging cells called (9) _____, which aid in defense and food capture. Most cnidarian life cycles have a (10) _____ larval stage. Of the two body types, the (11) _____ is the sexual stage, in which simple sex organs produce eggs or sperms. (12) _____ _____ are biradial predatory animals that appear to be made of jelly.

Labeling

Identify each indicated part of the illustration below.

13. _____ _____

14. _____ _____

15. _____ _____

16. _____

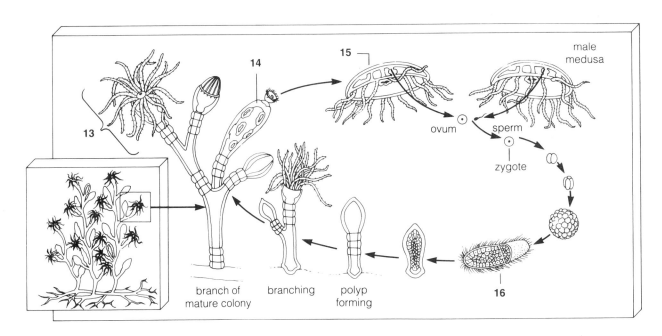

Identify the groups in the family tree shown by
writing the group names in the appropriate blank.

17. _____

18. _____

19. _____

20. _____

21. _____

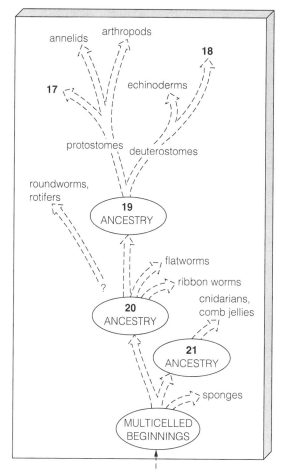

single-celled, protistanlike ancestors

26-II. FLATWORMS (pp. 418–419)
ROUNDWORMS (pp. 419–421)
RIBBON WORMS (p. 422)
ROTIFERS (pp. 422–423)

Selected Italicized Words

primary host, intermediate host, schistosomiasis, elephantiasis, "rotifer"

Boldfaced, Page-Referenced Terms

(418) flatworms _____

_____

(418) organs _____

_____

(418) organ systems _____

_____

(418) hermaphrodite _____

_____

(419) scolex _____

_____

(419) proglottids _____

_____

(419) roundworms _____

_____

(419) cuticle _____

_____

(422) ribbon worms _____

_____

Fill-in-the-Blanks

A mutant (1) _____ larva is believed to be the ancestor of the free-living flatworms. A shift from radial
to bilateral symmetry could have led to (2) _____ _____ of the sort seen in many flatworms; for
example, turbellarians have units called (3) _____ that regulate the volume and salt concentrations of
their body fluid. Flatworms have no (4) _____ systems, and their (5) _____ system is saclike.
Examples of parasitic flatworms are *Schistosoma*, a blood (6) _____ that causes schistosomiasis, and
Taenia saginata, a tapeworm that attaches to the intestine with a (7) _____ and releases eggs that
develop into proglottids. (8) _____ have complete digestive tracts, no circular muscles, and only a few
longitudinal muscles. Between the gut and body wall of a nematode is a (9) _____ _____ ,which
contains (10) _____ organs and serves as both a circulatory system and a hydrostatic (11) _____.
(12) _____ are tiny, abundant, aquatic pseudocoelomates that have ciliary crowns attracting their food,
single-celled organisms, to them.

Labeling

Identify the parts of the animals shown dissected in the accompanying drawings.

13. _____ _____

14. _____

15. _____

16. _____ _____

17. _____

18. _____

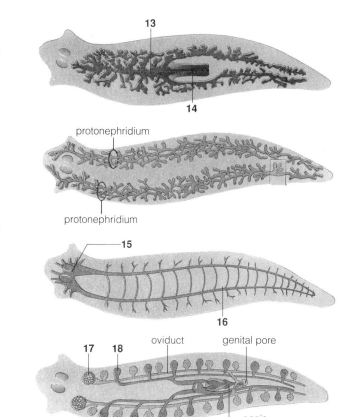

Answer exercises 19–22 for the drawing of the accompanying dissected animal.

19. What is the common name (or genus) of the animal dissected? _____
20. Is the animal parasitic? _____
21. Is the animal hermaphroditic? _____
22. Does the animal have a true coelom? _____

Answer exercises 23–25 for the drawing of the dissected animal below.

23. What is the common name of the animal dissected? _____
24. Is the animal hermaphroditic? _____
25. Does the animal have any kind of coelom? _____

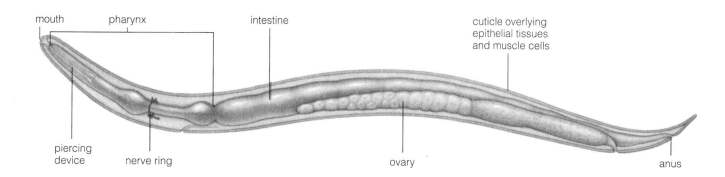

26-III. A MAJOR DIVERGENCE (p. 423)
MOLLUSKS (pp. 424–427)

Selected Italicized Words

closed circulatory system

Boldfaced, Page-Referenced Terms

(423) protostomes _____

(423) deuterostomes _____

_____

(423) radial cleavage _____

_____

(423) spiral cleavage _____

_____

(424) mollusks _____

_____

(424) mantle _____

_____

(425) chitons _____

_____

(426) bivalves _____

_____

(426) cephalopods _____

_____

Fill-in-the-Blanks

(1) _____ include echinoderms and chordates; in this group, the first opening to the gut becomes the

(2) _____, and the second one to appear becomes the (3) _____. The situation is reversed in the

(4) _____, which includes annelids [such as (5) _____], arthropods [such as (6) _____ and

crabs], and (7) _____ (such as abalones, limpets, squids, and chambered nautiluses).

The most highly evolved invertebrates are generally considered to be the (8) _____, which include

squids and octopuses; in terms of sheer size and complexity, the (9) _____ of these animals approach

those of mammals. Like vertebrates, these animals have acute (10) _____ and refined (11) _____

control, which is well integrated with the activities of the nervous system. In less highly evolved mollusks,

a structure known as the (12) _____ secretes one or more pieces of calcareous armor that protect these

soft-bodied animals from predation. In the cephalopods, the mantle has become a conical cloak that sur-

rounds the internal organs and the much-reduced shell; seawater moves in and out of the (13) _____

_____ in a jet-propulsive manner.

Matching

Identify the animals pictured below by matching each with the appropriate description.

14. _____ Animal A
15. _____ Animal B
16. _____ Animal C

   I. Bivalve
  II. Cephalopod
 III. Gastropod

Labeling

Identify each numbered part in the drawings below by writing its name in the appropriate blank.

17. _____
18. _____
19. _____
20. _____
21. _____

22. _____
23. _____
24. _____
25. _____
26. _____

27. _____
28. _____
29. _____
30. _____

31. _____ _____
32. _____
33. _____
34. _____

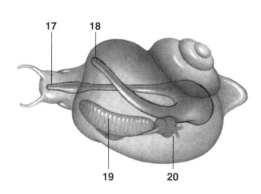

Animal A

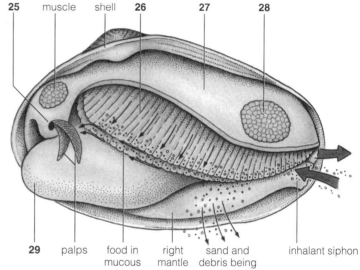

25 muscle   shell   **26**   **27**   **28**

29   palps   food in   right   sand and   inhalant siphon
             mucous   mantle   debris being
             string   rejected

Animal B

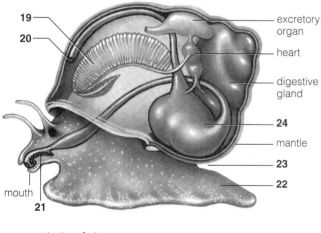

Animal A

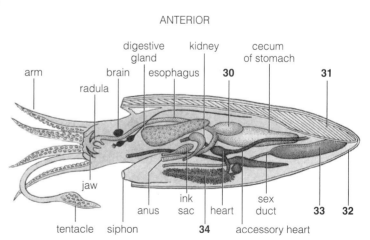

Animal C

26-IV. ANNELIDS (pp. 428–431)

Selected Italicized Words

setae, polychaetes

Boldfaced, Page-Referenced Terms

(428) annelids _____

_____

(428) nerve cord _____

_____

(428) ganglion, -glia _____

_____

(429) nephridium, -dia _____

_____

(430) leeches _____

_____

Fill-in-the-Blanks

(1) _____ include truly segmented worms such as earthworms, (2) _____, and leeches; they differ from flatworms in having (first) a complete digestive system with a mouth and (3) _____ and (second) a (4) _____, a fluid-filled space between the gut and body wall. In a circulatory system, (5) _____ provides a means for transporting materials between internal and external environments. In some annelids, (6) _____, a protein component of blood, dramatically increases the blood's oxygen-carrying capacity. (7) _____, which is a repeating series of body parts, is well developed in annelids. In each segment, there are swollen regions of the (8) _____ _____ that control local activity and a pair of (9) _____ that act as kidneys, as well as bristles embedded in the body wall. Many polychaetes (marine worms) have fleshy, paddle-shaped lobes called (10) _____, which project from the body wall. Earthworms (11) _____ soil and help to make nutrients available to plants; polychaetes live in (12) _____ habitats, where they are important food for fishes and birds.

Labeling

Identify each indicated part of the illustration.

13. _____

14. _____

15. _____ _____

16. _____

17. _____ _____

18. _____

19. _____
20. Name this animal. _____
21. Name this animal's phylum.
_____
22. Name two distinguishing characteristics
of this group. _____
_____
23. Protostome ❐ or deuterostome ❐ ?
24. Is this animal segmented? ❐ yes ❐ no
25. Symmetry of adult: ❐ radial ❐ bilateral
26. Does this animal have a true coelom?
❐ yes ❐ no

anus

intestine

coelum

14

13

mouth

16   17   19

15   18

26-V. ARTHROPODS (pp. 432–437)

Boldfaced, Page-Referenced Terms

(432) exoskeleton _____

_____

(432) molting _____

_____

(432) tracheas _____

_____

(432) metamorphosis _____

_____

(433) chelicerates _____

_____

(434) crustaceans _____

_____

(435) millipedes _____

_____

(435) centipedes _____

_____

(435) insects _____

_____

(435) Malpighian tubules _____

_____

Fill-in-the-Blanks

Arthropods developed a thickened (1) _____ and a hardened (2) _____. In freshwater and saltwater environments, arthropods known as (3) _____ came to be well represented. The first land arthropods were the (4) _____. Their descendants—(5) _____, scorpions, ticks, and mites—are still living. (6) _____ and millipedes arose later. The larger aquatic arthropods extract oxygen from water with (7) _____, whereas most insects utilize (8) _____ systems, which provide the basis for some of the highest metabolic rates known. Arthropods develop a lightweight armor called the (9) _____; it protects its owner from predators and supports a body deprived of water's buoyancy. In land-dwelling arthropods, it protects against (10) _____ loss. An arthropod sheds this structure by (11) _____.

Labeling

Identify each numbered part of the animal pictured at the right.

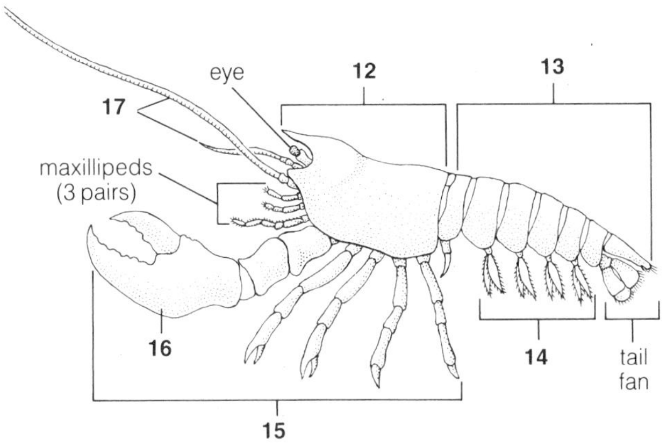

12. _____

13. _____

14. _____

15. _____

16. _____

17. _____

Answer exercises 18–24 for the animal pictured at the right.

18. Name the animal pictured. _____
19. Name the subgroup of arthropods to which this animal belongs. _____
20. Name two distinguishing characteristics of this group. _____
21. Protostome ❐ or deuterostome ❐ ?
22. Is this animal segmented? ❐ yes ❐ no
23. Symmetry of adult: ❐ radial ❐ bilateral
24. Does this animal have a true coelom? ❐ yes ❐ no

Identify each indicated body part in the illustration at the right.

25. _____

26. _____ _____

27. _____

28. _____

Answer exercises 29–30 for the animal pictured at the right.

29. Name the subgroup of arthropods to which the animal belongs. _____

30. Name the structure shown enlarged in the lower part of the picture. _____ _____

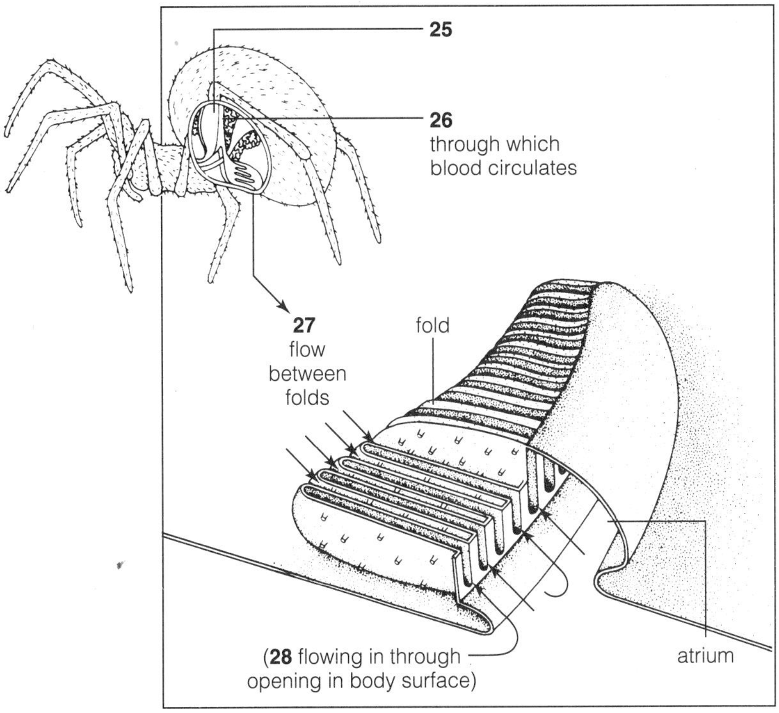

25

26
through which blood circulates

27
flow between folds

fold

(**28** flowing in through opening in body surface)

atrium

26-VI. ECHINODERMS (pp. 438–440)

Boldfaced, Page-Referenced Terms

(438) echinoderms _____

_____

(438) tube feet _____

_____

(438) water vascular system _____

_____

Fill-in-the-Blanks

Only two prominent groups of (1) _____ have survived to the present: the (2) _____ and the chordates. In echinoderms, (3) _____ symmetry has been overlaid on an earlier bilateral heritage; most echinoderms still go through a free-swimming, (4) _____ symmetrical larval stage. Echinoderm locomotion is based on constant circulation of seawater through a (5) _____-_____ system of canals and (6) _____ _____. (7) _____ _____ have a rounded, globose body with bristling spines. (8) _____ have many long feather-duster arms; both mouth and anus lie within the circlet of arms.

Labeling

Identify each indicated part of the illustration below.

9. _____ _____

10. _____ _____

11. _____

12. _____

13. _____

14. _____ _____

15. _____

16. _____ _____

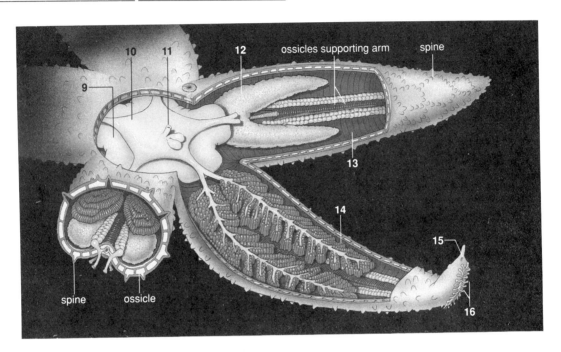

Answer exercises 17–19 for the animal pictured above.

17. Name the animal shown. _____
18. Does this animal have a true coelom? ☐ yes ☐ no
19. Protostome ☐ or deuterostome ☐ ?

Identifying

20. Name the system shown at right. _____

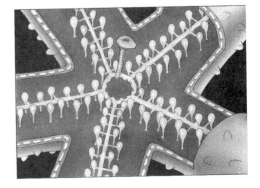

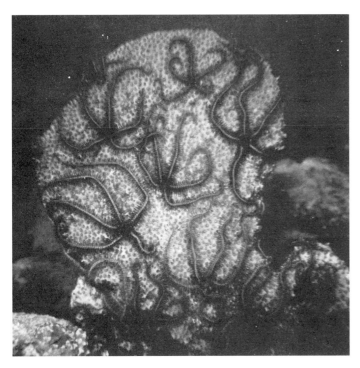

a. _____

b. _____

c. _____

d. _____

21. Identify each creature above by its common name.
22. Name the phylum of these animals pictured above. _____
23. Name two distinguishing characteristics of this group. _____
24. Symmetry of adult: ❏ radial ❏ bilateral

Self-Quiz

Multiple-Choice

___ 1. Which of the following is *not* true of sponges? They have no _____.
   a. distinct cell types
   b. nerve cells
   c. muscles
   d. gut

___ 2. Which of the following is *not* a protostome?
   a. earthworm
   b. crayfish or lobster
   c. sea star
   d. squid

___ 3. Bilateral symmetry is characteristic of _____.
   a. cnidarians
   b. sponges
   c. jellyfish
   d. flatworms

___ 4. Flukes and tapeworms are parasitic _____.
   a. leeches
   b. flatworms
   c. nematodes
   d. annelids

___ 5. Insects include _____.
   a. spiders, mites, and ticks
   b. centipedes and millipedes
   c. termites, aphids, and beetles
   d. all of the above

___ 6. Creeping behavior and a mouth located toward the "head" end of the body may have led, in some evolutionary lines, to _____.
   a. development of a circulatory system with blood
   b. sexual reproduction
   c. feeding on nutrients suspended in the water (filter feeding)
   d. concentration of sense organs in the head region

___ 7. Which of the following is associated with the shift from radial to bilateral body form?
   a. a circulatory system
   b. a one-way gut
   c. paired organs
   d. the development of a water-vascular system

___ 8. The _____ body plan is characterized by simple gas-exchange mechanisms, two-way traffic through a relatively unspecialized gut, and a thin body with all cells fairly close to the gut.
   a. annelid
   b. nematode
   c. echinoderm
   d. flatworm

___ 9. The _____ have a tough cuticle, longitudinal muscles, and a complete digestive system, and they can live under anaerobic conditions.
   a. nematodes
   b. cnidarians
   c. flatworms
   d. echinoderms

___ 10. _____ insulates various internal organs from the stresses of body-wall movement and bathes them in a liquid through which nutrients and waste products can diffuse.
   a. A coelom
   b. Mesoderm
   c. A mantle
   d. A water-vascular system

___ 11. The annelid _____ may resemble the ancestral structure from which the vertebrate kidney evolved.
   a. trachea
   b. nephridium
   c. mantle
   d. parapodium

Matching

Match each phylum below with the corresponding characteristics (a–k) and representatives (A–P). A phylum may match with more than one letter from the group of representatives.

___, ___12. Annelida

___, ___13. Arthropoda

___, ___14. Chordata

___, ___15. Cnidaria

___, ___16. Ctenophora

___, ___17. Echinodermata

___, ___18. Mollusca

___, ___19. Nematoda

___, ___20. Platyhelminthes

___, ___21. Porifera

___, ___22. Rotifera

a. radial (biradial) symmetry + no stinging cells + comb plates
b. choanocytes (= collar cells) + spicules
c. jointed legs + an exoskeleton
d. gill slits in pharynx + dorsal, tubular nerve cord + notochord
e. pseudocoelomate + wheel organ + soft body
f. soft body + mantle; may or may not have radula or shell
g. bilateral symmetry + blind-sac gut
h. radial symmetry + blind-sac gut; stinging cells
i. body compartmentalized into repetitive segments; coelom containing nephridia (= primitive kidneys)
j. tube feet + calcium carbonate structures in skin
k. complete gut + bilateral symmetry + cuticle; includes many parasitic species, some of which are harmful to humans

A. Dinosaurs
B. Corals, sea anemones, and *Hydra*
C. Salamanders and toads
D. Whales and opossums
E. Tapeworms and *Planaria*
F. Insects
G. Jellyfish and the Portuguese man-of-war
H. Sand dollars and starfishes

I. Earthworms and leeches
J. Lobsters, shrimp, and crayfish
K. Organisms with spicules and choanocytes
L. Scorpions and millipedes
M. Octopuses and oysters
N. Flukes
O. Hookworm, trichina worm
P. Comb jellies, sea gooseberries

Chapter Objectives/Review Questions

This section lists general and detailed chapter objectives that can be used as review questions. You can make maximum use of these items by writing answers on a separate sheet of paper. Fill in answers where blanks are provided. To check for accuracy, compare your answers with information given in the chapter or glossary.

Page	Objectives/Questions
(410–411)	1. Distinguish radial symmetry from bilateral symmetry and a true body cavity from a false body cavity.
(410)	2. Explain how radial symmetry might be more advantageous to floating or sedentary animals than bilateral symmetry.
(411)	3. List two benefits that the development of a coelom brings to an animal.
(412)	4. Be able to reproduce from memory a diagram that expresses the relationships between the major groups of animals.
(413)	5. List two characteristics that distinguish sponges from other animal groups.
(414)	6. State what nematocysts are used for and explain how they operate.
(414)	7. Describe the two cnidarian body types.

(418–419)   8. List the three main types of flatworms.

(418–419)   9. Describe the body plan of roundworms, comparing its various systems with those of the flatworm body plan.

(423)   10. Define *protostome* and *deuterostome* and give examples of each group.

(424–427) 11. Define *mantle* and tell what role it plays in the molluscan body.

(427)   12. Explain why you think cephalopods came to have such well-developed sensory and motor systems and are able to learn.

(428)   13. Name the three groups of annelids and give a specific example from each group.

(431)   14. Define *segmentation* and explain how it is related to the development of muscular and nervous systems.

(431)   15. State which ancestors are thought to have given rise to the arthropods.

(432)   16. List six different groups of arthropods.

(432)   17. Explain how the development of a thickened cuticle and a hardened exoskeleton affected the ways that arthropods lived.

(438)   18. Describe how locomotion occurs in echinoderms.

Integrating and Applying Key Concepts

Scan Table 26.2 to verify that most highly evolved animals have a complete gut, a closed blood-vascular system, both central and peripheral nervous systems, and are dioecious. Why do you suppose having two sexes in separate individuals is considered to be more highly evolved than the monoecious condition utilized by earthworms? Wouldn't it be more efficient if all individuals in a population could produce both kinds of gametes? Cross-fertilization would then result in both individuals being able to produce offspring.

Answers

Interactive Exercises

26-I. OVERVIEW OF THE ANIMAL KINGDOM
(pp. 408–411)
AN EVOLUTIONARY ROAD MAP (p. 412)
THE ONE PLACOZOAN (p. 412)
SPONGES (pp. 412–413)
CNIDARIANS (pp. 414–417)
COMB JELLIES (p. 417)

1. a. Porifera; b. Cnidaria; c. Platyhelminthes; d. Nematoda; e. Mollusca; f. Annelida; g. Arthropoda; h. Echinodermata; i. *Hydra* (*Obelia*, Portuguese man-of-war); j. Snails (nudibranchs, oysters); k. Earthworms (oligochaetes); 2. invertebrates; 3. placozoan; 4. Sponges; 5. collar cells; 6. Communication; 7. nerve; 8. radially; 9. nematocysts; 10. planula; 11. medusa; 12. Comb jellies; 13. feeding polyp; 14. reproductive polyp; 15. female medusa; 16. planula; 17. mollusks; 18. chordates; 19. coelomate; 20. bilateral; 21. radial.

26-II. FLATWORMS (pp. 418–419)
ROUNDWORMS (pp. 419–421)
RIBBON WORMS (p. 422)
ROTIFERS (pp. 422–423)

1. planuloid; 2. paired organs; 3. protonephridia; 4. respiratory (circulatory); 5. digestive; 6. fluke; 7. scolex; 8. Nematodes (Roundworms); 9. false coelom; 10. reproductive; 11. skeleton; 12. Rotifers; 13. branching gut; 14. pharynx; 15. brain; 16. nerve cord; 17. ovary; 18. testis; 19. planarian; 20. no; 21. yes; 22. no; 23. roundworm; 24. no; 25. yes (a "false" coelom).

26-III. A MAJOR DIVERGENCE (p. 423)
MOLLUSKS (pp. 424–427)

1. Deuterostomes; 2. anus; 3. mouth; 4. protostomes; 5. earthworms; 6. insects (spiders); 7. mollusks; 8. cephalopods; 9. brains (eyes); 10. vision; 11. motor (movement, muscular); 12. mantle; 13. mantle cavity; 14. III; 15. I; 16. II; 17. mouth; 18. anus; 19. gill; 20. heart; 21. radula; 22. foot; 23. shell; 24. stomach; 25. mouth; 26. gill; 27. mantle; 28. muscle; 29. foot;

30. stomach; 31. internal shell; 32. mantle; 33. gonad; 34. gill.

26-IV. ANNELIDS (pp. 428–431)
1. Annelids; 2. polychaetes (marine worms); 3. anus; 4. coelom; 5. blood; 6. hemoglobin; 7. Segmentation; 8. nerve cord; 9. nephridia; 10. parapodia; 11. aerate (burrow through); 12. marine; 13. brain; 14. pharynx; 15. nerve cord; 16. hearts; 17. blood vessel; 18. crop; 19. gizzard; 20. earthworm; 21. Annelida; 22. segmented; closed circulatory system; 23. protostome; 24. yes; 25. bilateral; 26. yes.

26-V. ARTHROPODS (pp. 432–437)
1. cuticle; 2. exoskeleton; 3. crustaceans; 4. arachnids; 5. spiders; 6. Centipedes; 7. gills; 8. tracheal; 9. exoskeleton; 10. water (fluid); 11. molting; 12. cephalothorax (carapace); 13. abdomen; 14. swimmerets; 15. legs; 16. cheliped; 17. antennae; 18. lobster; 19. crustaceans; 20. exoskeleton and jointed legs; 21. Protostome; 22. yes; 23. bilateral; 24. yes; 25. heart; 26. body cavity; 27. "blood" (body fluids); 28. air; 29. chelicerates; 30. book lung.

26-VI. ECHINODERMS (pp. 438–440)
1. deuterostomes; 2. echinoderms; 3. radial; 4. bilaterally; 5. water-vascular; 6. tube feet; 7. Sea urchins; 8. Crinoids; 9. lower stomach; 10. upper stomach; 11. anus; 12. gonad; 13. coelom; 14. digestive gland; 15. eyespot; 16. tube feet; 17. sea star, starfish; 18. yes; 19. deuterostome; 20. water-vascular system; 21. a. brittle stars; b. sea urchin; c. sea cucumber; d. feather star (crinoid); 22. Echinodermata; 23. water vascular system and body wall with spines, spicules, or plates; 24. radial.

Self-Quiz
1. a; 2. c; 3. d; 4. b; 5. c; 6. d; 7. c; 8. d; 9. a; 10. a; 11. b; 12. i, I; 13. c, F, J, L; 14. d, A, C, D; 15. h, B, G; 16. a, P; 17. j, H; 18. f, M; 19. k, O; 20. g, E, N; 21. b, K; 22. e.

27

ANIMALS: THE VERTEBRATES

Interactive Exercises

27-I. THE CHORDATE HERITAGE (pp. 441–446)
INVERTEBRATE CHORDATES (pp. 447–448)
ORIGIN OF VERTEBRATES (pp. 448–449)
EVOLUTIONARY TRENDS AMONG THE VERTEBRATES (pp. 450–451)
FISHES (pp. 452–456)

Selected Italicized Words

jawless fishes; jawed, armored fishes; amphibians; reptiles; birds; mammals; "tunic"

Boldfaced, Page-Referenced Terms

(446) chordates _____

_____

(446) notochord _____

_____

(446) nerve cord _____

(446) pharynx _____

(447) tunicates _____

(447) gill slits _____

(448) lancelets _____

(448) hemichordates _____

(450) vertebrae _____

(450) jaws _____

(450) fins _____

(451) gills _____

(451) lungs _____

(452) swim bladder _____

(452) ostracoderms _____

(453) placoderms _____

(453) lampreys _____

(453) hagfishes _____

(454) cartilaginous fishes _____

(454) scales _____

(454) bony fishes _____

(455) lobe-finned fishes _____

_____

Fill-in-the-Blanks

Four major features distinguish the embryos of chordates from those of all other animals: a hollow, dorsal (1) _____ _____; a(n) (2) _____ with slits in its wall; a(n) (3) _____; and a tail that extends past the anus during at least part of their lives. In some chordates, the (4) _____ chordates, the notochord is *not* divided into a skeletal column of separate, hard segments; in others, the (5) _____, it is.

Invertebrate chordates living today are represented by tunicates and (6) _____, which obtain their food by (7) _____ _____; they draw in plankton-laden water through the mouth and pass it over sheets of mucus, which trap the particulate food before the water exits through the (8) _____ _____ in the pharynx. (9) _____ are among the most primitive of all living chordates; when they are tiny, they look and swim like (10) _____. A rod of stiffened tissue, the (11) _____, runs the length of the larval body; it is the forerunner of the chordate's (12) _____.

As the ancestors of land vertebrates began spending less time immersed and more of their lives exposed to air, use of gills declined and (13) _____ evolved; more elaborate and efficient (14) _____ systems evolved along with more complex and efficient lungs.

Even though the adult forms of acorn worms, echinoderms, and chordates look very different, they have similar embryonic developmental patterns. Chordates may have developed from a mutated ancestral deuterostome (15) _____ of a sessile, filter-feeding adult. A larva is an immature, motile form of an organism, but if a (16) _____ occurred that caused (17) _____ _____ to become functional in the larval body, then a motile larva that could reproduce would have been more successful in finding food and avoiding (18) _____ than a sessile adult; in time, the sessile stages in the species would be eliminated.

The ancestors of the vertebrate line may, or may not have been mutated forms of modern forms they resemble, the (19) _____, in which the notochord became segmented and the segments became hardened (20) _____. The vertebral column was the foundation for fast-moving (21) _____, some of which were ancestral to all other vertebrates. The evolution of (22) _____ intensified the competition for prey and the competition to avoid being preyed upon; animals in which mutations expanded the nerve cord into a (23) _____ that enabled the animal to compete effectively survived more frequently than their duller-witted fellows and passed along their genes into the next generations. Fins became (24) _____; in some fishes, those became (25) _____ and equipped with skeletal supports; these forms set the stage for the development of legs, arms, and wings in later groups.

Exercises 26–27 refer to the illustrations on page 296.

26. Name the creature in illustration B. _____

27. Name the creature in illustration D. _____

Labeling

Name the structures numbered below.

28. _____

29. _____ with _____ _____

30. _____

31. _____ _____

32. _____

33. _____

34. _____

35. _____ _____

36. _____ _____

37. _____ _____

38. _____

39. _____ _____ _____

40. _____ _____ _____

41. _____

42. _____

43. _____ _____ _____

_____

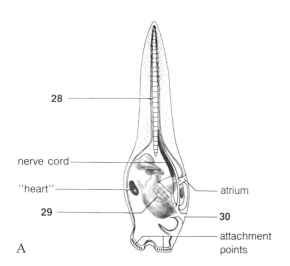

28

nerve cord

"heart"

29

A

atrium

30

attachment points

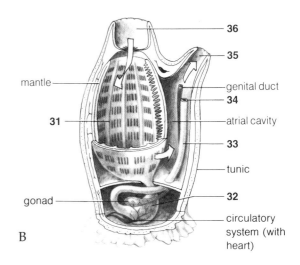

36

35

mantle

genital duct

34

31

atrial cavity

33

tunic

32

gonad

circulatory system (with heart)

B

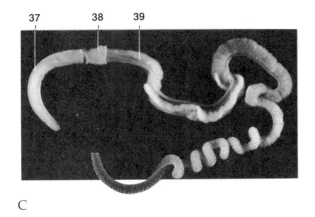

37   38   39

C

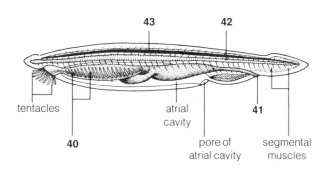

43   42

tentacles

atrial cavity

41

40

pore of atrial cavity

segmental muscles

D

44. Name the creature whose head is illustrated at the right. _____

45. What structure occupies most of the head space of the creature at the right? _____-_____ _____

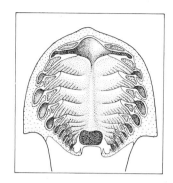

Complete the Table

Provide common names wherever (*) appears, and provide the class name wherever ❐ appears.

Phylum: Hemichordata   *Phylum: Chordata*

46*	*Subphylum: Urochordata*	*Subphylum: Cephalochordata*	*Subphylum: Vertebrata*	
	47*	48*	Class: *Agnatha*	49*
			Class: *Placodermi*	50*
			Class: *Chondrichthyes*	51*
			Class: *Osteichthyes*	52*
			Class: 53 ❐	amphibians
			Class: 54 ❐	reptiles
			Class: 55 ❐	birds
			Class: 56 ❐	mammals

Analysis and Short Answer

Fig. 27.10 mentions features that distinguish the amphibian-lungfish lineage from the lineage that leads to sturgeons and other bony fishes. At least one of those features may have developed near ⑤ in this evolutionary tree.

Exercises 57–67 refer to the evolutionary diagram illustrated below.

57. What single feature do the lampreys, hagfishes, and extinct ostracoderms have in common that is different from the placoderms? _____

58. How did ostracoderms feed? _____

59. A mutation in ostracoderm stock led to the development of what in all organisms that descended from ① ? _____

60. Mutation at ② led to the development of an endoskeleton made of what? _____

61. Mutations at ③ led to an endoskeleton of what? _____

62. Mutations at ④ led to which spectacularly diverse fishes that have delicate fins originating from the dermis? _____

63. Mutations at ⑤ led to which fishes whose fins incorporate fleshy extensions from the body?

_____

64. Which branch, ④ or ⑤ , gave rise to the amphibians? _____

65. Which branch gave rise to the modern bony fishes? _____

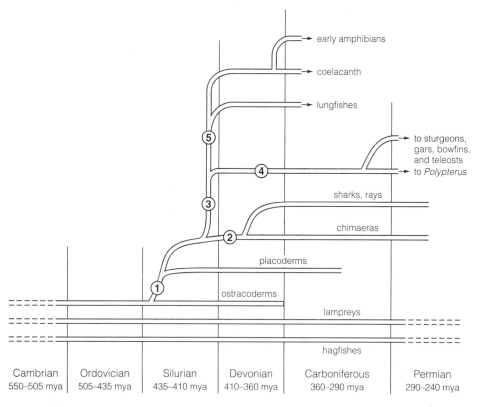

Cambrian	Ordovician	Silurian	Devonian	Carboniferous	Permian
550–505 mya	505–435 mya	435–410 mya	410–360 mya	360–290 mya	290–240 mya

mya = million years ago

66. In which period did three distinctly different lineages of *bony* fishes appear in the fossil record?

_____

67. Approximately how many million years ago did the fork in the evolutionary path that led to the amphibians occur? _____

27-II. AMPHIBIANS (pp. 456–457)
REPTILES (pp. 458–461)

Boldfaced, Page-Referenced Terms

(456) amphibian _____

_____

(456) salamanders _____

_____

(457) frogs _____

_____

(457) caecilians _____

_____

(458) reptiles _____

_____

(458) cerebral cortex _____

_____

(458) amniote egg _____

_____

(460) turtles _____

_____

(460) lizards _____

_____

(460) snakes _____

_____

(461) tuataras _____

_____

(461) crocodilians _____

_____

Fill-in-the-Blanks

Natural selection acting on lobe-finned fishes during the Devonian period favored the evolution of ever more efficient (1) _____ used in gas exchange and stronger (2) _____ used in locomotion. Without the buoyancy of water, an animal traveling over land must support its own weight against the pull of gravity. The (3) _____ of early amphibians underwent dramatic modifications that involved evaluating incoming signals related to vision, hearing, and (4) _____. Although fish have (5) _____-chambered hearts, amphibians have (6) _____-chambered hearts (see Fig. 27.9). Early in amphibian evolution, mutations may have created the third chamber, which added a second (7) _____ in addition to the already existing atrium and ventricle. The Carboniferous period brought humid, forested swamps with an abundance of aquatic invertebrates and (8) _____—ideal prey for amphibians.

There are three groups of existing amphibians: (9) _____, frogs and toads, and caecilians. Amphibians require free-standing (10) _____ or at least a moist habitat to (11) _____. Amphibian skin generally lacks scales but contains many glands, some of which produce (12) _____. In the late Carboniferous, (13) _____ began a major adaptive radiation into the lush habitats on land, and only amphibians that mutated and developed certain (14) _____ features were able to follow them and exploit an abundant food supply. Several features helped: modification of (15) _____ bones favored swiftness, modification of teeth and jaws enabled them to feed efficiently on a variety of prey items, and the development of a (16) _____ _____ egg protected the embryo inside from drying out, even in dry habitats.

(Consult Fig. 27.16 of the main text, the time line below, and the figure for the Analysis and Short Answer on the opposite page of this Study Guide.)

Today's reptiles include (17) _____, crocodilians, snakes, and (18) _____. All rely on (19) _____ fertilization, and most lay leathery eggs. Although amphibians originated during Devonian times, ancestral "stem" reptiles appeared during the (20) _____ period, about 340 million years ago. Reptilian groups living today that have existed on Earth longest are the (21) _____; their ancestral path diverged from that of the "stem" reptiles during the (22) _____ period. Crocodilian ancestors appeared in the early (23) _____ period, about 220 million years ago. Snake and lizard stocks diverged from the tuatara line during the late (24) _____ period, about 140 million years ago. (25) _____ are more closely related to extinct dinosaurs and crocodiles than to any other existing vertebrates; they, too, have a (26) _____-chambered heart. Mammals have descended from therapsids, which in turn are descended from the (27) _____ group of reptiles, which diverged earlier from the stem reptile group during the (28) _____ period, approximately 320 million years ago.

In blanks 29–35, arrange the following groups in sequence, from earliest to latest, according to their appearance in the fossil record:

A. birds   B. crocodilians   C. dinosaurs   D. early ancestors of mammals (= synapsid reptiles)
   E. early ancestors of turtles (= anapsid reptiles)   F. lizards and snakes   G. "stem" reptiles

29. ___ 30. ___ 31. ___ 32. ___ 33. ___ 34. ___ 35. ___

In blanks 36–42, select from the choices immediately below the geologic period in which each group first appeared and write each in the correct space.

Paleozoic Era			*Mesozoic Era*	
A. Carboniferous 360–290 mya	B. Permian 290–240 mya	C. Triassic 240–205 mya	D. Jurassic 205–138 mya	E. Cretaceous 138–65 mya

36. ___ 37. ___ 38. ___ 39. ___ 40. ___ 41. ___ 42. ___

Analysis and Short Answer

43. In this chapter of the text (p. 464) the authors mention the features that distinguish, say, the mammals from all of the remaining groups on the right-hand side of the diagram; therefore, at some time during the evolution of mammals, mutations that produced those features appeared. Consult the evolutionary diagram below and imagine what sort of mutation(s) occurred at the numbered places.

   a.   What may have occurred at ① ? _____
   b.   What may have occurred at ② ? _____
   c.   What may have occurred at ③ ? _____
   d.   What may have occurred at ④ ? _____
   e.   What may have occurred at ⑤ ? _____

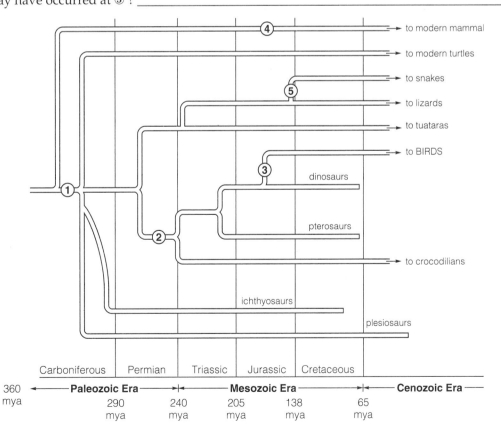

27-III. BIRDS (pp. 462–463)
MAMMALS (pp. 464–467)

Boldfaced, Page-Referenced Terms

(462) birds _____

_____

(462) feathers _____

_____

(464) mammals _____

_____

(464) therapsids _____

_____

(464) dentition _____

_____

(466) placenta _____

_____

Labeling

Label the structures pictured at the right.

1. _____

2. _____

3. _____ _____

Fill-in-the-Blanks

Birds descended from (4) _____ that ran around on two legs some 160 million years ago. All birds have (5) _____ that insulate them and help them get aloft. Generally, birds have a greatly enlarged (6) _____ to which flight muscles are attached. Bird bones contain (7) _____ _____, which reduce their weight. Air flows through sacs, not into and out of them, for gas exchange in the lungs. Birds also lay (8) _____ _____ eggs, have complex courtship behaviors, and generally nurture their offspring. Birds are able to regulate their body (9) _____, which is generally higher than that of mammals. Existing birds do not have socketed (10) _____ in their beaks.

There are three groups of existing mammals: those that lay eggs (examples are the (11) _____ and the spiny anteater), those that are (12) _____ (examples are the opossum and the kangaroo), and those that are (13) _____ mammals (there are more than 4,500 species of these). Mammals regulate their body temperature, have a (14) _____-chambered heart, and show a high degree of parental nurture. Most mammals have (15) _____ as a means of insulation, and mammalian mothers generally suckle their young with milk.

Self-Quiz

Multiple Choice

Select the best answer.

___1. Filter-feeding chordates rely on _____, which have cilia that create water currents and mucous sheets that capture nutrients suspended in the water.

a. notochords
b. differentially permeable membranes
c. filiform tongues
d. gill slits

___2. In true fishes, the gills serve primarily _____ function.

a. a gas-exchange
b. a feeding
c. a water-elimination
d. both a feeding and a gas-exchange

___3. The hearts in amphibians _____.

a. pumps blood more rapidly than the heart of fish
b. are efficient enough for amphibians but would not be efficient for birds and mammals
c. have three chambers (ventricle and two atria)
d. all of the above

___4. The feeding behavior of true fishes selected for highly developed _____.

a. parapodia
b. notochords
c. sense organs
d. gill slits

Identification

Provide the common name and the major chordate group to which each creature pictured below and on pages 304–305 belongs.

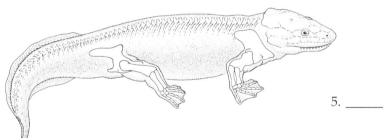

5. _____, _____

6. _____, _____

7. _____, _____

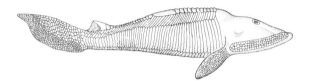

8. _____, _____

9. _____, _____

10. _____, _____

11. _____, _____

12. _____, _____

13. _____, _____

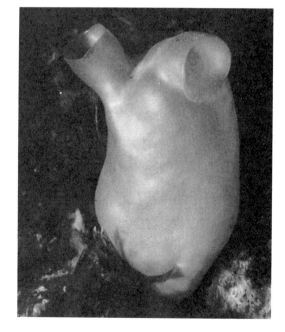

14. _____, _____

15. _____, _____

Matching

Match the following groups and classes with the corresponding characteristics (a–i) and representatives (A–I).

16. ___, ___ Amphibians

17. ___, ___ Birds

18. ___, ___ Bony fishes

19. ___, ___ Cartilaginous fishes

20. ___, ___ Cephalochordates

21. ___, ___ Jawless fishes

22. ___, ___ Mammals

23. ___, ___ Reptiles

24. ___, ___ Urochordates

a. hair + vertebrae
b. feathers + hollow bones
c. jawless + cartilaginous skeleton (in existing species)
d. two pairs of limbs (usually) + glandular skin + "jelly"-covered eggs
e. amniote eggs + scaly skin + bony skeleton
f. invertebrate + sessile adult that cannot swim
g. jaws + cartilaginous skeleton + vertebrae
h. in adult, notochord stretches from head to tail; mostly burrowed-in, adult can swim
i. bony skeleton + skin covered with scales and mucus

A. lancelet
B. loons, penguins, and eagles
C. tunicates, sea squirts
D. sharks and manta rays
E. lampreys and hagfishes (and ostracoderms)
F. true eels and sea horses
G. lizards and turtles
H. caecilians and salamanders
I. platypuses and opossums

Chapter Objectives/Review Questions

This section lists general and detailed chapter objectives that can be used as review questions. You can make maximum use of these items by writing answers on a separate sheet of paper. To check for accuracy, compare your answers with information given in the chapter or glossary.

Page	Objectives/Questions
(445–446)	1. List three characteristics found only in chordates.
(447–449)	2. Describe the adaptations that sustain the sessile or sedentary lifestyle seen in primitive chordates such as tunicates and lancelets.
(448–450)	3. State what sort of changes occurred in the primitive chordate body plan that could have promoted the emergence of vertebrates.
(450–456)	4. Describe the differences between primitive and advanced fishes in terms of skeleton, jaws, special senses, and brain.
(455–457)	5. Describe the changes that enabled aquatic fishes to give rise to land dwellers.
(451, 456, 458, 462, 464)	6. State what kind of heart each of the four groups of four-limbed vertebrates has, and list the principal skin structures that each produces.
(462–467)	7. Discuss the effects that increased parental nurture of offspring in birds and mammals has had on courtship behavior and reproductive physiology.

Integrating and Applying Key Concepts

Birds and mammals both have four-chambered hearts and high metabolic rates and regulate their body temperatures efficiently. Both groups evolved from reptiles, so one would think that those same traits would have developed in ancestral reptiles. Data suggest that most reptiles have a heart intermediate between three and four chambers, lower metabolic rates, and body temperatures that are not well regulated and tend to rise and fall in accord with the environmental temperature. If the three traits mentioned in the first sentence had developed in reptilian groups, how might their lives have been different?

Answers

Interactive Exercises

27-I. THE CHORDATE HERITAGE (p. 446)
   INVERTEBRATE CHORDATES (pp. 447–448)
   ORIGIN OF VERTEBRATES (pp. 448–449)
   EVOLUTIONARY TRENDS AMONG THE
   VERTEBRATES (pp. 450–451)
   FISHES (pp. 452–456)
1. nerve cord; 2. pharynx; 3. notochord; 4. invertebrate; 5. vertebrates; 6. lancelets; 7. filter feeding; 8. gill slits; 9. Tunicates (sea squirts); 10. tadpoles; 11. notochord; 12. spine (backbone); 13. lungs; 14. circulatory; 15. larva; 16. mutation; 17. sex organs; 18. predators; 19. cephalochordates (lancelets); 20. vertebrae; 21. predators; 22. jaws; 23. brain; 24. paired; 25. fleshy; 26. adult tunicate (sea squirt); 27. lancelet; 28. notochord; 29. pharynx with gill slits; 30. mouth; 31. sievelike pharynx; 32. stomach; 33. intestine; 34. anus; 35. excurrent siphon; 36. incurrent siphon; 37. conelike proboscis; 38. collar; 39. gill slit region; 40. pharyngeal gill slits; 41. anus; 42. notochord; 43. dorsal, tubular nerve cord; 44. ostracoderm; 45. food-straining pharynx; 46. acorn worms; 47. tunicates; 48. lancelets; 49. lampreys, hagfishes; 50. jawed, armored fishes; 51. cartilaginous fishes; 52. bony fishes; 53. Amphibia; 54. Reptilia; 55. Aves; 56. Mammalia; 57. all jawless; 58. filter feeders; 59. jaws; 60. cartilage; 61. bone; 62. ray-finned fishes; 63. lobe-finned fishes; 64. branch ⑤; 65. branch ④; 66. Devonian; 67. about 375 million years ago.

27-II. AMPHIBIANS (pp. 456–457)
   REPTILES (pp. 458–461)
1. lungs; 2. fins; 3. brain; 4. balance; 5. two; 6. three; 7. atrium; 8. insects; 9. salamanders; 10. water;

11. reproduce; 12. toxins; 13. insects; 14. reptilian; 15. limb; 16. shelled amniote; 17. turtles (lizards); 18. lizards (turtles); 19. internal; 20. Carboniferous; 21. turtles; 22. Carboniferous; 23. Triassic; 24. Jurassic; 25. Birds; 26. four; 27. synapsid; 28. Carboniferous; 29. G; 30. D; 31. E; 32. B; 33. C; 34. A; 35. F; 36. A; 37. A; 38. A; 39. C; 40. C; 41. D; 42. D; 43. a. dry, scaly skin; b. four-chambered heart; c. feather development; d. hair development; e. loss of limbs.

27-III. BIRDS (pp. 462–463)
   MAMMALS (pp. 464–467)
1. embryo (notochord); 2. albumin; 3. yolk sac; 4. reptiles; 5. feathers; 6. sternum (breastbone); 7. air cavities; 8. shelled amniote; 9. temperature; 10. teeth; 11. platypus; 12. pouched (marsupials); 13. placental; 14. four; 15. hair.

Self-Quiz
1. d; 2. a; 3. d; 4. c; 5. early amphibian, Amphibia; 6. Arctic fox, Mammalia; 7. soldier fish, Osteichthyes; 8. ostracoderm, Agnatha; 9. owl, Aves; 10. sea turtle, Reptilia; 11. shark, Chondrichthyes; 12. coelacanth (lobe-finned fish), Osteichthyes; 13. reef ray, Chondrichthyes; 14. tunicate, Urochordata; 15. lancelet, Cephalochordata; 16. d, H; 17. b, B; 18. i, F; 19. g, D; 20. h, A; 21. c, E; 22. a, I; 23. e, G; 24. f, C.

28

HUMAN EVOLUTION: A CASE STUDY

PRIMATE CLASSIFICATION

FROM PRIMATE TO HUMAN: KEY EVOLUTION-
ARY TRENDS
    Upright Walking
    Precision Grips and Power Grips
    Enhanced Daytime Vision
    Teeth for All Occasions
    Better Brains, Bodacious Behavior

PRIMATE ORIGINS

THE FIRST HOMINIDS

ON THE ROAD TO MODERN HUMANS
    Early *Homo*
    From *Homo erectus* to *H. sapiens*

Interactive Exercises

28-I. PRIMATE CLASSIFICATION (pp. 470–472)
    FROM PRIMATE TO HUMAN: KEY EVOLUTIONARY TRENDS (pp. 473–474)
    PRIMATE ORIGINS (p. 475)

Selected Italicized Words

prosimian, prehensile, opposable, Plesiadapis, Aegyptopithecus

Boldfaced, Page-Referenced Terms

(472) Primates _____

_____

(472) anthropoids _____

_____

(472) hominoids _____

_____

(472) hominids _____

_____

(473) savannas _____

_____

(473) bipedalism _____

_____

(474) culture _____

_____

(475) dryopiths _____

_____

Fill-in-the-Blanks

During the Cenozoic Era (65 million years ago to the present), birds, (1) _____, and flowering plants
evolved to dominate Earth's assemblage of organisms. (1) are warm-blooded vertebrates with
(2) _____ that began their evolution more than 200 million years ago. There are many groups
[(3) _____] within the class Mammalia; each order has its distinctive array of characteristics. Humans,
apes, monkeys, and prosimians are all (4) _____; members of this order have excellent (5) _____
perception as a result of their forward-directed eyes. They also have hands that are (6) _____; that is,
they are adapted for (7) _____ instead of running. Primates rely less on their sense of (8) _____
and more on daytime vision. Their (9) _____ became larger and more complex; this trend was accompanied by refined technologies and the development of (10) _____: the collection of behavior patterns
of a social group, passed from generation to generation by learning and by symbolic behavior (language).
Modification in the (11) _____ led to increased dexterity and manipulative skills. Changes in primate
(12) _____ indicate that there was a shift from eating insects to fruit and leaves and on to a mixed diet.
Primates began to evolve from ancestral mammals more than (13) _____ million years ago, during the
Paleocene. The first primates resembled small (14) _____ or tree shrews; they foraged at night for
(15) _____, seeds, buds, and eggs on the forest floor, and they could clumsily climb trees searching for
safety and sleep.

   During the (16) _____ (25–5 million years ago), the continents began to assume their current positions, and climates became cooler and (17) _____. Under these conditions, forests began to give way to
mixed woodlands and (18) _____; an adaptive radiation of apelike forms—the first hominoids—took
place as subpopulations of apes became reproductively isolated within the shrinking stands of trees. The
chimpanzee-sized "tree apes," or (19) _____, originated during this time; eventually, they ranged
throughout Africa, Europe, and Southern (20) _____. Between (21) _____ million and
(22) _____ million years ago, three divergences occurred that gave rise to the ancestors of modern
gorillas, chimpanzees, and humans.

Matching

Fill in the blanks of the evolutionary diagram below with the appropriate letter from the choice list.

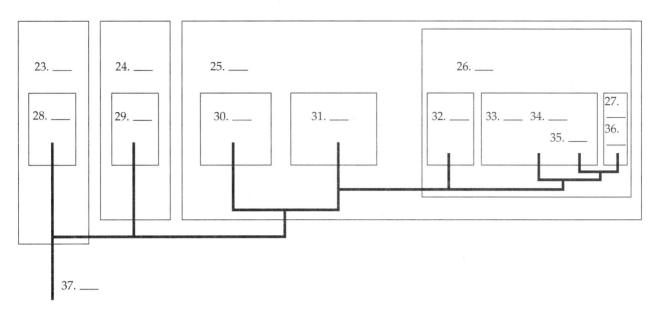

A. Anthropoids
B. Chimpanzee
C. Gibbon, siamang
D. Gorilla
E. Hominids
F. Hominoids
G. Humans and their most recent ancestors
H. Lemurs, lorises

I. New World monkeys (spider monkeys and so on)
J. Old World monkeys (baboons and so on)
K. Orangutan
L. Prosimians
M. Tarsiers
N. Tarsioids
O. Rodentlike primate of the Paleocene

28-II. THE FIRST HOMINIDS (pp. 476–477)
ON THE ROAD TO MODERN HUMANS (pp. 478–479)

Boldfaced, Page-Referenced Terms

(476) African savanna _____

_____

(476) australopiths _____

_____

(478) early *Homo* _____

_____

(478) *Homo erectus* _____

_____

(478) *Homo sapiens* _____

_____

Fill-in-the-Blanks

The family Hominidae (hominids) includes all species on the genetic path leading to humans since the time when that path diverged from the path leading to the (1) _____; hominids emerged between (2) _____ million and (3) _____ million years ago, during the late Miocene. (4) _____-million-year-old fossils of humanlike forms have been discovered in Africa, and they all were bipedal and omnivorous and had an expanded brain. Lucy was one of the earliest (5) _____, a collection of forms that combined ape and human features; they were fully two-legged, or (6) _____, with essentially human bodies and ape-shaped heads.

The oldest fossils of the genus *Homo*, makers of stone tools, date from approximately (7) _____ million years ago. Between 1.8 million and 400,000 years ago, during the Pleistocene periods of glaciation, there were also intermittent periods of warming; during these interglacial times, a larger-brained human species, (8) _____ _____, migrated out of Africa and into China, Southeast Asia, and Europe. Over time, (8) became better at toolmaking and learned how to control (9) _____ as members of the species became adapted to a wide range of habitats. Fossils of (8) are frequently associated with abundant (10) _____ artifacts. The (11) _____ were a distinct hominid population that appeared 130,000 years ago in southern France, central Europe, and the Near East; their cranial capacity was indistinguishable from our own, and they had a complex culture.

Anatomically modern humans evolved from (12) _____ _____. Analysis of (13) _____ and (14) _____ comparisons and specimens from the fossil record support the model of human origins that hypothesizes that *Homo erectus* migrated out of Africa, then formed distinctive subpopulations of modern humans as an outcome of genetic divergence in different geographic regions. From (15) _____ years ago to the present, human evolution has been almost entirely cultural rather than biological. By 30,000 years ago, there was only one remaining hominid species: (16) _____ _____.

Matching

Suppose you are a student who wants to be chosen to accompany a paleontologist who has spent forty years teaching and roaming the world in search of human ancestors. Above the desk in her office, she keeps reconstructions of the seven skulls shown on the opposite page, and you have heard that she chooses the graduate students who accompany her on her summer safaris on the basis of their ability on a short matching quiz. The quiz is presented below. *Without consulting your text (or any other)* while you are taking the quiz, match up all seven skulls with *all* applicable letters. Sixteen correctly placed letters win you a place on the expedition. Fourteen correct answers put you on a waiting list. Thirteen or fewer and she suggests that you may wish to investigate dinosaur fossils instead of human ancestry. Which will it be for you?

17.

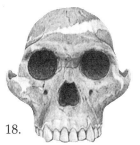

18.

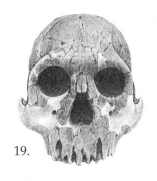

19.

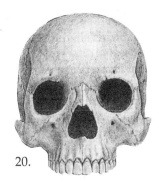

20.

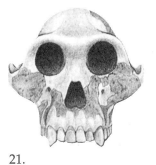

21.

A. Gracile forms of australopiths
B. Robust forms of australopiths
C. Fashioned stone tools and used them first
D. Has a chin
E. First controlled use of fire
F. Lived 4 million years ago
G. From a lineage that lived from approximately 3 million years ago until about 1.25 million years ago
H. Lived 2 million years ago
I. Lived 1.5 million years ago until at least 100,000 years ago
J. The most recent hominid to appear

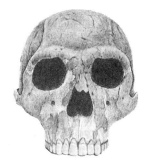

23.

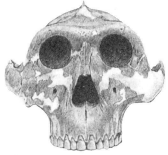

22.

17. ___
18. ___
19. ___
20. ___
21. ___
22. ___
23. ___

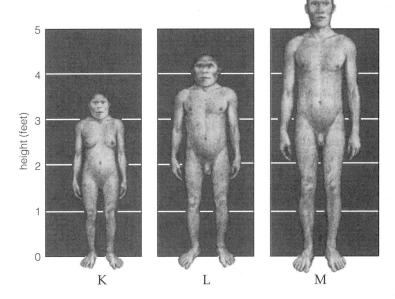

Crossword Puzzle: Primate Evolution

ACROSS

1. A group that includes prosimians, tarsioids, and anthropoids.
3. A group that includes apes and humans.
7. _____ evolution is a time of rapid branchings and adaptive radiation.
10. A kind of tall, showy flower; usually lavender or purple.
11. A cheek tooth that crushes and grinds.
12. All the behavior patterns of a social group passed by learning and language from generation to generation.
13. Pay great homage to; worship.
14. Forest apes that lived 13 million years ago in Africa, Europe, and southern Asia.
17. Pointed teeth that enable the tearing of flesh.
18. *Homo* _____ was the first hominid to control and use fire for heating and cooking.
19. The ability to adapt to a variety of agents of change in an animal's environment.

DOWN

2. Southern ape-humans, the fossils of which have dates from 3.7 to 1.25 million years ago.
4. A vertebrate with hair.
5. A flat chisel or conelike tooth that nips or cuts food.
6. A group that includes modern humans and their direct-line ancestors since divergence from the ape line.
8. The specific (species) name of modern humans.
9. Habitual two-legged method of locomotion.
15. _____ *Homo* used simple stone tools.
16. Hard structures that can provide clues about what an animal typically eats.

Self-Quiz

Multiple Choice

___ 1. Which of the following is *not* considered to have been a key character in early primate evolution?

    a. Eyes adapted for discerning color and shape in a three-dimensional field

    b. Body and limbs adapted for tree climbing

    c. Bipedalism and increased cranial capacity

    d. Eyes adapted for discerning movement in a three-dimensional field

___ 2. Primitive primates generally live
_____.

    a. in tropical and subtropical forest canopies

    b. in temperate savanna and grassland habitats

    c. near rivers, lakes, and streams in the East African Rift Valley

    d. in caves where there are abundant supplies of insects

___ 3. All the ancestral placental mammals apparently arose from ancestral forms of a group _____.

    a. that includes omnivorous shrews and moles

    b. that includes dogs, cats, and seals

    c. that includes mice and beavers

    d. that includes the koala ("teddy bear") and flying phalanger (resembles the flying squirrel)

___ 4. The hominid evolutionary line stems from a divergence from the ape line that apparently occurred _____.

    a. somewhere between 10 million and 5 million years ago

    b. about 3 million years ago

    c. during the Pliocene epoch

    d. less than 2 million years ago

___ 5. _____ was an Oligocene anthropoid that probably predated the divergence leading to Old World monkeys and the apes, with dentition more like that of dryopiths and less like that of the Paleocene primates with rodentlike teeth.

    a. *Aegyptopithecus*

    b. *Australopithecus*

    c. *Homo erectus*

    d. *Plesiadapis*

___ 6. Donald Johanson, from the University of California at Berkeley, discovered Lucy (named for the Beatles tune), who was a(n) _____.

    a. dryopith

    b. australopith

    c. member of *Homo*

    d. prosimian

___ 7. A hominid of Europe and Asia that became extinct nearly 30,000 years ago was _____.

    a. a dryopith

    b. *Australopithecus*

    c. *Homo erectus*

    d. Neanderthal

Matching

Choose the one most appropriate answer for each.

8. ___ anthropoids

9. ___ australopiths

10. ___ Cenozoic

11. ___ hominids

12. ___ hominoids

13. ___ Miocene

14. ___ primates

15. ___ prosimians

A. A group that includes apes and humans

B. Organisms in a suborder that includes New World and Old World monkeys, apes, and humans

C. An era that began 65 million to 63 million years ago; characterized by the evolution of birds, mammals, and flowering plants

D. A group that includes humans and their most recent ancestors

E. An epoch of the Cenozoic Era lasting from 25 million to 5 million years ago; characterized by the appearance of primitive apes, whales, and grazing animals of the grasslands

F. Organisms in a suborder that includes tree shrews, lemurs, and others

G. A group that includes prosimians, tarsioids, and anthropoids

H. Bipedal organisms living from about 4 million to 1 million years ago, with essentially human bodies and ape-shaped heads; brains no larger than those of chimpanzees

Chapter Objectives/Review Questions

This section lists general and detailed chapter objectives that can be used as review questions. You can make maximum use of these items by writing answers on a separate sheet of paper. Fill in answers where blanks are provided. To check for accuracy, compare your answers with information given in the chapter or glossary.

Page	Objectives/Questions
(472)	1. Where do prosimian survivors dwell today on Earth?
(472)	2. Beginning with the primates most closely related to humans, list the main groups of primates in order by decreasing closeness of relationship to humans.
(473–474)	3. Five key characteristics of primate evolution are _____, _____, _____, _____ and _____.
(475)	4. Describe the general physical features and behavioral patterns attributed to early primates.
(475)	5. Trace primate evolutionary development through the Cenozoic Era. Describe how Earth's climates were changing as primates changed and adapted. Be specific about times of major divergence.
(475–477)	6. State which anatomical features underwent the greatest changes along the evolutionary line from early anthropoids to humans.
(478–479)	7. Explain how you think *Homo sapiens sapiens* arose. Make sure your theory incorporates existing paleontological (fossil), biochemical, and morphological data.

Integrating and Applying Key Concepts

Suppose someone told you that, some time between 12 million and 6 million years ago, dryopiths were forced by larger predatory members of the cat family to flee the forests and take up residence in estuarine, riverine, and coastal marine habitats where they could take refuge in the nearby water to evade the tigers. Those that, through mutations, became naked, developed an upright stance, developed subcutaneous fat deposits as insulation, and developed a bridged nose that had advantages in watery habitats (features that other dryopiths that remained inland never developed) survived and expanded their populations. As time went on, predation by the big cats and competition with other animals for available food caused most of the terrestrial dryopiths to become extinct, but the water-habitat varieties survived as scattered remnant populations, adapting to easily available shellfish and fish; wild rice and oats; and various tubers, nuts, and fruits. It was in these aquatic habitats that the first food-getting tools (baskets, nets, and pebble tools) were developed, as well as the first words that signified different kinds of food. How does such a story fit with current speculations about and evidence of human origins? How could such a story be shown to be true or false?

Answers

Interactive Exercises

28-I. PRIMATE CLASSIFICATION (pp. 470–472)
FROM PRIMATE TO HUMAN: KEY EVOLUTIONARY TRENDS (pp. 473–474)
PRIMATE ORIGINS (p. 475)

1. mammals; 2. hair (mammary glands); 3. orders; 4. Primates; 5. depth; 6. prehensile; 7. grasping; 8. smell; 9. brains; 10. culture; 11. handbones; 12. teeth; 13. 60; 14. rodents; 15. insects; 16. Miocene; 17. drier; 18. grasslands; 19. dryopiths; 20. Asia; 21. 10; 22. 5; 23. L; 24. N; 25. A; 26. F; 27. E; 28. H; 29. M; 30. I; 31. J; 32. C; 33. K; 34. D; 35. B; 36. G; 37. O.

28-II. THE FIRST HOMINIDS (pp. 476–477)
ON THE ROAD TO MODERN HUMANS (pp. 478–479)

1. apes; 2. 10; 3. 5; 4. Four; 5. australopiths; 6. bipedal; 7. 2.5; 8. *Homo erectus*; 9. fire; 10. tool; 11. Neandertals; 12. *Homo erectus*; 13. biochemical; 14 immunological; 15. 40,000; 16. *Homo sapiens*; 17. B, G, H; 18. A, G, K; 19. C, H; 20. D, J; 21. A, F, M; 22. B, G, (H), L; 23. E, I.

Self-Quiz

1. c; 2. a; 3. a; 4. a; 5. a; 6. b; 7. d; 8. B; 9. H; 10. C; 11. D; 12. A; 13. E; 14. G; 15. F.

¹P	R	I	M	²A	T	E		³H	⁴O	M	I	N	⁵O	I	D
				U					A				N		
⁶H		⁷B	U	S	H	Y		⁸S	M		⁹B		C		
O		T						A	M		¹⁰I	R	I	S	
¹¹M	O	L	A	R				P	A		P	E	S	O	
I				A				I	L		E		O		
N		¹²C	U	L	T	U	R	E		¹³A	D	O	R	E	
I				O				N			A				
¹⁴D	R	Y	O	P	I	T	H	S		¹⁵E		L		¹⁶T	
				I					¹⁷C	A	N	I	N	E	S
¹⁸E	R	E	C	T	U	S				R			S		E
				H						L			M		T
	¹⁹P	L	A	S	T	I	C	I	T	Y				H	

29

PLANT TISSUES

Interactive Exercises

29-I. OVERVIEW OF THE PLANT BODY (pp. 482–485)

Selected Italicized Words

primary tissues, lateral meristems, secondary tissues, primary growth, secondary growth

Boldfaced, Page-Referenced Terms

(484) shoots _____

_____

(484) roots _____

_____

(485) ground tissue system _____

_____

(485) vascular tissue system _____

_____

(485) dermal tissue system _____

_____

(485) meristems _____

_____

(485) apical meristems _____

_____

(485) vascular cambium _____

_____

(485) cork cambium _____

_____

Label-Match

Identify each part of the accompanying illustration. Choose from dermal tissues, root system, ground tissues, shoot system, and vascular tissues. Complete the exercise by matching and entering the letter of the proper description in the parentheses following each label.

1. _____ _____ ( )
2. _____ _____ ( )
3. _____ _____ ( )
4. _____ _____ ( )
5. _____ _____ ( )

A. Typically consists of stems, leaves, and reproductive structures
B. Usually grows below ground, absorbs soil water and minerals, stores food, anchors the plant, and sometimes lends support
C. Plant tissues other than epidermal and vascular tissues
D. Protective covering for the plant body
E. The conduction tissues of the plant, xylem and phloem

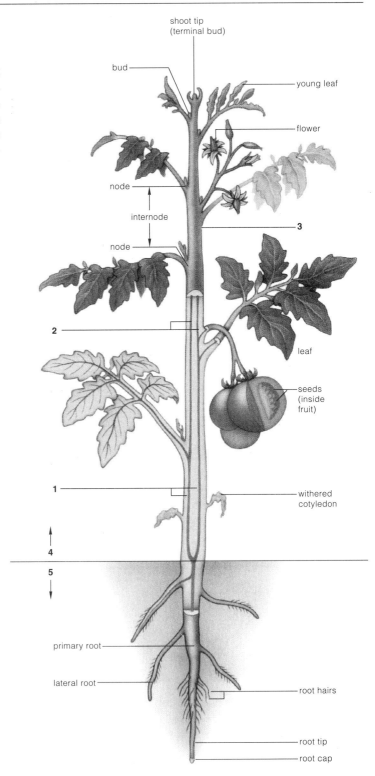

Matching

Choose the most appropriate answer to match with each term.

6. ___ secondary growth

7. ___ shoots

8. ___ vascular cambium and cork cambium

9. ___ meristems

10. ___ ground tissue system

11. ___ apical meristems

12. ___ primary growth

13. ___ vascular tissue system

14. ___ roots

15. ___ dermal tissue system

A. Lengthening of stems and roots; originates with cell divisions at apical meristems
B. Located inside dome-shaped tips of stems and roots; cell divisions and enlargements give rise to three primary meristems
C. The plant's descending parts; specialized for absorbing water and dissolved nutrients
D. Contains two kinds of conducting tissues that distribute water and solutes through the plant body
E. Two types of lateral meristems; growth there produces secondary tissues of the plant body
F. Tissues that make up the bulk of the plant body
G. Localized regions of embryonic cells that continue dividing to lengthen and thicken stems and roots
H. Thickening of stems and roots, originates at lateral meristems
I. Covers and protects the plant's surfaces
J. Consist of stems and their branchings, leaves, flowers, and other components

Label-Match

Identify each part of the accompanying illustration. Choose from the following and enter in the numbered blanks: vascular cambium, apical meristem, ground meristem, vascular bundle, protoderm, cork cambium, procambium. Complete the exercise by matching from the list of descriptions below and entering the correct letter in the parentheses following each label.

16. _____ _____ ( )

17. _____ ( )

18. _____ _____ ( )

19. _____ ( )

20. _____ _____ ( )

21. _____ _____ ( )

22. _____ _____ ( )

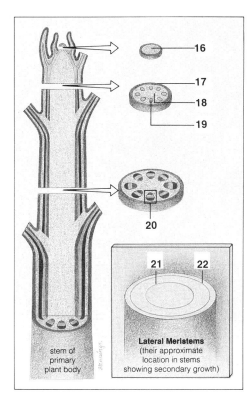

stem of primary plant body

Lateral Meristems
(their approximate location in stems showing secondary growth)

A. Originates secondary growth; increases diameter of older roots and stems
B. Cell lineage of apical meristem that gives rise to epidermis
C. A cord of primary xylem and phloem threading lengthwise through ground tissue
D. Cell lineage of apical meristem giving rise to primary xylem and phloem
E. Dome-shaped cell masses at root and shoot tips
F. Cell lineage of apical meristem giving rise to ground tissues
G. Lateral meristem forming periderm in stems and roots

29-II. SIMPLE TISSUES (p. 486)

Selected Italicized Words

fibers, sclereids

Boldfaced, Page-Referenced Terms

(486) parenchyma _____

_____

(486) collenchyma _____

_____

(486) sclerenchyma _____

_____

(486) lignin _____

_____

Labeling

Place one of these three labels under each of the following illustrations: collenchyma, parenchyma, or sclerenchyma.

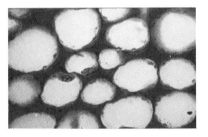

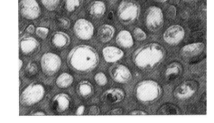

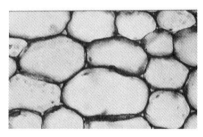

1. _____   2. _____   3. _____

Label-Match

Identify each of the ground tissues below by entering the correct name in the blank adjacent to the sketch. Complete the exercise by matching and entering the letter of the correct description in the parentheses following each label.

4. _____ ( )

5. _____ ( )

6. _____ ( )

A. Usually thick, lignified secondary walls; fibers and sclereids
B. Living, thin-walled cells with ample air spaces between them
C. Living cells; primary walls thickened with cellulose and pectin; cell corners appear thickened

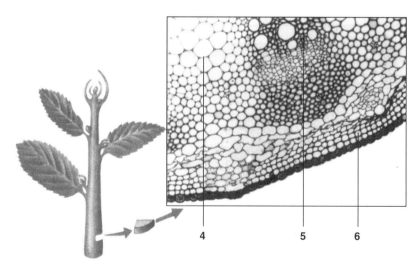

Choice

For questions 7–17, choose from the following:

a. parenchyma   b. collenchyma   c. sclerenchyma

___ 7. Patches or cylinders of this tissue are found near the surface of lengthening stems.

___ 8. The cells have thick, lignin-impregnated walls.

___ 9. Cells are alive at maturity and retain the capacity to divide.

___ 10. Some types specialize in storage, secretion, and other tasks.

___ 11. Cells are mostly elongated, with unevenly thickened walls.

___ 12. Provides flexible support for primary tissues.

___ 13. Abundant air spaces are found around the cells.

___ 14. Supports mature plant parts and often protects seeds.

___ 15. Cells are thin-walled, pliable, and many-sided.

___ 16. Mesophyll is specialized for photosynthesis.

___ 17. Fibers and sclereids.

29-III. COMPLEX TISSUES (pp. 486–487)

Boldfaced, Page-Referenced Terms

(486) xylem _____

_____

(486) phloem _____

_____

(487) epidermis _____

_____

(487) cuticle _____

_____

(487) stoma _____

_____

(487) periderm _____

_____

Complete the Table

1. Two kinds of vascular tissues occur in strands (vascular bundles) within the plant body. Carefully study text Figs. 29.6 and 29.7 to note distinctions in the types of conducting cells found in xylem and phloem (each also has fibers and parenchyma). Complete the table below, which summarizes more information about these vascular tissues.

Vascular Tissue	Major Conducting Cells	Cells Alive	Function(s)
a. Xylem			
b. Phloem			

2. Two types of dermal tissues cover the plant body. Complete the table below, which summarizes information about the dermal tissues.

Dermal Tissue	Primary/Secondary Plant Body	Function(s)
a. Epidermis		
b. Periderm		

29-IV. MONOCOTS AND DICOTS COMPARED (p. 488)

Selected Italicized Words

cotyledons ·

Boldfaced, Page-Referenced Terms

(488) monocots _____

_____

(488) dicots _____

_____

Complete the Table

1. There are two classes of flowering plants, monocots and dicots. Complete the table below, which summarizes information about the two groups of flowering plants.

Class	Number of Cotyledons	Number of Floral Parts	Leaf Venation	Pollen Grains	Vascular Bundles
a. Monocots					
b. Dicots					

29-V. SHOOT PRIMARY STRUCTURE (pp. 488–493)

Selected Italicized Words

terminal bud, lateral bud, Atropa belladonna, Digitalis purpurea, Aloe vera, Agave, Nicotiana rusticum, Cannabis sativa, internode, medicago, Zea mays, Phaseolus, Coleus

Boldfaced, Page-Referenced Terms

(488) vascular bundles _____

_____

(489) cortex _____

_____

(489) pith _____

_____

(490) leaves _____

_____

(491) mesophyll _____

_____

(491) veins _____

_____

(492) node _____

_____

(492) bud _____

_____

Matching

Choose the most appropriate answer to match with each term.

1. ___ terminal bud
2. ___ dicot stems
3. ___ leaves
4. ___ lateral bud
5. ___ vascular bundles
6. ___ veins
7. ___ internode
8. ___ monocot stems
9. ___ evergreen
10. ___ mesophyll
11. ___ simple leaf
12. ___ compound leaf
13. ___ pith
14. ___ cortex
15. ___ petiole
16. ___ leaflets
17. ___ node
18. ___ deciduous
19. ___ bud

A. Located along the side of a branch in the upper angles where leaves attach to stems
B. The parts of the blade of a compound leaf
C. Stem location where one or more leaves are attached
D. Distinct vascular bundles that lace through the interior of a leaf
E. Portion of a stem between two successive nodes
F. Species of plants that do not drop their leaves all at the same time
G. An undeveloped shoot of mostly meristematic tissue, often covered and protected by modified leaves (scales); gives rise to new stems and leaves, flowers, or both
H. Leaf composed of many leaflets
I. In dicot stems; the part of the ground tissue between the vascular bundles and the epidermis
J. Most have a ring of vascular bundles that divides ground tissue into an outer cortex and an inner pith
K. Photosynthetic cell layers within a leaf
L. A plant that drops its leaves as winter approaches
M. The stalk of a leaf
N. Located at the end of a branch
O. A leaf with a single broad blade, attached to the stem by a stalk or petiole
P. Most have vascular bundles scattered throughout the ground tissue
Q. Strands of primary xylem and phloem threading lengthwise through the ground tissue; phloem faces the stem surface, xylem faces the stem center
R. Metabolic food factories equipped with photosynthetic cells; form as bulges from meristems
S. In dicots, the ground tissue inside the ring of vascular bundles (center); in monocots, vascular bundles are scattered through this ground tissue

Label-Match

Identify each indicated part of the accompanying illustration (Fig. 29.12, text). Complete the exercise by matching and entering the letter of the proper function description in the parentheses following each label.

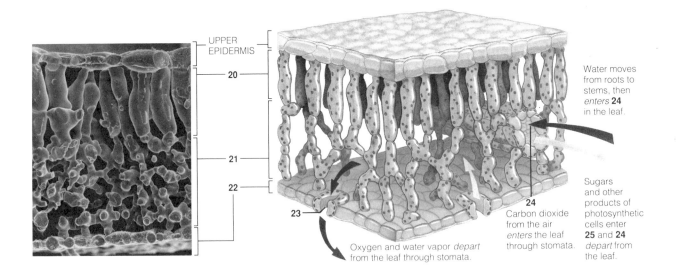

UPPER
EPIDERMIS

— 20 —

— 21 —

— 22 —

23 —

24

Water moves from roots to stems, then *enters* **24** in the leaf.

Sugars and other products of photosynthetic cells enter **25** and **24** *depart* from the leaf.

Carbon dioxide from the air *enters* the leaf through stomata.

Oxygen and water vapor *depart* from the leaf through stomata.

20. _____ _____ ( )
21. _____ _____ ( )
22. _____ _____ ( )
23. _____ ( )
24. _____ ( )

A. Lowermost cuticle-covered cell layer
B. Loosely packed photosynthetic parenchyma cells just above the lower epidermal layer
C. Allow movement of oxygen and water vapor out of leaves and allows carbon dioxide to enter
D. Photosynthetic parenchyma cells just beneath the upper epidermis
E. Move water and solutes to photosynthetic cells and carry products away from them

Labeling

Name the structures numbered in the illustrations of the monocot stem below. Choose from air space, epidermis, sclerenchyma cells, vessel, vascular bundle, sieve tube, ground tissue, and companion cell.

25. _____

26. _____ _____

27. _____ _____

28. _____ _____

29. _____ _____

30. _____

31. _____ _____

32. _____ _____

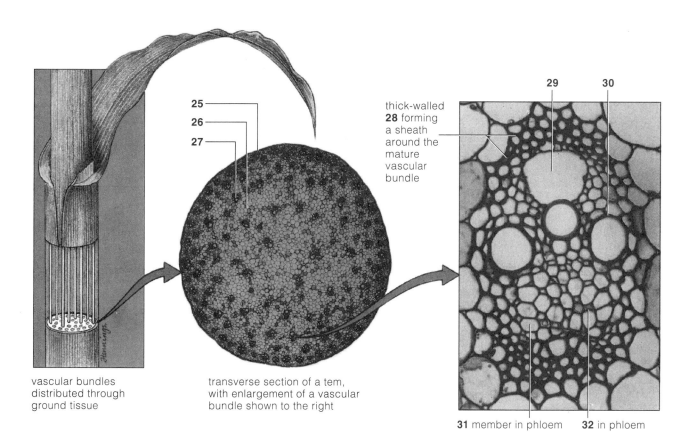

vascular bundles distributed through ground tissue

transverse section of a tem, with enlargement of a vascular bundle shown to the right

thick-walled **28** forming a sheath around the mature vascular bundle

31 member in phloem   **32** in phloem

Labeling

Name the structures numbered in illustrations of the herbaceous dicot stem below. Choose from sieve tube and companion cells in phloem, vascular bundle, xylem vessel, cortex, vascular cambium, pith, phloem fibers, and epidermis.

33. _____

34. _____

35. _____ _____

36. _____

37. _____

38. _____ _____

39. _____ _____

40. _____

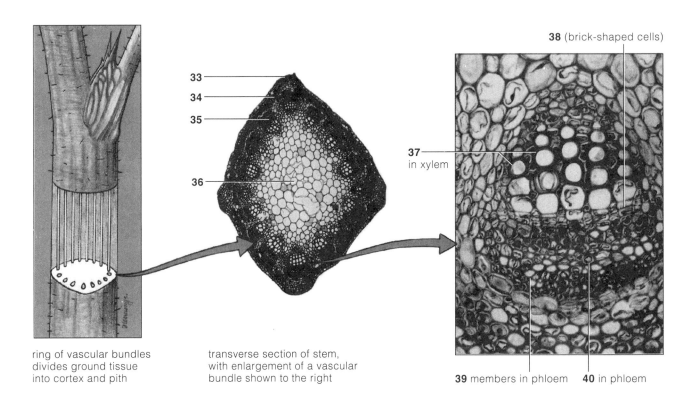

ring of vascular bundles divides ground tissue into cortex and pith

transverse section of stem, with enlargement of a vascular bundle shown to the right

38 (brick-shaped cells)

37 in xylem

39 members in phloem **40** in phloem

29-VI. ROOT PRIMARY STRUCTURE (pp. 494–495)

Selected Italicized Words

adventitious roots

Boldfaced, Page-Referenced Terms

(494) primary root _____

_____

(494) lateral roots _____

_____

(494) taproot system _____

_____

(494) fibrous root system _____

_____

(494) root hairs _____

_____

(494) vascular cylinder _____

_____

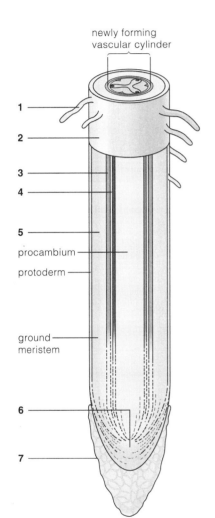

Label-Match

Identify each indicated part of the accompanying illustration. Complete the exercise by matching the letter of the proper description in the parentheses following each label. Some choices are used more than once.

1. _____ _____ ( )
2. _____ ( )
3. _____ ( )
4. _____ ( )
5. _____ ( )
6. _____ _____ _____ ( )
7. _____ _____ ( )
8. _____ ( )
9. _____ ( )
10. _____ _____ ( )

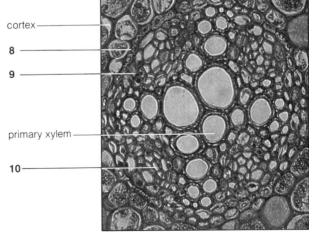

Vascular Cylinder

A. Dome-shaped cell mass produced by the apical meristem
B. Part of the vascular cylinder; gives rise to lateral roots
C. Part of the vascular cylinder; transports photosynthetic products
D. Ground tissue region surrounding the vascular cylinder
E. The absorptive interface with the root's environment
F. The region of dividing cells
G. Innermost part of the root cortex; helps control water and mineral movement into the vascular column
H. Epidermal cell extension; greatly increases the surface available for taking up water and solutes

Fill-in-the-Blanks

Oak trees, carrots, and dandelions are examples of plants whose primary root and its lateral branchings represent a(n) (11) _____ system. In monocots such as grasses, the primary root is short-lived; in its place, numerous (12) _____ roots arise from the stem of the young plant. Such roots and their branches are somewhat alike in length and diameter and form a (13) _____ root system. Vascular tissues form a (14) _____ cylinder, a central column inside the root. Ground tissues surrounding the cylinder are called the root (15) _____. Water entering the root moves from cell to cell until it reaches the (16) _____, the innermost part of the root cortex. This is a sheetlike layer, one cell thick, around the vascular cylinder. Abutting walls of its cells are waterproof, so they force incoming water to pass through the cytoplasm of (17) _____ cells. This arrangement helps (18) _____ the movement of water and dissolved substances into the vascular cylinder. Just inside the endodermis is the (19) _____. This part of the vascular cylinder gives rise to (20) _____ roots.

29-VII. WOODY PLANTS (pp. 496–497)

Selected Italicized Words
early wood, late wood

Boldfaced, Page-Referenced Terms

(496) annuals _____

_____

(496) biennials _____

_____

(496) perennials _____

_____

(496) lateral meristems _____

_____

(497) annual growth layers _____

_____

Choose the most appropriate answer to match with each term.

1. ___ "tree rings"
2. ___ lateral meristem
3. ___ early wood
4. ___ biennial
5. ___ girdling
6. ___ perennial
7. ___ cork cambium
8. ___ late wood
9. ___ annual
10. ___ bark

A. These cells produce periderm—a corky replacement for lost epidermis
B. Plant life cycle completed in one growing season
C. All living and nonliving tissues between the vascular cambium and the stem or root surface
D. As the growing season progresses, the xylem cell diameters become smaller and the walls become thinner
E. Vegetative growth and seed formation continue year after year; some have secondary tissues, others do not
F. Stripping off a band of phloem all the way around a tree's circumference; interrupts phloem transport
G. A plant life cycle completed in two growing seasons
H. Alternating bands representing annual growth layers
I. A cylinder of cells in older stems and roots that give rise to secondary xylem and phloem; xylem forms on the inner face, phloem on the outer
J. The first xylem cells produced at the start of the growing season; tend to have large diameters and thin walls

Label-Match

Identify each indicated part of the accompanying illustration. Complete the exercise by matching and entering the letter of the proper description in the parentheses following each label.

11. _____ _____ ( )
12. _____ ( )
13. _____ ( )
14. _____ ( )
15. _____ ( )

A. Corky replacement for epidermis
B. Vascular tissue that conducts water and dissolved minerals absorbed from soil; gives mechanical support to the plant
C. Has meristematic cells that give rise to secondary xylem and phloem tissues
D. The vascular tissue that transports sugars and other solutes through the plant body
E. All living and nonliving tissues between the vascular cambium and the stem or root surface

Self-Quiz

___ 1. _____ develops into the plant's surface layers.
   a. Ground tissue
   b. Dermal tissue
   c. Vascular tissue
   d. Pericycle

___ 2. Which of the following is *not* considered a ground cell type?
   a. Epidermis

   b. Parenchyma
   c. Collenchyma
   d. Sclerenchyma

___ 3. The _____ produces secondary xylem growth.
   a. apical meristem
   b. lateral meristem
   c. cork cambium
   d. endodermis

___ 4. The _____ is a leaflike structure that is part of the embryo; monocot embryos have one, dicot embryos have two.

a. shoot tip
b. root tip
c. cotyledon
d. apical meristem

___ 5. Leaves are differentiated and buds develop at specific points along the stem called _____.

a. nodes
b. internodes
c. vascular bundles
d. cotyledons

___ 6. Which of the following structures is *not* considered to be meristematic?

a. Vascular cambium
b. Lateral meristem
c. Cork cambium
d. Endodermis

___ 7. Which of the following statements about monocots is *false*?

a. They are usually herbaceous.
b. They develop one cotyledon in their seeds.
c. Their vascular bundles are scattered throughout the ground tissue of their stems.
d. They have a single central vascular cylinder in their stems.

___ 8. New plants grow and older plant parts lengthen through cell divisions at _____ meristems present at root and shoot tips; older roots and stems of woody plants increase in diameter through cell divisions at _____ meristems.

a. lateral; lateral
b. lateral; apical
c. apical; apical
d. apical; lateral

___ 9. Vascular bundles called _____ form a network through a leaf blade.

a. xylem
b. phloem
c. veins
d. stomata

___ 10. A primary root and its lateral branchings represent a _____ system.

a. lateral root
b. adventitious root
c. taproot
d. branch root

___ 11. Plants whose vegetative growth and seed formation continue year after year are _____ plants.

a. annual
b. perennial
c. biennial
d. herbaceous

___ 12. The _____ layer of a root divides to produce lateral roots.

a. endodermis
b. pericycle
c. xylem
d. cortex

Chapter Objectives/Review Questions

This section lists general and detailed chapter objectives that can be used as review questions. You can make maximum use of these items by writing answers on a separate sheet of paper. To check for accuracy, compare your answers with information given in the chapter or glossary.

Page   *Objectives/Questions*

(484)   1. The aboveground parts of flowering plants are called _____; the plants' descending parts are called _____.

(485)   2. Distinguish between the ground tissue system, the vascular tissue system, and the dermal tissue system.

(485)   3. Plants grow at localized regions of self-perpetuating embryonic cells called _____.

(485)   4. Lengthening of stems and roots originates at _____ meristems and at dividing tissues derived from them; this is called _____ growth.

(485)   5. Increases in the diameter of a plant originate at _____ meristems.

(485)   6. Describe the role of vascular cambium and cork cambium in producing secondary tissues of the plant body.

(486)   7. Name and generally describe the simple tissues called parenchyma, collenchyma, and sclerenchyma.

(486)   8. What cell wall compound was necessary for the evolution of rigid and erect land plants?

(486)   9. _____ tissue conducts soil water and dissolved minerals, and it mechanically supports the plant.

(486) 10. _____ tissue transports sugars and other solutes.

(486) 11. Name and describe the functions of the conducting cells in xylem and phloem.

(487) 12. All surfaces of primary plant parts are covered and protected by a dermal tissue system called _____.

(487) 13. What is the function of guard cells and stomata found within the epidermis of young stems and leaves?

(487) 14. The cork cells of _____ replace the epidermis of stems and roots showing secondary growth.

(488) 15. Distinguish between monocots and dicots by listing their characteristics and citing examples of each group.

(488) 16. The primary xylem and phloem develop as vascular _____.

(489) 17. The _____ is the ground tissue between the vascular bundles and the epidermis of stems; the _____ is the ground tissue inside the vascular bundles.

(490) 18. How does the simplest type of leaf differ from "compound" leaves?

(491) 19. State the major functions of leaf mesophyll and vein tissue.

(492) 20. Relate the origin of leaves to the node regions of stems.

(492) 21. A _____ is an undeveloped shoot of mostly meristematic tissue, often covered and protected by scales.

(492) 22. Distinguish the location and function of terminal and lateral buds.

(494) 23. The _____ root is the first to poke through the coat of a germinating seed; later, _____ roots erupt through the epidermis.

(494) 24. How does a taproot system differ from a fibrous root system?

(494) 25. What are the origin and function of root hairs?

(494–495) 26. Describe the passage of soil water through root epidermis to the xylem of the vascular cylinder; include the role of the endodermis.

(496) 27. Define these three categories of flowering plants: *annuals*, *biennials*, and *perennials*.

(496) 28. Each growing season, new tissues that increase the girth of woody plants originate at their _____ meristems.

(497) 29. Distinguish early wood from late wood.

(497) 30. Explain the origin of the annual growth layers (tree rings) seen in a cross-section of a tree trunk.

Integrating and Applying Key Concepts

Try to imagine the specific behavioral restrictions that might be imposed if the human body resembled the plant body in having (1) open growth with apical meristematic regions, (2) stomata in the epidermis, (3) cells with chloroplasts, (4) excess carbohydrates stored primarily as starch rather than as fat, and (5) dependence on the soil as a source of water and inorganic compounds.

Answers

Interactive Exercises

29-I. OVERVIEW OF THE PLANT (pp. 484–485)
1. ground tissues (C); 2. vascular tissues (E); 3. dermal tissues (D); 4. shoot system (A); 5. root system (B); 6. H; 7. J; 8. E; 9. G; 10. F; 11. B; 12. A; 13. D; 14. C; 15. I; 16. apical meristem (E); 17. protoderm (B); 18. ground meristem (F); 19. procambium (D); 20. vascular bundle (C); 21. vascular cambium (A); 22. cork cambium (G).

29-II. SIMPLE TISSUES (p. 486)
1. collenchyma; 2. sclerenchyma; 3. parenchyma; 4. parenchyma (B); 5. sclerenchyma (A); 6. collenchyma (C); 7. b; 8. c; 9. a; 10. a; 11. b; 12. b; 13. a; 14. c; 15. a; 16. a; 17. c.

29-III. COMPLEX TISSUES (pp. 486–487)
1. a. Vessel members and tracheids; No; Conduct water and dissolved minerals absorbed from soil, mechanical support; b. Sieve tube members and companion cells; Yes; Transports sugar and other solutes; 2. a. Primary plant body; Cutin in the cuticle layer over epidermal cells restricts water loss and resists microbial attack; openings (stomates) permit water vapor and gases to enter and leave the plant; b. Secondary plant body; Replaces epidermis to cover roots and stems.

29-IV. MONOCOTS AND DICOTS COMPARED (p. 488)
1. a. One; In threes or multiples thereof; Usually parallel; One pore or furrow; Distributed throughout ground stem tissue; b. Two; In fours or fives or multiples thereof; Usually netlike; Three pores or pores with furrows; Positioned in a ring in the stem.

29-V. SHOOT PRIMARY STRUCTURE (pp. 488–493)
1. N; 2. J; 3. R; 4. A; 5. Q; 6. D; 7. E; 8. P; 9. F; 10. K; 11. O; 12. H; 13. S; 14. I; 15. M; 16. B; 17. C; 18. L; 19. G; 20. palisade mesophyll (D); 21. spongy mesophyll (B); 22. lower epidermis (A); 23. stomata (C); 24. vein(s) (E); 25. epidermis; 26. ground tissue; 27. vascular bundle; 28. sclerenchyma cells; 29. air space; 30. vessel; 31. sieve tube; 32. companion cell; 33. epidermis; 34. cortex; 35. vascular bundle; 36. pith; 37. vessel; 38. vascular cambium; 39. sieve tube; 40. fibers.

29-VI. ROOT PRIMARY STRUCTURE (pp. 494–495)
1. root hair (H); 2. endodermis (G); 3. pericycle (B); 4. epidermis (E); 5. cortex (D); 6. root apical meristem (F); 7. root cap (A); 8. endodermis (G); 9. pericycle (B); 10. primary phloem (C); 11. taproot; 12. adventitious; 13. fibrous; 14. vascular; 15. cortex; 16. endodermis; 17. endodermal; 18. control; 19. pericycle; 20. lateral.

29-VII. WOODY PLANTS (pp. 496–497)
1. H; 2. I; 3. J; 4. G; 5. F; 6. E; 7. A; 8. D; 9. B; 10. C; 11. vascular cambium (C); 12. bark (E); 13. periderm (A); 14. phloem (D); 15. xylem (B).

Self-Quiz
1. b; 2. a; 3. b; 4. c; 5. a; 6. d; 7. d; 8. d; 9. c; 10. c; 11. b; 12. b.

30

PLANT NUTRITION
AND TRANSPORT

UPTAKE OF WATER AND NUTRIENTS
  Nutritional Requirements
  Specialized Absorptive Structures
CONTROL OF NUTRIENT UPTAKE
  Focus on the Environment: Putting Down Roots
WATER TRANSPORT
CONTROL OF WATER LOSS
  The Water-Conserving Cuticle
  Controlled Water Loss at Stomata

TRANSPORT OF ORGANIC SUBSTANCES
  Storage and Transport Forms of Organic
    Compounds
  Translocation
  Pressure Flow Theory

Interactive Exercises

30-I. UPTAKE OF WATER AND NUTRIENTS (pp. 500–503)

Boldfaced, Page-Referenced Terms

(502) essential elements _____

_____

(502) macronutrients _____

_____

(502) micronutrients _____

_____

(502) mutualism _____

_____

(502) symbiosis _____

_____

(502) nitrogen fixation _____

_____

(503) root nodules _____

_____

(503) mycorrhizae _____

(503) root hairs _____

Fill-in-the-Blanks

The three essential elements that plants use as their main metabolic building blocks are oxygen, carbon, and

(1) _____ . The thirteen essential elements available to plants are known as dissolved (2) _____

_____ . Of these, six are present in easily detectable concentrations in plant tissues and are known as

(3) _____ . The remainder occur in very small amounts in plant tissues and are known as

(4) _____ .

Complete the Table

5. Thirteen essential elements are available to plants as mineral ions. Complete the table below, which summarizes information about these important plant nutrients. Refer to Table 30.1 (p. 502) in the text.

Mineral Element	Macronutrient or Micronutrient	Known Functions
a.		Roles in chlorophyll synthesis, electron transport
b.		Activation of enzymes, role in maintaining water–solute balance
c.		Component of chlorophylls; activation of enzymes
d.		Role in root, shoot growth; role in photolysis
e.		Role in chlorophyll synthesis; coenzyme activity
f.		Component of enzyme used in nitrogen metabolism
g.		Component of proteins, nucleic acids, coenzymes, chlorophyll
h.		Component of most proteins, two vitamins
i.		Roles in flowering, germination, fruiting, cell division, nitrogen metabolism
j.		Role in formation of auxin, chloroplasts, and starch; enzyme component
k.		Roles in cementing cell walls, regulation of many cell functions
l.		Component of several enzymes
m.		Component of nucleic acids, phospholipids, ATP

Dichotomous Choice

Circle one of two possible answers given between parentheses in each statement.

6. A form of mutualism, (symbiosis/parasitism), refers to permanent and intimate interactions between species in which two-way benefits pass between species.
7. (Gaseous nitrogen/Nitrogen "fixed" by bacteria) represents the chemical form of nitrogen plants can use in their metabolism.
8. Nitrogen-fixing, mutualistic bacteria reside in localized swellings on legume plants known as (root hairs/root nodules).
9. Mycorrhizae represent symbiotic relationships in which fungi and the roots they cover both benefit; the roots receive (sugars and nitrogen-containing compounds/scarce minerals).
10. (Root nodules/Root hairs) greatly increase a plant's capacity for absorbing water and nutrients from the soil.

30-II. CONTROL OF NUTRIENT UPTAKE (pp. 504–505)

Boldfaced, Page-Referenced Terms

(504) vascular cylinder _____

_____

(504) endodermis _____

_____

(504) Casparian strip _____

_____

(504) exodermis _____

_____

Matching

Consider the information in the short paragraph below (text p. 504). Then match each event in the left column with the event in the right column that would immediately *precede* it.

> Once nutrients enter a vascular cylinder, they are distributed to all living tissues in coordinated, interrelated ways that affect plant growth. Thus living cells expend ATP energy and actively transport certain nutrients at membrane proteins. In photosynthetic cells, the required ATP is produced during photosynthesis and aerobic respiration. In other cells, ATP forms mostly by aerobic respiration.

1. ___ photosynthesis

2. ___ absorption of minerals and water by roots

3. ___ respiration of sucrose by roots

4. ___ transport of minerals and water to leaves

5. ___ ATP formation by roots

6. ___ transport of sucrose to roots

A. photosynthesis
B. absorption of minerals and water by roots
C. respiration of sucrose by roots
D. transport of minerals and water to leaves
E. ATP formation by roots
F. transport of sucrose to roots

Sequence

Arrange in correct chronological sequence the path that nutrients take from the soil to cells in living plant tissues. Write the letter of the first step next to 7, the letter of the second step next to 8, and so on.

7. ___   A. Cortex cells lacking Casparian strips
8. ___   B. Endodermis cells with Casparian strips
9. ___   C. Root hairs
10. ___   D. ATP energy (from photosynthesis or respiration or both) actively transporting nutrients at
11. ___   membrane proteins
12. ___   E. Exodermis cells with Casparian strips
               F. Vascular cylinder

Label-Match

Identify each indicated part of the illustration below that deals with the control of nutrient uptake by plant roots. Choose from the following: water movement, cytoplasm, vascular cylinder, endodermal cell wall, exodermis, endodermis, and Casparian strip. Complete the exercise by matching from the list below and entering the correct letter in the parentheses following each label.

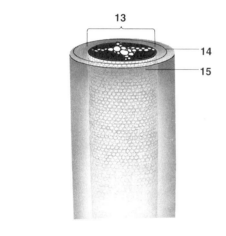

13. _____ _____ ( )   17. _____ _____ ( )

14. _____ ( )   18. _____ _____ ( )

15. _____ ( )   19. _____ _____ _____ ( )

16. _____ ( )

A. Cellular area through which water and dissolved nutrients must move due to Casparian strips
B. A layer of cortex cells just inside the epidermis; also equipped with Casparian strips
C. Waxy band acting as an impermeable barrier between the walls of abutting endodermal cells; forces water and dissolved nutrients through the cytoplasm of endodermal cells
D. Specific location of the waxy strips known as Casparian strips
E. Substance whose diffusion occurs through the cytoplasm of endodermal cells due to Casparian strips
F. Sheetlike layer of single cortex cells wrapped around the vascular cylinder
G. Tissues include the xylem, phloem, and pericycle

30-III. WATER TRANSPORT (pp. 506–507)

Boldfaced, Page-Referenced Terms

(506) transpiration _____

_____

(506) xylem _____

_____

(506) tracheids _____

_____

(506) vessel members _____

_____

(506) cohesion-tension theory _____

_____

Fill-in-the-Blanks

The cohesion-tension theory explains (1) _____ transport to the tops of plants. The water travels
inside tubelike strands of (2) _____, which are formed by hollow, dead cells called tracheids and vessel
members. The process begins with the drying power of air, which causes (3) _____, the evaporation of
water from plant parts exposed to air. (4) _____ of hydrogen-bonded water molecules in the xylem of
roots, stems, and leaves provides a continuous column of water. This places the xylem water in a state of
(5) _____ that extends from veins in leaves, down through the stems, to roots. As water continues to
escape from plant surfaces, water enters the roots to replace that which is lost.

30-IV. CONTROL OF WATER LOSS (pp. 508–509)

Selected Italicized Words
inward and outward diffusion, Opuntia, Commelina communis

Boldfaced, Page-Referenced Terms

(508) cuticle _____

_____

(508) cutin _____

_____

(509) stomata _____

_____

(509) guard cells _____

_____

(509) CAM plants _____

_____

True-False

If the statement is true, write a T in the blank. If the statement is false, make it correct by changing the underlined word(s) and writing the correct word(s) in the answer blank.

_____ 1. Of the water moving into a leaf, 2 percent or more is lost by transpiration.

_____ 2. When evaporation exceeds water uptake by roots, plant tissues wilt and water-dependent activities are seriously disrupted.

_____ 3. The surfaces of all plant epidermal cell walls have an outer layer of waxes embedded in cutin, the cuticle, which reduces rate of water loss, limits inward diffusion of carbon dioxide, and limits outward diffusion of oxygen.

_____ 4. Plant epidermal layers are peppered with tiny openings called guard cells through which water leaves the plant and carbon dioxide enters.

_____ 5. When a pair of guard cells swells with water, the opening between them closes.

_____ 6. In most plants, stomata remain open during the daylight photosynthetic period; water is lost but they gain carbon dioxide.

_____ 7. Stomata stay closed at night in most plants.

_____ 8. Photosynthesis starts when the sun comes up; as the morning progresses, carbon dioxide levels increase in cells, including guard cells.

_____ 9. A drop in carbon dioxide level within guard cells triggers an inward active transport of potassium ions; water follows by osmosis and the fluid pressure closes the stoma.

_____ 10. When the sun goes down and photosynthesis stops, carbon dioxide levels rise; stomata close when potassium, then water, move out of the guard cells.

_____ 11. CAM plants such as cacti and other succulents open stomata during the day when they fix carbon dioxide by way of a special C4 metabolic pathway.

_____ 12. CAM plants use carbon dioxide the following night when stomata are closed.

30-V. TRANSPORT OF ORGANIC SUBSTANCES (pp. 510–511)

Selected Italicized Words

source, sink, Sonchus

Boldfaced, Page-Referenced Terms

(510) translocation _____

_____

(510) sieve-tube members _____

_____

(510) companion cells _____

_____

(511) pressure flow theory _____

_____

Fill-in-the-Blanks

Sucrose and other organic compounds resulting from (1) _____ are used throughout the plant. Most plant cells store their carbohydrates as (2) _____. Quantities of (3) _____ become stored in some fruits; (4) _____ store proteins and fats. (5) _____ molecules are too large to cross cell membranes and too insoluble to be transported to other regions of the plant body. (6) _____ are largely insoluble in water and cannot be transported from storage sites. (7) _____ proteins do not lend themselves to transport. Specific chemical reactions within plant cells, such as (8) _____, convert storage forms of organic compounds to their subunits, which are transportable forms. For example, the hydrolysis of starch liberates glucose units, which combine with fructose to form (9) _____, a transport form for sugars. Experiments with phloem-embedded aphid mouthparts revealed that (10) _____ was the most abundant carbohydrate being forced out of those conducting tubes.

Matching

Match the appropriate letter with its numbered partner.

11. ___ translocation
12. ___ sieve tube members
13. ___ companion cells
14. ___ aphids
15. ___ source
16. ___ sink
17. ___ pressure flow theory

A. Any region where organic compounds are being loaded into the sieve tube system
B. Nonconducting cells adjacent to sieve tube members that supply energy to load sucrose at the source
C. Any region of the plant where organic compounds are being unloaded from the sieve tube system and used or stored
D. Process occurring in phloem that distributes sucrose and other organic compounds through the plant (apparently under pressure)
E. States that pressure builds up at the source end of a sieve tube system and pushes solutes toward a sink, where they are removed
F. Passive conduits for translocation within vascular bundles; water and organic compounds flow rapidly through large pores on their end walls
G. Insects used to verify that in most plant species sucrose is the main carbohydrate translocated under pressure

Self-Quiz

___ 1. The _____ theory of water transport states that hydrogen bonding allows water molecules to maintain a continuous fluid column as water is pulled from roots to leaves.

    a. pressure flow
    b. cohesion-tension
    c. evaporation
    d. abscission

___ 2. The three elements that are present in carbohydrates, lipids, proteins, and nucleic acids are _____.

    a. oxygen, carbon, and nitrogen
    b. oxygen, hydrogen, and nitrogen
    c. oxygen, carbon, and hydrogen
    d. carbon, nitrogen, and hydrogen

___ 3. Macronutrients are the six dissolved mineral ions that _____.

    a. play vital roles in photosynthesis and other metabolic events
    b. occur in only small traces in plant tissues
    c. become detectable in plant tissues
    d. can function only without the presence of micronutrients
    e. both a and c

___ 4. Without _____, plants would rapidly wilt and die during hot, dry spells.

    a. a cuticle
    b. mycorrhizae
    c. phloem
    d. cotyledons

___ 5. Gaseous nitrogen is converted to a plant-usable form by _____.

    a. root nodules
    b. mycorrhizae
    c. nitrogen-fixing bacteria
    d. Venus flytraps

___ 6. _____ prevent(s) inward-moving water from moving past the abutting walls of the root endodermal cells.

    a. Cytoplasm
    b. Plasma membranes
    c. Osmosis
    d. Casparian strips

___ 7. Most of the water moving into a leaf is lost through _____.

    a. osmotic gradients being established
    b. transpiration
    c. pressure-flow forces
    d. translocation

___ 8. Stomata remain _____ during daylight, when photosynthesis occurs, but remain _____ during the night when carbon dioxide accumulates through aerobic respiration.

    a. open; open
    b. closed; open
    c. closed; closed
    d. open; closed

___ 9. By control of _____ levels inside the guard cells of stomata, the activity of stomata is controlled when leaves are losing more water than roots can absorb.

    a. oxygen
    b. potassium
    c. carbon dioxide
    d. ATP

___ 10. Leaves represent _____ regions; growing leaves, stems, fruits, seeds, and roots represent _____ regions.

    a. source; source
    b. sink; source
    c. source; sink
    d. sink; sink

Chapter Objectives/Review Questions

This section lists general and detailed chapter objectives that can be used as review questions. You can make maximum use of these items by writing answers on a separate sheet of paper. To check for accuracy, compare your answers with information given in the chapter or glossary.

Page	Objectives/Questions
(502)	1. Plants generally require _____ (number) essential elements.
(502)	2. Distinguish between macronutrients and micronutrients in relation to their role in plant nutrition.
(503)	3. Describe the roles of root nodules and mycorrhizae in plant nutrition.
(504)	4. Differentiate between the endodermis and the exodermis of the root cortex.
(504)	5. Due to the presence of the _____ strip in the walls of endodermal cells, water can move into the vascular cylinder only by crossing the plasma membrane and diffusing through the cytoplasm.
(506)	6. Water evaporation from leaves, stems, and other plant parts is called _____.
(506)	7. Water moves through pipelines in the vascular tissue called _____.
(506)	8. Henry Dixon's _____-_____ theory explains how water moves upward in an unbroken column through xylem to the tops of tall trees.
(506)	9. Be able to give the key points of Dixon's explanation of upward water transport in plants.
(508)	10. Even mildly stressed plants would rapidly wilt and die if it were not for the _____ covering their parts; describe additional functions of this structure.
(508)	11. Describe the chemical compounds found in cutin.
(509)	12. Transpiration occurs mostly at _____, tiny epidermal passageways of leaves and stems.
(509)	13. Explain the mechanism by which stomata open during daylight and close during the night.
(509)	14. Describe the mechanisms by which CAM plants conserve water.
(510)	15. Carbohydrates are stored as _____ in most plants.
(510)	16. _____ is the main form in which sugars are transported through most plants.
(510)	17. Describe the role of sieve tube members and companion cells in translocation.
(510–511)	18. A _____ is any region of the plant where organic compounds are loaded into the sieve tube system; a _____ is any region where organic compounds are unloaded from the sieve tube system and used or stored.
(511)	19. According to the _____ _____ theory, pressure builds up at the source end of a sieve tube system and pushes solutes toward a sink, where they are removed.
(511)	20. Companion cells supply the _____ that loads sucrose at the source.
(511)	21. Name the gradients responsible for continuous flow of organic compounds through phloem.

Integrating and Applying Key Concepts

How do you think maple syrup is made from maple trees? Which specific systems of the plant are involved, and why are maple trees tapped only at certain times of the year?

Answers

Interactive Exercises

Interactive Exercises

30-I. UPTAKE OF WATER AND NUTRIENTS
(pp. 502–503)

1. hydrogen; 2. mineral ions; 3. macronutrients; 4. micronutrients; 5. a. iron, micronutrient; b. potassium, macronutrient; c. magnesium, macronutrient; d. chlorine, micronutrient; e. manganese, micronutrient; f. molybdenum, micronutrient; g. nitrogen, macronutrient; h. sulfur, macronutrient; i. boron, micronutrient; j. zinc, micronutrient; k. calcium, macronutrient; l. copper, micronutrient; m. phosphorus, macronutrient; 6. symbiosis; 7. Nitrogen "fixed" by bacteria; 8. root nodules; 9. scarce minerals; 10. Root hairs.

30-II. CONTROL OF NUTRIENT UPTAKE
(pp. 504–505)

1. D; 2. E; 3. F; 4. B; 5. C; 6. A; 7. C; 8. E; 9. A; 10. B; 11. F; 12. D; 13. vascular cylinder (G); 14. exodermis (B); 15. endodermis (F); 16. cytoplasm (A); 17. water movement (E); 18. Casparian strip (C); 19. endodermal cell wall (D).

30-III. WATER TRANSPORT (pp. 506–507)

1. water; 2. xylem; 3. transpiration; 4. Cohesion; 5. tension.

30-IV. CONTROL OF WATER LOSS (pp. 508–509)

1. 90 percent; 2. T; 3. T; 4. stomata; 5. opens; 6. T; 7. T; 8. decrease; 9. opens; 10. T; 11. night; 12. day.

30-V. TRANSPORT OF ORGANIC SUBSTANCES
(pp. 510–511)

1. photosynthesis; 2. starch; 3. fats; 4. seeds; 5. Starch; 6. Fats; 7. Storage; 8. hydrolysis; 9. sucrose; 10. sucrose; 11. D; 12. F; 13. B; 14. G; 15. A; 16. C; 17. E.

Self-Quiz

1. b; 2. c; 3. e; 4. a; 5. c; 6. d; 7. b; 8. d; 9. b; 10. c.

31

PLANT REPRODUCTION

Interactive Exercises

31-I. REPRODUCTIVE MODES (pp. 514–516)
FLORAL STRUCTURE (pp. 516–517)

Selected Italicized Words

sexual reproduction, asexual reproduction, angiosperm, perfect flowers, imperfect flowers

Boldfaced, Page-Referenced Terms

(516) sporophyte _____

_____

(516) flowers _____

_____

(516) gametophytes _____

_____

(517) stamens _____

_____

(517) pollen grains _____

_____

(517) carpels _____

_____

Label-Match

Study the generalized life cycle for flowering plants as shown in Fig. 31.2a in the text. Identify each indicated part of the accompanying illustration. Complete the exercise by matching and entering the letter of the proper description in the parentheses following each label.

1. _____ ( )

2. _____ ( )

3. _____ ( )

4. _____ ( )

5. _____ ( )

6. _____ ( )

A. An event that produces a young sporophyte
B. A reproductive shoot produced by the sporophyte
C. The "plant"; a vegetative body that develops from a zygote
D. Cellular division event occurring within flowers to produce spores
E. Produces haploid eggs by mitosis
F. Produces haploid sperm by mitosis

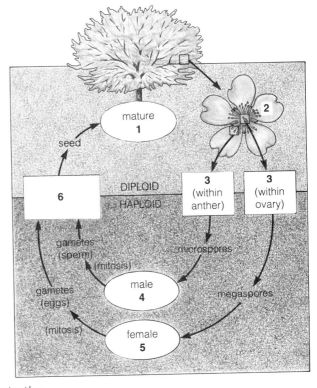

Labeling

Identify each indicated part of the accompanying illustration.

7. _____

8. _____

9. _____

10. _____

11. _____

12. _____

13. _____

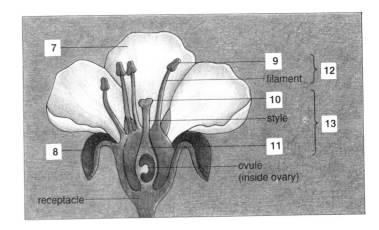

Matching

Choose the one most appropriate answer for each.

14. ___ sepals
15. ___ petals
16. ___ stamens
17. ___ ovule
18. ___ pollen sacs
19. ___ carpels
20. ___ ovary
21. ___ perfect flowers
22. ___ imperfect flowers

A. Have both male and female parts
B. Four chambers within an anther where pollen grains develop
C. Collectively, the flower's "corolla"
D. Have male or female parts, but not both
E. Female reproductive parts; include stigma, style, and ovary
F. Structure that matures to become a seed
G. Found just inside the flower's corolla; the male reproductive parts
H. Lower portion of the carpel where egg formation, fertilization, and seed development occur
I. Outermost leaflike whorl of floral organs; collectively, the calyx

31-II. A NEW GENERATION BEGINS (pp. 518–521)

Boldfaced, Page-Referenced Terms

(518) microspores _____

_____

(518) ovule _____

_____

(518) megaspores _____

_____

(518) endosperm _____

_____

(518) pollination _____

_____

(519) double fertilization _____

_____

Fill-in-the-Blanks

The numbered items on the illustration below represent missing information; complete the numbered blanks of the narrative below to supply the missing information on the illustration.

Within each (1) _____, mitotic divisions produce four masses of spore-forming cells, each mass forming within a(n) (2) _____ _____. Each one of these diploid cells is known as a(n) (3) _____ _____ cell and undergoes (4) _____ to produce four haploid (5) _____. Mitosis within each haploid (5) results in a two-celled haploid body, the immature male gametophyte. One of these cells will give rise to a (6) _____ _____; the other cell will develop into a (7) _____-_____ cell. Mature microspores are eventually released from the pollen sacs of the anther as (8) _____. Pollination occurs, and after the pollen lands on a(n) (9) _____ of a carpel, the pollen tube develops from one of the cells in the pollen grain; the other cell within the pollen grain divides to form two sperm cells. As the pollen tube grows through the carpel tissues, it contains the two sperm cells and a tube nucleus. The pollen tube, with its two sperm cells and the tube nucleus, is known as the mature (10) _____ _____.

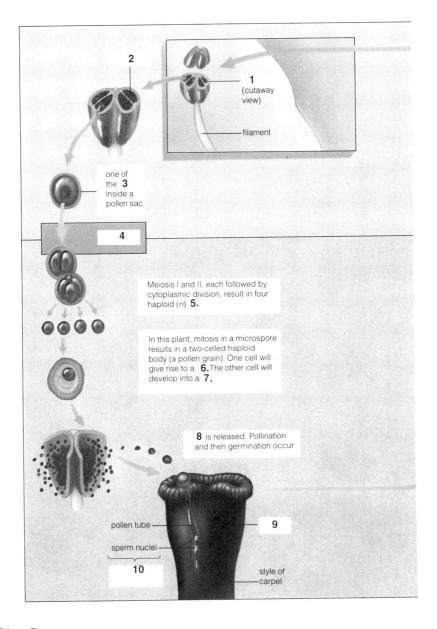

Fill-in-the-Blanks

The numbered items on the illustration below represent missing information; complete the numbered blanks in the narrative below to supply the missing information on the illustration.

In the carpel of the flower, one or more dome-shaped, diploid tissue masses develop on the inner wall of the ovary. Each mass is the beginning of a(n) (11) _____. A tissue forms inside a domed mass as it grows, and one or two protective layers called (12) _____ form around it. Inside each mass, a cell (the megaspore mother cell) divides by (13) _____ to form four haploid spores, the (14) _____. Commonly, three (15) _____ disintegrate. The remaining (16) _____ undergoes (17) _____ three times without cytoplasmic division. At first, this structure is a cell with (18) _____ haploid nuclei. Cytoplasmic division results in a seven-cell (19) _____ _____, which represents the mature (20) _____ _____. One of these cells, the (21) _____ mother cell, has two nuclei and will

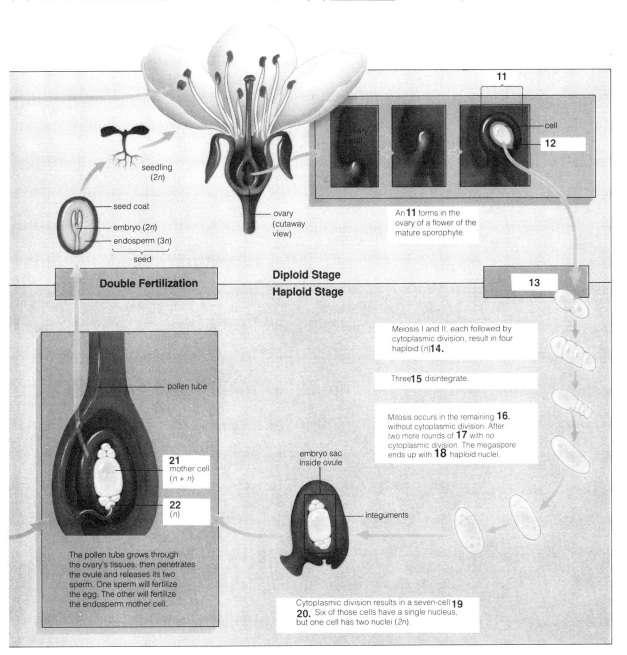

seedling
(2n)

seed coat
embryo (2n)
endosperm (3n)
seed

ovary
(cutaway
view)

Double Fertilization

Diploid Stage
Haploid Stage

11

cell

12

An **11** forms in the ovary of a flower of the mature sporophyte.

13

Meiosis I and II, each followed by cytoplasmic division, result in four haploid (n) **14.**

Three **15** disintegrate.

Mitosis occurs in the remaining **16.** without cytoplasmic division. After two more rounds of **17** with *no* cytoplasmic division. The megaspore ends up with **18** haploid nuclei.

pollen tube

21
mother cell
(n + n)

22
(n)

embryo sac
inside ovule

integuments

The pollen tube grows through the ovary's tissues, then penetrates the ovule and releases its two sperm. One sperm will fertilize the egg. The other will fertilize the endosperm mother cell.

Cytoplasmic division results in a seven-cell **19 20.** Six of those cells have a single nucleus, but one cell has two nuclei (2n).

help form the 3*n* endosperm, a nutritive tissue for the forthcoming embryo. Another haploid cell within the embryo sac is the (22) _____, which will combine with one sperm to form the diploid embryo.

Short Answer

23. What guides the growth of the pollen tube down through the female floral tissues toward the chamber holding the egg? _____

_____

24. Describe the site of double fertilization in flowering plants. _____

_____

Complete the Table

25. Complete the table below to summarize the unique double fertilization occurring only in flowering plant life cycles. Refer to Figure 31.5 in the text.

Double Fertilization Products	Origin	Produces	Function
a. Zygote (2*n*) nucleus			
b. Endosperm (3*n*) nucleus			

31-III. FROM ZYGOTE TO SEED (pp. 522–523)

Selected Italicized Words

Capsella

Boldfaced, Page-Referenced Terms

(522) cotyledons _____

_____

(522) seed _____

_____

(522) fruit _____

_____

Complete the Table

1. Complete the following table, which summarizes concepts associated with seeds and fruits.

Structure	Origin	Function(s)
a. Cotyledons		
b. Seeds		
c. Seed coat		
d. Fruit		

Labeling

Identify each indicated part of the illustration below.

2. _____

3. _____

4. _____

5. _____

6. _____

7. _____

8. _____ _____

9. _____

10. _____

11. _____ _____

12. _____

13. _____

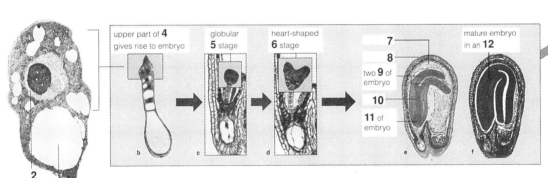

upper part of **4** gives rise to embryo

globular **5** stage

heart-shaped **6** stage

7

8

two **9** of embryo

10

11 of embryo

mature embryo in an **12**

13 (ovary) partially cut open to show mature ovules; embryos inside are at stage shown in (**f**).

2

3

Sequence

Arrange the following developmental events in correct chronological sequence from first to last.

14. ___   A. Early on, cell divisions produce a single row of cells.
15. ___   B. The ovule's integuments thicken and harden into a seed coat.
16. ___   C. The lower cells (of the single row) develop into a stalklike structure that anchors the embryo
17. ___   and absorbs endosperm nutrients for it; the cells above the stalklike structure develop into
18. ___   the embryo proper.
19. ___   D. The embryo and food reserves are now a self-contained package. The ovule has become a
20. ___   seed.
               E. Following fertilization, the newly formed zygote embarks on a course of cell divisions leading to a mature embryo sporophyte.
               F. Primary meristems begin to form and two cotyledons (monocot embryos have one) have begun to develop from two tissue lobes.
               G. Food reserves accumulate in the expanding endosperm or cotyledons.

Choice

For questions 21–27, choose from the following:

> a. simple; wall dry; splits at maturity   b. simple; wall dry; intact at maturity
> c. simple; fleshy; often leathery   d. aggregate
> e. multiple   f. accessory; simple   g. accessory; aggregate

___ 21. apple, pear

___ 22. pea, magnolia, mustard

___ 23. strawberry

___ 24. sunflower, wheat, rice, maple

___ 25. pineapple, fig, mulberry

___ 26. grape, banana, lemon, cherry

___ 27. blackberry, raspberry

31-IV. FRUIT AND SEED DISPERSAL (pp. 524–525)

Fill-in-the-Blanks

Long before humans began eating them, (1) _____ were undergoing natural selection as seed-dispersing mechanisms. They have coevolved with air currents, water currents, and with particular (2) _____. The wings of a dropping maple fruit spin sideways with air currents to arrive far enough that its enclosed embryo will not compete with the (3) _____ plant for soil water, minerals, and sunlight. Many (4) _____ have hooks, spines, hairs, and sticky surfaces that allow animals to move them to new locations. Fleshy fruits such as blueberries and cherries have their seeds dispersed as they are eaten and expelled by (5) _____. Digestive enzymes in the animal guts remove enough of the hard seed coats to increase the chance of successful (6) _____ after expulsion from the body.

31-V. ASEXUAL REPRODUCTION OF FLOWERING PLANTS (pp. 526–527)

Selected Italicized Words

Populus tremuloides, Larrea divericata, Daucus carota

Boldfaced, Page-Referenced Terms

(526) vegetative growth _____

_____

(526) parthenogenesis _____

_____

(527) tissue culture propagation _____

_____

Matching

Match the following asexual reproductive modes of flowering plants.

1. ___ corm
2. ___ bulb
3. ___ parthenogenesis
4. ___ runner
5. ___ vegetative propagation on modified stems
6. ___ rhizome
7. ___ tuber
8. ___ tissue culture propagation (induced propagation)
9. ___ vegetative growth

A. In a general sense, new plants develop from tissues or organs that drop or separate from parent plants
B. New shoots arise from axillary buds (enlarged tips of slender underground rhizomes)
C. New plants arise from cells in parent plant that were not irreversibly differentiated; a laboratory technique
D. New plants arise at nodes of underground horizontal stem
E. New plant arises from axillary bud on short, thick, vertical underground stem
F. New plants arise at nodes on an aboveground horizontal stem
G. New structure arises from axillary bud on short underground stem
H. Involves asexual reproduction utilizing runners, rhizomes, corms, tubers, and bulbs
I. Embryo develops without nuclear or cellular fusion

More Matching

Choose the one most appropriate example for each modified stem.

10. ___ bulb
11. ___ rhizome
12. ___ tuber
13. ___ runner
14. ___ corm

A. Potato
B. Strawberry
C. Gladiolus
D. Onion, lily
E. Bermuda grass

Self-Quiz

___ 1. A stamen is _____.
   a. composed of a stigma
   b. the mature male gametophyte
   c. the site where microspores are produced
   d. part of the vegetative phase of an angiosperm

___ 2. A gametophyte is _____.
   a. a gamete-producing plant
   b. haploid
   c. both a and b
   d. the plant produced by the fusion of gametes

___ 3. The phase in the life cycle of plants that gives rise to spores is known as the _____.
   a. gametophyte
   b. embryo
   c. sporophyte
   d. seed

___ 4. The ovary of a flower is the site of _____.
   a. egg development
   b. fertilization
   c. seed maturation
   d. all of these

___ 5. A characteristic of a seed is that it _____.
   a. contains an embryo sporophyte
   b. represents an arrested growth stage
   c. is covered by hardened and thickened integuments
   d. all of these

___ 6. An immature fruit is a(n) _____ and an immature seed is a(n) _____.
   a. ovary; megaspore
   b. ovary; ovule

   c. megaspore; ovule
   d. ovule; ovary

___ 7. The joint evolution of flowers and their pollinators is known as _____.
   a. adaptation
   b. coevolution
   c. joint evolution
   d. covert evolution

___ 8. In flowering plants, one sperm nucleus fuses with that of an egg, and a zygote forms that develops into an embryo. Another sperm fuses with _____.
   a. a primary endosperm cell to produce three cells, each with one nucleus
   b. a primary endosperm cell to produce one cell with one triploid nucleus
   c. both nuclei of the endosperm mother cell, forming a primary endosperm cell with a single triploid nucleus
   d. one of the smaller megaspores to produce what will eventually become the seed coat

___ 9. "Simple, aggregate, multiple, and accessory" refer to types of _____.
   a. carpels
   b. seeds
   c. fruits
   d. ovaries

___ 10. "When a leaf falls or is torn away from a jade plant, a new plant can develop from the leaf, from meristematic tissue." This statement refers to _____.
   a. parthenogenesis
   b. runners
   c. tissue culture propagation
   d. vegetative propagation

Chapter Objectives/Review Questions

This section lists general and detailed chapter objectives that can be used as review questions. You can make maximum use of these items by writing answers on a separate sheet of paper. To check for accuracy, compare your answers with information given in the chapter or glossary.

Page	Objectives/Questions
(516)	1. _____ reproduction requires formation of gametes, followed by fertilization.
(516)	2. The production of individuals in clones describes _____ reproduction.
(516)	3. State the relationship of a flower to a sporophyte.
(516)	4. Distinguish between sporophytes and gametophytes.
(516)	5. _____ have several means of reproducing themselves asexually.
(516)	6. _____ are shoots on sporophytes that are specialized for reproduction.
(516–517)	7. Be able to identify the various parts of a typical flower and give their functions.
(517)	8. The male floral reproductive parts are called _____; list the two parts of one such structure.
(517)	9. _____ are the female reproductive parts of a flower; list the three parts of one such structure.
(517)	10. Distinguish between a flower that is perfect and one that is imperfect.
(518)	11. Relate the sequence of events giving rise to microspores and megaspores.
(518)	12. Describe the development and contents of the male gametophyte.
(518)	13. Describe the development and contents of the embryo sac.
(518)	14. What structures represent the male gametophyte and female gametophyte in flowering plants?
(518)	15. The gamete found in the embryo sac is the _____.
(518)	16. The endospore mother cell in the embryo sac is composed of two _____.
(518)	17. _____ is the transfer of pollen grains to a receptive stigma.
(519)	18. Describe the double fertilization that occurs uniquely in the flowering plant life cycle.
(519)	19. How is endosperm formed? What is the function of endosperm?
(520–521)	20. Provide one explanation as to why the flowering plants have been so successful on a worldwide basis; cite examples of coevolution.
(522)	21. _____, or "seed leaves," develop as part of the embryo.
(522)	22. An ovule is an immature _____.
(522)	23. An ovary containing ovule(s) develops into a _____.
(522)	24. Review the general types of fruits produced by flowering plants (text, Table 31.1).
(524)	25. From the plant's perspective, what is the function of fruits? List common dispersal mechanisms.
(526)	26. Review Table 31.2 in the text, which lists major types of asexual reproductive modes among flowering plants.
(526–527)	27. Distinguish between parthenogenesis, vegetative propagation, and tissue culture propagation.
(526)	28. List representative plant examples of a runner, a rhizome, a corm, a tuber, and a bulb.

Integrating and Applying Key Concepts

Unlike the growth pattern of animals, the vegetative body of vascular plants continues growing throughout its life by means of meristematic activity. Explain why most representatives of the plant kingdom probably would not have survived and grown as well as they do if they had followed the growth pattern of animals.

Answers

Interactive Exercises

31-I. REPRODUCTIVE MODES (p. 516)
FLORAL STRUCTURE (pp. 516–517)

1. sporophyte (C); 2. flower (B); 3. meiosis (D); 4. gametophyte (F); 5. gametophyte (E); 6. fertilization (A); 7. petal; 8. sepal; 9. anther; 10. stigma; 11. ovary; 12. stamen; 13. carpel; 14. I; 15. C; 16. G; 17. F; 18. B; 19. E; 20. H; 21. A; 22. D.

31-II. A NEW GENERATION BEGINS (pp. 518–521)

1. anther; 2. pollen sac; 3. microspore mother; 4. meiosis; 5. microspores; 6. pollen tube; 7. sperm-producing; 8. pollen; 9. stigma; 10. male gametophyte; 11. ovule; 12. integuments; 13. meiosis; 14. megaspores; 15. megaspores; 16. megaspore; 17. mitosis; 18. eight; 19. embryo sac; 20. female gametophyte; 21. endosperm; 22. egg; 23. Chemical and molecular cues guide a pollen tube's growth through tissues of the ovary, toward the egg chamber and sexual destiny; 24. The embryo sac within the ovule is the site of double fertilization; 25. a. Fusion of one egg nucleus (*n*) with one sperm nucleus (*n*); The plant embryo (2*n*); Eventually develops into a new sporophyte plant; b. Fusion of one sperm nucleus (*n*) with the endosperm mother cell (2*n*); Endosperm tissues (3*n*); Nourishes the embryo within the seed.

31-III. FROM ZYGOTE TO SEED (pp. 522–523)

1. a. "Seed leaves" that develop as part of the embryo; Plants with large cotyledons absorb endosperm and store food; thin cotyledons may produce enzymes for transferring food from the embryo to the germinating seedling; b. Ovules within the ovary; Contains a 2*n* embryo sporophyte, 3*n* endosperm; these are wrapped within a seed coat; c. Integuments of ovule; protection for embryo with its endosperm (stored food); d. Usually a mature, ripened ovary; seed protection and dispersal in specific environments; 2. nucleus; 3. vacuole; 4. zygote; 5. embryo; 6. embryo; 7. endosperm; 8. seed coat; 9. cotyledons; 10. embryo; 11. root tip; 12. ovule; 13. fruit; 14. E; 15. A; 16. C; 17. F; 18. G; 19. B; 20. D; 21. f; 22. a; 23. g; 24. b; 25. e; 26. c; 27. d.

31-IV. FRUIT AND SEED DISPERSAL (pp. 524–525)

1. fruits; 2. animals; 3. parent; 4. fruits; 5. animals; 6. germination.

31-V. ASEXUAL REPRODUCTION OF FLOWERING PLANTS (pp. 526–527)

1. E; 2. G; 3. I; 4. F; 5. H; 6. D; 7. B; 8. C; 9. A; 10. D; 11. E; 12. A; 13. B; 14. C.

Self-Quiz

1. c; 2. c; 3. c; 4. d; 5. d; 6. b; 7. b; 8. c; 9. c; 10. d.

32

PLANT GROWTH AND DEVELOPMENT

PATTERNS OF GROWTH AND DEVELOPMENT
    Seed Germination
    Prescribed Growth Patterns

PLANT HORMONES
    Auxins
    Gibberellins
    Focus on Health: About Those Herbicides
    Cytokinins
    Abscisic Acid
    Ethylene
    Other Hormones and Hormone-Like Substances

THE TROPISMS
    Phototropism
    Gravitropism
    Thigmotropism

RESPONSES TO MECHANICAL STRESS

BIOLOGICAL CLOCKS AND THEIR EFFECTS
    Circadian Rhythms
    Photoperiodism

THE FLOWERING PROCESS

VERNALIZATION

SENESCENCE

DORMANCY
    Entering Dormancy
    Breaking Dormancy
    Focus on the Environment: The Rise and Fall of a Giant

Interactive Exercises

32-I. PATTERNS OF GROWTH AND DEVELOPMENT (pp. 530–533)

Boldfaced, Page-Referenced Terms

(532) germination _____

_____

(532) imbibition _____

_____

(532) primary root _____

_____

Fill-in-the-Blanks

Before or after seed dispersal, the growth of the (1) _____ sporophyte idles. For seeds, (2) _____ is the resumption of growth after a time of arrested embryonic development. The development process is governed by (3) _____ and the environment. Environmental factors influencing this process include

(4) _____ temperature, moisture and oxygen levels, and the number of daylight hours. Usually germination coincides with the return of spring (5) _____. In a process called (6) _____, water molecules move into the seed. As more and more water molecules move inside it, the seed swells and its (7) _____ ruptures. Once the seed coat splits, more oxygen reaches the embryo, and (8) _____ respiration moves into high gear. Cells then rapidly divide to provide growth of the embryo. Cells of the embryonic (9) _____ are activated, divide, elongate, and give rise to a(n) (10) _____ root. Germination is concluded when the primary root breaks through the seed coat.

Growth involves cell (11) _____ and cell (12) _____. There are two basic patterns of growth and development among flowering plants, those of a(n) (13) _____ and a(n) (14) _____. Basic growth and development patterns are (15) _____ and dictated by the plant's genes. Plant development requires cell (16) _____, as brought about by selective gene expression. The growth patterns can be adjusted in response to (17) _____ pressures. Enzymes, hormones, and other gene products within the cells carry out the (18) _____ response.

32-II. PLANT HORMONES (pp. 534–535)

Selected Italicized Words

Agent Orange, Dioscorea

Boldfaced, Page-Referenced Terms

(534) hormone _____

_____

(534) auxins _____

_____

(534) coleoptile _____

_____

(534) gibberellins _____

_____

(535) herbicide _____ _____

_____

(535) cytokinins _____

_____

(535) abscisic acid _____

_____

(535) apical dominance _____

_____

Choice

For questions 1–20, choose from the following:

a. auxins b. gibberellins c. cytokinins
d. abscisic acid e. ethylene f. saponins g. florigen

___ 1. Chinese burning incense to hurry fruit ripening

___ 2. Natural and synthetic versions are used to prolong the shelf life of cut flowers, lettuces, mushrooms, and other vegetables

___ 3. Orchardists spray trees with IAA to thin out overcrowded seedlings in the spring

___ 4. Inhibits cell growth, promotes bud dormancy, and prevents seeds from germinating prematurely

___ 5. IAA, the most important naturally occurring compound

___ 6. Causes stomata to close and help conserve plant water

___ 7. Not plant hormones; they are toxic substances that may kill or repel herbivores or pathogenic organisms

___ 8. Arbitrary designation for as-yet-unidentified hormone (or hormones) thought to cause flowering

___ 9. Exposed to oranges and other citrus fruits to brighten the color of their rind before they are displayed in the market

___ 10. Their complex ring structures are similar to human sex hormones

___ 11. Stimulate stem lengthening; influence plant responses to gravity and light and promote coleoptile elongation

___ 12. Make stems lengthen; help buds and seeds break dormancy and resume growth in the spring

___ 13. Used to prevent premature fruit drop so all fruit can be picked at the same time

___ 14. Stimulate cell division

___ 15. Used by food distributors to ripen green fruit after shipment

___ 16. Most abundant in root and shoot meristems and in the tissues of maturing fruits

___ 17. Used to synthesize birth control pills

___ 18. Under its influence, flowers, fruits, and leaves drop away from plants at prescribed times of the year

___ 19. Promote leaf expansion and retard leaf aging

___ 20. Wild yams contain a modified form; used by rural Mexican natives for contraception

Choose the one most appropriate answer for each.

21. ___ 2,4-D
22. ___ hormone
23. ___ apical dominance
24. ___ herbicide
25. ___ abscission
26. ___ IAA
27. ___ 2,4,5-T
28. ___ target cell

A. Agent Orange; used to defoliate forests during the Vietnam conflict
B. Dropping of flowers, leaves, fruits, or other plant parts
C. A cell with receptors for a given signaling molecule
D. Hormonal effect that inhibits lateral bud growth, which promotes stem elongation
E. Synthetic auxin used as an herbicide
F. A signaling molecule released from one cell that changes the activity of target cells
G. The most important naturally occurring auxin
H. Any compound used to selectively kill plants

32-III. THE TROPISMS (pp. 536–537)

Boldfaced, Page-Referenced Terms

(536) phototropism _____

_____

(536) flavoprotein _____

_____

(536) gravitropism _____

_____

(536) statoliths _____

_____

(537) thigmotropism _____

_____

(537) vines _____

_____

(537) tendrils _____

_____

Choice

For questions 1–5, choose from the following:

a. phototropism   b. gravitropism   c. thigmotropism

___ 1. More intense sunlight on one side of a plant; stems curve toward the light

___ 2. Vines climbing around a fencepost as they grow upward

___ 3. A root turned on its side will curve downward

___ 4. A potted seedling turned on its side in the dark; the growing stem curves upward

___ 5. Leaves turn until their flat surfaces face light

32-IV. RESPONSES TO MECHANICAL STRESS (p. 537)
BIOLOGICAL CLOCKS AND THEIR EFFECTS (p. 538)

Boldfaced, Page-Referenced Terms

(538) biological clocks _____

_____

(538) circadian rhythms _____

_____

(538) phytochrome _____

_____

(538) photoperiodism _____

_____

Matching

1. ___ photoperiodism
2. ___ Pr
3. ___ a plant response to mechanical stress
4. ___ phytochrome activation
5. ___ Pfr deficiency
6. ___ circadian rhythms
7. ___ Pfr
8. ___ biological clocks
9. ___ rhythmic leaf movements
10. ___ phytochrome

A. Biological activities that recur in cycles of twenty-four hours or so
B. Full sun plants grown in the dark put less resources into leaf expansion or branching
C. Internal time-measuring mechanisms with roles in adjusting daily activities
D. Any biological response to a change in the relative length of daylight and darkness in a cycle of twenty-four hours
E. A blue-green pigment that absorbs red or far-red wavelengths, with different results
F. An example of a circadian rhythms
G. Active form of phytochrome
H. Inhibition of plant growth due to strong winds, grazing animals, farm machinery, and shaking plants daily
I. Inactive form of phytochrome
J. May stimulate plant cells to take up calcium ions, or it may induce certain plant cell organelles to release them

Labeling

Identify each indicated part of the accompanying illustration.

11. _____
12. _____
13. _____
14. _____
15. _____

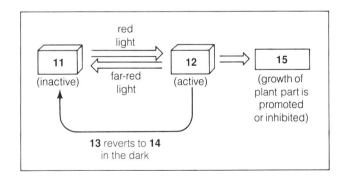

32-V. THE FLOWERING PROCESS (p. 539)

Selected Italicized Words

long-day plants, short-day plants

Fill-in-the-Blanks

Photoperiodism is especially apparent in the (1) _____ process, which is often keyed to daylength changes throughout the year. "Long-day plants" flower in spring when daylength becomes (2) (choose one) ❑ shorter ❑ longer than some critical value. "Short-day plants" flower in late summer or early autumn when daylength becomes (3) (choose one) ❑ shorter ❑ longer than some critical value. "Day-neutral plants" flower whenever they become (4) _____ enough to do so without regard to daylength. Spinach is a (5) _____ - _____ plant because it will not flower and produce seeds unless it is exposed to fourteen hours of light every day for two weeks. Cocklebur is termed a (6) _____-_____ plant because it flowers after a single night that is longer than eight and a half hours.

32-VI. VERNALIZATION (p. 540)
SENESCENCE (p. 540)
DORMANCY (pp. 541–543)

Selected Italicized Words

Secale cereale, Quercus agrifolia

Boldfaced, Page-Referenced Terms

(540) vernalization _____

_____

(540) abscission _____

_____

(540) senescence _____

_____

(541) dormancy _____

_____

Choice

For questions 1–15, choose from the following:

a. vernalization   b. senescence   c. abscission   d. entering dormancy   e. breaking dormancy

___ 1. Dropping of leaves or other parts from a plant

___ 2. A process at work between fall and spring; temperatures become milder, and rain and nutrients become available again

___ 3. Strong cues are short days; long, cold nights; and dry, nitrogen-deficient soil

___ 4. The process occurs at tissues in the base of leaves, flowers, fruits, or other plant parts; a special zone is involved

___ 5. The sum of processes leading to the death of plant parts or the whole plant

___ 6. Low temperature stimulation of flowering

___ 7. A cue is the funneling of nutrients into reproductive parts

___ 8. When a plant stops growing under conditions that appear quite suitable for growth

___ 9. The forming of ethylene in cells near the breakpoints may trigger the process

___ 10. Unless buds of some biennials and perennials are exposed to low winter temperatures, flowers do not form on stems in the spring

___ 11. The term comes from a Latin root meaning "to make springlike"

___ 12. Depending on the species, the process probably involves gibberellins and abscisic acid and requires exposure to low winter temperatures for specific periods

___ 13. Abscisic acid may cause cells near break points to produce the required ethylene

___ 14. Postponement can be effected when gardeners remove flower buds from plants to maintain vegetative growth

___ 15. As an outcome of the redistribution of nutrients to newly forming flowers, fruits, and seeds, leaves usually wither and die

Self Quiz

___ 1. Promoting fruit ripening and abscission of leaves, flowers, and fruits is a function ascribed to _____.

   a. gibberellins
   b. ethylene
   c. abscisic acid
   d. auxins

___ 2. Auxins _____.

   a. cause flowering
   b. promote stomatal closure
   c. promote cell division
   d. promote cell elongation in coleoptiles and stems

___ 3. _____ is demonstrated by a germinating seed whose first root always curves down while the stem always curves up.

   a. Phototropism
   b. Photoperiodism
   c. Gravitropism
   d. Thigmotropism

___ 4. Light of _____ wavelengths is the main stimulus for phototropism.

   a. blue
   b. yellow
   c. red
   d. green

___ 5. Plants whose leaves are open during the day but folded at night are exhibiting a _____.

   a. growth movement
   b. circadian rhythm
   c. biological clock
   d. both b and c are correct

___ 6. 2,4-D, a potent dicot weed killer, is a synthetic _____.

   a. auxin
   b. gibberellin
   c. cytokinin
   d. phytochrome

___ 7. All the processes that lead to the death of a plant or any of its organs are called _____.

   a. dormancy
   b. vernalization
   c. abscission
   d. senescence

___ 8. Phytochrome is converted to an active form, _____, at sunrise and reverts to an inactive form, _____, at sunset, at night, or in the shade.

   a. Pr; Pfr
   b. Pfr; Pfr
   c. Pr; Pr
   d. Pfr; Pr

___ 9. Which of the following is not promoted by the active form of phytochrome?

   a. Seed germination
   b. Flowering
   c. Leaf expansion
   d. Stem elongation

___ 10. When a perennial or biennial plant stops growing under conditions suitable for growth, it has entered a state of _____.

   a. senescence
   b. vernalization
   c. dormancy
   d. abscission

Chapter Objectives/Review Questions

This section lists general and detailed chapter objectives that can be used as review questions. You can make maximum use of these items by writing answers on a separate sheet of paper. To check for accuracy, compare your answers with information given in the chapter or glossary.

Page		Objectives/Questions
(532)	1.	_____ is a resumption of growth after a time of arrested embryonic development.
(532)	2.	Define *imbibition* and describe its role in germination.
(532)	3.	List the environmental factors that influence germination.
(532)	4.	What portion of the embryo emerges from the seed first?
(532)	5.	The basic patterns of growth and development are heritable, dictated by the plant's _____.
(532)	6.	Growth involves cell divisions and _____.
(532)	7.	What determines the directions in which plant embryo cells will divide and how they will differentiate?
(533)	8.	Contrast the growth and development patterns of monocot and dicot plants.
(534)	9.	Describe the general role of plant hormones and target cells.
(534)	10.	Be able to list the five types of hormones found in most flowering plants.
(534)	11.	_____ promote cell elongation in coleoptiles and stems and are involved in phototropism and gravitropism.
(534–535)	12.	_____ promote stem elongation, might help break dormancy of seeds and buds, and stimulate starch breakdown.
(535)	13.	_____ promote cell division and leaf expansion and retard leaf aging.
(535)	14.	_____ promotes stomatal closure and bud and seed dormancy.
(535)	15.	_____ promotes fruit ripening and abscission of leaves, flowers, and fruits.
(535)	16.	_____ is an arbitrary designation for the as-yet-unidentified hormone (or hormones) thought to cause flowering.
(535)	17.	Name a natural source of saponins and describe their value.
(535)	18.	Define *herbicide* and cite examples.
(536–537)	19.	Define *phototropism*, *gravitropism*, and *thigmotropism* and cite examples of each.
(536)	20.	Plants make the strongest phototropic response to light of _____ wavelengths; _____ is a yellow pigment molecule that absorbs blue wavelengths.
(536)	21.	Explain how statoliths form the basis for gravity-sensing mechanisms.
(537)	22.	Provide an example of how mechanical stress can affects plants.
(538)	23.	Plants have internal time-measuring mechanisms called biological _____.
(538)	24.	What are circadian rhythms? Give an example.
(538)	25.	_____ is a biological response to a change in the relative length of daylight and darkness in a twenty-four-hour cycle.
(538)	26.	Phytochrome is converted to an active form, _____, at sunrise, when red wavelengths dominate the sky. It reverts to an inactive form, _____, at sunset, at night, or even in shade, where far-red wavelengths predominate.
(538)	27.	List the plant activities influenced by Pfr.
(538)	28.	_____ serves as a switching mechanism in the biological clock governing these responses.
(539)	29.	Describe the photoperiodic responses of "long-day," "short-day," and "day-neutral" plants.
(540)	30.	Low-temperature stimulation of flowering is called _____.
(540)	31.	_____ is the dropping of leaves, flowers, fruits, or other plant parts.
(540)	32.	Describe the events that signal plant senescence.
(541)	33.	What condition exists when a plant is in dormancy.
(541)	34.	List environmental cues that send a plant into dormancy?
(541)	35.	What conditions are instrumental in the dormancy-breaking process?

Integrating and Applying Key Concepts

An oak tree has grown up in the middle of a forest. A lumber company has just cut down all of the surrounding trees except for a narrow strip of woods that includes the oak. How will the oak be likely to respond as it adjusts to its changed environment? To what new stresses will it be exposed? Which hormones will most probably be involved in the adjustment?

Answers

Interactive Exercises

32-I. PATTERNS OF GROWTH AND DEVELOPMENT (pp. 532–533)
1. embryo; 2. germination; 3. genes; 4. soil; 5. rains; 6. imbibition; 7. coat; 8. aerobic; 9. root; 10. primary; 11. divisions (enlargements); 12. enlargements (divisions); 13. dicot (monocot); 14. monocot (dicot); 15. heritable; 16. differentiation; 17. environmental; 18. growth.

32-II. PLANT HORMONES (pp. 534–535)
1. e; 2. c; 3. a; 4. d; 5. a; 6. d; 7. f; 8. g; 9. e; 10. f; 11. a; 12. b; 13. a; 14. c; 15. e; 16. c; 17. f; 18. e; 19. c; 20. f; 21. E; 22. F; 23. D; 24. H; 25. B; 26. G; 27. A; 28. C.

32-III. THE TROPISMS (pp. 536–537)
1. a; 2. c; 3. b; 4. b; 5. a.

32-IV. RESPONSES TO MECHANICAL STRESS (p. 537)
BIOLOGICAL CLOCKS AND THEIR EFFECTS (p. 538)
1. D; 2. I; 3. H; 4. J; 5. B; 6. A; 7. G; 8. C; 9. F; 10. E; 11. Pr; 12. Pfr; 13. Pfr; 14. Pr; 15. response.

32-V. THE FLOWERING PROCESS (p. 539)
1. flowering; 2. longer; 3. shorter; 4. mature; 5. long-day; 6. short-day.

32-VI. VERNALIZATION (p. 540)
SENESCENCE (p. 540)
DORMANCY (pp. 541–543)
1. c; 2. e; 3. d; 4. c; 5. b; 6. a; 7. b; 8. d; 9. c; 10. a; 11. a; 12. c; 13. c; 14. b; 15. b.

Self-Quiz
1. b; 2. d; 3. c; 4. a; 5. d; 6. a; 7. d; 8. d; 9. d; 10. c.

33

TISSUES, ORGAN SYSTEMS, AND HOMEOSTASIS

ANIMAL STRUCTURE AND FUNCTION: AN
OVERVIEW

EPITHELIAL TISSUE
    General Characteristics
    Cell-to-Cell Contacts
    Glandular Epithelium

CONNECTIVE TISSUE
    Connective Tissue Proper
    Specialized Connective Tissue

MUSCLE TISSUE

NERVOUS TISSUE
    Focus on Science: Frontiers in Tissue Research

ORGAN SYSTEMS
    Tissue and Organ Formation
    Overview of the Major Organ Systems

HOMEOSTASIS AND SYSTEMS CONTROL
    The Internal Environment
    Mechanisms of Homeostasis

Interactive Exercises

33-I. ANIMAL STRUCTURE AND FUNCTION: AN OVERVIEW (pp. 545–547) EPITHELIAL TISSUE (pp. 548–549)

Selected Italicized Words

anatomy, physiology; simple, stratified, epithelium

Boldfaced, Page-Referenced Terms

(547) tissue _____

_____

(547) organ _____

_____

(547) organ system _____

_____

(548) epithelium, epithelia _____

_____

(549) exocrine glands _____

_____

(549) endocrine glands _____

_____

True-False

If the statement in true, write a T in the blank. If false, make it true by changing the underlined word.

_____ 1. <u>Physiology</u> is the study of the way body parts are arranged in an organism.

_____ 2. Groups of <u>like</u> cells that work together to perform a task are known as an organ.

_____ 3. Most animals are constructed of only four types of <u>tissue</u>: epithelial, connective, nervous, and muscle tissues.

_____ 4. Mammalian skin contains <u>squamous epithelium</u> and other tissues.

_____ 5. The more tight junctions there are in a tissue, the more <u>permeable</u> the tissue will be.

_____ 6. <u>Endocrine</u> glands secrete their products through ducts that empty onto an epithelial surface.

_____ 7. <u>Endocrine cell</u> products include digestive enzymes, saliva, and mucus.

Fill-in-the-Blanks

Groups of like cells that work together to perform a task are known as a(n) (8) _____. Groups of *different* types of tissues that interact to carry out a task are known as a(n) (9) _____. Each cell engages in basic (10) _____ activities that assure its own survival. The combined contributions of cells, tissues, organs, and organ systems help maintain a stable (11) _____ _____ that is required for individual cell survival.

While a specific kind of tissue (for example, simple squamous epithelium) is composed of cells that look very similar and do similar jobs, a specific (12) _____ (for example, a kidney) is composed of different tissues that cooperate to do a specific job (in this case, to remove waste products from blood, produce urine, and maintain the composition of body fluids). Kidneys, a urinary bladder, and various tubes are grouped together into a(n) (13) _____ _____, which adds the functions of urine storage and elimination to the jobs that the kidneys do. A multicellular (14) _____ is most often composed of organ systems that cooperate to keep activities running smoothly in a coordinated fashion; this maintenance of stable operating conditions in the internal environment is known as (15) _____.

All the diverse body parts found in different animals can be assembled from a few (16) _____ types through variations in the way they are combined and arranged. Somatic cells compose the physical structure of the animal body; they become differentiated into four main types of specialized tissue: (17) _____, (18) _____, muscle, and nervous tissues.

33-II. CONNECTIVE TISSUE (pp. 550–551)
MUSCLE TISSUE (p. 552)
NERVOUS TISSUE (p. 553)

Selected Italicized Words

"*ground substance,*" *laboratory-grown epidermis, designer organs*

Boldfaced, Page-Referenced Terms

(550) loose connective tissue _____

_____

(550) dense, irregular connective tissue _____

_____

(550) dense, regular connective tissue _____

_____

(550) cartilage _____

_____

(550) bone _____

_____

(551) adipose tissue _____

_____

(551) blood _____

_____

(552) skeletal muscle tissue _____

_____

(552) smooth muscle tissue _____

_____

(552) cardiac muscle tissue _____

_____

(553) nervous tissue _____

_____

(553) neurons _____

_____

True-False

If the statement is true, write a T in the blank. If false, make it true by changing the underlined word.

_____ 1. Muscle bundles are identical to <u>individual</u> skeletal muscle cells.

_____ 2. Both skeletal and cardiac muscle tissues are <u>striated</u>.

_____ 3. Cardiac muscle cells are fused end-to-end, but each cell contracts <u>independently</u> <u>of</u> other cardiac muscle cells.

_____ 4. Smooth muscle tissue is involuntary and not <u>striated</u>.

_____ 5. Smooth muscle tissue is located in the walls of the <u>intestine</u>.

_____ 6. Neurons conduct messages to other <u>neurons</u> or to muscles or glands.

Fill-in-the-Blanks

Cells of connective tissues are scattered throughout an extensive extracellular (7) _____ _____.
(8) _____ connective tissue contains a weblike scattering of strong, flexible protein fibers and a few
highly elastic protein fibers and serves as a packing material that holds in place blood vessels, nerves, and
internal organs. (9) _____ and (10) _____ are examples of dense, regular connective tissue that
help connect elements of the skeletal and muscular systems.

Label-Match

Identify each of the illustrations below by labeling it with one of the following: connective, epithelial, muscle, nervous, or gametes. Complete the exercise by matching and entering *all* appropriate letters and numbers from each group on the next page in the parentheses following each label.

11. _____ ( )

12. _____ ( )

13. _____ ( )

14. _____ ( )

15. _____ ( )

16. _____ ( )

17. _____ ( )

18. _____ ( )

19. _____ ( )

20. _____ ( )

21. _____ ( )

22. _____ ( )

23. _____ ( )

A. Adipose
B. Bone
C. Cardiac
D. Dense, regular
E. Loose
F. Simple columnar
G. Simple cuboidal
H. Simple squamous
I. Smooth
J. Skeletal

1. Absorption
2. Maintain diploid number of chromosomes in sexually reproducing populations
3. Communication by means of electrochemical signals
4. Energy reserve
5. Contraction for voluntary movements
6. Diffusion
7. Padding
8. Contract to propel substances along internal passageways; not striated
9. Attaches muscle to bone and bone to bone
10. In vertebrates, provides the strongest internal framework of the organism
11. Elasticity
12. Secretion
13. Pumps circulatory fluid; striated
14. Insulation
15. Transport of nutrients and waste products to and from body cells

21

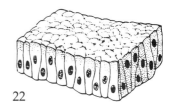
22

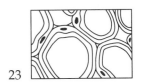
23

33-III. ORGAN SYSTEMS (pp. 554–555)

Selected Italicized Words

ventral, dorsal

Boldfaced, Page-Referenced Terms

(554) ectoderm _____

_____

(554) mesoderm _____

_____

(554) endoderm _____

_____

(554) circulatory system _____

_____

(554) endocrine system _____

_____

(554) integumentary system _____

_____

(554) muscular system _____

_____

(554) nervous system _____

_____

(554) skeletal system _____

_____

(555) lymphatic system _____

_____

(555) digestive system _____

_____

(555) reproductive system _____

_____

(555) respiratory system _____

_____

(555) urinary system _____

_____

Fill-in-the-Blanks

In (1) _____ _____ (immature reproductive cells that later develop into (2) _____), a special form of cell division known as (3) _____ occurs. In all other body tissues that consist of (4) _____ cells, the usual form of cell division, (5) _____, occurs. The life of almost any animal begins with two gametes merging to form a fertilized egg, which undergoes reorganization and then divides by mitosis to form first undifferentiated cells, then three types of (6) _____ (groups of similar cells that perform similar activities) in the early embryo. Eventually, the outer layer of skin and the tissues of the nervous system are formed from the relatively unspecialized embryonic tissue known as (7) _____ located on the embryo's surface. The inner lining of the gut and the major organs formed from the embryonic gut develop from the internal embryonic tissue known as (8) _____. Most of the internal skeleton; muscle; the circulatory, reproductive, and urinary systems; and the connective tissue layers of the gut and body covering are formed from (9) _____, the embryonic tissue composed of cells that can move about like amoebae.

Complete the Table

Supply the name of the "primary" tissue of the embryo that does the job indicated by becoming specialized in particular ways.

Primary Tissue	Functions
10.	Forms internal skeleton and muscle, circulatory, reproductive, and urinary systems
11.	Forms inner lining of gut and linings of major organs formed from the embryonic gut
12.	Forms outer layer of skin and the tissues of the nervous system

Labeling

Supply the name of the correct organ system.

13. _____ Picks up nutrients absorbed from gut and transports them to cells throughout body

14. _____ Helps cells use nutrients by supplying them with oxygen and relieving them of CO_2 wastes

15. _____ Helps maintain the volume and composition of body fluids that bathe the body's cells

16. _____ Provides basic framework for the animal and supports other organs of the body

17. _____ Uses chemical messengers to control and guide body functions

18. _____ Protects the body from viruses, bacteria, and other foreign agents

19. _____ Produces younger, temporarily smaller versions of the animal

20. _____ Breaks down larger food molecules into smaller nutrient molecules that can be absorbed by body fluids and transported to body cells

21. _____ Consists of contractile parts that move the body through the environment and propel substances about in the animal

22. _____ Serves as an electrochemical communications system in the animal's body

23. _____ In the meerkat, served as a heat catcher in the morning and protective insulation at night

Matching

Match the most appropriate function with each system shown on the next page.

24. ___ Male: production and transfer of sperm to the female. Female: production of eggs; provision of a protected nutritive environment for developing embryo and fetus. Both systems have hormonal influences on other organ systems.

25. ___ Ingestion of food, water; preparation of food molecules for absorption; elimination of food residues from the body.

26. ___ Movement of internal body parts; movement of whole body; maintenance of posture; heat production.

27. ___ Detection of external and internal stimuli; control and coordination of responses to stimuli; integration of activities of all organ systems.

28. ___ Protection from injury and dehydration; body temperature control; excretion of some wastes; reception of external stimuli; defense against microbes.

29. ___ Provisioning of cells with oxygen; removal of carbon dioxide wastes produced by cells; pH regulation.

30. ___ Support, protection of body parts; sites for muscle attachment; blood cell production, and calcium and phosphate storage.

31. ___ Hormonal control of body functioning; works with nervous system in integrative tasks.

32. ___ Maintenance of the volume and composition of extracellular fluid.

33. ___ Rapid internal transport of many materials to and from cells; helps stabilize internal temperature and pH.

34. ___ Return of some extracellular fluid to blood; roles in immunity (defense against specific invaders of the body).

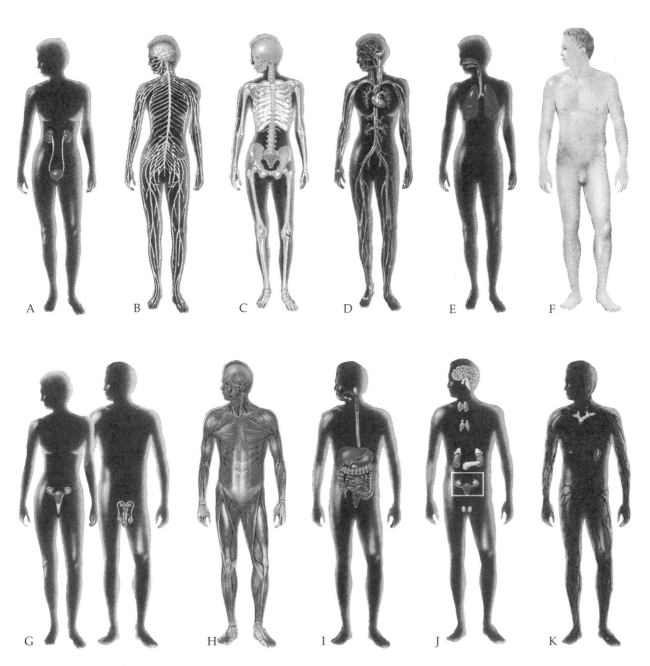

A   B   C   D   E   F

G   H   I   J   K

Fill-in-the-Blanks

There are four major body cavities in humans. Lungs are located in the (35) _____ cavity; the brain is in the (36) _____ cavity; and, in the female, ovaries and the urinary bladder are in the (37) _____ cavity. In humans, the (38) _____ plane divides the body into right and left halves. The (39) _____ plane divides the body into (40) _____ (front) and posterior (back) parts. In humans, the (41) _____ plane divides the body into dorsal (upper) and (42) _____ (lower) parts. The urinary system of an animal is responsible for the disposal of (43) _____ wastes; fecal material is not considered in the category. The endocrine system is generally responsible for internal (44) _____ control; together with the (45) _____ system, it integrates physiological processes.

33-IV. HOMEOSTASIS AND SYSTEMS CONTROL (pp. 556–557)

Selected Italicized Words

interstitial, plasma

Boldfaced, Page-Referenced Terms

(556) extracellular fluid _____

_____

(556) homeostasis _____

_____

(556) sensory receptors _____

_____

(556) stimulus _____

_____

(556) integrator _____

_____

(556) effectors _____

_____

(556) negative feedback mechanism _____

_____

(557) positive feedback mechanism _____

_____

True-False

If the statement is true, write a T in the blank. If false, make it true by changing the underlined word or giving an explanation.

_____ 1. The process of childbirth is an example of a <u>negative</u> feedback mechanism.

_____ 2. The human body's <u>effectors</u> are, for the most part, muscles and glands.

_____ 3. An integrator is constructed in such a way that it is informed of specific energy changes in the environment and relays messages about them to <u>a receptor</u>.

_____ 4. <u>Interstitial fluid</u> is a synonym for plasma.

Fill-in-the-Blanks

In a(n) (5) _____ feedback mechanism, a chain of events is set in motion that intensifies the original condition before returning to a set point; sexual arousal and childbirth are two examples. Generally, physiological controls work by means of (6) _____ feedback, in which an activity changes some condition in the internal environment, and the change causes the condition to be reversed; the maintenance of body temperature close to a set point is an example. Your brain is a(n) (7) _____, a control point where different bits of information are pulled together in the selection of a response. Muscles and (8) _____ are examples of effectors. Internal temperature control of the human body is achieved by (9) _____ in the skin and elsewhere sensing a temperature change at the body's surface and relaying the neural information to an integrator. In this example, the (10) _____ in the brain (see Fig. 33.13) compares neural input against a set point. This part of the brain then sends output signals to (11) _____. Examples of these are the (12) _____ glands, which excrete water from the skin.

Self-Quiz

___ 1. Which of the following is *not* included in connective tissues?

    a. Bone
    b. Blood
    c. Cartilage
    d. Skeletal muscle

___ 2. A surrounding material within which something originates, develops, or is contained is known as a _____.

    a. lamella
    b. ground substance
    c. plasma
    d. lymph

___ 3. Blood is considered to be a(n) _____ tissue.

    a. epithelial
    b. muscular
    c. connective
    d. none of these

___ 4. _____ are abundant in tissues of the heart and liver where they promote diffusion of ions and small molecules from cell to cell.

    a. Adhesion junctions
    b. Filter junctions
    c. Gap junctions
    d. Tight junctions

___ 5. Muscle that is not striped and is involuntary is _____.

    a. cardiac
    b. skeletal
    c. striated
    d. smooth

___ 6. Chemical and structural bridges link groups or layers of like cells, uniting them in structure and function as a cohesive _____.

    a. organ
    b. organ system
    c. tissue
    d. cuticle

___ 7. A fish embryo was accidentally stabbed by a graduate student in developmental biology. Later, the embryo developed into a creature that could not move and had no supportive or circulatory systems. Which embryonic tissue had suffered the damage?

    a. ectoderm
    b. endoderm
    c. mesoderm
    d. protoderm

___ 8. A tissue whose cells are striated and fused at the ends by cell junctions so that the cells contract as a unit is called _____ tissue.

    a. smooth muscle
    b. dense fibrous connective
    c. supportive connective
    d. cardiac muscle

___ 9. The secretion of tears, milk, sweat, and oil are functions of _____ tissues.

    a. epithelial
    b. loose connective
    c. lymphoid
    d. nervous

___ 10. Memory, decision making, and issuing commands to effectors are functions of _____ tissue.
a. connective
b. epithelial
c. muscle
d. nervous

___ 11. An animal that feels heated from the sun moves to an environment that tends to cool its body. This is an example of _____.

a. intensifying an original condition
b. positive feedback mechanism
c. positive phototropic response
d. negative feedback mechanism

___ 12. Which group is arranged correctly from smallest structure to largest?
a. muscle cells, muscle bundle, muscle
b. muscle cells, muscle, muscle bundle
c. muscle bundle, muscle cells, muscle
d. none of the above

Matching

Choose the most appropriate answer to match with each of the following terms.

13. ___ circulatory system

14. ___ digestive system

15. ___ endocrine system

16. ___ immune system

17. ___ integumentary system

18. ___ muscular system

19. ___ nervous system

20. ___ reproductive system

21. ___ respiratory system

22. ___ skeletal system

23. ___ urinary system

A. Picks up nutrients absorbed from gut and transports them to cells throughout body
B. Helps cells use nutrients by supplying them with oxygen and relieving them of CO_2 wastes
C. Helps maintain the volume and composition of body fluids that bathe the body's cells
D. Provides basic framework for the animal and supports other organs of the body
E. Uses chemical messengers to control and guide body functions
F. Protects the body from viruses, bacteria, and other foreign agents
G. Produces younger, temporarily smaller versions of the animal
H. Breaks down larger food molecules into smaller nutrient molecules that can be absorbed by body fluids and transported to body cells
I. Consists of contractile parts that move the body through the environment and propel substances about in the animal
J. Serves as an electrochemical communications system in the animal's body
K. In the meerkat, served as a heat catcher in the morning and protective insulation at night

Chapter Objectives/Review Questions

This section lists general and detailed chapter objectives that can be used as review questions. You can make maximum use of these items by writing answers on a separate sheet of paper. Fill in answers where blanks are provided. To check for accuracy, compare your answers with information given in the chapter or glossary.

Page   *Objectives/Questions*

(546–547)   1. Explain how the meerkat maintains a rather constant internal environment in spite of changing external conditions.

(547)   2. Cells are the basic units of life; in a multicellular animal, like cells are grouped into a(n) _____.

(548–553)   3. Know the characteristics of the various types of tissues. Know the types of cells that compose each tissue type, and be able to cite some examples of organs that contain significant amounts of each tissue type.

(548–549)   4. _____ tissues cover the body surface of all animals and line internal organs from gut cavities to vertebrate lungs; this tissue always has one _____ surface; the opposite surface adheres to a(n) _____ _____.

(548) 5. List the functions carried out by epithelial tissue and state the general location of each type.

(549) 6. Explain the nature of three different cell-to-cell junctions, and state the types of tissues in which these junctions occur.

(549) 7. Explain the meaning of the term *gland*, cite three examples of glands, and state the extracellular products secreted by each.

(550–551) 8. Describe the basic features of connective tissue, and explain how they enable connective tissue to carry out its various tasks.

(550) 9. Connective tissue cells and fibers are surrounded by a(n) _____ _____.

(551) 10. List three functions of blood.

(552) 11. Distinguish among skeletal, cardiac, and smooth muscle tissues in terms of location, structure, and function.

(552) 12. Muscle tissues contain specialized cells that can _____.

(553) 13. Neurons are organized as lines of _____.

(554) 14. Explain how, if each cell can perform all its basic activities, organ systems contribute to cell survival.

(554–555) 15. List each of the eleven principal organ systems in humans and match each to its main task.

(556) 16. Describe the ways by which extracellular fluid helps cells survive.

(556–557) 17. Draw a diagram that illustrates the mechanism of homeostatic control.

(556–557) 18. Describe the relationships among receptors, integrators, and effectors in a negative feedback system.

Interpreting and Applying Key Concepts

Explain why, of all places in the body, marrow is located on the interior of long bones. Explain why your bones are remodeled after you reach maturity. Why does your body not keep the same mature skeleton throughout life?

Answers

Interactive Exercises

33-I. ANIMAL STRUCTURE AND FUNCTION: AN OVERVIEW (p. 547)
EPITHELIAL TISSUE (pp. 548–549)
1. Anatomy; 2. unlike tissues; 3. T; 4. T; 5. impermeable; 6. Exocrine; 7. Exocrine; 8. tissue; 9. organ; 10. metabolic; 11. internal environment; 12. organ; 13. organ system (urinary system, excretory system); 14. organism; 15. homeostasis; 16. tissue; 17. epithelial (connective); 18. connective (epithelial).

33-II. CONNECTIVE TISSUE (pp. 550–551)
MUSCLE TISSUE (p. 552)
NERVOUS TISSUE (p. 553)
1. groups of; 2. T; 3. synchronously with; 4. T; 5. T; 6. T; 7. ground substance; 8. Loose; 9. Tendons (Ligaments); 10. ligaments (tendons); 11. connective, D, 9, 11; 12. epithelial, G, 1, 6, 12; 13. muscle, I, 8, (11); 14. muscle, J, 5, (11); 15. connective, E, 7, 11, (14); 16. gametes, 2; 17. connective, B, 10; 18. epithelial, H, 1, 6, 12; 19. connective, 1, 6, 15; 20. nervous, 3; 21. muscle, C, 13; 22. epithelial, F, 1, 6, 12; 23. connective, A, 4, 14.

33-III. ORGAN SYSTEMS (pp. 554–555)
1. germ cells; 2. gametes; 3. meiosis; 4. somatic; 5. mito-sis; 6. tissues; 7. ectoderm; 8. endoderm; 9. mesoderm; 10. mesoderm; 11. endoderm; 12. ectoderm; 13. circulatory; 14. respiratory; 15. urinary (= excretory); 16. skeletal; 17. endocrine; 18. immune; 19. reproductive; 20. digestive; 21. muscular; 22. nervous; 23. integumentary; 24. G; 25. I; 26. H; 27. B; 28. F; 29. E; 30. C; 31. J; 32. A; 33. D; 34. K; 35. thoracic; 36. cranial; 37. pelvic; 38. midsagittal; 39. transverse; 40. anterior; 41. frontal; 42. ventral; 43. fluid; 44. hormonal (chemical); 45. nervous.

33-IV. HOMEOSTASIS AND SYSTEMS CONTROL (pp. 556–557)
1. positive; 2. T; 3. an effector; 4. *Plasma* is the fluid portion of blood located *in* blood vessels. *Interstitial fluid* is located outside of blood vessels, lymphatic vessels, and body cells, but within the body proper; 5. positive; 6. negative; 7. integrator; 8. glands; 9. receptors; 10. hypothalamus; 11. effectors; 12. sweat.

Self-Quiz
1. d; 2. b; 3. c; 4. c; 5. d; 6. c; 7. c; 8. d; 9. a; 10. d; 11. d 12. a; 13. A; 14. H; 15. E; 16. F; 17. K; 18. I; 19. J; 20. G; 21. B; 22. D; 23. C.

34

INFORMATION FLOW
AND THE NEURON

Interactive Exercises

34-I. CELLS OF THE NERVOUS SYSTEM (pp. 560–561)
HOW THE NEURON RESPONDS TO STIMULATION (pp. 562–563)

Selected Italicized Words

input, conducting, trigger, output zones

Boldfaced, Page-Referenced Terms

(561) neurons _____

_____

(561) sensory neuron _____

_____

(561) interneurons _____

_____

(561) motor neurons _____

_____

(561) neuroglia _____

_____

(562) dendrites _____

_____

(562) axon _____

_____

(562) resting membrane potential _____

_____

(562) action potential _____

_____

(563) sodium-potassium pumps _____

_____

Fill-in-the-Blanks

Nerve cells that conduct messages are called (1) _____. (2) _____ cells, which support and nurture the activities of neurons, make up more than half the volume of the nervous system. (3) _____
neurons are receptors for environmental stimuli, (4) _____ connect different neurons in the central
nervous system, and (5) _____ neurons are linked with muscles or glands. All neurons have a
(6) _____ _____ that contains the nucleus and the metabolic means to carry out protein synthesis.
(7) _____ are short, slender extensions of (6), and together these two neuronal parts are the neurons'
"input zone" for receiving (8) _____. The (9) _____ is a single, long, cylindrical extension of the
(10) _____ _____; in motor neurons, the (11) _____ has finely branched (12) _____ that
terminate on muscle or gland cells and are "output zones," where messages are sent on to other cells.

A neuron at rest establishes unequal electric charges across its plasma membrane, and a(n)
(13) _____ _____ is maintained. Another name for (13) is the (14) _____ _____
_____; it represents a tendency for activity to happen along the membrane. Weak disturbances of the
neuronal membrane might set off only slight changes across a small patch, but strong disturbances can
cause a(n) (15) _____ _____, which is an abrupt, short-lived reversal in the polarity of charge
across the plasma membrane of the neuron. For a fraction of a second, the cytoplasmic side of a bit of membrane becomes positive with respect to the outside. The (16) _____ that travels along the neuronal
membrane is nothing more than short-lived changes in the membrane potential. When action potentials
reach the end of a motor neuron, they cause (17) _____ to be released that serve as chemical signals to
adjacent muscle cells. Muscles (18) _____ in response to the signals.

How is the resting membrane potential established, and what restores it between action potentials? The
concentrations of (19) _____ ions (K^+), sodium ions (20) (___), and other charged substances are not the
same on the inside and outside of the neuronal membrane. (21) _____ proteins that span the mem-

brane affect the diffusion of specific types of ions across it. (22) _____ proteins that span the membrane pump sodium and potassium ions against their concentration gradients by using energy stored in ATP. A neuronal membrane has many more positively charged (23) _____ ions inside than out and many more positively charged (24) _____ ions outside than inside. An electrical gradient also exists across the neuronal membrane; compared with the outside, the inside of a neuron at rest has an overall (25) _____ charge. For many neurons in most animals, the difference in charge across the neuronal membrane is about 70 (26) _____. There are about (27) _____ (number) times more potassium ions on the cytoplasmic side as outside, and there are about (28) _____ (number) times more sodium ions outside as inside. These ions can cross the membrane only by traveling along passages through (29) _____ proteins. Some (29) proteins leak ions through them all the time; others have (30) _____ that open only when stimulated. Transport proteins called (31) _____-_____ _____ counter the leakage of ions across the neuronal membrane and maintain the resting membrane potential.

Labeling

Identify the parts of the neuron illustrated below.

32. _____ _____

33. _____ _____

34. _____

35. _____-_____ _____

36. _____ _____

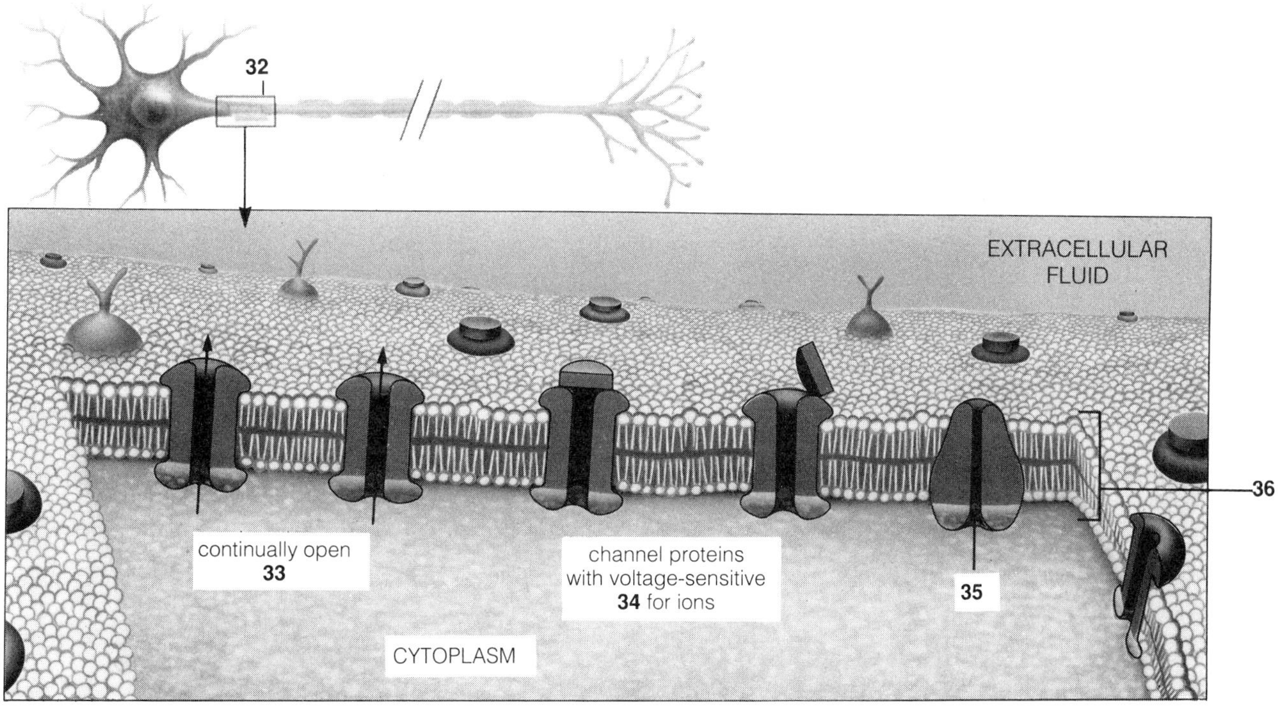

continually open
33

channel proteins
with voltage-sensitive
34 for ions

EXTRACELLULAR
FLUID

36

35

CYTOPLASM

34-II. ACTION POTENTIALS (pp. 564–565)
NODE-TO-NODE HOPPING ALONG SHEATHED AXONS (p. 566)

Selected Italicized Words

graded, local signal; all-or-nothing event; saltatory conduction; multiple sclerosis

Boldfaced, Page-Referenced Terms

(564) threshold level _____

_____

(566) myelin sheath _____

_____

Fill-in-the-Blanks

In all neurons, stimulation at an input zone produces (1) _____ signals that do not spread very far (half a millimeter or less). (2) _____ means that signals can vary in magnitude—small or large—depending on the intensity and (3) _____ of the stimulus. When stimulation is intense or prolonged, graded signals can spread into an adjacent (4) _____ _____ of the membrane—the site where action potentials can be initiated.

A(n) (5) _____ _____ is an all-or-nothing, brief reversal in membrane potential. Once an action potential has been achieved, it is a(n) (6) _____-_____-_____ event; its amplitude will not change even if the strength of the stimulus changes. The minimum change in membrane potential needed to achieve an action potential is the (7) _____ value. Each action potential is followed by a time of insensitivity to stimulation. Some narrow-diameter neurons are wrapped in lipid-rich (8) _____ produced by specialized neuroglial cells called Schwann cells; each of these is separated from the next by a(n) (9) _____ _____—a small gap where the axon is exposed to extracellular fluid. An action potential jumps from one node to the next in line and is called (10) _____ conduction. In the largest myelinated axon, signals travel (11) _____ (number) meters per second.

Labeling

Identify the parts of the illustrations.

12. _____ _____

13. _____

14. _____ _____ _____

15. _____

16. _____

17. _____ _____

18. _____ _____

34-III. CHEMICAL SYNAPSES (pp. 566–567)
SYNAPTIC INTEGRATION (pp. 568–569)
PATHS OF INFORMATION FLOW (pp. 570–571)

Selected Italicized Words

presynaptic, postsynaptic, excitatory, inhibitory, neuromuscular junction, depolarizing, hyperpolarizing, spatial, temporal, botulism, tetanus, divergent, convergent, reverberating, Clostridium botulinum, C. tetani

Boldfaced, Page-Referenced Terms

(566) neurotransmitters _____

_____

(566) chemical synapses _____

_____

(566) acetylcholine (ACh) _____

_____

(568) neuromodulators _____

_____

(568) EPSP _____

_____

(568) IPSP _____

_____

(568) synaptic integration _____

_____

(570) nerves _____

_____

(570) reflex arc _____

_____

(570) muscle spindles _____

_____

Fill-in-the-Blanks

The junction specialized for transmission between a neuron and another cell is called a(n) (1) _____

_____. Usually, the signal being sent to the receiving cell is carried by chemical messengers called

(2) _____. (3) _____ is an example of this type of chemical messenger that diffuses across the

synaptic cleft, combines with protein receptor molecules on the muscle cell membrane, and soon thereafter

is rapidly broken down by enzymes. At a(n) (4) _____ synapse, the membrane potential is driven

toward the threshold value and increases the likelihood that an action potential will occur. At a(n)

(5) _____ synapse, the membrane potential is driven away from the threshold value, and the receiving

neuron is less likely to achieve an action potential. A specific neurotransmitter can have either excitatory or

inhibitory effects depending on which type of protein channel it opens up in the (6) _____ membrane.

A(n) (7) _____ _____ is a synapse between a motor neuron and muscle cells.

(8) _____ acts on brain cells that govern sleeping, sensory perception, temperature regulation, and

emotional states. (9) _____ are neuromodulators that inhibit perceptions of pain and may have roles in

memory and learning, emotional depression, and sexual behavior. (10) _____ _____ at the cellu-

lar level is the moment-by-moment tallying of all excitatory and inhibitory signals acting on a neuron.

Incoming information is (11) _____ by cell bodies, and the charge differences across the membranes are

either enhanced or inhibited. A(n) (12) _____ postsynaptic potential (EPSP) brings the membrane

closer to threshold and has a depolarizing effect. An inhibitory postsynaptic potential (IPSP) drives the

membrane away from threshold and either has a(n) (13) _____ effect or maintains the membrane at its

resting level. Cordlike communication lines called (14) _____ _____ connect neurons from one

region to neurons in different regions within the brain and spinal cord.

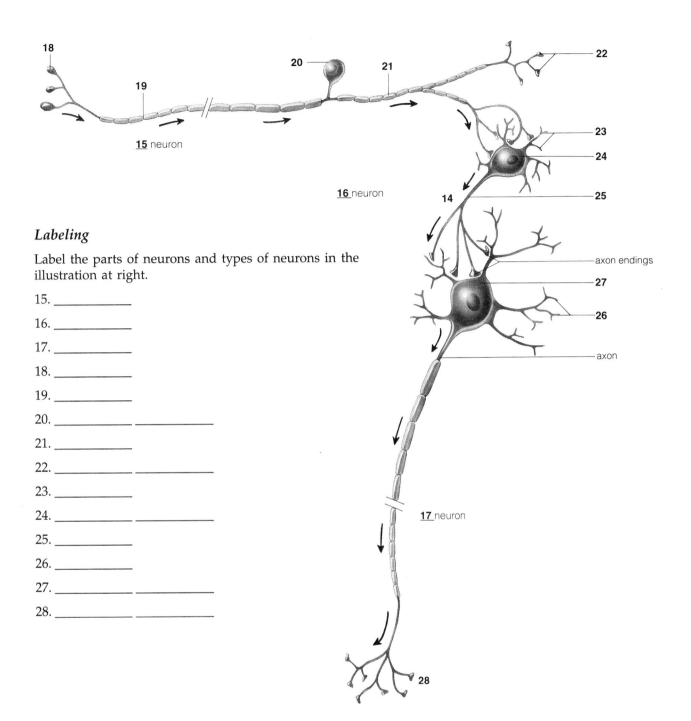

15 neuron

16 neuron

17 neuron

axon endings

axon

Labeling

Label the parts of neurons and types of neurons in the illustration at right.

15. _____

16. _____

17. _____

18. _____

19. _____

20. _____ _____

21. _____

22. _____ _____

23. _____

24. _____ _____

25. _____

26. _____

27. _____ _____

28. _____ _____

Label the parts of the nerve illustrated here.

29. _____

30. _____ _____

31. _____ _____

32. _____

29

30

outer wrapping
of the nerve

31

a nerve fascicle
(many **32**
bundled in
connective
tissue)

Fill-in-the-Blanks

A(n) (33) _____ is an involuntary sequence of events elicited by a stimulus. During a(n)
(34) _____ _____, a muscle contracts involuntarily whenever conditions cause a stretch in length;
many of these help you maintain an upright posture despite small shifts in balance. Located within skeletal
muscles are length-sensitive organs called (35) _____ _____, which generate action potentials
when stretched beyond a critical point; these potentials are conducted rapidly to the (36) _____
_____, where they are communicated to motor neurons leading right back to the muscle that was
stretched.

Imbalances can occur at chemical synapses; a neurotoxin produced by *Clostridium tetani* interferes with
the effect of (37) _____ on motor neurons, which may cause tetanus—a prolonged, spastic paralysis
that can lead to death.

Matching

Match the choices below with the correct number in the diagram. Two of the numbers match with two lettered choices.

38. _____   43. _____

39. _____   44. _____

40. _____   45. _____

41. _____   46. _____

42. _____

A. Response
B. Action potentials generated in motor neuron and propagated along its axon toward muscle
C. Motor neuron synapses with muscle cells
D. Muscle cells contract
E. Local signals in receptor endings of sensory neuron
F. Muscle spindle stretches
G. Action potentials generated in all muscle cells innervated by motor neuron
H. Stimulus
I. Axon endings synapse with motor neuron
J. Spinal cord
K. Action potential propagated along sensory neuron toward spinal cord

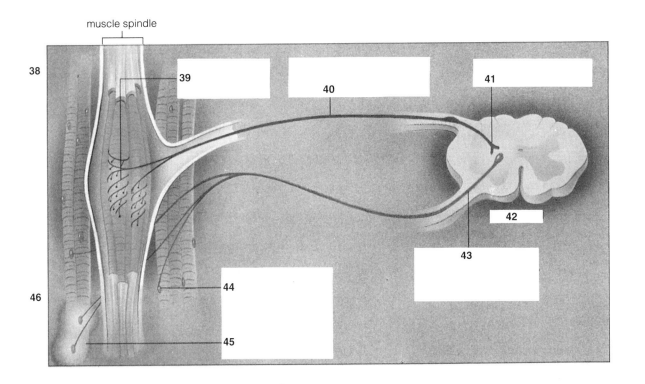

Self-Quiz

___ 1. Which of the following is *not* true of an action potential?

    a. It is a short-range message that can vary in size.

    b. It is an all-or-none brief reversal in membrane potential.

    c. It doesn't decay with distance.

    d. It is self-propagating.

___ 2. The conducting zone of a neuron is the _____.

    a. axon

    b. axonal terminals

    c. cell body

    d. dendrite

___ 3. The integrative zone of a neuron is the _____.

    a. axon

    b. axonal terminals

    c. cell body

    d. dendrite

___ 4. In the nervous system, a cotransport mechanism involves _____.

    a. the inactivation of signals along the parasympathetic nerves by signals from the sympathetic division

    b. acetylcholine being reciprocally inactivated by cholinesterase in the synaptic zone

    c. the exchange of signals between the peripheral and central nervous systems

    d. two different substances being exchanged across a membrane by the same enzyme system

___ 5. An action potential is brought about by _____.

    a. a sudden membrane impermeability

    b. the movement of negatively charged proteins through the neuronal membrane

    c. the movement of lipoproteins to the outer membrane

    d. a local change in membrane permeability caused by a greater-than-threshold stimulus

___ 6. The resting membrane potential _____.

    a. exists as long as a charge difference sufficient to do work exists across a membrane

    b. occurs because there are more potassium ions outside the neuronal membrane than there are inside

    c. occurs because of the unique distribution of receptor proteins located on the dendrite exterior

    d. is brought about by a local change in membrane permeability caused by a greater-than-threshold stimulus

___ 7. The phrase "all or none" used in conjunction with discussion about an action potential means that _____.

    a. a resting membrane potential has been received by the cell

    b. an impulse does not decay or dissipate as it travels away from the stimulus point

    c. the membrane either achieves total equilibrium or remains as far from equilibrium as possible

    d. propagation along the neuron is saltatory

___ 8. Endorphins _____.

    a. are neuromodulators

    b. block perceptions of pain

    c. may play a role in causing emotional depression

    d. are involved in all of the above roles

___ 9. An action potential passes from neuron to neuron across a synaptic cleft by _____.

    a. saltatory conduction

    b. the resting membrane potential

    c. neurotransmitter substances

    d. neuromodulator substances

___ 10. _____ are responsible for integration in the nervous system.

    a. Interneurons

    b. Schwann cells

    c. Motor neurons

    d. Sensory neurons

Chapter Objectives/Review Questions

This section lists general and detailed chapter objectives that can be used as review questions. You can make maximum use of these items by writing answers on a separate sheet of paper. To check for accuracy, compare your answers with information given in the chapter or glossary.

Page   *Objectives/Questions*

(562)   1. Draw a neuron and label it according to its three general zones, its specific structures, and the specific function(s) of each structure.

(562)   2. Define *resting membrane potential*; explain what establishes it and how it is used by the cell neuron.

(563–565)   3. Describe the distribution of the invisible array of large proteins, ions, and other molecules in a neuron, both at rest and as a neuron experiences a change in potential.

(563)   4. Define *sodium-potassium pump* and state how it helps maintain the resting membrane potential.

(563)   5. Define *action potential* by stating its three main characteristics.

(562–565)   6. Explain the chemical basis of the action potential. Look at Fig. 34.6 in your text and determine which part of the curve represents the following:
   a. the point at which the stimulus was applied
   b. the events prior to achievement of the threshold value
   c. the opening of the ion gates and the diffusing of the ions
   d. the change from net negative charge inside the neuron to net positive charge and back again to net negative charge
   e. the active transport of sodium ions out of and potassium ions into the neuron

(564)   7. Explain how graded signals differ from action potentials.

(565)   8. Define *period of insensitivity* and state what causes it.

(566)   9. Define *Schwann cell*, *unsheathed nodes*, and *myelin sheath*, and explain how each helps narrow-diameter neurons conduct nerve impulses quickly.

(566)   10. Distinguish the way excitatory synapses function from the way inhibitory synapses function.

(566–567)   11. Understand how a nerve impulse is received by a neuron, conducted along a neuron, and transmitted across a synapse to a neighboring neuron, muscle, or gland.

(568–570)   12. Outline some of the ways by which information flow is regulated and integrated in the human body.

(570–571)   13. Explain what a reflex is by drawing and labeling a diagram and telling how it functions.

(570–571)   14. Explain what the stretch reflex is and tell how it helps an animal survive.

Integrating and Applying Key Concepts

What do you think might happen to human behavior if inhibitory postsynaptic potentials did not exist and if the threshold stimulus necessary to provoke an EPSP were much higher?

Answers

Interactive Exercises

34-I. CELLS OF THE NERVOUS SYSTEM (pp. 560–561)
HOW THE NEURON RESPONDS TO STIMULATION (pp. 562–563)

1. neurons; 2. Neuroglial; 3. Sensory; 4. interneurons; 5. motor; 6. cell body; 7. Dendrites; 8. signals (stimuli); 9. axon; 10. cell body; 11. axon; 12. endings; 13. voltage differential; 14. resting membrane potential; 15. action potential (nerve impulse); 16. disturbance; 17. molecules (chemicals); 18. contract; 19. potassium; 20. Na$^+$; 21. Channel; 22. Transport; 23. potassium; 24. sodium; 25. negative; 26. millivolts; 27. 30; 28. 10; 29. channel; 30. gates; 31. sodium-potassium pumps; 32. axonal membrane; 33. channel proteins; 34. gates; 35. sodium-potassium pump; 36. lipid bilayer.

34-II. ACTION POTENTIALS (pp. 564–565)
NODE-TO-NODE HOPPING ALONG SHEATHED AXONS (p. 566)

1. localized; 2. Graded; 3. duration; 4. trigger zone; 5. action potential; 6. all-or-nothing; 7. threshold; 8. myelin; 9. unsheathed node; 10. saltatory; 11. 120; 12. action potential; 13. threshold; 14. resting membrane potential; 15. milliseconds; 16. millivolts; 17. unsheathed node; 18. Schwann cell (myelin sheath).

34-III. CHEMICAL SYNAPSES (pp. 566–567)
SYNAPTIC INTEGRATION (pp. 568–569)
PATHS OF INFORMATION FLOW (pp. 570–571)

1. chemical synapse; 2. neurotransmitters; 3. Acetylcholine (ACh); 4. excitatory; 5. inhibitory; 6. postsynaptic; 7. neuromuscular junction; 8. Serotonin; 9. Endorphins; 10. Synaptic integration; 11. summed; 12. excitatory; 13. hyperpolarizing; 14. nerve tracts; 15. sensory; 16. inter; 17. motor; 18. receptor; 19. dendrite; 20. cell body; 21. axon; 22. axon endings; 23. dendrites; 24. cell body; 25. axon; 26. dendrites; 27. cell body; 28. axon endings; 29. axon; 30. myelin sheath; 31. blood vessels; 32. axons; 33. reflex; 34. stretch reflex; 35. muscle spindles; 36. spinal cord; 37. acetylcholine; 38. H, F; 39. E; 40. K; 41. I; 42. J; 43. B; 44. C; 45. G; 46. A, D.

Self-Quiz

1. a; 2. a; 3. c; 4. d; 5. d; 6. a; 7. b; 8. d; 9. c; 10. a.

35

INTEGRATION AND CONTROL: NERVOUS SYSTEMS

Interactive Exercises

35-I. INVERTEBRATE NERVOUS SYSTEMS (pp. 574–577)

Selected Italicized Words

crack, radial, bilateral, cephalization

Boldfaced, Page-Referenced Terms

(576) nervous system _____

_____

(576) nerve net _____

_____

(576) reflex pathways _____

_____

(576) ganglion, ganglia _____

_____

(576) planula _____

_____

Fill-in-the-Blanks

The (1) _____ system enables animals to sense specific information about external and internal conditions, to (2) _____ or evaluate that information, and to issue commands to the muscles and glands so that a response can be made. The simplest nervous systems are the (3) _____ _____ of cnidarians, which have (4) _____ symmetry; a nerve net is arranged about a central axis like spokes of a bike wheel.

Probably the earliest response mechanism that developed in invertebrates was that of the (5) _____, a simple, stereotyped movement made in response to a specific kind of stimulus; in the simplest of these pathways, a(n) (6) _____ neuron directly signals a(n) (7) _____ neuron, which acts on muscle cells that respond to the stimulus. Radially symmetrical organisms are adapted for a life that is either fixed to some surface or floating through water, but a crawling existence might get an organism more resources. In crawling animals, the leading, forward end is the part most likely to encounter (8) _____ or food; it would be more advantageous if (9) _____ _____ were located there rather than in the tail section that follows, if the animal is interested in surviving. Organisms in which sensory nerve cells became concentrated as a sort of brain in the forward, head end are said to have undergone the evolutionary process known as (10) _____; such animals are (11) _____ symmetrical, with right and left sides. Invertebrate animals tend to have two (12) _____ _____, which are located on the lower side and receive information from and issue commands to the right and left halves of the animal.

As invertebrate body structure became more complex, the cell bodies of neurons became clustered to form integrative and evaluative centers known as (13) _____; the fiberlike axons of their neurons were grouped together and encased in connective tissue and became known as (14) _____. As brain regions and sense organs evolved and became more complex, they exerted more and more control over the more basic (15) _____ behaviors; brain regions were able to store and compare information about experiences and use that information to override reflex activities and initiate innovative actions. These neural pathways within the more recent evolutionary additions to brains are the bases of memory, (16) _____, and reasoning. Octopuses, for example, can be taught symbols that guide them to a crab feast.

Labeling

Label each numbered part of the illustration below.

17. _____

18. _____

19. _____

20. _____

21. _____

22. _____

23. _____

24. _____

FLATWORM

17

18

EARTHWORM

19

nerve cord

20

CRAYFISH

21

nerve cord

nerve cord

24

GRASSHOPPER

22

23

nerve cord

35-II. VERTEBRATE NERVOUS SYSTEMS (pp. 578–579) PERIPHERAL NERVOUS SYSTEM (pp. 580–581)

Selected Italicized Words

afferent, efferent, spinal nerves, cranial nerves

Boldfaced, Page-Referenced Terms

(578) central nervous system _____

_____

(578) peripheral nervous system _____

_____

(580) somatic nerves _____

_____

(580) autonomic nerves _____

_____

(580) parasympathetic nerves _____

_____

(580) sympathetic nerves _____

_____

(581) fight-flight response _____

_____

Fill-in-the-Blanks

A shift from radial to bilateral symmetry could have led, in some evolutionary lines of animals, to paired (1) _____ and (2) _____, paired sensory structures such as eyes, and paired (3) _____ regions. The (4) _____ nervous system consists of the brain and spinal cord (or paired cords). The (5) _____ nervous system includes cell bodies of sensory neurons and all nerves (bundles of axons) that lead to and from the central nervous system.

In addition to cephalization and bilateral symmetry, the first vertebrates also had a (6) _____ _____, a hollow, tubular structure running dorsally above the (7) _____, and it was the forerunner of the (8) _____ _____ and brain. As time passed, the brain expanded and became divided into three specialized parts: the (9) _____, midbrain, and hindbrain.

We call the nerve cord that develops in all vertebrate embryos the (10) _____ _____; it undergoes expansion and regional modification and becomes enclosed within the (11) _____ _____. Adjacent tissues in the embryo form (12) _____ that thread through all body regions and connect with the spinal cord and brain.

All motor-nerves-to-skeletal-muscle pathways and all sensory pathways make up the (13) _____ nervous system. The remaining nerve tissue, which generally is not under conscious control, is collectively known as the (14) _____ nervous system; it is subdivided into two parts: (15) _____ nerves, which respond to emergency situations, and (16) _____ nerves, which oversee the restoration of normal body functioning. A (17) _____ encloses the brain, and the (18) _____ _____ encloses and protects the spinal cord in vertebrates. In humans, thirty-one pairs of spinal nerves connect with the spinal cord and are grouped by anatomical region; twelve pairs of (19) _____ nerves connect parts of the head and neck with brain centers.

Matching

Choose the most appropriate letter to match with each numbered blank.

20. ___ cervical
21. ___ coccygeal
22. ___ lumbar
23. ___ sacral
24. ___ thoracic

A. Chest
B. Neck
C. Pelvic
D. Tail
E. Waist

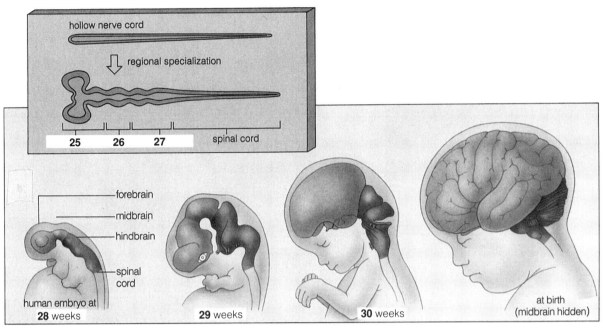

Labeling

Label each numbered part of the accompanying illustrations.

25. _____

26. _____

27. _____

28. _____

29. _____

30. _____

    Identify the divisions of the nervous system in the posterior view at right.

31. _____ _____

32. _____ nerves

33. _____ nerves

34. _____ nerves

35. _____ nerves

36. _____ nerves

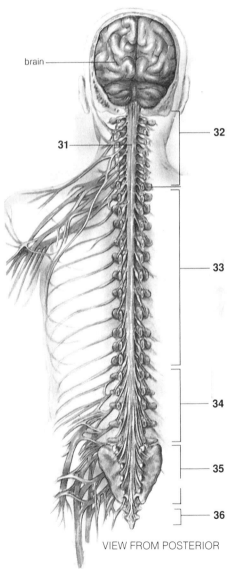

VIEW FROM POSTERIOR

Label each numbered part of the accompanying illustration.

37. _____   40. _____

38. _____   41. _____

39. _____

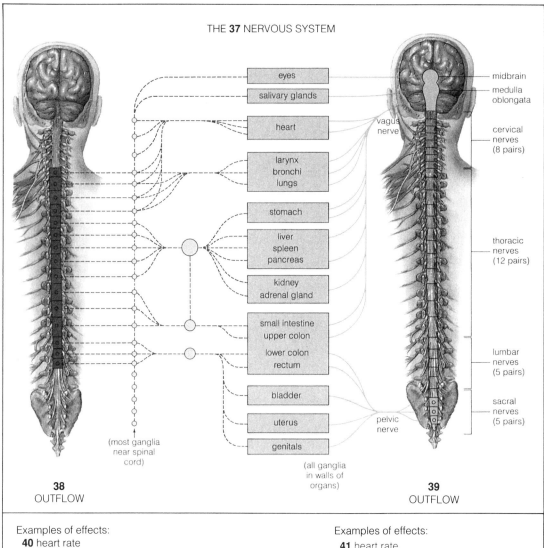

THE **37** NERVOUS SYSTEM

eyes

salivary glands

heart

larynx
bronchi
lungs

stomach

liver
spleen
pancreas

kidney
adrenal gland

small intestine
upper colon

lower colon
rectum

bladder

uterus

genitals

midbrain

medulla
oblongata

vagus
nerve

cervical
nerves
(8 pairs)

thoracic
nerves
(12 pairs)

lumbar
nerves
(5 pairs)

sacral
nerves
(5 pairs)

pelvic
nerve

(most ganglia
near spinal
cord)

(all ganglia
in walls of
organs)

38
OUTFLOW

39
OUTFLOW

Examples of effects:
 40 heart rate
 Dilation of pupil in eyes
 Inhibits bronchial glandular secretion
 Inhibits stomach, intestinal movements
 Contracts sphincters

Examples of effects:
 41 heart rate
 Contraction of pupil in eyes
 Stimulates bronchial glandular secretion
 Stimulates stomach, intestinal movements
 Relaxes sphincters

35-III. CENTRAL NERVOUS SYSTEM: THE SPINAL CORD (p. 581)
CENTRAL NERVOUS SYSTEM: THE VERTEBRATE BRAIN (pp. 582–583)

Selected Italicized Words

meningitis, "white matter," "gray matter," extensor

Boldfaced, Page-Referenced Terms

(581) spinal cord _____

_____

(581) meninges _____

_____

(582) brain _____

_____

(582) brain stem _____

_____

(582) hindbrain _____

_____

(582) medulla oblongata _____

_____

(582) cerebellum _____

_____

(582) pons _____

_____

(582) midbrain _____

_____

(582) tectum _____

_____

(583) forebrain _____

_____

(583) cerebrum _____

_____

(583) thalamus _____

_____

(583) hypothalamus _____

_____

(583) reticular formation _____

_____

(583) cerebrospinal fluid _____

_____

Fill-in-the-Blanks

The (1) _____ _____ is a region of local integration and reflex connections with nerve pathways leading to and from the brain; its (2) _____ _____, which contains myelinated sensory and motor axons, is the through-conducting zone. The (3) _____ _____ includes neuronal cell bodies, dendrites, nonmyelinated axon terminals, and neuroglial cells; this is the (4) _____ zone, where sensory input is linked to motor output.

The hindbrain is an extension and enlargement of the upper spinal cord; it consists of the (5) _____ _____, which contains the control centers for the heartbeat rate, blood pressure, and breathing reflexes. The hindbrain also includes the (6) _____, which helps coordinate motor responses associated with refined limb movements, maintenance of posture, and spatial orientation. The (7) _____ is a major routing station for nerve tracts passing between brain centers.

The (8) _____ evolved as a center that coordinates visual and auditory (hearing) input with reflex responses. The midbrain, pons, and medulla oblongata make up the brain stem; within its core, the (9) _____ _____ is a major network of interneurons that extends its entire length and governs an organism's level of alertness.

The (10) _____ contains two cerebral hemispheres that contain most of the brain's gray matter. Underlying the cerebral hemisphere is the (11) _____, a region that monitors internal organs and influences hunger, thirst, and (12) _____ behaviors, and the (13) _____, where some motor pathways converge and which is the principal pathway to and from the cerebral hemispheres. Both the spinal cord and the brain are bathed in (14) _____ _____, which (15) _____ the vital nervous tissue from sudden jarring movements.

Labeling

Identify the vertebrae in the side view of the spinal cord shown.

16. _____ vertebrae

17. _____ vertebrae

18. _____ vertebrae

19. _____ vertebrae

20. _____ vertebrae

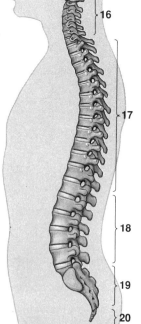

Identify the numbered parts of the illustrations.

21. _____ _____

22. _____

23. _____

24. _____ _____

25. _____ _____

26. _____

27. _____ _____

28. _____ _____

29. _____ _____

30. _____ _____

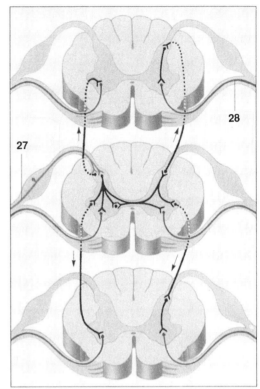

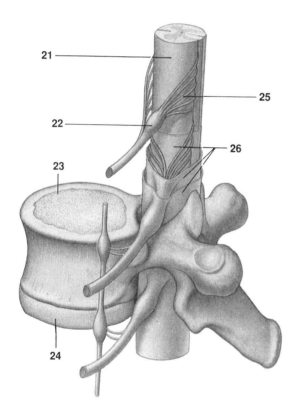

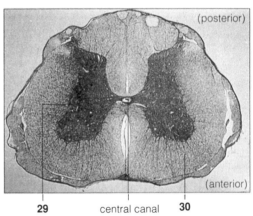

35-IV. A CLOSER LOOK AT THE HUMAN CEREBRUM (pp. 584–586)
MEMORY (p. 587)
STATES OF CONSCIOUSNESS (p. 587)
THE BRAIN AND BEHAVIOR (pp. 588–589)

Selected Italicized Words

occipital lobe, temporal lobe, parietal lobe, frontal lobe, epilepsy, "split-brain," retrograde amnesia, Alzheimer's disease, alpha rhythm, slow-wave sleep, REM sleep, EEG arousal, short-term memory, long-term memory

Boldfaced, Page-Referenced Terms

(584) cerebral hemispheres _____

_____

(584) cerebral cortex _____

_____

(587) memory _____

_____

(587) association _____

_____

(588) limbic system _____

_____

(588) drug addiction _____

_____

(588) stimulants _____

_____

(589) depressants _____

_____

(589) hypnotics _____

_____

(589) analgesics _____

_____

(589) psychedelics _____

_____

(589) hallucinogens _____

_____

Matching

Match the named part with the letter that describes its function.

1. ___ cerebellum
2. ___ corpus callosum
3. ___ hypothalamus
4. ___ limbic system
5. ___ medulla oblongata
6. ___ meninges
7. ___ primary motor cortex
8. ___ primary auditory cortex
9. ___ visual cortex
10. ___ primary somatosensory cortex
11. ___ thalamus

A. Monitors visceral activities; influences behaviors related to thirst, hunger, reproductive cycles, and temperature control
B. Protective coverings of brain and spinal cord
C. Receives inputs from cochleae of inner ears
D. Issues commands to muscles
E. Relays and coordinates sensory signals to the cerebrum
F. Receives inputs from retinas of eyeballs
G. Receives and processes input from body feeling areas
H. Broad channel of white matter that keeps the two cerebral hemispheres communicating with each other
I. Coordinates nerve signals for maintaining balance, posture, and refined limb movements
J. Connects pons and spinal cord; contains reflex centers involved in respiration, stomach secretion, and cardiovascular function
K. Contains brain centers that coordinate activities underlying emotional expression

Labeling

Identify each numbered part of the accompanying illustration.

12. _____
13. _____ _____
14. _____
15. _____ _____

16. _____ _____
17. _____
18. _____
19. _____ _____

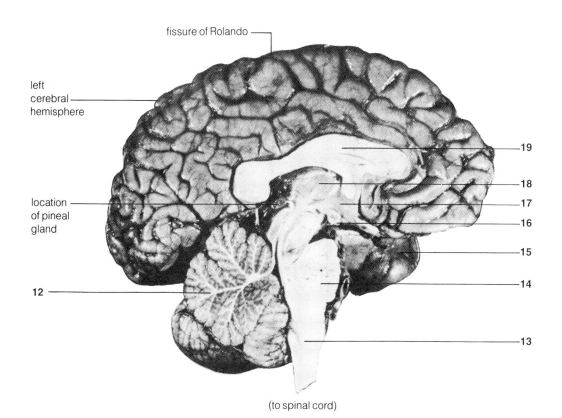

fissure of Rolando

left cerebral hemisphere

location of pineal gland

12

19
18
17
16
15
14
13

(to spinal cord)

Fill-in-the-Blanks

The storage of individual bits of information somewhere in the brain is called (20) _____. Experiments suggest that at least two stages are involved in its formation. One is a(n) (21) _____-_____ _____ period, lasting only a few minutes; information then becomes spatially and temporally organized in neural pathways. The other is a(n) (22) _____-_____ _____; information then is put in a different neural representation and is permanently filed in the brain.

The principal wave pattern for someone who is relaxed, with eyes closed, is a(n) (23) _____ _____. The (24) _____-_____ _____ pattern occupies about 80 percent of the total sleeping time for adults. When individuals shift from sleep to a conscious focus on external stimuli or on their own thoughts, the pattern is called (25) _____ _____. (26) _____ _____ accompanies vivid dreaming periods. Activities in the (27) _____ _____ determine whether you are awake or asleep. (28) _____ are analgesics produced by the brain that inhibit regions concerned with our emotions and perception of (29) _____. High (30) _____ levels in the brain stem's core bring about drowsiness and sleep.

Matching

Match the most appropriate letter with each numbered item.

31. ___ amphetamines   A. Depressant or hypnotic drug
32. ___ caffeine   B. Narcotic analgesic drug
       C. Psychedelic or hallucinogenic drug
33. ___ cocaine   D. Stimulant
34. ___ codeine
35. ___ ethyl alcohol
36. ___ heroin
37. ___ LSD
38. ___ marijuana
39. ___ nicotine

Self-Quiz

___ 1. All nerves that lead away from the central nervous system are _____.
     a. efferent nerves
     b. sensory nerves
     c. afferent nerves
     d. spinal nerves
     e. peripheral nerves

___ 2. _____ nerves generally dominate internal events when environmental conditions permit normal body functioning.

     a. Ganglia
     b. Pacemaker
     c. Sympathetic
     d. Parasympathetic
     e. All of the above

___ 3. The center of consciousness and intelligence is the _____.
     a. medulla
     b. thalamus
     c. hypothalamus
     d. cerebellum
     e. cerebrum

___ 4. The _____ are the protective coverings of the brain.

a. ventricles
b. meninges
c. tectums
d. olfactory bulbs
e. pineal glands

___ 5. The _____ monitors internal organs; acts as gatekeeper to the limbic system; helps the reasoning centers of the brain to dampen rage and hatred; and governs hunger, thirst, and sex drives.

a. medulla
b. pons
c. thalamus
d. hypothalamus
e. reticular formation

___ 6. The left hemisphere of the brain is responsible for _____.

a. music
b. artistic ability and spatial relationships
c. language skills
d. abstract abilities

___ 7. The part of the brain that controls the basic responses necessary to maintain life processes (breathing, heartbeat) is _____.

a. the cerebral cortex
b. the cerebellum
c. the corpus callosum
d. the medulla

___ 8. To produce a split-brain individual, an operation would be required to sever the _____.

a. pons
b. fissure of Rolando
c. hypothalamus
d. reticular formation
e. corpus callosum

___ 9. The center for balance and coordination in the human brain is the _____.

a. cerebrum
b. pons
c. cerebellum
d. hypothalamus
e. thalamus

___ 10. The sleep center of the human brain is the _____.

a. medulla
b. pons
c. thalamus
d. hypothalamus
e. reticular formation

Chapter Objectives/Review Questions

This section lists general and detailed chapter objectives that can be used as review questions. You can make maximum use of these items by writing answers on a separate sheet of paper. To check for accuracy, compare your answers with information given in the chapter or glossary.

Page   *Objectives/Questions*

(575–579) 1. Contrast invertebrate and vertebrate nervous systems in terms of neural patterns.
(576–577) 2. Describe how the shift from radial to bilateral symmetry influenced the complexity of nervous systems.
(578–581) 3. Describe the basic structural and functional organization of the spinal cord. In your answer, distinguish spinal cord from vertebral column.
(578–579) 4. Define and contrast the central and peripheral nervous systems.
(580–581) 5. Explain how parasympathetic nerve activity balances sympathetic nerve activity. List activities of the sympathetic and parasympathetic nerves in regulating pupil diameter, rate of heartbeat, activities of the gut, and elimination of urine.
(581, 584) 6. Compare the structures of the spinal cord and brain with respect to white matter and gray matter.
(582–583) 7. List the parts of the brain found in the hindbrain, midbrain, and forebrain and tell the basic functions of each.
(584–585) 8. Describe how the cerebral hemispheres are related to the other parts of the forebrain.
(586)   9. State what the results of the "split-brain" experiments suggest about the functioning of the cerebral hemispheres.

(587) 10. Explain what an electroencephalogram is and what EEGs can tell us about the levels of conscious experience. Describe three typical EEG patterns and tell which level of consciousness each characterizes.

(583, 587) 11. Locate and identify the function of the reticular formation.

(587, 589) 12. Explain the relationship between transmitter substances and analgesics.

(588–589) 13. List the major classes of psychoactive drugs and provide an example of each class.

Integrating and Applying Key Concepts

Suppose that anger is eventually determined to be caused by excessive amounts of specific transmitter substances in the brains of angry people. Also suppose that an inexpensive antidote to anger that neutralizes these anger-producing transmitter substances is readily available. Can violent murderers now argue that they have been wrongfully punished because they were victimized by their brain's transmitter substances and could not have acted in any other way? Suppose an antidote is prescribed to curb violent tempers in an easily angered person. Suppose also that the person forgets to take the pill and subsequently murders a family member. Can the murderer still claim to be victimized by transmitter substances?

Answers

Interactive Exercises

35-I. INVERTEBRATE NERVOUS SYSTEMS
(pp. 574–577)
1. Nervous; 2. integrate; 3. nerve nets; 4. radial;
5. reflex; 6. sensory; 7. motor; 8. danger (trouble);
9. sensory cells (sense organs); 10. cephalization;
11. bilaterally; 12. nerve cords; 13. ganglia; 14. nerves;
15. reflex; 16. learning; 17. rudimentary "brain";
18. nerve cord; 19. rudimentary "brain"; 20. segmental ganglia; 21. brain; 22. optic lobe; 23. brain; 24. segmental ganglia.

35-II. VERTEBRATE NERVOUS SYSTEMS
(pp. 578–579)
PERIPHERAL NERVOUS SYSTEM (pp. 580–581)
1. nerves (muscles); 2. muscles (nerves); 3. brain;
4. central; 5. peripheral; 6. nerve cord; 7. notochord;
8. spinal cord; 9. forebrain; 10. neural tube; 11. vertebral column; 12. nerves; 13. peripheral (somatic); 14. autonomic; 15. sympathetic; 16. parasympathetic; 17. skull;
18. vertebral column; 19. cranial; 20. B; 21. D; 22. E;
23. C; 24. A; 25. forebrain; 26. midbrain; 27. hindbrain;
28. three; 29. seven; 30. nine; 31. spinal cord; 32. cervical; 33. thoracic; 34. lumbar; 35. sacral; 36. coccygeal;
37. autonomic; 38. sympathetic; 39. parasympathetic;
40. Increases; 41. Decreases.

35-III. CENTRAL NERVOUS SYSTEM: THE SPINAL CORD (p. 581)
CENTRAL NERVOUS SYSTEM: THE VERTEBRATE BRAIN (pp. 582–583)
1. spinal cord; 2. white matter; 3. gray matter; 4. integrative; 5. medulla oblongata; 6. cerebellum; 7. pons;

8. midbrain; 9. reticular formation; 10. forebrain;
11. hypothalamus; 12. sexual; 13. thalamus; 14. cerebrospinal fluid; 15. cushions; 16. cervical; 17. thoracic;
18. lumbar; 19. sacral; 20. coccygeal; 21. spinal cord;
22. ganglion; 23. vertebra; 24. intervertebral disk;
25. spinal nerve; 26. meninges; 27. sensory axon;
28. motor axon; 29. gray matter; 30. white matter.

35-IV. A CLOSER LOOK AT THE HUMAN CEREBRUM (pp. 584–586)
MEMORY (p. 587)
STATES OF CONSCIOUSNESS (p. 587)
THE BRAIN AND BEHAVIOR (pp. 588–589)
1. I; 2. H; 3. A; 4. K; 5. J; 6. B; 7. D; 8. C; 9. F; 10. G;
11. E; 12. cerebellum; 13. medulla oblongata; 14. pons;
15. temporal lobe; 16. optic chiasm; 17. hypothalamus;
18. thalamus; 19. corpus callosum; 20. memory;
21. short-term memory; 22. long-term memory;
23. alpha rhythm; 24. slow-wave sleep; 25. EEG arousal;
26. REM sleep; 27. reticular formation (sleep centers);
28. Endorphins (Enkephalins); 29. pain; 30. serotonin;
31. D; 32. D; 33. D; 34. B; 35. A; 36. B; 37. C; 38. C;
39. D.

Self-Quiz
1. a; 2. d; 3. e; 4. b; 5. d; 6. c; 7. d; 8. e; 9. c; 10. e.

36

SENSORY RECEPTION

Interactive Exercises

36-I. SENSORY SYSTEMS (pp. 592–595)

Selected Italicized Words

amplitude, frequency

Boldfaced, Page-Referenced Terms

(592) echolocation _____

_____

(594) sensory systems _____

_____

(594) sensation _____

_____

(594) chemoreceptors _____

_____

(594) mechanoreceptors _____

_____

(594) photoreceptors _____

_____

(594) thermoreceptors _____

_____

(594) nociceptors _____

_____

(595) sensory adaptation _____

_____

(595) somatic sensations _____

_____

(595) special senses _____

_____

Fill-in-the-Blanks

Finely branched peripheral endings of sensory neurons that detect specific kinds of stimuli are

(1) _____. A(n) (2) _____ is any form of energy change in the environment that the body actually

detects. (3) _____ detect impinging chemical energy; (4) _____ detect mechanical energy associ-

ated with changes in pressure, position, or acceleration; (5) _____ detect the energy of visible and

ultraviolet light; (6) _____ detect radiant energy associated with temperature changes. A(n)

(7) _____ is conscious awareness of change in internal or external conditions; this is not to be confused

with (8) _____, which is an understanding of what sensation means. A sensory system consists of

sensory receptors for specific stimuli, (9) _____ _____ that conduct information from those recep-

tors to the brain, and (10) _____ _____ where information is evaluated.

Signals from receptors in the skin and joints travel to the primary (11) _____ _____

_____, which is a strip little more than an inch wide running from the top of the (12) _____ to just

above the (13) _____ on the surface of each (14) _____ _____. The largest portion of the

somatic sensory cortex is A (see figure at right), which

receives signals coming from the (15) _____. The

second largest region is C, which receives signals com-

ing from the (16) _____. Every type of sensation

is caused by (17) _____ _____ arriving from

particular nerve pathways activating specific neurons

in the (18) _____. Besides sensing a stimulus, the

brain also interprets variations in (19) _____

_____. Interpretation is based on the

(20) _____ of action potentials propagated along

single axons and the (21) _____ of axons carrying

action potentials from a given tissue.

Matching

Select the best match for each item below.

22. ___ Vision is associated with _____.

23. ___ Pain is associated with _____.

24. ___ Odors are detected by _____.

25. ___ Hearing is detected by _____.

26. ___ CO_2 concentration in the blood is detected by _____.

27. ___ Environmental temperature is detected by _____.

28. ___ Internal body temperature is detected by _____.

29. ___ Touch is detected by _____.

30. ___ Rods and cones

31. ___ Hair cells in the ear's organ of Corti

32. ___ Pacinian corpuscles in the skin

33. ___ Olfactory receptors in nose

34. ___ Sensation of pain due to tissue damage

35. ___ The movement of fluid in the inner ear is associated with _____.

A. Chemoreceptors
B. Mechanoreceptors
C. Nociceptors
D. Photoreceptors
E. Thermoreceptors

36-II. SOMATIC SENSATIONS (pp. 596–597)
TASTE AND SMELL (p. 598)

Selected Emphasized Words

"referred pain"

Boldfaced, Page-Referenced Terms

(596) somatosensory cortex _____

_____

(596) pain _____

_____

(598) taste receptors _____

_____

(598) olfactory receptors _____

_____

(598) pheromones _____

_____

Fill-in-the-Blanks

The somatic sensations [awareness of (1) _____, pressure, heat, (2) _____, and pain] start with receptor endings that are embedded in (3) _____ and other tissues at the body's surfaces, in (4) _____ muscles, and in the walls of internal organs. All skin (5) _____ are easily deformed by pressure on the skin's surface; these make you aware of touch, vibrations, and pressure. (6) _____ nerve endings serve as "heat" receptors, and their firing of action potentials increases with increases in temperature.

(7) _____ is the perception of injury to some body region; the perception begins when (8) _____, which include free nerve endings, send signals via the thalamus to the parietal lobe of the brain, where they are interpreted. When the (9) _____ _____ pass a certain threshold, the signals generated are translated into sensations of pain. The brain sometimes gets confused and may associate perceived pain with a tissue some distance from the damaged area; this phenomenon is called (10) _____ _____. Usually the nerve pathways to both the injured and the mistaken areas pass through the same segment of spinal cord. (11) _____ in skeletal muscle, joints, tendons, ligaments, and (12) _____ are responsible for awareness of the body's position in space and of limb movements.

Label-Match

Identify each indicated part of the illustration below. Complete the exercise by entering the appropriate letter in the parentheses that follow the labels.

13. _____ _____
    _____ ( )
14. _____ _____ ( )
15. _____ _____ ( )
16. _____
17. _____
18. _____ _____ ( )

A. React continually to ongoing stimuli
B. Contribute to sensations of vibrations
C. Involved in sensing heat, light pressure, and pain
D. Stimulated at the beginning and end of sustained pressure

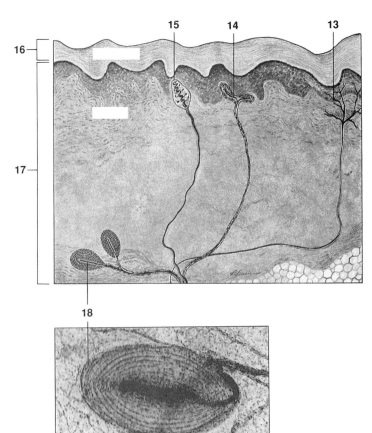

Fill-in-the-Blanks

(19) _____receptors detect molecules that become dissolved in fluid next to some body surface. Receptors are the modified dendrites of (20) _____ neurons. In the case of taste, these receptors, when located on animal tongues, are often part of sensory organs, (21) _____ _____, that are enclosed by circular papillae.

Animals smell substances by means of (22)_____receptors, such as the ones in your (23) _____; humans have about (24) _____ million of these in a nose. Sensory nerve pathways lead from the nasal cavity to the region of the brain where odors are identified and associated with their sources—the (25) _____ bulb and nerve tract. (26) _____receptors sampling odors from food in the mouth are important for our sense of taste. When we suffer from the common cold and have a runny nose, odor molecules from food have difficulty reaching the olfactory receptors and contributing their information signals. Our senses of taste and (27) _____ are both dulled.

36-III. BALANCE (p. 599)
HEARING (pp. 600–601)

Selected Italicized Words

dynamic equilibrium, static equilibrium, motion sickness, acoustical organs, frequency, pitch, outer ear, middle ear, inner ear, organ of Corti

Boldfaced, Page-Referenced Terms

(599) hair cells _____

_____

(599) cristae _____

_____

(599) otolithic organs _____

_____

(600) ears _____

_____

(600) acoustical receptors _____

_____

Fill-in-the-Blanks

The (1) _____ (perceived loudness) of sound depends on the height of the sound wave. The (2) _____ (perceived pitch) of sound depends on how fast the wave changes occur. The faster the vibrations, the (3) (choose one) ❏ higher ❏ lower the sound. Hair cells are (4) (choose one) ❏ nociceptors ❏ mechanoreceptors ❏ thermoreceptors that detect vibrations. The hammer, anvil, and stirrup are located in the (5) (choose one) ❏ inner ❏ middle ear. The (6) _____ is a coiled tube that resembles a snail shell and contains the (7) _____ _____ _____—the organ that changes vibrations into electrochemical impulses. Structures that detect rotational acceleration in humans are (8)_____

_____.

Labeling

Identify each indicated part of the accompanying illustrations.

9. _____ _____

10. _____

11. _____ _____

12. _____ _____

13. _____ _____

14. _____ _____

15. _____ _____

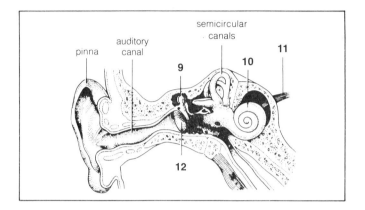

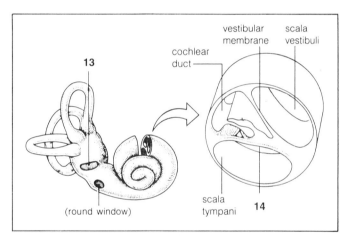

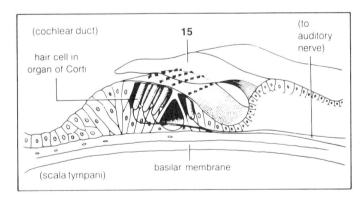

36-IV. VISION (p. 602)
INVERTEBRATE EYES (pp. 602–603)
VERTEBRATE EYES (pp. 604–607)
FROM SIGNALING TO VISUAL PERCEPTION (pp. 608–609)

Selected Italicized Words

nearsightedness, farsightedness, red-green color blindness, astigmatism, histoplasmosis, trachoma, glaucoma, retinal detachment, corneal transplant surgery, radial keratotomy, laser coagulation, bipolar, ganglion, amacrine, receptive field

Boldfaced, Page-Referenced Terms

(602) vision _____

_____

(602) eyes _____

_____

(602) eyespots _____

_____

(602) cornea _____

_____

(602) retina _____

_____

(602) lens _____

_____

(602) iris _____

_____

(602) pupil _____

_____

(602) compound eyes _____

_____

(604) accommodation _____

_____

(608) rod cells _____

_____

(608) cone cells _____

_____

Fill-in-the-Blanks

Light is a stream of (1) _____—discrete energy packets. (2) _____ is a process in which photons are absorbed by pigment molecules and photon energy is transformed into the electrochemical energy of a nerve signal. (3) _____ requires precise light focusing onto a layer of photoreceptive cells that are dense enough to sample details of the light stimulus, followed by image formation in the brain. (4) _____ are simple clusters of photosensitive cells, usually arranged in a cuplike depression in the epidermis. (5) _____ are well-developed photoreceptor organs that allow at least some degree of image formation. The (6) _____ is a transparent cover of the lens area, and the (7) _____ consists of tissue containing densely packed photoreceptors. Compound eyes contain several thousand photosensitive units known as (8) _____.

In the vertebrate eye, lens adjustments assure that the (9) _____ _____ for a specific group of light rays lands on the retina. (10) _____ refers to the lens adjustments that bring about precise focusing onto the retina. (11) _____ people focus light from nearby objects posterior to the retina. (12) _____ cells are concerned with daytime vision and, usually, color perception. A (13) _____ is a funnel-shaped pit on the retina that provides the greatest visual acuity.

Labeling

Identify each indicated part of the accompanying illustration.

14. _____ _____
15. _____
16. _____
17. _____
18. _____ _____
19. _____
20. _____
21. _____
22. _____ _____
23. _____ _____
24. _____

Self-Quiz

___ 1. According to the mosaic theory,
_____.

    a. the basement membrane's pigment molecules prevent the scattering of light
    b. light falling on the inner area of an "on-center" field activates firing of the cells
    c. hair cells in the semicircular canals cooperate to detect rotational acceleration
    d. each ommatidium detects information about only one small region of the visual field; many ommatidia contribute "bits" to the total image
    e. all of the above

___ 2. The principal place in the human ear where sound waves are amplified is
_____.

    a. the pinna
    b. the ear canal
    c. the middle ear
    d. the organ of Corti
    e. none of the above

___ 3. The place where vibrations are translated into patterns of nerve impulses is the
_____.

    a. pinna
    b. ear canal
    c. middle ear
    d. organ of Corti
    e. none of the above

For questions 4–8, choose from the following answers:

    a. fovea
    b. cornea
    c. iris
    d. retina
    e. sclera

___ 4. The white protective fibrous tissue of the eye is the _____.

___ 5. Rods and cones are located in the
_____.

___ 6. The highest concentration of cones is in the
_____.

___ 7. The adjustable ring of contractile and connective tissues that controls the amount of light entering the eye is the _____.

___ 8. The outer transparent protective covering of part of the eyeball is the _____.

___ 9. Accommodation involves the ability to
_____.

    a. change the sensitivity of the rods and cones by means of transmitters
    b. change the width of the lens by relaxing or contracting certain muscles
    c. change the curvature of the cornea
    d. adapt to large changes in light intensity
    e. all of the above

___ 10. Nearsightedness is caused by _____.

    a. eye structure that focuses an image in front of the retina
    b. uneven curvature of the lens
    c. eye structure that focuses an image posterior to the retina
    d. uneven curvature of the cornea
    e. none of the above

Chapter Objectives/Review Questions

This section lists general and detailed chapter objectives that can be used as review questions. You can make maximum use of these items by writing answers on a separate sheet of paper. To check for accuracy, compare your answers with information given in the chapter or glossary.

Page   *Objectives/Questions*

(594)   1. Define and distinguish among *chemoreceptors, mechanoreceptors, photoreceptors,* and *thermoreceptors*. Name at least one example of each type that appears in an animal.

(594)   2. Distinguish the types of stimuli detected by tactile and stretch receptors from those detected by hearing and equilibrium receptors.

(594, 596, 597, 598)   3. Explain how a taste bud works and distinguish the types of stimuli it detects from those detected by touch or stretch receptors.

(599)   4. Explain how the three semicircular canals of the human ear detect changes of position and acceleration in a variety of directions.

(600)   5. State how low- and high-pitch sounds affect the organ of Corti.

(600–601)   6. Follow a sound wave from pinna to organ of Corti; mention the name of each structure it passes and state where the sound wave is amplified and where the pattern of pressure waves is translated into electrochemical impulses.

(600–601)   7. State how low- and high-amplitude sounds affect the organ of Corti.

(602)   8. Explain what a visual system is and list four of the five aspects of a visual stimulus that are detected by different components of a visual system.

(602–604)   9. Contrast the structure of compound eyes with the structures of invertebrate eyespots and of the human eye.

(606–607)   10. Define *nearsightedness* and *farsightedness* and relate each to eyeball structure.

(608–609)   11. Describe how the human eye perceives color and black and white.

(609)   12. Explain the general principles that affect how light is detected by photoreceptors and changed into electrochemical messages.

Integrating and Applying Key Concepts

How might human behavior be changed if human eyes were compound eyes composed of ommatidia and if humans perceived only vibrations—as fish do—rather than sounds?

Answers

Interactive Exercises

36-I. SENSORY SYSTEMS (pp. 592–595)
1. receptors; 2. stimulus; 3. Chemoreceptors; 4. mechanoreceptors; 5. photoreceptors; 6. thermoreceptors; 7. sensation; 8. perception; 9. nerve pathways; 10. brain regions; 11. somatic sensory cortex; 12. head; 13. ear; 14. cerebral hemisphere; 15. mouth; 16. hand; 17. action potentials; 18. brain; 19. stimulus intensity; 20. frequency; 21. number; 22. D; 23. C; 24. A; 25. B; 26. A; 27. E; 28. E; 29. B; 30. D; 31. B; 32. B; 33. A; 34. C; 35. B.

36-II. SOMATIC SENSATIONS (pp. 596–597)
   TASTE AND SMELL (p. 598)
1. touch (cold); 2. cold (touch); 3. skin; 4. skeletal; 5. mechanoreceptors; 6. Free; 7. Pain; 8. nociceptors; 9. action potentials; 10. referred pain; 11. Mechanoreceptors; 12. skin; 13. free nerve endings (C); 14. Ruffini endings (A); 15. Meissner corpuscle (D); 16. epidermis; 17. dermis; 18. Pacinian corpuscle (B); 19. Chemo; 20. sensory; 21. taste buds; 22. chemo; 23. nose; 24. 10; 25. olfactory; 26. Chemo; 27. smell.

36-III. BALANCE (p. 599)
   HEARING (pp. 600–601)
1. amplitude; 2. frequency; 3. higher; 4. mechanoreceptors; 5. middle; 6. cochlea; 7. organ of Corti; 8. semicircular canals; 9. middle earbones (malleus, incus, stapes); 10. cochlea; 11. auditory nerve; 12. tympanic membrane (eardrum); 13. oval window; 14. basilar membrane; 15. tectorial membrane.

36-IV. VISION (p. 602)
   INVERTEBRATE EYES (pp. 602–603)
   VERTEBRATE EYES (pp. 604–607)
   FROM SIGNALING TO VISUAL PERCEPTION (pp. 608–609)
1. photons; 2. Photoreception; 3. Vision; 4. Eyespots; 5. Eyes; 6. cornea; 7. retina; 8. ommatidia; 9. focal point; 10. Accommodation; 11. Farsighted; 12. Cone; 13. fovea; 14. vitreous humor; 15. cornea; 16. iris; 17. lens; 18. aqueous humor; 19. suspensory ligament (fiberlike ligaments); 20. retina; 21. fovea; 22. optic nerve; 23. blind spot (optic disk); 24. sclera.

Self-Quiz
1. d; 2. c; 3. d; 4. e; 5. d; 6. a; 7. c; 8. b; 9. b; 10. a.

37

INTEGRATION AND CONTROL: ENDOCRINE SYSTEMS

Interactive Exercises

37-I. THE ENDOCRINE SYSTEM (pp. 612–615)

Selected Italicized Words

target cells, neurotransmitters, local signaling molecules, pheromones

Boldfaced, Page-Referenced Terms

(614) hormones _____

_____

(614) endocrine system _____

_____

Matching

Choose the one most appropriate answer for each.

1. ___ hormones
2. ___ neurotransmitters
3. ___ target cells
4. ___ local signaling molecules
5. ___ pheromones

A. Signaling molecules released from neurons and acting swiftly on abutting target cells
B. Released by many types of body cells; alter conditions within localized regions of tissues
C. Exocrine gland secretions; signaling molecules that act on cells of other animals of the same species and help integrate social behavior
D. Secretions from endocrine glands, endocrine cells, and some neurons that the bloodstream distributes to nonadjacent target cells
E. Cells that have receptors for any given type of signaling molecule

Complete the Table

6. Complete the table below by identifying the numbered components of the endocrine system shown in the illustration on the facing page as well as the hormones produced by each gland (see text Fig. 37.2).

Gland Name	Number	Hormones Produced
a. Hypothalamus		
b. Pituitary, anterior lobe		
c. Pituitary, posterior lobe		
d. Adrenal glands (cortex)		
e. Adrenal glands (medulla)		
f. Ovaries (two)		
g. Testes (two)		
h. Pineal		
i. Thyroid		
j. Parathyroids (four)		
k. Thymus		
l. Pancreatic islets		

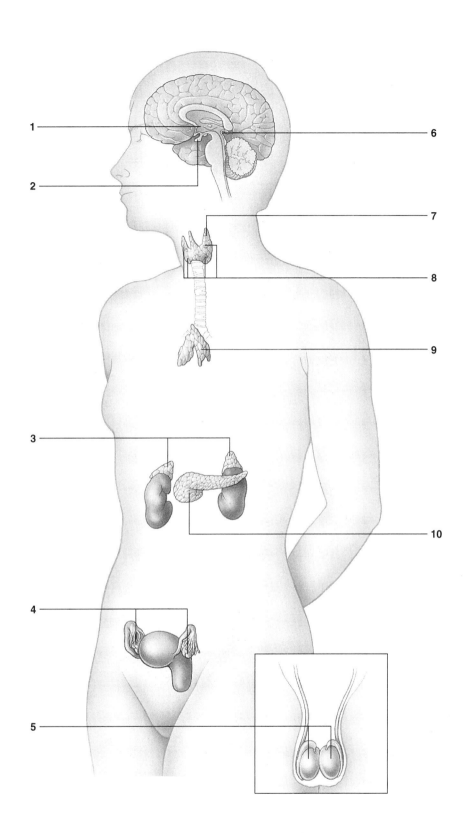

37-II. SIGNALING MECHANISMS (pp. 616–617)

Selected Italicized Words

testicular feminization syndrome, protein hormones

Boldfaced, Page-Referenced Terms

(616) steroid hormones _____

_____

(617) nonsteroid hormones _____

_____

(617) second messengers _____

_____

Choice

For questions 1–8, choose from the following:

a. steroid hormones   b. nonsteroid hormones

___ 1. Hormones synthesized from cholesterol, are lipid-soluble, and diffuse directly across the lipid bilayer of a target cell's plasma membrane.

___ 2. Include amines, peptides, proteins, and glycoproteins.

___ 3. A defective receptor for one of these results in a condition called testicular feminization.

___ 4. Hormones that activate second messengers.

___ 5. Hormones that bind to receptors at the plasma membrane with the complex moving into the cell by endocytosis to cause further action.

___ 6. Lipid-soluble hormones that move through the target cell's plasma membrane to the nucleus, where they bind to a receptor. The hormone-receptor complex moves into the nucleus and interacts with specific DNA regions to stimulate or inhibit transcription of mRNA.

___ 7. A hormone-receptor complex that activates transport proteins or triggers the opening of channel proteins across the membrane.

___ 8. Involves molecules such as cyclic AMP that activate many enzymes in cytoplasm that, in turn, cause alteration in some cell activity.

37-III. THE HYPOTHALAMUS AND THE PITUITARY GLAND (pp. 618–619)
EXAMPLES OF ABNORMAL PITUITARY OUTPUT (p. 620)

Selected Italicized Words

posterior pituitary lobe, anterior pituitary lobe, releasers, inhibitors, gigantism, pituitary dwarfism, acromegaly

Boldfaced, Page-Referenced Terms

(618) hypothalamus _____

_____

(618) pituitary gland _____

_____

Choice-Match

Label each hormone given below with an "A" if it is secreted by the anterior lobe of the pituitary or a "P" if it is released from the posterior pituitary. Complete the exercise by entering the letter of the corresponding action in the parentheses following each label.

1. ___ ( ) ACTH

2. ___ ( ) ADH

3. ___ ( ) FSH

4. ___ ( ) GH (STH)

5. ___ ( ) LH

6. ___ ( ) oxytocin

7. ___ ( ) PRL

8. ___ ( ) TSH

A. Acts on ovaries and testes to produce gametes
B. Acts on mammary glands to develop milk supplies
C. Acts on ovaries and testes to release gametes and to stimulate the production of testosterone in males and formation of the corpus luteum in females
D. Induces uterine contractions and milk movement into secretory ducts
E. Acts on the thyroid gland
F. Acts on the kidneys to conserve water
G. Acts on the adrenal cortex to increase adrenal steroid hormone production
H. Acts on most cells to induce protein synthesis and cell division; also plays a role in the metabolism of glucose and protein in adults

Dichotomous Choice

Circle one of two possible answers given between parentheses in each statement.

9. The (hypothalamus/pituitary gland) region of the brain monitors internal organs and activities related to their functioning, such as eating and sexual behavior; it also secretes some hormones.

10. The (posterior/anterior) lobe of the pituitary stores and secretes two of the hypothalamic hormones.

11. The (posterior/anterior) lobe of the pituitary produces and secretes its own hormones, which govern the release of hormones from other endocrine glands.

12. Most hypothalamic hormones acting in the posterior pituitary lobe are (releasers/inhibitors) and cause target cells there to secrete hormones of their own.

13. Some hypothalamic hormones slow down secretion from their targets; these are classed as (releasers/inhibitors).

14. (Pituitary dwarfism/Gigantism) results when not enough somatotropin is produced during childhood.

15. Production of excessive amounts of somatotropin during childhood results in (pituitary dwarfism/gigantism).

16. Excess somatotropin production during adulthood results in thicker bone, cartilage, and connective tissues of hands, feet, jaws, and epithelia; this condition is known as (gigantism/acromegaly).

17. (ACTH/TSH) is an anterior pituitary hormone acting on adrenal glands.

18. In addition to LH, the anterior pituitary hormone having a role in reproduction is (FSH/TSH).

Complete the Table

19. Complete the table below to summarize examples of abnormal pituitary output.

Condition	Hormone/Abnormality	Characteristics
a.	Excessive somatotropin produced during childhood	Affected adults are proportionally similar to a normal person but larger
b.	Insufficient somatotropin produced during childhood	Affected adults are proportionally similar to a normal person but much smaller
c.	Excessive somatotropin output during adulthood when long bones can no longer lengthen	Abnormal thickening of bone, cartilage, and other connective tissues in the hands, feet, and jaws

37-IV. EXAMPLES OF ABNORMAL PITUITARY OUTPUT (p. 620)

Selected Italicized Words

gigantism, pituitary dwarfism, acromegaly

SELECTED EXAMPLES OF HORMONAL CONTROL (p. 621)

Complete the Table

1. Complete the table on the facing page by matching the gland/organ and the hormone(s) produced by it to the descriptions of hormone action.

Gland/Organ

A. adrenal cortex
B. adrenal medulla
C. thyroid
D. parathyroids
E. testes
F. ovaries
G. pancreas (alpha cells)
H. pancreas (beta cells)
I. pancreas (delta cells)
J. thymus
K. pineal

Hormones

a. thyroxine and triiodothyronine
b. glucagon
c. PTH
d. androgens (includes testosterone)
e. somatostatin
f. thymosins
g. glucocorticoids
h. estrogens (includes progesterone)
i. epinephrine and norepinephrine
j. melatonin
k. insulin
l. progesterone
m. mineralocorticoids (including aldosterone)
n. calcitonin
o. norepinephrine

Gland/Organ	Hormone	Hormone Action
a.		Stimulates bone cells to release calcium phosphate and the kidneys to conserve it; also helps activate vitamin D
b.		Influences carbohydrate metabolism by control of food digestion; can block secretion of insulin and glucagon
c.		Required in egg maturation and release; prepares and maintains the uterine lining for pregnancy; influences growth and development
d.		Helps maintain the blood level of glucose; cortisol is an example; also suppresses inflammatory responses
e.		Lowers blood sugar level by stimulating glucose uptake by liver, muscle, and adipose cells; promotes protein and fat synthesis; inhibits protein conversion to glucose
f.		Required in sperm formation, genital development, and maintenance of sexual traits; influences growth and development
g.		Regulates metabolism; roles in growth and development
h.		Influences daily biorhythms and influences gonad development and reproductive cycles
i.		Helps adjust blood circulation and carbohydrate metabolism when the body is excited or stressed, the "fight-flight" response
j.		Roles in immunity
k.		Raises blood sugar level by causing glycogen and amino acid conversion to glucose in the liver
l.		Prepares, maintains uterine lining for pregnancy; stimulates breast development
m.		Promote sodium reabsorption; control salt, water balance
n.		Promotes constriction or dilation of blood vessels
o.		Lowers calcium levels in blood

37-V. PANCREATIC ISLETS (pp. 622–623)

Selected Italicized Words

exocrine cells, *endocrine cells*, *alpha cells*, *glucagon*, *glucose level*, *beta cells*, *insulin*, *delta cells*, *diabetes mellitus*

Boldfaced, Page-Referenced Terms

(622) pancreatic islet _____

Complete the Table

1. Complete the table below to summarize function of the pancreatic islets.

Pancreatic Islet Cells	Hormone Secreted	Hormone Action
a. Alpha cells		
b. Beta cells		
c. Delta cells		

Dichotomous Choice

Circle one of two possible answers given between parentheses in each statement.

2. Insulin deficiency can lead to diabetes mellitus, a disorder in which the blood glucose level (rises/decreases) and glucose accumulates in the urine.
3. In a person with diabetes mellitus, urination becomes (reduced/excessive), so the body's water–solute balance becomes disrupted.
4. Lacking a steady glucose supply, body cells of a person with diabetes mellitus begin breaking down fats and proteins for (energy/water).
5. (Insulin/Glucagon) deficiency can lead to diabetes mellitus.
6. After a meal, blood glucose rises; pancreatic beta cells secrete (glucagon/insulin); targets use glucose or store it as glycogen.
7. Blood glucose levels decrease between meals. (Glucagon/Insulin) is secreted by stimulated pancreas alpha cells; targets convert glycogen back to glucose, which then enters the blood.
8. In (type 1 diabetes/type 2 diabetes) the body mounts an immune response against its own insulin-secreting beta cells and destroys them.
9. Juvenile-onset diabetes is also known as (type 1 diabetes/type 2 diabetes).
10. In (type 1 diabetes/type 2 diabetes), insulin levels are close to or above normal, but target cells fail to respond to insulin.
11. (Type 1 diabetes/Type 2 diabetes) usually is manifested during middle age and is less dramatically dangerous than the other type.

37-VI. ADRENAL GLANDS (p. 623)

Selected Italicized Words

hypoglycemia, stress response, fight-flight response

Boldfaced, Page-Referenced Terms

(623) adrenal medulla _____

_____

Matching

Choose the most appropriate answer to match with each term.

1. ___ adrenal medulla
2. ___ ACTH
3. ___ nervous system
4. ___ glucocorticoids
5. ___ CRH
6. ___ cortisol
7. ___ hypoglycemia
8. ___ cortisol-like drugs
9. ___ adrenal cortex
10. ___ homeostatic feedback loops

A. Occurs when the glucose blood level falls below a set point, and a negative feedback mechanism kicks in
B. Outer portion of each adrenal gland; some of its cells secrete glucocorticoids and other hormones
C. Govern the secretion of the hypothalamus and the pituitary
D. Initiates a stress response during severe stress, painful injury, or prolonged illness
E. Inner portion of the adrenal gland; neurons located here release epinephrine and norepinephrine
F. Stimulates the adrenal cortex to secrete cortisol; this helps raise the level of glucose by preventing muscle cells from taking up more blood glucose
G. Used to counter asthma and other chronic inflammatory disorders
H. Help maintain blood glucose level and help suppress inflammatory responses
I. Secreted by the hypothalamus in response to falling blood glucose level
J. Blocks the uptake and use of glucose by muscle cells; also stimulates liver cells to form glucose from amino acids

37-VII. THE THYROID AND PARATHYROID GLANDS (p. 624)

Selected Italicized Words

goiter, hypothyroidism, hyperthyroidism, rickets

Boldfaced, Page-Referenced Terms

(624) thyroid gland _____

_____

(624) parathyroid glands _____

_____

Fill-in-the-Blanks

Thyroxine and triiodothyronine are the main hormones secreted by the human (1) _____ gland. They are critical for normal development of many tissues, and they control overall (2) _____ rates in humans and other warm-blooded animals. The synthesis of thyroid hormones requires (3) _____, which is obtained from the diet. In the absence of iodine, blood levels of these hormones decrease. The anterior pituitary responds by secreting (4) _____. When thyroid hormones cannot be synthesized, the feedback signal continues—and so does TSH secretion. TSH secretion continues, overstimulates the thyroid gland, and causes (5) _____, an enlargement of the thyroid gland. (6) _____ results from insufficient concentrations of thyroid hormones. Affected (7) _____ are often overweight, sluggish, dry-skinned, intolerant of cold, and sometimes confused and depressed. (8) _____ results from excess concentrations of thyroid hormones. Affected (9) _____ show an increased heart rate, elevated blood pressure, weight loss despite normal caloric intake, heat intolerance, and profuse sweating. They are also nervous, agitated, and have trouble sleeping.

Four (10) _____ glands are positioned next to the back of the human thyroid; they secrete (11) _____ hormone when the concentration of (12) _____ ions in their surroundings decreases. This hormone stimulates bone cells to release calcium and phosphate and the (13) _____ to conserve it; it also helps to activate vitamin D. The activated form is a hormone that enhances (14) _____ absorption from ingested food. In vitamin D deficiency, too little calcium and phosphorus are absorbed, so bones develop improperly. This ailment is called (15) _____.

37-VIII. PINEAL GLAND (p. 625)
SOME FINAL EXAMPLES OF INTEGRATION AND CONTROL (pp. 626–627)

Selected Italicized Words

Platysamia cecropia

Boldfaced, Page-Referenced Terms

(625) pineal gland _____

_____

(626) thymus gland _____

_____

(626) gonads _____

_____

(626) local signaling molecules _____

_____

(626) molting _____

_____

Choice

For questions 1–15, choose from the following:

a. pineal gland   b. thymus gland   c. gonads   d. local signaling molecules   e. invertebrate hormone

___ 1. The primary reproductive organs

___ 2. Controls molting, the periodic discarding and replacement of a hardened cuticle

___ 3. A photosynthetic organ in the brain

___ 4. Actions are confined to the immediate vicinity of change

___ 5. Located behind the breastbone, between the two lungs

___ 6. Production of melatonin

___ 7. Ecdysone, a steroid hormone active in molting

___ 8. Secretes hormones known as thymosins

___ 9. Produce gametes and secrete sex hormones

___ 10. Examples are prostaglandins and growth factors

___ 11. Hamsters produce high melatonin levels in winter (long nights), suppressing sexual activity

___ 12. Epidermal growth factor (EGF) and nerve growth factor (NGF)

___ 13. Produces hormones that may induce the formation and early development of stem cells that give rise to T lymphocytes, central players in the human body's immune system

___ 14. Produce sex hormones that also influence secondary sexual traits

___ 15. Responsible for decreased melatonin secretion, which may trigger human puberty

---

Self-Quiz

Multiple-Choice

___ 1. The _____ governs the release of hormones from other endocrine glands; it is controlled by the _____.
   a. pituitary; hypothalamus
   b. pancreas; hypothalamus
   c. thyroid; parathyroid glands
   d. hypothalamus; pituitary
   e. pituitary; thalamus

___ 2. Neurons of the _____ produce ADH and oxytocin that are stored within axon endings of the _____.
   a. anterior pituitary, posterior pituitary
   b. adrenal cortex, adrenal medulla
   c. posterior pituitary, hypothalamus
   d. posterior pituitary, thyroid
   e. hypothalamus, posterior pituitary

___ 3. If you were lost in the desert and had no fresh water to drink, the level of _____ in your blood would increase as a means to conserve water.
   a. insulin
   b. corticotropin
   c. oxytocin
   d. antidiuretic hormone
   e. salt

For questions 4–6, choose from the following answers:

    a. Estrogen
    b. PTH
    c. FSH
    d. Somatotropin
    e. Prolactin

___ 4. _____ stimulates bone cells to release calcium and phosphate and the kidneys to conserve it.

___ 5. _____ stimulates and sustains milk production in mammary glands.

___ 6. Protein synthesis and cell division are activities stimulated by _____.

For questions 7–9, choose from the following answers:

    a. adrenal medulla
    b. adrenal cortex
    c. thyroid
    d. anterior pituitary
    e. posterior pituitary

___ 7. The _____ produces glucocorticoids that help maintain the blood level of glucose and suppress inflammatory responses.

___ 8. The gland that is most closely associated with emergency situations is the _____.

___ 9. The _____ gland regulates the basic metabolic rate.

___ 10. If all sources of calcium were eliminated from your diet, your body would secrete more _____ in an effort to release calcium stored in your body and send it to the tissues that require it.

    a. parathyroid hormone
    b. aldosterone
    c. calcitonin
    d. mineralocorticoids
    e. none of the above

Matching

11. ___ ACTH
12. ___ ADH
13. ___ thymosins
14. ___ oxytocin
15. ___ cortisol
16. ___ epinephrine and norepinephrine
17. ___ estrogen
18. ___ glucagon
19. ___ insulin
20. ___ melatonin
21. ___ parathyroid hormone
22. ___ STH (GH)
23. ___ calcitonin
24. ___ testosterone
25. ___ thyroxine
26. ___ progesterone
27. ___ TSH

A. Raises the glucose level in the blood
B. Influences daily biorhythms, gonad development, and reproductive cycles
C. Affects development of male sexual traits
D. Increases heart rate and controls blood volume; the "emergency hormones"
E. Produced by gonads; essential for egg maturation and maintenance of secondary sex characteristics in the female
F. The water conservation hormone; released from posterior pituitary
G. Lowers blood sugar by encouraging cells to take in glucose; responsible for synthesis of proteins and fats
H. Stimulates adrenal cortex to secrete cortisol
I. Elevates calcium levels in blood by stimulating calcium reabsorption from bone and kidneys and calcium absorption from gut
J. Influences overall metabolic rate, growth, and development
K. Roles in immunity
L. Triggers uterine contractions during labor and causes milk release during nursing
M. Prepares and maintains uterine lining for pregnancy; stimulates breast development
N. Inhibits uptake of more blood glucose by muscle cells
O. Lowers calcium levels in blood; bone is the target
P. Secreted by anterior pituitary; stimulates release of thyroid hormones
Q. Secreted by anterior pituitary; enhances growth in young animals, especially of cartilage and bone

Chapter Objectives/Review Questions

This section lists general and detailed chapter objectives that can be used as review questions. You can make maximum use of these items by writing answers on a separate sheet of paper. To check for accuracy, compare your answers with information given in the chapter or glossary.

Page   *Objectives/Questions*

(614)   1. Hormones, transmitter substances, local signaling molecules, and pheromones are all known as _____ molecules that carry out integration.

(614)   2. _____ cells are any cells that have receptors for a specific signaling molecule and that may alter their behavior in response to it.

(614)   3. Collectively, sources of hormones came to be viewed as the _____ system.

(615)   4. Be able to locate and name the components of the human endocrine systems on a diagram such as text Fig. 37.2.

(616–617) 5. Contrast the proposed mechanisms of hormonal action on target cell activities by (a) steroid hormones and (b) hormones that are proteins or are derived from proteins.

(618)   6. State how, even though the anterior and posterior lobes of the pituitary are compounded as one gland, the tissues of each part differ in character.

(618)   7. The _____ and the pituitary gland interact as a major neural-endocrine control center.

(618)   8. Identify the hormones released from the posterior lobe of the pituitary and state their target tissues.

(618–619) 9. Identify the hormones produced by the anterior lobe of the pituitary and tell which target tissues or organs each acts on.

(619)   10. Most hypothalamic hormones acting in the anterior lobe are _____; they cause target cells to secrete hormones of their own. Some are _____; they slow down secretion from their targets.

(620)   11. Pituitary dwarfism, gigantism, and acromegaly are all associated with abnormal secretion of _____ by the pituitary gland.

(621)   12. Be familiar with the major human hormone sources and their secretions, main targets, and primary actions as shown on text Table 37.3.

(621–622) 13. Be able to name the hormones secreted by alpha, beta, and delta pancreatic cells; list the effect of each.

(623)   14. Describe the symptoms of diabetes mellitus and distinguish between type 1 and type 2 diabetes.

(623)   15. The adrenal _____ secretes glucocorticoids.

(623)   16. The adrenal _____ contains neurons that secrete epinephrine and norepinephrine.

(623)   17. Define *hypoglycemia*; relate the role of the hypothalamus and the anterior pituitary in this condition.

(623)   18. The _____ system initiates the stress response.

(623)   19. List the features of the *fight-flight response*.

(624)   20. Describe the characteristics of hypothyroidism and hyperthyroidism.

(624)   21. Name the glands that secrete PTH and give the function of this hormone.

(624)   22. Describe an ailment called rickets and its cause.

(625)   23. The pineal gland secretes the hormone _____; relate two examples of the action of this hormone.

(626)   24. Thymosins are secreted by the _____ gland; they appear to be related to the proper functioning of the _____ system.

(626)   25. Give two examples that illustrate the effects of local signaling molecules.

(626–627) 26. The invertebrate hormone _____ is related to the phenomenon known as _____ that occurs among crustaceans and insects.

Integrating and Applying Key Concepts

Suppose you suddenly quadruple your already high daily consumption of calcium. State which body organs would be affected and tell how they would be affected. Name two hormones whose levels would most probably be affected and tell whether your body's production of them would increase or decrease. Suppose you continue this high rate of calcium consumption for ten years. Can you predict the organs that would be subject to the most stress as a result?

Answers

Interactive Exercises

37-I. THE ENDOCRINE SYSTEM (pp. 614–615)
1. D; 2. A; 3. E; 4. B; 5. C; 6. a. 1; Six releasing and inhibiting hormones; synthesizes ADH, oxytocin; b. 2; ACTH, TSH, FSH, LH, GSH; c. 2; Stores and secretes two hypothalamic hormones, ADH and oxytocin; d. 3; Sex hormones of opposite sex, cortisol, aldosterone; e. 3; Epinephrine, norepinephrine; f. 4; Estrogens, progesterone; g. 5; Testosterone; h. 6; Melatonin; i. 7; Thyroxine and triiodothyronine; j. 8; Parathyroid hormone (PTH); k. 9; Thymosins; l. 10; Insulin, glucagon, somatostatin.

37-II. SIGNALING MECHANISMS (pp. 616–617)
1. a; 2. b; 3. a; 4. b; 5. b; 6. a; 7. b; 8. b.

37-III. THE HYPOTHALAMUS AND THE PITUITARY GLAND (pp. 618–619) EXAMPLES OF ABNORMAL PITUITARY OUTPUT (p. 620)
1. A(G); 2. P(F); 3. A(A); 4. A(H); 5. A(C); 6. P(D); 7. A(B); 8. A(E); 9. hypothalamus; 10. posterior; 11. anterior; 12. releasers; 13. inhibitors; 14. Pituitary dwarfism; 15. gigantism; 16. acromegaly; 17. ACTH; 18. FSH; 19. a. Gigantism; b. Pituitary dwarfism; c. Acromegaly.

37-IV. SELECTED EXAMPLES OF HORMONAL CONTROL (p. 621)
1. a. D(c); b. I(e); c. F(h); d. A(g); e. H(k); f. E(d); g. C(a); h. K(j); i. B(i); j. J(f); k. G(b); l. F (l); m. A (m); n. B (o); o. C (n).

37-V. PANCREATIC ISLETS (pp. 622–623)
1. a. Glucagon; Causes glycogen (a storage polysaccharide) and amino acids to be converted to glucose in the liver (glucagon raises the glucose level); b. Insulin; Stimulates glucose uptake by liver, muscle, and adipose cells; promotes synthesis of proteins and fats, and inhibits protein conversion to glucose (lowers the glucose level); c. Somatostatin; Helps control digestion; can block secretion of insulin and glucagon; 2. rises; 3. excessive; 4. energy; 5. Insulin; 6. insulin; 7. Glucagon; 8. type 1 diabetes; 9. type 1 diabetes; 10. type 2 diabetes; 11. Type 2 diabetes.

37-VI. ADRENAL GLANDS (p. 623)
1. E; 2. F; 3. D; 4. H; 5. I; 6. J; 7. A; 8. G; 9. B; 10. C.

37-VII. THE THYROID AND PARATHYROID GLANDS (p. 624)
1. thyroid; 2. metabolic; 3. iodine; 4. TSH; 5. goiter; 6. Hypothyroidism; 7. adults; 8. Hyperthyroidism; 9. adults; 10. parathyroid; 11. parathyroid (PTH); 12. calcium; 13. kidneys; 14. calcium; 15. rickets.

37-VIII. PINEAL GLAND (p. 625) SOME FINAL EXAMPLES OF INTEGRATION AND CONTROL (pp. 626–627)
1. c; 2. e; 3. a; 4. d; 5. b; 6. a; 7. e; 8. b; 9. c; 10. d; 11. a; 12. d; 13. b; 14. c; 15. a.

Self-Quiz
1. a; 2. e; 3. d; 4. b; 5. e; 6. d; 7. b; 8. a; 9. c; 10. a; 11. H; 12. F; 13. K; 14. L; 15. N; 16. D; 17. E; 18. A; 19. G; 20. B; 21. I; 22. Q; 23. O; 24. C; 25. J; 26. M; 27. P.

38

PROTECTION, SUPPORT, AND MOVEMENT

Interactive Exercises

38-I. INTEGUMENTARY SYSTEM (pp. 630–633)
A CLOSER LOOK AT THE STRUCTURE OF SKIN (pp. 634–635)

Selected Italicized Words

blister, cold sweats, acne, hirsutism, cold sores, epidermal skin cancers, basal cell carcinomas, "stratum corneum," "stretch marks," "split ends"

Boldfaced, Page-Referenced Terms

(631) integumentary system _____

_____

(631) muscle system _____

_____

(631) skeletal system _____

_____

(632) integument _____

_____

(632) cuticle _____

_____

(632) epidermis _____

_____

(632) dermis _____

_____

(634) keratinocytes _____

_____

(634) melanocytes _____

_____

(634) sweat glands _____

_____

(635) oil glands _____

_____

(635) hairs _____

_____

Fill-in-the-Blanks

Human skin is an organ system that consists of two layers: the outermost (1) _____, which contains mostly dead cells, and the (2)_____, which contains hair follicles, nerves, tiny muscles associated with the hairs, and various types of glands. The (3)_____ layer, with its loose connective tissue and store of fat in (4) _____ tissue, lies beneath the skin.

Labeling

Label the numbered parts of the illustration.

5. _____
6. _____ _____ _____
7. _____ _____
8. _____ _____
9. _____ _____
10. _____ _____
11. _____ _____
12. _____
13. _____
14. _____

Match the following proteins (or protein derivatives) with the particular ability that each substance lends to the skin. For questions 15–21, choose from these letters:

a. collagen   b. elastin   c. keratin   d. melanin   e. hemoglobin

___ 15. Fibers that run through the dermis and lend a flexible but substantive structure to it

___ 16. Protects against loss of moisture

___ 17. Protects against ultraviolet radiation and sunburn

___ 18. Helps skin to be stretchable yet return to its previous shape

___ 19. Beaks, hooves, hair, fingernails, and claws contain a lot of this

___ 20. Helps ward off bacterial attack by making the skin surface rather impermeable

___ 21. Located in red blood cells; binds with O_2

38-II. INVERTEBRATE AND VERTEBRATE SKELETONS (pp. 636–637)
CHARACTERISTICS OF BONE (pp. 638–639)
HUMAN SKELETAL SYSTEM (pp. 640–641)

Selected Italicized Words

hydraulic, compact bone tissue, "Haversian canals," spongy bone tissue, cartilage, osteoporosis, herniated disks, strain, sprain, osteoarthritis, rheumatoid arthritis

Boldfaced, Page-Referenced Terms

(636) hydrostatic skeleton _____

_____

(636) exoskeleton _____

_____

(636) endoskeleton _____

_____

(638) bones _____

_____

(638) red marrow _____

_____

(638) yellow marrow _____

_____

(639) osteocytes _____

_____

(640) intervertebral disks _____

_____

(640) skeletal joint _____

_____

(640) ligaments _____

_____

Fill-in-the-Blanks

All motor systems are based on muscle cells that are able to (1) _____ and (2) _____ and on the presence of a medium against which the (3) _____ force can be applied. Longitudinal and radial muscle layers work as a(n) (4) _____ muscle system, in which the action of one motor element opposes the action of another. A membrane filled with fluid resists compression and can act as a(n) (5) _____ skeleton. Arthropods have antagonistic muscles attached to an (6) (choose one) ❒ endoskeleton ❒ exoskeleton. (7) _____ are hard compartments that enclose and protect the brain, spinal cord, heart, lungs, and other vital organs of vertebrates. Bones support and anchor (8) _____ and soft organs, such as eyes. (9) _____ systems are found in the long bones of mammals and contain living bone cells that receive their nutrients from the blood. (10) _____ _____ is a major site of blood cell formation. Bone tissue serves as a "bank" for (11) _____, (12) _____, and other mineral ions; depending on metabolic needs, the body deposits ions into and withdraws ions from this "bank."

Bones develop from (13) _____ secreting material inside the shaft and on the surface of the cartilage model. Bone can also give ions back to interstitial fluid as osteoclasts dissolve out component minerals and remodel bone in response to (14) _____ signals, lack of exercise, and calcium deficiencies in the diet. Extreme decreases in bone density result in (15) _____, particularly among older women.

The (16) _____ skeleton includes the skull, vertebral column, ribs, and breastbone; the (17) _____ skeleton includes the pectoral and pelvic girdles and the forelimbs and hindlimbs, when they exist, in vertebrates.

(18) _____ joints are freely movable and are lubricated by a fluid secreted into the capsule of dense connective tissue that surrounds the bones of the joint. Bones are often tipped with (19) _____; as a person ages, the cartilage at (20) _____ joints may simply wear away, a condition called (21) _____. By contrast, in (22) _____ _____, the synovial membrane becomes inflamed, cartilage degenerates, and bone becomes deposited in the joint.

Labeling

Identify each indicated part of the accompanying illustrations below and on the next page.

23. _____ _____

24. _____ _____

25. _____ _____

26. _____ _____

27. _____ _____ _____

28. _____ _____

29. _____ _____

30. _____ _____

31. _____

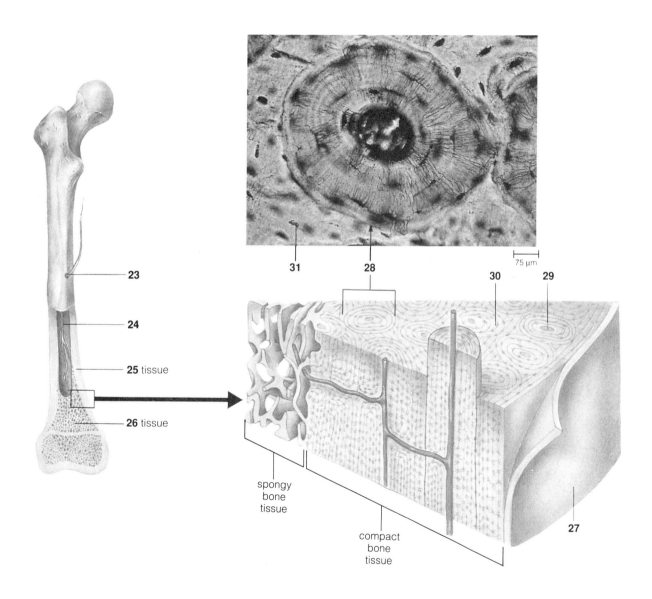

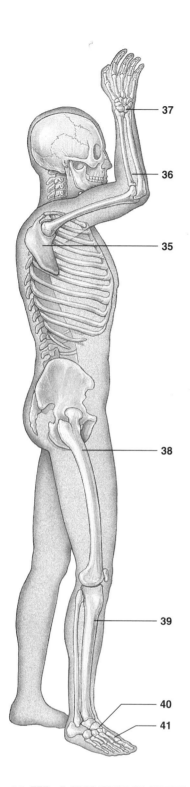

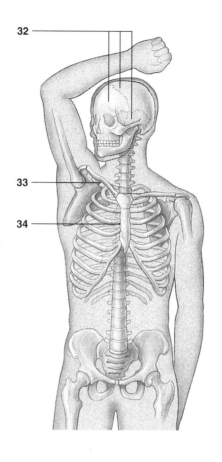

32. _____

33. _____

34. _____

35. _____

36. _____

37. _____ _____

38. _____

39. _____

40. _____ _____

41. _____

38-III. MUSCLE SYSTEMS (pp. 642–643)
MUSCLE FUNCTION (pp. 644–645)
CONTROL OF CONTRACTION (pp. 646–647)
MUSCLE TENSION, STRENGTH, AND FATIGUE (pp. 648–649)

Selected Italicized Words

myofibril, "excitable," biceps brachii, deltoid, gluteus maximus, biceps femoris, gastrocnemius, pectoralis major, rectus abdominis, quadriceps femoris

Boldfaced, Page-Referenced Terms

(642) skeletal muscle _____

(642) tendons _____

(644) sarcomeres _____

(644) actin _____

(644) myosin _____

(645) sliding-filament model _____

(645) cross-bridge formation _____

(646) action potential _____

(646) sarcoplasmic reticulum _____

(647) creatine phosphate _____

(648) muscle tension _____

(648) motor unit _____

(648) muscle twitch _____

(648) tetanus _____

(648) muscle fatigue _____

(648) anabolic steroids _____

Sequence

Arrange in order of decreasing size.

1. ___
2. ___
3. ___
4. ___
5. ___
6. ___

A. Muscle fiber (muscle cell)
B. Myosin filament
C. Muscle bundle
D. Muscle
E. Myofibril
F. Actin filament

Labeling

Identify each indicated part of the accompanying illustration.

7. _____

8. _____ _____

9. _____ _____

10. _____ _____

11. _____ _____

12. _____ _____

13. _____ _____

14. (two of the _____

_____)

15. _____

16. _____ _____

17. _____ _____

18. _____ _____

19. _____ _____

20. _____ _____

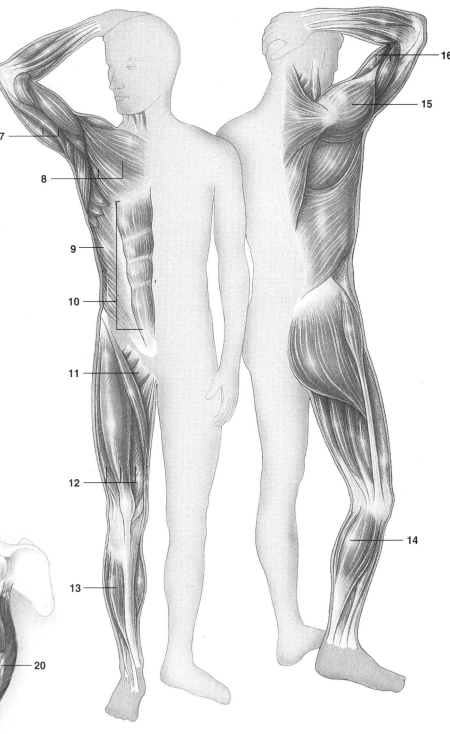

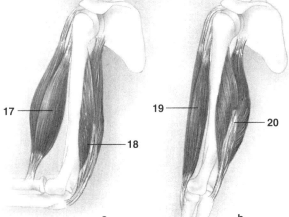

Opposing Muscle Groups

Fill-in-the-Blanks

There are three types of muscle tissue: (21) _____, which is striated, involuntary, and located in the heart; smooth, which is involuntary, not striated, and mostly located in the wall of internal organs in vertebrates; and (22) _____, which is striated, largely voluntary, and generally attached to (23) _____ or cartilage by means of (24) _____. (25) _____ muscle interacts with the skeleton to create positional changes of body parts and to move the animal through its environment.

Together, the skeleton and its attached muscles are like a system of levers in which rigid rods, (26) _____, move about at fixed points, called (27) _____. Most attachments are close to joints, so a muscle has to shorten only a small distance to produce a large movement of some body part. When the (28) _____ contracts, the elbow joint bends (flexes). As it relaxes and as its partner, the (29) _____, contracts, the forelimb extends and straightens.

Muscle cells have three properties in common: contractility, elasticity, and (30) _____. Like a(n) (31) _____, a muscle cell is "excited" when a wave of electrical disturbance, a(n) (32) _____ _____, travels along its cell membrane. Neurons that send signals to muscles are (33) _____ neurons.

Each muscle cell contains (34) _____, threadlike structures packed together in parallel array. Every striated (34) is functionally divided into (35) _____, which appear to be striped and are arranged one after another along its length. Each myofibril contains (36) _____ filaments, which have cross-bridges, and (37) _____ filaments, which are thin and lack cross-bridges.

According to the (38) _____-_____ model, (39) _____ filaments physically slide along actin filaments and pull them toward the center of a(n) (40) _____ during a contraction. The energy that drives the forming and breaking of the cross-bridges comes immediately from (41) _____, which obtained its phosphate group from (42) _____ _____ that is stored in muscle cells. (43) _____ of an entire muscle is brought about by the combined decreases in length of the individual sarcomeres that make up the myofibrils of the muscle cells. The parallel orientation of a muscle's component parts directs the force of contraction toward a(n) (44) _____ that must be pulled in some direction.

When excited by incoming signals from motor neurons, muscle cell membranes cause (45) _____ ions to be released from the (46) _____ _____. These ions remove all obstacles that might interfere with myosin heads binding to the sites along (47) _____ filaments. When muscle cells relax, calcium ions are sent back to the sarcoplasmic reticulum by (48) _____ _____. By controlling the (49) _____ _____ that reach the (50) _____ _____ in the first place, the nervous system controls muscle contraction by controlling calcium ion levels in muscle tissue.

Complete the Table

For each item in the table below, state its specific role in muscle contraction.

Component	Role in Contraction
Aerobic respiration	51.
ATP	52.
Calcium ions	53.
Creatine phosphate	54.
Glycogen	55.
Lactate fermentation	56.
Motor neuron	57.
Myosin heads	58.
Sarcomere	59.
Sarcoplasmic reticulum	60.

Fill-in-the-Blanks

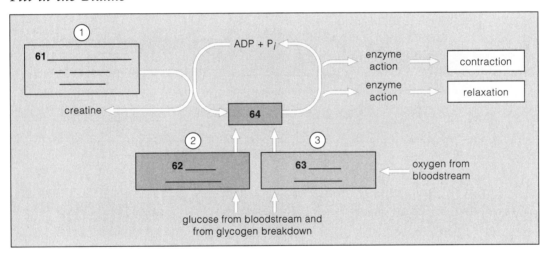

More Fill-in-the-Blanks

The larger the (65) _____ of a muscle, the greater its strength. A (66) _____ neuron and the muscle cells under its control are called a (67) _____ _____. A(n) (68) _____ _____ is a response in which a muscle contracts briefly when reacting to a single, brief stimulus and then relaxes. The strength of the muscular contraction depends on how far the (69) _____ response has proceeded by the time another signal arrives. (70) _____ is the state of contraction in which a motor unit that is being stimulated repeatedly is maintained. In a(n) (71) _____ contraction, the nervous system activates only a small number of motor units; in a stronger contraction, a larger number are activated at a high (72) _____ of stimulation.

Labeling

Label the numbered parts of the accompanying illustrations.

73. _____ bundle

74. _____ _____

75. _____

76. _____

77. _____ _____

78. _____ _____

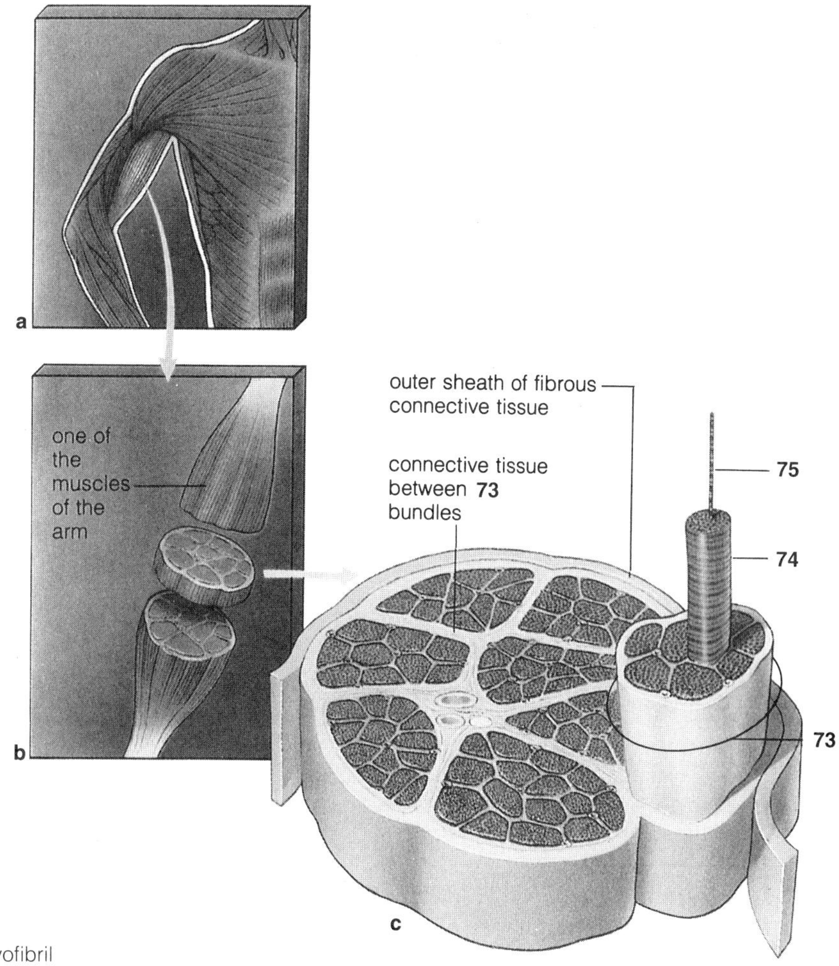

a

b

one of
the
muscles
of the
arm

outer sheath of fibrous
connective tissue

connective tissue
between **73**
bundles

75

74

73

c

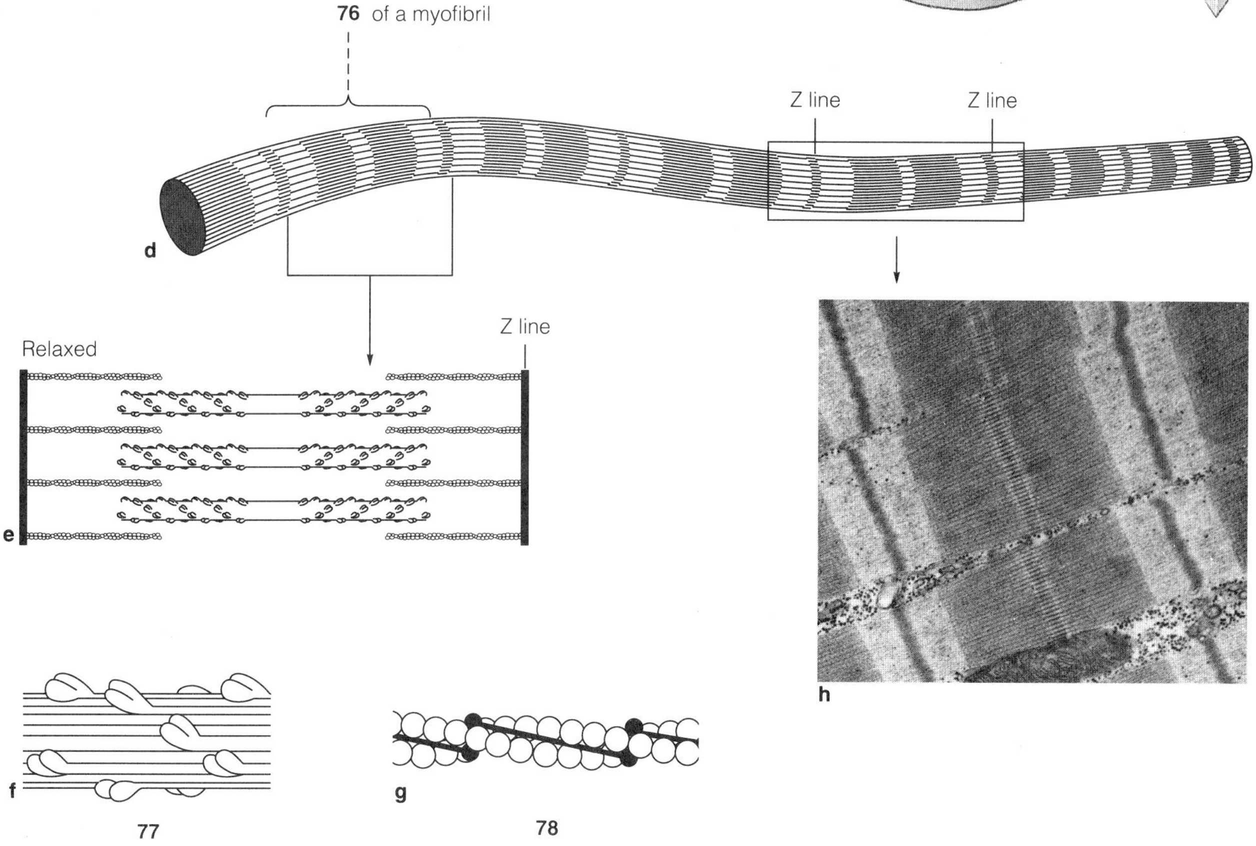

76 of a myofibril

Z line   Z line

d

Relaxed

Z line

e

f

77

g

78

h

Labeling

Identify the numbered parts of the accompanying illustrations.

79. _____

80. _____ _____

81. _____

82. _____ _____

    _____

83. _____

84. _____ _____

85. _____ _____

86. _____ (_____)

87. _____ _____

    _____

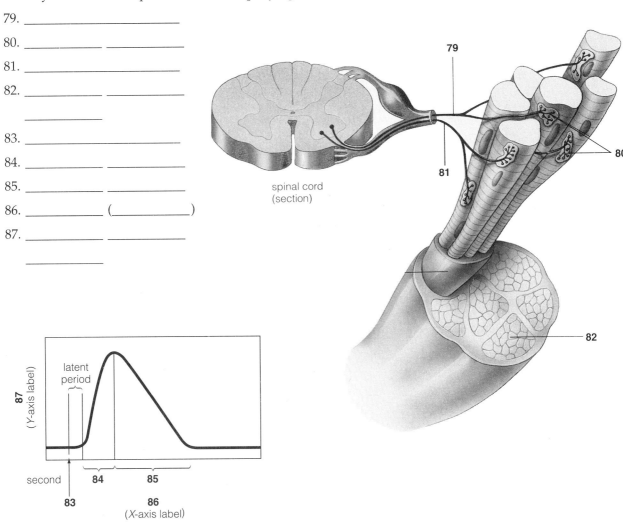

spinal cord
(section)

Fill-in-the-Blanks

(88) _____ _____ are synthetic hormones that mimic the effects of testosterone in building greater

(89) _____ mass in both men and women. It is illegal for competitive athletes to use them because of

unfair advantage and because of the side effects. In men, (90) _____, baldness, shrinking

(91) _____, and infertility are the first signs of damage. In women, (92) _____ hair becomes more

noticeable, (93) _____ _____ become irregular, breasts may shrink, and the (94) _____ may

become grossly enlarged. Aside from these physical side effects, some men experience uncontrollable

(95) _____, delusions, and wildly manic behavior.

Self-Quiz

For questions 1–7, choose from the following answers:

    a. bone
    b. cartilage
    c. epidermis
    d. dermis
    e. hypodermis

___ 1. Fat cells in adipose tissue are most likely to be located in this.

___ 2. Keratinized squamous cells are most likely to be located in this.

___ 3. Melanin in melanocytes is most likely to be here.

___ 4. Smooth muscles attached to hairs are probably here.

___ 5. This makes up the original "model" of the skeletal framework.

___ 6. The receiving ends of sensory receptors are most likely here.

___ 7. This serves as a "bank" for withdrawing and depositing calcium and phosphate ions.

For questions 8–11, choose from the following answers:

    a. Ligaments
    b. Osteoblasts
    c. Osteoclasts
    d. Red marrow
    e. Tendons

___ 8. _____ secrete bone-dissolving enzymes.

___ 9. Major site of blood cell formation.

___ 10. _____ remove Ca^{++} and $PO_4^{\equiv}$ ions from blood and build bone.

___ 11. _____ attach muscles to bone.

For questions 12–16, choose from the following answers:

    a. An action potential
    b. Cross-bridge formation
    c. The sliding-filament model
    d. Tension
    e. Tetanus

___ 12. _____ is a mechanical force that causes muscle cells to shorten if it is not exceeded by opposing forces.

___ 13. A wave of electrical disturbance that moves along a neuron or muscle cell in response to a threshold stimulus.

___ 14. _____ is a large contraction caused by repeated stimulation of motor units that are not allowed to relax.

___ 15. _____ is assisted by calcium ions and ATP.

___ 16. _____ explains how myosin filaments move to the centers of sarcomeres and back.

For questions 17–20, choose from the following answers:

    a. actin
    b. myofibril
    c. myosin
    d. sarcomere
    e. sarcoplasmic reticulum

___ 17. A(n) _____ contains many repetitive units of muscle contraction.

___ 18. The repetitive unit of muscle contraction.

___ 19. Thin filaments that depend on calcium ions to clear their binding sites so that they can attach to parts of thick filaments.

___ 20. _____ stores calcium ions and releases them in response to an action potential.

Chapter Objectives/Review Questions

This section lists general and detailed chapter objectives that can be used as review questions. You can make maximum use of these items by writing answers on a separate sheet of paper. To check for accuracy, compare your answers with information given in the chapter or glossary.

Page	Objectives/Questions
(632)	1. Name the four functions of human skin.
(632)	2. Describe the two-layered structure of human skin and identify the items located in each layer.
(636–637)	3. Compare invertebrate and vertebrate motor systems in terms of skeletal and muscular components and their interactions.
(638–639)	4. Explain the various roles of osteoblasts, osteoclasts, cartilage models, and long bones in the development of human bones.
(641)	5. Identify human bones by name and location.
(643)	6. Refer to Fig. 38.14 of your main text and indicate (a) a muscle used in sit-ups, (b) another used in dorsally flexing and inverting the foot, and (c) another used in flexing the elbow joint.
(644–647)	7. Explain in detail the structure of muscles, from the molecular level to the organ systems level. Then explain how biochemical events occur in muscle contractions and how antagonistic muscle action refines movements.
(644–645)	8. Describe the fine structure of a muscle fiber; use terms such as *myofibril*, *sarcomere*, *motor unit*, *actin*, and *myosin*.
(644–645)	9. List, in sequence, the biochemical and fine structural events that occur during the contraction of a skeletal muscle fiber and explain how the fiber relaxes.
(648)	10. Distinguish twitch contractions from tetanic contractions.

Integrating and Applying Key Concepts

If humans had an exoskeleton rather than an endoskeleton, would they move differently from the way they do now? Name any advantages or disadvantages that having an exoskeleton instead of an endoskeleton would present in human locomotion.

Answers

Interactive Exercises

38-I. INTEGUMENTARY SYSTEM (pp. 630–633)
A CLOSER LOOK AT THE STRUCTURE OF SKIN (pp. 634–635)

1. epidermis; 2. dermis; 3. hypodermis; 4. adipose; 5. hair; 6. sensory nerve ending; 7. sebaceous gland; 8. smooth muscle; 9. hair follicle; 10. sweat gland; 11. blood vessels; 12. epidermis; 13. dermis; 14. hypodermis; 15. a; 16. c; 17. d; 18. b; 19. c; 20. c; 21. e.

38-II. INVERTEBRATE AND VERTEBRATE SKELETONS (pp. 636–637)
CHARACTERISTICS OF BONE (pp. 638–639)
HUMAN SKELETAL SYSTEM (pp. 640–641)

1. contract (relax); 2. relax (contract); 3. contractile; 4. antagonistic; 5. hydrostatic; 6. exoskeleton; 7. Bones; 8. muscles; 9. Haversian; 10. Red marrow; 11. calcium (phosphate); 12. phosphate (calcium); 13. osteoblasts; 14. hormonal; 15. osteoporosis; 16. axial; 17. appendicular; 18. Synovial; 19. cartilage; 20. synovial; 21. osteoarthritis; 22. rheumatoid arthritis; 23. nutrient canal; 24. yellow marrow; 25. compact bone; 26. spongy bone; 27. connective tissue covering (periosteum); 28. Haversian system; 29. Haversian canal (blood vessel); 30. mineral deposits (calcium, phosphate); 31. osteocyte (bone cell); 32. cranium; 33. clavicle; 34. sternum; 35. scapula; 36. radius; 37. carpal bones; 38. femur; 39. tibia; 40. tarsal bones; 41. metatarsals.

38-III. MUSCLE SYSTEMS (pp. 642–643)
MUSCLE FUNCTION (pp. 644–645)
CONTROL OF CONTRACTION (pp. 646–647)
MUSCLE TENSION, STRENGTH, AND FATIGUE (pp. 648–649)

1. D; 2. C; 3. A; 4. E; 5. B; 6. F; 7. triceps brachii; 8. pectoralis major; 9. external oblique; 10. rectus abdominis; 11. adductor longus; 12. quadriceps femoris; 13. tibialis anterior; 14. gastronemius; 15. deltoid; 16. biceps brachii; 17. biceps contracts; 18. triceps relaxes; 19. biceps relaxes; 20. triceps contracts; 21. cardiac; 22. skeletal; 23. bones; 24. tendons; 25. Skeletal; 26. bones; 27. joints; 28. biceps; 29. triceps; 30. excitability; 31. neuron; 32. action potential; 33. motor; 34. myofibrils; 35. sarcomeres; 36. myosin; 37. actin; 38. sliding-filament; 39. myosin; 40. sarcomere; 41. ATP;

42. creatine phosphate; 43. Contraction; 44. bone; 45. calcium; 46. sarcoplasmic reticulum; 47. actin; 48. active transport; 49. action potentials; 50. sarcoplasmic reticulum; 51. Oxygen-requiring reactions that provide most of the ATP needed for muscle contraction during prolonged, moderate exercise; 52. Provides the energy to make myosin filaments slide along actin filaments; 53. Used to clear the actin binding sites of any obstacles to cross-bridge formation with myosin heads; 54. Supplies phosphate to ADP → ATP, which powers muscle contraction for a short time because creatine phosphate supplies are limited; 55. Stored in muscles and in the liver; glucose is stored by the animal body in this form of starch; 56. An anaerobic pathway in which glucose is broken down to lactate with a small yield of ATP; this pathway operates during intense exercise; 57. Supplies commands (signals) to muscle cells to contract and relax; 58. Projections from the thick filaments of myosin that bind to actin sites and form temporary cross-bridges; making and breaking these cross-bridges causes myosin filaments to be pulled to the center of a sarcomere; 59. The repetitive unit of muscle contraction; many sarcomeres constitute a myofibril; many myofibrils constitute a muscle cell; 60. The endoplasmic reticulum of a muscle cell; stores calcium ions and releases them in response to incoming signals from motor neurons; uses active transport to bring the calcium ions back inside; 61. dephosphorylation of creatine phosphate; 62. lactate fermentation; 63. aerobic respiration; 64. ATP; 65. diameter; 66. motor; 67. motor unit; 68. muscle twitch; 69. twitch; 70. Tetanus; 71. weak; 72. frequency (rate); 73. muscle; 74. muscle cell (muscle fiber); 75. myofibril; 76. sarcomere; 77. myosin filament; 78. actin filament; 79. axon of motor neuron serving one motor unit; 80. neuromuscular junction; 81. axon of another motor neuron serving another motor unit; 82. individual muscle cells; 83. time that stimulus is applied; 84. contraction phase; 85. relaxation phase; 86. time (seconds); 87. strength of contraction; 88. Anabolic steroids; 89. muscle; 90. acne; 91. testes; 92. facial; 93. menstrual periods; 94. clitoris; 95. aggression.

Self-Quiz

1. e; 2. c; 3. c; 4. d; 5. b; 6. d; 7. a; 8. c; 9. d; 10. b; 11. e; 12. d; 13. a; 14. e; 15. b; 16. c; 17. b; 18. d; 19. a; 20. e.

39

CIRCULATION

Interactive Exercises

39-I. CIRCULATORY SYSTEMS—AN OVERVIEW (pp. 652–655)

Selected Italicized Words

open circulatory system, closed circulatory system

Boldfaced, Page-Referenced Terms

(654) circulatory system _____

_____

(654) interstitial fluid _____

_____

(654) blood _____

_____

(654) heart _____

_____

(654) capillary beds _____

_____

(654) capillaries _____

_____

(654) lymphatic system _____

_____

Fill-in-the-Blanks

Cells survive by taking in from their surroundings what they need, (1) _____, and giving back to their surroundings materials what they don't need, (2) _____. In most animals, substances move rapidly to and from living cells by way of a (3) _____ circulatory system. (4) _____, a fluid connective tissue within the (5) _____ and blood vessels, is the transport medium.

Most of the cells of animals are bathed in a(n) (6) _____ _____; blood is constantly delivering nutrients and removing wastes from that fluid. The (7) _____ generates the pressure that keeps blood flowing. Blood flows (8) (choose one) ❐ rapidly ❐ slowly through large-diameter vessels to and from the heart, but where the exchange of nutrients and wastes occurs, in the (9) _____ beds, the blood is divided up into vast numbers of smaller-diameter vessels with tremendous surface area that enable the exchange to occur by diffusion. An elaborate network of drainage vessels attracts excess interstitial fluid and reclaimable (10) _____ and returns them to the circulatory system. This network is part of the (11) _____ system, which also helps clean the blood of disease agents.

Nutrients are absorbed into the blood from the (12) _____ and (13) _____ systems. Carbon dioxide is given to the (14) _____ system for elimination, and excess water, solutes, and wastes are eliminated by the (15) _____ system.

Labeling

Label the numbered parts in the illustrations below.

16. _____

17. _____ _____

18. _____

Describe the kind of circulatory system in:

19. Creature A. _____

20. Creature B. _____

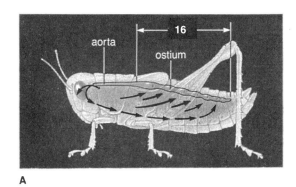

A

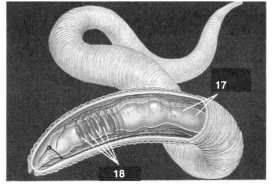

B

39-II. CHARACTERISTICS OF BLOOD (pp. 656–657)

Boldfaced, Page-Referenced Terms

(656) plasma _____

_____

(656) red blood cells _____

_____

(656) white blood cells _____

_____

(656) platelets _____

_____

(656) stem cells _____

_____

(657) cell count _____

_____

Complete the Table

1. Complete the following table, which describes the components of blood.

Components	Relative Amounts	Functions
Plasma Portion (50%–60% of total volume):		
Water	91%–92% of plasma volume	Solvent
a. (albumin, globulins, fibrinogen, and so on)	7%–8%	Defense, clotting, lipid transport, roles in extracellular fluid volume, and so forth
Ions, sugars, lipids, amino acids, hormones, vitamins, dissolved gases	1%–2%	Roles in extracellular fluid volume, pH, and so forth
Cellular Portion (40%–50% of total volume):		
b.	4,800,000–5,400,000 per microliter	O_2, CO_2 transport
White blood cells:		
c.	3,000–6,750	Phagocytosis
d.	1,000–2,700	Immunity
Monocytes (macrophages)	150–720	Phagocytosis
Eosinophils	100–360	Roles in inflammatory response, immunity
Basophils	25–90	Roles in inflammatory response, anticlotting
e.	250,000–300,000	Roles in clotting

Fill-in-the-Blanks

Blood is a highly specialized, fluid (2) _____ tissue that helps stabilize internal (3) _____ and equalize internal temperature throughout an animal's body. Organisms with (4) _____ circulatory systems generally also have a supplementary (5) _____ _____ _____ that recovers and purifies interstitial fluid and returns it to the major blood vessels. Oxygen binds with the (6) _____ atom in a hemoglobin molecule. The red blood (7) _____ _____ in males is about 5.4 million cells per microliter of blood; in females, it is 4.8 million per microliter. The plasma portion constitutes approximately (8) _____ to _____ percent of the total blood volume. Erythrocytes are produced in the (9) _____ _____ _____. (10) _____ _____ are immature cells not yet fully differentiated. (11) _____ and monocytes are highly mobile and phagocytic; they chemically detect, ingest, and destroy bacteria, foreign matter, and dead cells. (12) _____ (thrombocytes) are cell fragments that aid in forming blood clots.

In humans, red blood cells lack their (13) _____, but they contain enough to sustain them for about (14) _____ months. Platelets also have no (15) _____, but they last a maximum of (16) _____ days in the human bloodstream.

Label-Match

Identify the numbered cell types in the illustration below. Complete the exercise by matching and entering the letter of the appropriate function in the parentheses following the given cell types. A letter may be used more than once.

17. _____ ( )

18. _____

_____ ( )

19. _____ ( )

20. _____ ( )

21. _____ ( )

22. _____ ( )

23. _____ ( )

24. _____ ( )

A. Phagocytosis
B. Play a role in the inflammatory response
C. Play a role in clotting
D. Immunity
E. O_2, CO_2 transport
F. Play a role in producing blood cells

18 ___ cells

eosinophils   20 ___   basophils

(mature in bone marrow)   (mature in thymus)

21 ___ lymphocytes   22 ___ lymphocytes

17 ___ cells

19 ___

23 ___

wandering 24 ___

39-III. HUMAN CARDIOVASCULAR SYSTEM (pp. 658–659)
THE HUMAN HEART (pp. 660–661)

Selected Italicized Words

endothelium, "coronary circulation"

Boldfaced, Page-Referenced Terms

(658) artery, arteries _____

_____

(658) arterioles _____

_____

(658) venules _____

_____

(658) veins _____

_____

(658) pulmonary circuit _____

_____

(658) systemic circuit _____

_____

(660) atrium, atria _____

_____

(660) ventricle _____

_____

(660) aorta _____

_____

(661) cardiac cycle _____

_____

(661) cardiac muscle tissue _____

_____

(661) cardiac pacemaker _____

_____

Fill-in-the-Blanks

A(n) (1) _____ carries blood away from the heart. A(n) (2) _____ is a blood vessel with such a small diameter that red blood cells must flow through it single file; its wall consists of no more than a single layer of (3) _____ cells resting on a basement membrane. In each (4) _____ _____, small molecules move between the bloodstream and the (5) _____ fluid. (6) _____ are in the walls of veins and prevent backflow. Both (7) _____ and (8) _____ serve as temporary reservoirs for blood volume.

(9) _____ are pressure reservoirs that keep blood flowing smoothly away from the heart while the (10) _____ are relaxing. (11) _____ are control points where adjustments can be made in the volume of blood flow to be delivered to different capillary beds. They offer great resistance to flow, so there is a major drop in (12) _____ in these tubes.

In the pulmonary circuit, the heart pumps (13) _____-poor blood to the lungs; then the (14) _____-enriched blood flows back to the (15) _____. The (16) _____ _____ is the cardiac pacemaker. During a cardiac cycle, contraction of the (17) _____ is the driving force for blood circulation; (18) _____ contraction helps fill the ventricles.

A receiving zone of a vertebrate heart is called a(n) (19) _____; a departure zone is called a(n) (20)_____. Each contraction period is called (21) _____; each relaxation period is (22) _____.

The heart is a pumping station for two major blood transport routes: the (23) _____ circulation to and from the lungs, and the (24) _____ circulation to and from the rest of the body.

Fill-in-the-Blanks

Look at Fig. 39.8 in your text and use a red pen to redden all tubes in this illustration (except the pulmonary artery) that are indicated by "aorta" or "artery." Memorize the names, then fill in all of the blanks. Also, redden the parts of the heart that contain oxygen-rich blood.

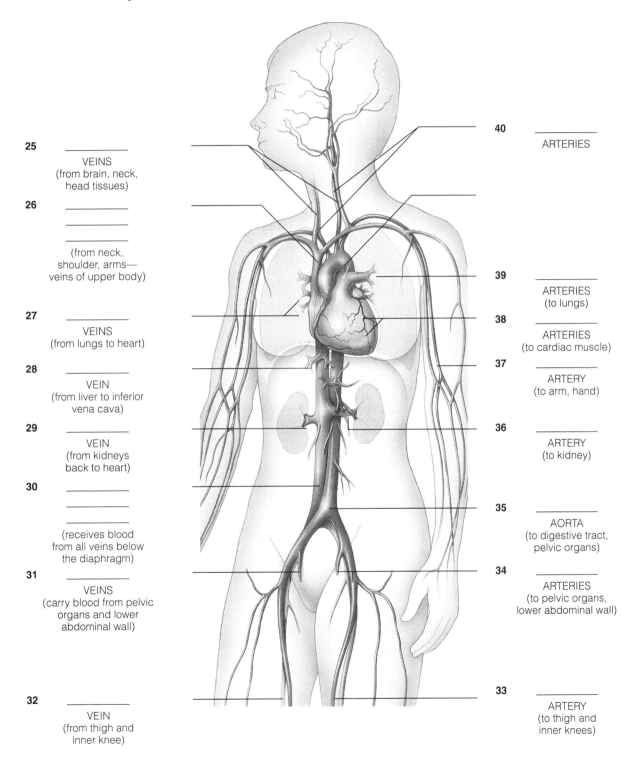

25 _____
VEINS
(from brain, neck, head tissues)

26 _____
_____
_____
(from neck, shoulder, arms—veins of upper body)

27 _____
VEINS
(from lungs to heart)

28 _____
VEIN
(from liver to inferior vena cava)

29 _____
VEIN
(from kidneys back to heart)

30 _____
_____
_____
(receives blood from all veins below the diaphragm)

31 _____
VEINS
(carry blood from pelvic organs and lower abdominal wall)

32 _____
VEIN
(from thigh and inner knee)

40 _____
ARTERIES

39 _____
ARTERIES
(to lungs)

38 _____
ARTERIES
(to cardiac muscle)

37 _____
ARTERY
(to arm, hand)

36 _____
ARTERY
(to kidney)

35 _____
AORTA
(to digestive tract, pelvic organs)

34 _____
ARTERIES
(to pelvic organs, lower abdominal wall)

33 _____
ARTERY
(to thigh and inner knees)

Labeling

Identify each indicated part of the accompanying illustration.

41. _____

42. _____ _____

    _____

43. _____ _____

44. _____ _____

45. _____

    _____

46. _____

47. _____ _____

    _____

48. _____ _____

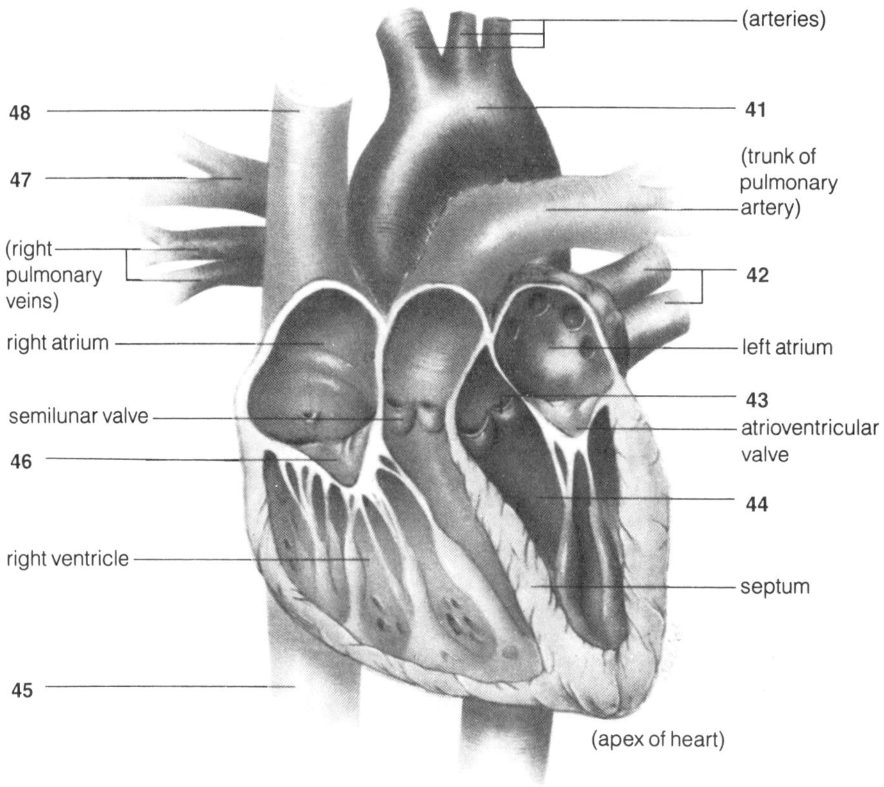

    _____

39-IV. BLOOD PRESSURE IN THE CARDIOVASCULAR SYSTEM (pp. 662–663)
FROM CAPILLARY BEDS BACK TO THE HEART (pp. 664–667)

Selected Italicized Words

edema, hypertension, atherosclerosis, heart attack, stroke, low-density lipoproteins (LDLs), high-density lipoproteins (HDLs), thrombus, embolus, angiography, diffusion zones, tachycardia

Boldfaced, Page-Referenced Terms

(662) blood pressure _____

_____

(663) carotid arteries _____

_____

(663) vasodilation _____

_____

(663) vasoconstriction _____

_____

Fill-in-the-Blanks

Blood pressure is normally high in the (1) _____ immediately after leaving the heart, but then it drops as the fluid passes along the circuit through different kinds of blood vessels. Energy in the form of (2) _____ is lost as it overcomes (3) _____ to the flow of blood. Arterial walls are thick, muscular, and (4) _____ and have large diameters. Arteries present (5) (choose one) ❑ much ❑ little resistance to blood flow, so pressure (6) (choose one) ❑ drops a lot ❑ does not drop much in the arterial portion of the systemic and pulmonary circuits.

The greatest drop in pressure occurs at (7) (choose one) ❑ capillaries ❑ arterioles ❑ veins. With this slow-down, blood flow can now be allotted in different amounts to different regions of the body in response to signals from the (8) _____ system and endocrine system or even changes in local chemical conditions.

When a person is resting, blood pressure is influenced most by reflex centers in the (9) _____ _____. When the resting level of blood pressure increases, reflex centers command the heart to (10) (choose one) ❑ beat more slowly ❑ beat faster and smooth muscle cells in arteriole walls to (11) (choose one) ❑ contract ❑ relax, which results in (12) (choose one) ❑ vasodilation ❑ vasoconstriction.

Capillary beds are (13) _____ zones where substances are exchanged between blood and interstitial fluid. Capillaries have the thinnest walls across which (14) _____ _____ drives fluid, forcing mole-cules of water and solutes to move in the same direction. Veins contain (15) _____ that prevent blood from flowing backward; they contain (16) (choose one) ❑ 20–30 ❑ 35–45 ❑ 50–60 percent of the total blood volume.

Labeling

Identify each indicated part of the accompanying illustrations.

17. _____   20. _____

18. _____   21. _____ _____

19. _____   22. _____

outer coat   21   basement membrane   endothelium

a. 17   22

outer coat   smooth muscle rings over elastic layer   basement membrane   endothelium

c. 19

outer coat   smooth muscle between elastic layers   basement membrane   endothelium

b. 18

basement membrane   endothelium

d. 20

True-False

If the statement in true, write a T in the blank. If false, make it true by changing the underlined word.

_____ 23. The pulse <u>rate</u> is the difference between the systolic and the diastolic pressure readings.

_____ 24. Because the total volume of blood remains constant in the human body, blood pressure must <u>also</u> <u>remain</u> <u>constant</u> throughout the circuit.

Fill-in-the-Blanks

One cause of (25) _____ is the rupture of one or more blood vessels in the brain. A(n) (26) _____ _____ blocks a coronary artery. (27)_____ is a term for a formation that can include cholesterol, calcium salts, and fibrous tissue. It is not healthful to have a high concentration of (28) _____-density lipoproteins in the bloodstream. A clot that stays in place is a(n) (29) _____, but a clot that travels in the bloodstream is an embolus.

39-V. HEMOSTASIS (p. 668)
BLOOD TYPING (pp. 668–669)
LYMPHATIC SYSTEM (pp. 670–671)

Selected Italicized Words

transfusion, erythroblastosis fetalis, "valves," flaplike

Boldfaced, Page-Referenced Terms

(668) hemostasis _____

_____

(668) antibodies _____

_____

(668) ABO blood typing _____

_____

(668) agglutination _____

_____

(668) Rh blood typing _____

_____

(670) lymph _____

_____

(671) lymph capillaries _____

_____

(671) lymph vessels _____

(671) lymph nodes _____

(671) spleen _____

(671) thymus _____

Fill-in-the-Blanks

Bleeding is stopped by several mechanisms that are referred to as (1) _____; the mechanisms include blood vessel spasm, (2) _____ _____ _____, and blood (3) _____. Once the platelets reach a damaged vessel, through chemical recognition they adhere to exposed (4) _____ fibers in damaged vessel walls. Reactions cause rod-shaped proteins to assemble into long (5) _____ fibers. These trap blood cells and components of plasma. Under normal conditions, a clot eventually forms at the damaged site.

If you are blood type (6) _____, you have no antibodies against A or B markers in your plasma. If you are type (7) _____, you have antibodies against A and B markers in your plasma. People who are (8) (choose one) ❒ Rh⁺ ❒ Rh⁻ (9) (choose one) ❒ women ❒ men have to be careful so that their fetuses don't develop erythroblastosis fetalis.

(10) _____ is a response in which antibodies act against "foreign" cells bearing specific markers and cause them to clump together.

Identification/Fill-in-the-Blanks

11. _____

12. _____ gland

13. _____ duct

14. _____

15. _____ _____

16. organized arrays of _____ and

    _____

17. _____ _____

18. _____ vessels reclaim fluid lost

    from the bloodstream, purify the blood

    of microorganisms, and transport (19)

    _____ from the

    (20) _____ _____ to the

    bloodstream.

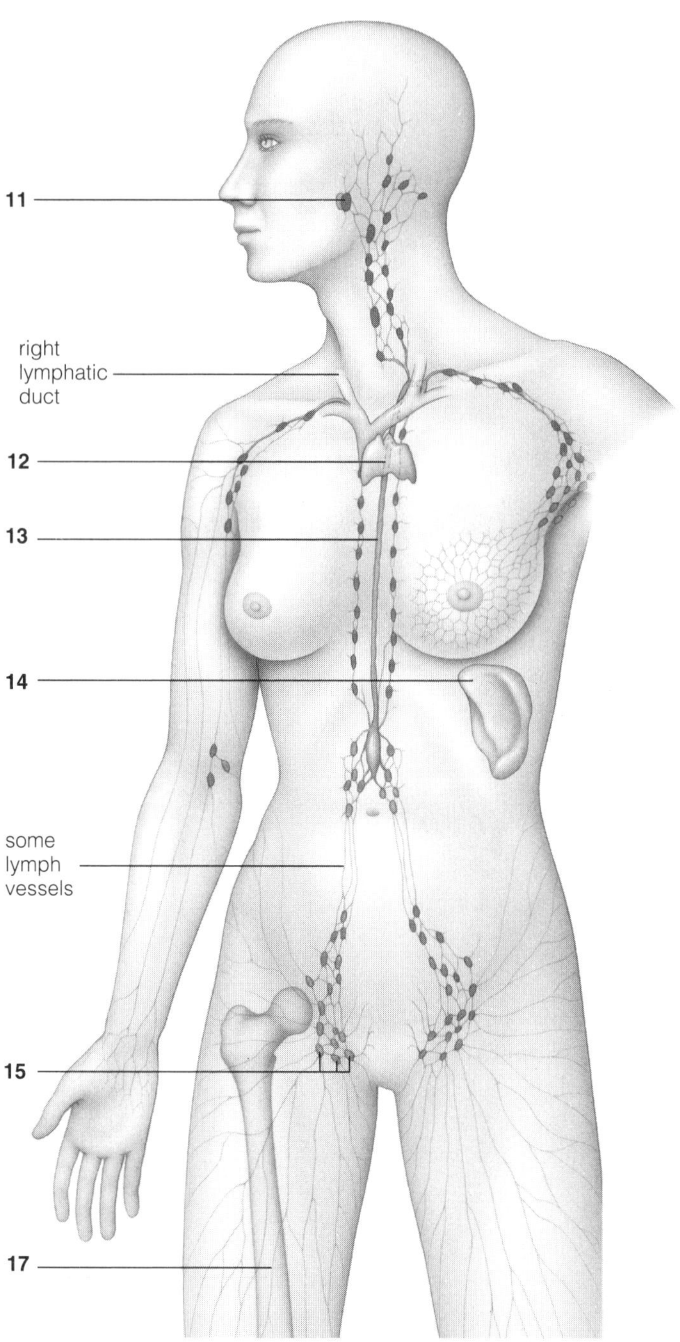

11

right
lymphatic
duct

12

13

14

some
lymph
vessels

15

17

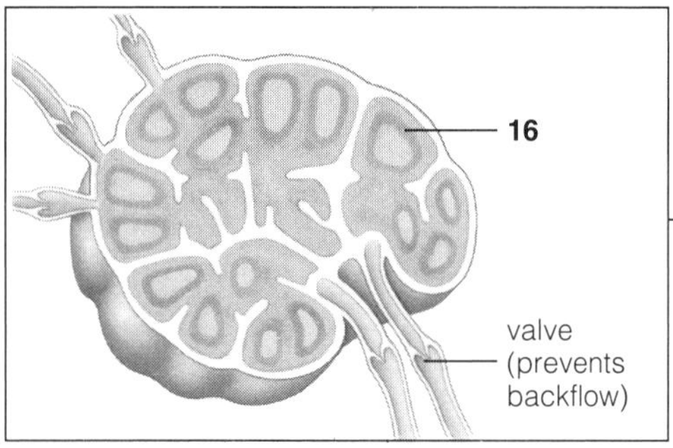

16

valve
(prevents
backflow)

Self-Quiz

Multiple Choice

___ 1. Most of the oxygen in human blood is transported by _____.

    a. plasma
    b. serum
    c. platelets
    d. hemoglobin
    e. leukocytes

___ 2. Of all the different kinds of white blood cells, two classes of _____ are the ones that respond to specific invaders and confer immunity to a variety of disorders.

    a. basophils
    b. eosinophils
    c. monocytes
    d. neutrophils
    e. lymphocytes

___ 3. Open circulatory systems generally lack _____.

    a. a heart
    b. arterioles
    c. capillaries
    d. veins
    e. arteries

___ 4. Red blood cells originate in the _____.

    a. liver
    b. spleen
    c. yellow bone marrow
    d. thymus gland
    e. red bone marrow

___ 5. Hemoglobin contains _____.

    a. copper
    b. magnesium
    c. sodium
    d. calcium
    e. iron

___ 6. The pacemaker of the human heart is the _____.

    a. sinoatrial node
    b. semilunar valve
    c. inferior vena cava
    d. superior vena cava
    e. atrioventricular node

___ 7. During systole, _____.

    a. oxygen-rich blood is pumped to the lungs
    b. the heart muscle tissues contract
    c. the atrioventricular valves suddenly open
    d. oxygen-poor blood from all parts of the human body, except the lungs, flows toward the right atrium
    e. none of the above

___ 8. _____ are reservoirs of blood pressure in which resistance to flow is low.

    a. Arteries
    b. Arterioles
    c. Capillaries
    d. Venules
    e. Veins

___ 9. Begin with a red blood cell located in the superior vena cava and travel with it in proper sequence as it goes through the following structures. Which will be *last* in sequence?

    a. aorta
    b. left atrium
    c. pulmonary artery
    d. right atrium
    e. right ventricle

___ 10. The lymphatic system is the principal avenue in the human body for transporting _____.

    a. fats
    b. wastes
    c. carbon dioxide
    d. amino acids
    e. interstitial fluids

Matching

11. ___ agglutination

12. ___ angiogram, angiography

13. ___ atherosclerosis

14. ___ carotid arteries

15. ___ coronary arteries

16. ___ edema

17. ___ embolism, embolus

18. ___ erythroblastosis fetalis

19. ___ hypertension

20. ___ inferior vena cava

21. ___ jugular veins

22. ___ renal arteries

23. ___ stroke

24. ___ tachycardia

25. ___ thrombosis, thrombus

A. Receive(s) blood from the brain, tissues of the head, and neck
B. The heart's own blood supplier(s)
C. Excessively fast rate of heartbeat; occurs during heavy exercising
D. Damage to brain caused by damaged blood vessels
E. High blood pressure
F. A blood clot that is on the move from one place to another
G. Diagnostic technique that uses opaque dyes and x-rays to discover blocked blood vessels
H. Deliver(s) blood to the head, neck, brain
I. The clumping of red blood cells, or of antibodies with antigens
J. Delivers blood to kidneys, where its composition and volume are adjusted
K. A blood clot that is lodged in a blood vessel and is blocking it
L. Receive(s) blood from all veins below the diaphragm
M. Progressive thickening of the arterial wall and narrowing of the arterial lumen (space)
N. Blood cell destruction caused by mismatched Rh blood types (Rh$^-$ mother and Rh$^+$ fetus)
O. Accumulation of excess fluid in interstitial spaces; extreme in elephantiasis

Chapter Objectives/Review Questions

This section lists general and detailed chapter objectives that can be used as review questions. You can make maximum use of these items by writing answers on a separate sheet of paper. To check for accuracy, compare your answers with information given in the chapter or glossary.

Page   *Objectives/Questions*

(654–655)   1. Distinguish between open and closed circulatory systems.
(656)   2. Describe the composition of human blood, using percentages of volume.
(657)   3. Distinguish the five types of leukocytes from each other in terms of structure and functions.
(657)   4. State where erythrocytes, leukocytes, and platelets are produced.
(658–659)   5. Trace the path of blood in the human body. Begin with the aorta and name all major components of the circulatory system through which the blood passes before it returns to the aorta.
(660–661)   6. Explain what causes a heart to beat. Then describe how the rate of heartbeat can be slowed down or sped up.
(662–667)   7. List the factors that cause blood to leave the heart and the factors that cooperate to return blood to the heart.
(662–663)   8. Explain what causes high pressure and low pressure in the human circulatory system. Then show where major drops in blood pressure occur in humans.
(662)   9. Describe how the structures of arteries, capillaries, and veins differ.
(664)   10. Explain how veins and venules can act as reservoirs of blood volume.
(666, 667, 11. Distinguish a stroke from a coronary artery blockage or occlusion.
 672)
(666)   12. Describe how hypertension develops, how it is detected, and whether it can be corrected.
(667)   13. State the significance of high- and low-density lipoproteins to cardiovascular disorders.
(668)   14. List in sequence the events that occur in the formation of a blood clot.
(668–669)   15. Describe how blood is typed for the ABO blood group and for the Rh factor.
(670–671)   16. Describe the composition and function of the lymphatic system.

Integrating and Applying Key Concepts

You observe that some people appear as though fluid has accumulated in their lower legs and feet. Their lower extremities resemble those of elephants. You inquire about what is wrong and are told that the condition is caused by the bite of a mosquito that is active at night. Construct a testable hypothesis that would explain (1) why the fluid was not being returned to the torso, as normal, and (2) what the mosquito did to its victims.

Answers

Interactive Exercises

39-I. CIRCULATORY SYSTEMS—AN OVERVIEW
(pp. 654–655)

1. nutrients (food); 2. wastes; 3. closed; 4. Blood;
5. heart; 6. interstitial fluid; 7. heart; 8. rapidly; 9. capillary; 10. solutes; 11. lymphatic; 12. digestive (respiratory); 13. respiratory (digestive); 14. respiratory;
15. urinary; 16. heart(s); 17. blood vessels; 18. hearts;
19. Open circulatory system; blood is pumped into short tubes that open into spaces in the body's tissues, mingles with tissue fluids, then is reclaimed by open-ended tubes that lead back to the heart; 20. Closed circulatory system; blood flow is confined within blood vessels that have continuously connected walls and is pumped by five pairs of "hearts."

39-II. CHARACTERISTICS OF BLOOD (pp. 656–657)

1. a. Plasma proteins; b. Red blood cells; c. Neutrophils;
d. Lymphocytes; e. Platelets; 2. connective; 3. pH;
4. closed; 5. lymph vascular system; 6. iron; 7. cell count; 8. 50 to 60; 9. red bone marrow; 10. Stem cells;
11. Neutrophils; 12. Platelets; 13. nucleus; 14. four;
15. nucleus; 16. nine; 17. stem (F); 18. red blood (E);
19. platelets (C); 20. neutrophils (A); 21. B (D); 22. T (D);
23. monocytes (A); 24. macrophages (A).

39-III. HUMAN CARDIOVASCULAR SYSTEM
(pp. 658–659)
THE HUMAN HEART (pp. 660–661)

1. artery; 2. capillary; 3. endothelial; 4. capillary bed (diffusion zone); 5. interstitial; 6. Valves; 7. veins (venules); 8. venules (veins); 9. Arteries; 10. ventricles;
11. Arterioles; 12. pressure; 13. oxygen; 14. oxygen;
15. heart; 16. sinoatrial (SA) node; 17. ventricles;
18. atrial; 19. atrium; 20. ventricle; 21. systole; 22. diastole; 23. pulmonary; 24. systemic; 25. jugular; 26. superior vena cava; 27. pulmonary; 28. hepatic; 29. renal;

30. inferior vena cava; 31. iliac; 32. femoral; 33. femoral;
34. iliac; 35. dorsal; 36. renal; 37. brachial; 38. coronary;
39. pulmonary; 40. carotid; 41. aorta; 42. left pulmonary veins; 43. semilunar valve; 44. left ventricle; 45. inferior vena cava; 46. atrioventricular valve; 47. right pulmonary artery; 48. superior vena cava.

39-IV. BLOOD PRESSURE IN THE CARDIOVASCULAR SYSTEM (pp. 662–663)
FROM CAPILLARY BEDS BACK TO THE HEART
(pp. 664–667)

1. aorta; 2. pressure; 3. resistance; 4. elastic; 5. little;
6. does not drop much; 7. arterioles; 8. nervous;
9. medulla oblongata; 10. beat more slowly; 11. relax;
12. vasodilation; 13. diffusion; 14. bulk flow (blood pressure); 15. valves; 16. 50–60; 17. vein; 18. artery;
19. arteriole; 20. capillary; 21. smooth muscle (elastic fibers); 22. valve; 23. pressure; 24. cannot remain constant (because it passes through various kinds of vessels that have varied structures); 25. stroke; 26. coronary occlusion; 27. Plaque; 28. low; 29. thrombus.

39-V. HEMOSTASIS (p. 668)
BLOOD TYPING (pp. 668–669)
LYMPHATIC SYSTEM (pp. 670–671)

1. hemostasis; 2. platelet plug formation; 3. coagulation;
4. collagen; 5. insoluble; 6. AB; 7. O; 8. Rh⁻; 9. women;
10. Agglutination (Clumping); 11. tonsils; 12. thymus;
13. thoracic; 14. spleen; 15. lymph node(s); 16. macrophages, lymphocytes; 17. bone marrow; 18. Lymph;
19. fats; 20. small intestine.

Self-Quiz

1. d; 2. e; 3. c; 4. e; 5. e; 6. a; 7. b; 8. a; 9. a; 10. a; 11. I;
12. G; 13. M; 14. H; 15. B; 16. O; 17. F; 18. N; 19. E;
20. L; 21. A; 22. J; 23. D; 24. C; 25. K.

40

IMMUNITY

Interactive Exercises

40-I. THREE LINES OF DEFENSE (pp. 674–676)
COMPLEMENT PROTEINS (p. 677)
INFLAMMATION (pp. 678–679)

Selected Italicized Words

"vaccination," pasteurization, athlete's foot, nonspecific response, specific response, fever

Boldfaced, Page-Referenced Terms

(676) pathogens _____

_____

(676) lysozyme _____

_____

(677) complement system _____

_____

(677) lysis _____

_____

(678) neutrophils _____

_____

(678) eosinophils _____

_____

(678) basophils _____

_____

(678) macrophages _____

_____

(678) acute inflammation _____

_____

(678) histamine _____

_____

(678) interleukins _____

_____

Fill-in-the-Blanks

The best way to deal with damaging foreign agents is to prevent their entry. Several barriers prevent pathogens from crossing the boundaries of your body. Intact skin and (1)_____ membranes are effective barriers. (2) _____ is an enzyme that destroys the cell walls of many bacteria. (3) _____ fluid destroys many food-borne pathogens in the gut. Normal (4) _____ residents of the skin, gut, and vagina outcompete pathogens for resources and help keep their numbers under control.

Even simple aquatic invertebrates defend themselves with (5) _____ cells and antimicrobial substances, including lysozymes that cause bacteria to burst open. In most animals, when a sharp object cuts through the skin and foreign microbes enter, some plasma proteins come to the rescue and seal the wound with a (6) _____ mechanism. (7) _____ white blood cells engulf bacteria soon thereafter.

Also, in vertebrates there are about twenty plasma proteins [collectively referred to as the (8) _____ _____] that are activated one after another in a "cascade" of reactions to help destroy invading microorganisms. When the complement system is activated, circulating basophils and mast cells in tissues release (9) _____, which dilates (10) _____ and makes them "leaky," so fluid seeps out and causes the inflamed area to become swollen and warm.

Vertebrates have (11) _____, general defenses such as those mentioned above that isolate and destroy threatening agents identified as "nonself." (11) defenses do not identify which specific foreign agent needs to be destroyed; they identify only that there is some foreign agent that needs to be destroyed. If foreign agents penetrate the boundaries of your body, (11) defenses constitute a fast way to exterminate them. If (11) defenses fail to control invading pathogens, (12) _____ defenses come to the rescue, in which well-coordinated armies of white blood cells named (13) _____ make highly focused counterat- tacks against pathogens whose specific identity has been identified. These armies and the organs of your body that house them constitute our (14) _____ system.

Complete the Table

Complete the table by providing the specific functions carried out by each of the four different kinds of white blood cells listed below.

Type of Cells	Functions
Basophils	15.
Eosinophils	16.
Neutrophils	17.
Macrophages	18.

40-II. THE IMMUNE SYSTEM (pp. 680–681)
CELL-MEDIATED RESPONSES (pp. 682–683)
ANTIBODY-MEDIATED RESPONSES (pp. 684–685)

Selected Italicized Words

specificity, memory, self, nonself, effector cells, memory cells, cell-mediated, antibody-mediated, primary immune response, secondary immune response, clone, immunotherapy, monoclonal antibodies, IgM, IgG, IgA, IgE

Boldfaced, Page-Referenced Terms

(680) B lymphocytes (=B cells) _____

_____

(680) T lymphocytes (=T cells) _____

_____

(680) immune system _____

_____

(680) antigen _____

_____

(680) MHC markers _____

_____

(680) antigen-MHC complexes _____

_____

(680) antigen-presenting cell _____

_____

(680) helper T cells _____

_____

(680) cytotoxic T cells _____

_____

(680) antibodies _____

_____

(683) perforins _____

_____

(683) natural killer cells (NK cells) _____

_____

(685) immunoglobulins (Igs) _____

_____

Fill-in-the-Blanks

If the (1) _____ defenses (fast-acting white blood cells such as neutrophils, eosinophils, and basophils; slower-acting macrophages; and the plasma proteins involved in clotting and complement) fail to repel the microbial invaders, then the body calls on its (2) _____ _____, which identifies *specific* targets to kill and *remembers* the identities of its targets. Your own unique (3) _____ _____ patterns identify your cells as "self" cells. Any other surface pattern is, by definition, (4) _____ and doesn't belong in your body.

The principal actors of the immune system are (5) _____ descended from stem cells (consult text-book Fig. 39.6) in the bone marrow that have two different strategies of action to deal with their different kinds of enemies. (6) _____ _____ clones secrete antibodies and act principally against the extra-cellular (ones that stay outside the body cells) enemies that are pathogens in blood or on the cell surfaces of body tissues. (7) _____ _____ clones descend from lymphocytes that matured in the (8) _____ where they acquired specific markers on their cell surfaces; they defend principally against intracellular pathogens, such as (9) _____, and against any cells [(10) _____ cells and grafts of foreign tissue] that are perceived as abnormal or foreign.

Each kind of cell, virus, or substance bears unique molecular configurations (patterns) that give it a unique (11) _____. A(n) (12) _____ is any molecular configuration that causes the formation of lymphocyte armies. Any cell that processes and displays (12) together with a suitable MHC molecule is known as a(n) (13) _____-_____ cell that can activate lymphocytes to undergo rapid cell divi-sions. Lymphocyte subpopulations that fight and destroy enemies are known as (14) _____ cells; among these are (15) _____ _____ cells that promote the formation of large populations of lym-phocytes and (16) _____ _____ cells that eliminate infected body cells and tumor cells by a lethal hit. Together with other "killers," they execute the (17) _____-_____ immune responses. By con-trast, (6) produce Y-shaped antigen-binding receptor molecules called (18) _____. (6) and (18) carry out the (19) _____-_____ immune responses.

Other lymphocyte subpopulations, (20) _____ cells, enter a resting phase, but they "remember" the specific agent that was conquered and will undertake a larger, more rapid response if it shows up again.

Labeling

Identify each numbered part in the accompanying illustration.

21. _____ - _____ _____

22. _____ _____ _____ _____

23. _____ _____ _____

24. _____

25. _____ _____ _____

26. _____ _____ _____

27. _____

28. _____

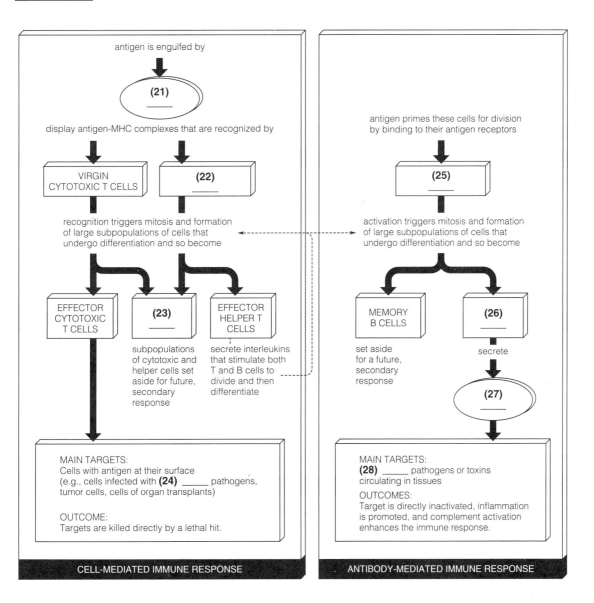

antigen is engulfed by

(21)
_____

display antigen-MHC complexes that are recognized by

VIRGIN CYTOTOXIC T CELLS   **(22)** _____

recognition triggers mitosis and formation of large subpopulations of cells that undergo differentiation and so become

EFFECTOR CYTOTOXIC T CELLS   **(23)** _____   EFFECTOR HELPER T CELLS

subpopulations of cytotoxic and helper cells set aside for future, secondary response

secrete interleukins that stimulate both T and B cells to divide and then differentiate

MAIN TARGETS:
Cells with antigen at their surface (e.g., cells infected with **(24)** _____ pathogens, tumor cells, cells of organ transplants)

OUTCOME:
Targets are killed directly by a lethal hit.

CELL-MEDIATED IMMUNE RESPONSE

antigen primes these cells for division by binding to their antigen receptors

(25) _____

activation triggers mitosis and formation of large subpopulations of cells that undergo differentiation and so become

MEMORY B CELLS   **(26)** _____

set aside for a future, secondary response

secrete

(27) _____

MAIN TARGETS:
(28) _____ pathogens or toxins circulating in tissues

OUTCOMES:
Target is directly inactivated, inflammation is promoted, and complement activation enhances the immune response.

ANTIBODY-MEDIATED IMMUNE RESPONSE

Choice

Match each of the white blood cell types with the appropriate function (29–33).

a. effector cytotoxic T cells   b. effector helper T cells   c. macrophages
d. memory cells   e. effector B cells

__ 29. Lymphocytes that directly destroy body cells already infected by certain viruses or by parasitic fungi

__ 30. A portion of B and T cell populations that were set aside as a result of a first encounter now circulate freely and respond rapidly to any later attacks by the same type of invader

__ 31. Lymphocytes and their progeny that produce antibodies

__ 32. Lymphocytes that serve as master switches of the immune system; stimulate the rapid division of B cells and cytotoxic T cells

__ 33. Nonlymphocytic white blood cells that develop from monocytes, engulf anything perceived as foreign, and alert helper T cells to the presence of specific foreign agents

Fill-in-the-Blanks

In addition to effector B cells and cytotoxic T cells (executioner lymphocytes that mature in the thymus), (34) _____ _____ cells mature in other lymphoid tissues and search out any cell that is either coated with complement proteins or antibodies, or bears any foreign molecular pattern. When cytotoxic T cells find foreign cells, they secrete (35) _____ and other toxic substances to poison and lyse the offenders.

Labeling

Identify each numbered part in the accompanying illustrations.

36. _____

37. _____ _____

38. _____ _____

39. _____-_____

40. _____

41. _____

42. _____ _____

43. _____ _____

Below: Example of a(n) (36) _____-mediated immune response to a bacterial invasion.

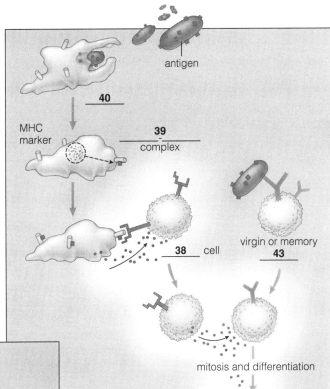

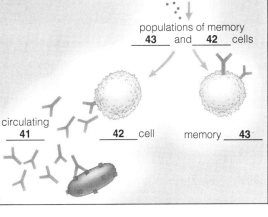

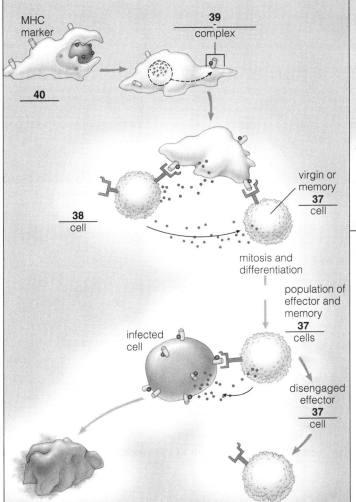

Above: Example of a(n) (41) _____-mediated immune response to a bacterial invasion.

Note: If the same number is used more than once, that is because the label is the same in all such situations. If you have labeled it correctly on the lefthand illustration, then its identification should also be correct on the one above.

The (44) _____ _____ _____ is the route taken during a first-time contact with an antigen; the secondary immune response is more rapid because patrolling battalions of (45) _____ _____ are in the bloodstream on the lookout for enemies they have conquered before. When these (45) meet up with recognizable (46) _____, they divide at once, producing large clones of B or T cells within two to three days.

(47) _____ refers to cells that have lost control over cell division. Milstein and Kohler developed a means of producing large amounts of (48) _____ _____, which are produced by clones of proliferating (49) _____ cells: Mouse B cells fused with tumor cells that divide nonstop.

Complete the Table

Immunoglobulin Type	Function
50.	enter mucus-coated surfaces of the respiratory, digestive, and reproductive tracts where they neutralize pathogens.
51.	trigger inflammation when parasitic worms attack the body; play a role in allergic responses.
52.	activate complement proteins; neutralize many toxins; long-lasting; can cross placenta and protect developing fetus; also present in colostrum from mammary glands.
53.	first to be secreted during immune responses; after binding to antigen, trigger complement cascade; also tag invaders and bind them in clumps for later phagocytosis.

40-III. LYMPHOCYTE BATTLEGROUNDS (p. 685)
SPECIFICITY AND MEMORY (pp. 686–687)
IMMUNIZATION (p. 688)

Selected Italicized Words

"immunological memory," active immunization, passive immunization

Boldfaced, Page-Referenced Terms

(687) clonal selection theory _____

_____

(688) immunization _____

_____

(688) vaccine _____

_____

Fill-in-the-Blanks

Deliberately provoking the production of memory lymphocytes is known as (1) _____. In a(n) (2) _____ immunization, a vaccine containing antigens is injected into the body or taken orally. The first one elicits a(n) (3) _____ _____ _____, and the second one (a booster shot) elicits a secondary immune response, which causes the body to produce more antibodies and (4) _____ _____ to provide long-lasting protection.

The (5) _____ _____ _____ explains how an individual has immunological memory, which is the basis of a secondary immune response; the theory also explains in part how self cells are distinguished from (6) _____ cells in the vertebrate immune response.

Antibodies are plasma proteins that are part of the (7) _____ group of proteins. Some of these circulate in blood; others are present in other body fluids or bound to B cells. (8) _____ are Y-shaped proteins that lock onto specific foreign targets and thereby tag them for destruction by phagocytes or by activating the complement system. While B or T cells are maturing, different regions of the genes that code for antigen receptors are shuffled at random into one of millions of possible combinations; this process of (9) _____ produces a gene that codes for just one of millions of possible antigen receptors.

Labeling

Identify each numbered part in the accompanying illustration and its legend.

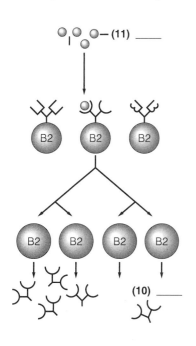

10. _____

11. _____

12. _____

13. _____ _____

14. _____ _____

Clonal selection of a (12) _____ cell, the descendants of which produced the specific (10) _____ that can combine with a specific antigen.

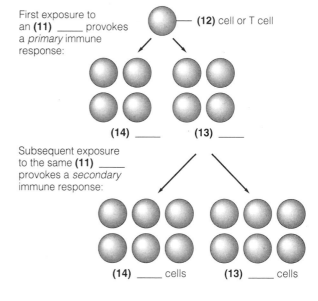

First exposure to an (11) ____ provokes a *primary* immune response:

(12) cell or T cell

(14) ____   (13) ____

Subsequent exposure to the same (11) ____ provokes a *secondary* immune response:

(14) ____ cells   (13) ____ cells

Complete the Table

Indicate with a check (√) the type of vaccine used (oral vs. combined injected vaccine) and the age(s) of vaccination.

Disease	(a) CIV	(b) O	(c) 0–8 months after birth	(d) 12–20 months after birth	(e) 4–6 years after birth	(f) 11–12 years after birth
				Age Vaccination Is Administered		
15. Diphtheria						
16. *Hemophilus influenzae*						
17. Hepatitis B						
18. Measles						
19. Mumps						
20. Polio						
21. Rubella						
22. Tetanus						
23. Whooping cough						

40-IV. ABNORMAL OR DEFICIENT IMMUNE RESPONSES (pp. 689–691)

Selected Italicized Words

asthma, hay fever, anaphylactic shock, Grave's disorder, myasthenia gravis, rheumatoid arthritis, AIDS

Boldfaced, Page-Referenced Terms

(689) allergens _____

_____

(689) allergy _____

_____

(689) autoimmune response _____

_____

Fill-in-the-Blanks

(1) _____ is an altered secondary response to a normally harmless substance and may actually cause injury to tissues. (2) _____ _____ is a disorder in which the body mobilizes its forces against certain of its own tissues. (3) _____ _____ is an example of this kind of disorder in which anti-

bodies tag acetylcholine receptors on skeletal muscle cells and cause progressive weakness. (4) _____

_____ is a similar kind of disorder in which skeletal joints are chronically inflamed.

AIDS is a constellation of disorders that follow infection by the (5) _____ _____ _____.

In the United States, transmission has occurred most often among intravenous drug abusers who share

needles and among (6) _____ _____. HIV is a(n) (7) _____; its genetic material is RNA

rather than DNA, and it has several copies of the enzyme (8) _____ _____, which uses the viral

RNA as a template for making DNA, which is then inserted into a host chromosome.

HIV is transmitted when (9) _____ _____ of an infected person enter another person's tis-

sues. The virus cripples the immune system by attacking (10) _____ _____ cells and

(11) _____. Approximately (12) _____ (number) Americans are now infected with HIV. Worldwide,

an estimated (13) _____ (number) are infected and (14) _____ (number) are already dead. By the

year 2000, the number of infected people may reach an estimated 40 million to (15) _____ million.

Self-Quiz

Multiple Choice

___ 1. All the body's phagocytes are derived from stem cells in the _____.
   a. spleen
   b. liver
   c. thymus
   d. bone marrow
   e. thyroid

___ 2. The plasma proteins that are activated when they contact a bacterial cell are collectively known as the _____ system.
   a. shield
   b. complement
   c. IgG
   d. MHC
   e. HIV

___ 3. _____ are divided into two groups: T cells and B cells.
   a. Macrophages
   b. Lymphocytes
   c. Platelets
   d. Complement cells
   e. Cancer cells

___ 4. _____ produce and secrete antibodies that set up bacterial invaders for subsequent destruction by macrophages.
   a. B cells
   b. Phagocytes
   c. T cells
   d. Bacteriophages
   e. Thymus cells

___ 5. Antibodies are shaped like the letter _____.
   a. Y
   b. W
   c. Z
   d. H
   e. E

___ 6. The markers for every cell in the human body are referred to by the letters _____.
   a. HIV
   b. MBC
   c. RNA
   d. DNA
   e. MHC

___ 7. Effector B cells _____.
   a. fight against extracellular pathogens and toxins circulating in tissues
   b. develop from virgin or memory B cells
   c. manufacture and secrete antibodies
   d. do not divide and form clones
   e. all of the above

___ 8. Clones of B or T cells are _____.
   a. being produced continually
   b. sometimes known as memory cells if they keep circulating in the bloodstream
   c. only produced when their surface proteins recognize other specific proteins previously encountered
   d. produced and mature in the bone marrow
   e. both (b) and (c)

___ 9. Whenever the body is reexposed to a specific sensitizing agent, IgE antibodies cause _____.
   a. prostaglandins and histamine to be produced
   b. clonal cells to be produced
   c. histamine to be released
   d. the immune response to be suppressed
   e. none of the above

___ 10. The clonal selection theory explains _____.
   a. how a few distinct lymphocytes are "chosen" to proliferate into many millions of identical effector cells
   b. how B cells differ from T cells
   c. how so many different kinds of antigen-specific receptors can be produced by lymphocytes
   d. how memory cells are set aside from effector cells
   e. how antigens differ from antibodies

Matching

Choose the most appropriate description for each term.

11. ___ allergy
12. ___ antibody
13. ___ antigen
14. ___ macrophage
15. ___ clone
16. ___ complement
17. ___ histamine
18. ___ MHC marker
19. ___ effector B cell
20. ___ T cell

A. Begins its development in bone marrow, but matures in the thymus gland
B. Cells that have directly or indirectly descended from the same parent cell
C. A potent chemical that causes blood vessels to dilate and let protein pass through the vessel walls
D. Y-shaped immunoglobulin
E. A nonself marker
F. The progeny of turned-on B cells
G. A group of about fifteen proteins that participate in the inflammatory response
H. An altered secondary immune response to a substance that is normally harmless to other people
I. The basis for self-recognition at the cell surface
J. Principal perpetrator of phagocytosis

Chapter Objectives/Review Questions

This section lists general and detailed chapter objectives that can be used as review questions. You can make maximum use of these items by writing answers on a separate sheet of paper. To check for accuracy, compare your answers with information given in the chapter or glossary.

Page		Objectives/Questions
(676)	1.	Describe typical external barriers that organisms present to invading organisms.
(676)	2.	List and discuss four nonspecific defense responses that serve to exclude microbes from the body.
(677)	3.	Explain how the complement system is related to an inflammatory response.
(680)	4.	Understand how vertebrates (especially mammals) recognize and discriminate between self and nonself tissues.
(680–681)	5.	Distinguish the roles of T cells from the roles of B cells.
(681)	6.	List the three general types of cells that form the basis of the vertebrate immune system.
(681)	7.	Explain what is meant by *primary immune response* as contrasted with *secondary immune response*.
(682–685)	8.	Distinguish between the antibody-mediated response pattern and the cell-mediated response pattern.
(683; 688)	9.	Explain what monoclonal antibodies are and tell how they are currently being used in passive immunization and cancer treatment.
(685–687)	10.	Describe how recognition proteins and antibodies are made. State how they are used in immunity.
(686–687)	11.	Describe the clonal selection theory and tell what it helps to explain.
(688)	12.	Describe two ways that people can be immunized against specific diseases.
(689)	13.	Distinguish allergy from autoimmune disorder.
(689–691)	14.	Describe some examples of immune failures and identify as specifically as you can which weapons in the immunity arsenal failed in each case.
(690–691)	15.	Describe specifically how AIDS interferes with the human immune system.

Integrating and Applying Key Concepts

Suppose you wanted to get rid of forty-seven warts that you have on your hands by treating them with monoclonal antibodies. Outline the steps you would have to take.

Answers

Interactive Exercises

40-I. THREE LINES OF DEFENSE (pp. 674–676)
COMPLEMENT PROTEINS (p. 677)
INFLAMMATION (pp. 678–679)

1. mucous; 2. Lysozyme; 3. Gastric; 4. bacterial; 5. phagocytic; 6. clotting; 7. Phagocytic (Macrophage); 8. complement system; 9. histamine; 10. capillaries; 11. nonspecific; 12. specific; 13. lymphocytes; 14. immune; 15. secrete histamine and prostaglandins that change permeability of blood vessels in damaged or irritated tissues; 16. attack parasitic worms by secreting corrosive enzymes; 17. the most abundant white blood cells; they quickly phagocytize bacteria and reduce them to molecules that can be used for other purposes; 18. slow, "big eaters"; engulf and digest foreign agents, and clean up dead and damaged tissues.

40-II. THE IMMUNE SYSTEM (pp. 680–681)
CELL-MEDIATED RESPONSES (pp. 682–683)
ANTIBODY-MEDIATED RESPONSES (pp. 684–685)

1. nonspecific; 2. immune system; 3. MHC marker; 4. nonself; 5. lymphocytes; 6. B cell; 7. T cell; 8. thymus; 9. viruses; 10. cancer; 11. identity; 12. antigen; 13. antigen-presenting; 14. effector; 15. helper T; 16. cytotoxic T; 17. cell-mediated; 18. antibodies; 19. antibody-mediated; 20. memory; 21. antigen-presenting cells (macrophages); 22. virgin helper T cells; 23. memory T cells; 24. intracellular; 25. virgin B cells; 26. effector B cells; 27. antibodies; 28. Extracellular; 29. a; 30. d; 31. e; 32. b; 33. c; 34. natural killer; 35. perforins; 36. cell; 37. cytotoxic T; 38. helper T; 39. antigen-MHC;

40. macrophage; 41. antibody; 42. effector B; 43. B cell; 44. primary immune response; 45. memory cells; 46. antigens; 47. Cancer; 48. monoclonal (pure) antibodies; 49. hybrid; 50. IgA; 51. IgE; 52. IgG; 53. Ig M.

40-III. LYMPHOCYTE BATTLEGROUNDS (p. 685)
SPECIFICITY AND MEMORY (pp. 686–687)
IMMUNIZATION (p. 688)

1. immunization; 2. active; 3. primary immune response; 4. memory cells; 5. clonal selection theory; 6. nonself; 7. immunoglobulin; 8. Antibodies; 9. recombination; 10. antibodies; 11. antigen; 12. B; 13. memory cells; 14. effector cells; 15. a, c, d, e; 16. a, c, d; 17. a, c, d; 18. a, d, f; 19. a, d, f; 20. b, c, d; 21. a, d, f; 22. a, c, d, e; 23. a, c, d, e.

40-IV. ABNORMAL OR DEFICIENT IMMUNE RESPONSES (pp. 689–691)

1. Allergy; 2. Autoimmune disease; 3. Myasthenia gravis; 4. Rheumatoid arthritis; 5. human immunodeficiency virus; 6. male homosexuals; 7. retrovirus; 8. reverse transcriptase; 9. body fluids; 10. helper T (antigen-presenting); 11. macrophages; 12. 1 million; 13. 13 million; 14. 2 million; 15. 110.

Self-Quiz

1. d; 2. b; 3. b; 4. a; 5. a; 6. e; 7. e; 8. e; 9. a; 10. a; 11. H; 12. D; 13. E; 14. J; 15. B; 16. G; 17. C; 18. I; 19. F; 20. A.

41

RESPIRATION

Interactive Exercises

41-I. THE NATURE OF RESPIRATION (pp. 694–697)
MODES OF RESPIRATION (pp. 698–699)

Selected Italicized Words

"altitude sickness," "respiration," pressure gradient, "partial pressure," external gills, internal gills

Boldfaced, Page-Referenced Terms

(696) respiratory system _____

_____

(696) respiratory surface _____

_____

(697) ventilation _____

_____

(697) hemoglobin _____

_____

(698) integumentary exchange _____

_____

(698) tracheal respiration _____

_____

(698) gills _____

_____

(699) countercurrent flow _____

_____

(699) lungs _____

_____

Fill-in-the-Blanks

A(n) (1) _____ is an outfolded, thin, moist membrane endowed with blood vessels; it may be protected (as in fish) by bony covering or (as in aquatic insects) it may be naked. Gas transfer is enhanced by (2) _____ _____, in which water flows past the bloodstream in the opposite direction. Insects have (3) _____, which are chitin-lined air tubes leading from the body surface to the interior. At sea level, atmospheric pressure is approximately 760 mm Hg, and oxygen represents about (4) _____ percent of the total volume.

The energy to drive animal activities comes mainly from (5) _____ _____, which uses (6) _____ and produces (7) _____ _____ wastes. In a process called (8) _____, animals move (6) into their internal environment and give up (7) to the external environment.

All respiratory systems make use of the tendency of any gas to diffuse down its (9) _____ _____. Such a (9) exists between (6) in the atmosphere [(10) _____ pressure] and the metabolically active cells in body tissues [where (6) is used rapidly; pressure is (11) (choose one) ❑ highest ❑ lowest here]. Another (9) exists between (7) in body tissues, where (12) (choose one) ❑ high ❑ low pressure exists, and the atmosphere, with its (13) (choose one) ❑ higher ❑ lower amount of (7).

The more extensive the (14) _____ _____ of a respiratory surface membrane and the larger the differences in (15) _____ _____ across it, the faster a gas diffuses across the membrane. (16) _____ is an important transport pigment, each molecule of which can bind loosely with as many as four O_2 molecules in the lungs.

A(n) (17) _____ is an internal respiratory surface in the shape of a cavity or sac. In all lungs, (18) _____ carry gas to and from one side of the respiratory surface, and (19) _____ in blood vessels carries gas to and from the other side.

Labeling

Identify the numbered parts of the accompanying illustration, which shows the respiratory system of many fishes.

20. _____

21. _____

    _____

22. _____-

    _____

    _____

23. _____-

    _____

    _____

24. _____

25. _____

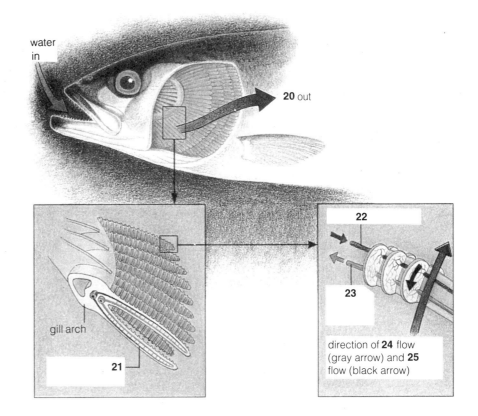

water in

20 out

gill arch

21

22

23

direction of **24** flow (gray arrow) and **25** flow (black arrow)

41-II. HUMAN RESPIRATORY SYSTEM (pp. 700–701)
VENTILATION (pp. 702–703)

Selected Italicized Words

nasal cavity, trachea, pleurisy, inhale, exhale, "tidal volume," "vital capacity," laryngitis

Boldfaced, Page-Referenced Terms

(700) pharynx _____

_____

(700) larynx _____

_____

(700) epiglottis _____

_____

(700) bronchus _____

_____

(701) diaphragm _____

_____

(701) bronchioles _____

_____

(701) alveolus, alveoli _____

_____

(703) vocal cords _____

_____

(703) glottis _____

_____

Fill-in-the-Blanks

During inhalation, the (1) _____ moves downward and flattens, and the (2) _____ _____

moves outward and upward; when these things happen, the chest cavity volume (3) (choose one)

❏ increases ❏ decreases, and the internal pressure (4) (choose one) ❏ rises ❏ drops ❏ stays the same. Every

time you take a breath, you are (5) _____ the respiratory surfaces of your lungs. The (6) _____

_____ surrounds each lung. In succession, air passes through the nasal cavities, pharynx, and

(7) _____, past the epiglottis into the (8) _____ (the space between the true vocal cords), into the

trachea, and then to the (9) _____, (10) _____, and alveolar ducts. Exchange of gases occurs

across the epithelium of the (11) _____.

Labeling

Identify each indicated part of the accompanying illustration.

12. _____ _____

13. _____

14. _____

15. _____

16. _____ _____

17. _____

18. _____

19. _____

20. _____ _____

21. _____ _____

22. _____ _____

23. _____

24. _____ (sing.),

_____ (plural)

25. _____

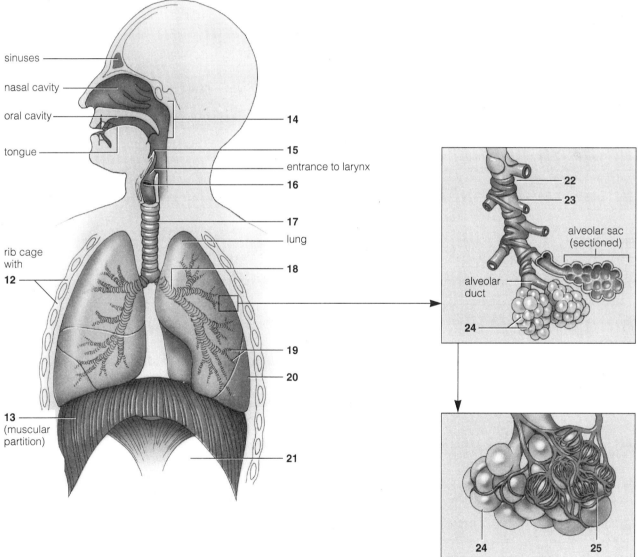

sinuses

nasal cavity

oral cavity

tongue

rib cage with

12

13 (muscular partition)

14

15

entrance to larynx

16

17

lung

18

19

20

21

22

23

alveolar sac (sectioned)

alveolar duct

24

24   **25**

Complete the Table

26. Complete the following table with the structures that carry out the functions listed.

Structure	Function
a.	Thin-walled sacs where O_2 diffuses into body fluids and CO_2 diffuses out
b.	Increasingly branched airways that connect the trachea and alveoli
c.	Muscle sheet that contracts during inhalation
d.	Airway where breathing is blocked while swallowing and where sound is produced
e.	Airway that enhances speech sounds; connects nasal cavity with larynx

41-III. GAS EXCHANGE AND TRANSPORT (pp. 704–707)
RESPIRATION IN UNUSUAL ENVIRONMENTS (pp. 708–711)

Selected Italicized Words

bronchitis, emphysema, "smoker's cough," hypoxia, hyperventilation, decompression sickness

Boldfaced, Page-Referenced Terms

(704) oxyhemoglobin _____

_____

(705) carbonic anhydrase _____

_____

(705) respiratory centers _____

_____

(709) myoglobin _____

_____

Fill-in-the-Blanks

Oxygen is said to exert a(n) (1) _____ _____ of $760 \times 21/100$ or 160 mm Hg. (2) _____ alone moves oxygen from the alveoli into the bloodstream, and it is enough to move (3) _____ _____ in the reverse direction. (4) _____ is the medical name for oxygen deficiency; it is characterized by faster breathing, faster heart rate, and anxiety at altitudes of 8,000 feet above sea level. About 70 percent of the carbon dioxide in the blood is transported as (5) _____. Without (6) _____, the plasma would be able to carry only about 2 percent of the oxygen that whole blood carries. When oxygen-rich blood reaches a(n) (7) _____ tissue capillary bed, oxygen diffuses outward, and carbon dioxide moves from tissues into the capillaries. When the (8) _____ _____ of carbon dioxide is lower in the alveoli than in the neighboring blood capillaries, carbonic acid dissociates to form water and carbon dioxide. The rate of breathing is governed by clusters of cells that make up a respiratory center in the (9) _____

_____, which monitors and coordinates signals coming in from arterial walls, from blood vessels, and from other brain regions. The respiratory center regulates contractions of the diaphragm and intercostal muscles associated with inhalation and exhalation.

(10) _____ is the distention of lungs and the loss of gas exchange efficiency such that running, walking, and even exhaling are painful experiences. At least 90 percent of all (11) _____ _____ deaths are the result of cigarette smoking; only about 10 percent of afflicted individuals will survive.

When a diver ascends, (12) _____ tends to move out of the tissues and into the bloodstream. If the ascent is too rapid, many bubbles of (12) collect at the (13) _____, hence the common name "the bends" for what is otherwise known as (14) _____ sickness.

Self-Quiz

___ 1. Most forms of life depend on _____ to obtain oxygen and eliminate carbon dioxide.
   a. active transport
   b. bulk flow
   c. diffusion
   d. osmosis
   e. muscular contractions

___ 2. _____ is the most abundant gas in Earth's atmosphere.
   a. Water vapor
   b. Oxygen
   c. Carbon dioxide
   d. Hydrogen
   e. Nitrogen

___ 3. With respect to respiratory systems, countercurrent flow is a mechanism that explains how _____.
   a. oxygen uptake by blood capillaries in the lamellae of fish gills occurs
   b. ventilation occurs
   c. intrapleural pressure is established
   d. sounds originating in the vocal cords of the larynx are formed
   e. all of the above

___ 4. _____ have the most efficient respiratory system.
   a. Amphibians
   b. Reptiles
   c. Birds
   d. Mammals
   e. Humans

___ 5. Immediately before reaching the alveoli, air passes through the _____.
   a. bronchioles
   b. glottis
   c. larynx
   d. pharynx
   e. trachea

___ 6. During inhalation, _____.
   a. the pressure in the thoracic cavity is less than the pressure within the lungs
   b. the pressure in the chest cavity is greater than the pressure within the lungs
   c. the diaphragm moves upward and becomes more curved
   d. the thoracic cavity volume decreases
   e. all of the above

___ 7. Hemoglobin _____.

    a. releases oxygen more readily in tissues with high rates of cellular respiration

    b. tends to release oxygen in places where the temperature is lower

    c. tends to hold on to oxygen when the pH of the blood drops

    d. tends to give up oxygen in regions where partial pressure of oxygen exceeds that in the lungs

    e. all of the above

___ 8. Oxygen moves from alveoli to the blood-stream _____.

    a. whenever the concentration of oxygen is greater in alveoli than in the blood

    b. by means of active transport

    c. by using the assistance of carbaminohemoglobin

    d. principally due to the activity of carbonic anhydrase in the red blood cells

    e. by all of the above

___ 9. Oxyhemoglobin releases O_2 when _____.

    a. carbon dioxide concentrations are high

    b. body temperature is lowered

    c. pH values are high

    d. CO_2 concentrations are low

    e. all of the above occur

___ 10. Nonsmokers live an average of _____ longer than people in their mid-twenties who smoke two packs of cigarettes each day.

    a. 6 months

    b. 1–2 years

    c. 3–5 years

    d. 7–9 years

    e. over 12 years

Matching

11. ___ bronchioles

12. ___ bronchitis

13. ___ carbonic anhydrase

14. ___ emphysema

15. ___ glottis

16. ___ hyperventilation

17. ___ intercostal muscles

18. ___ larynx

19. ___ myoglobin

20. ___ pharynx

21. ___ pleurisy

22. ___ rete mirabile

23. ___ tidal volume

24. ___ ventilation

25. ___ vital capacity

A. membrane that encloses human lung becomes inflamed and swollen; painful breathing generally results

B. an oxygen-binding protein associated with muscles

C. the amount of air inhaled and exhaled during normal breathing of a human at rest; generally about 500 ml

D. throat passageway that connects to both the respiratory tract below and the digestive tract

E. greatly increases gas concentrations in a swim bladder

F. inflammation of the two principal passageways that lead air into the human lungs

G. contract when air is leaving the lungs; relax when lungs are filling with air

H. the opening into the "voicebox"

I. finer and finer branchings that lead to alveoli

J. maximum volume of air that can move out of your lungs after a single, maximal inhalation

K. an enzyme that increases the rate of production of H_2CO_3 from CO_2 and H_2O

L. lungs have become distended and inelastic so that walking, running, and even exhaling are difficult

M. where sound is produced by vocal cords

N. movements that keep air or water moving across a respiratory surface

O. breathing much faster and deeper than normal in order to compensate for oxygen deficiency

Chapter Objectives/Review Questions

This section lists general and detailed chapter objectives that can be used as review questions. You can make maximum use of these items by writing answers on a separate sheet of paper. To check for accuracy, compare your answers with information given in the chapter or glossary.

Page	Objectives/Questions
(696–699)	1. Understand the behavior of gases and the types of respiratory surfaces that participate in gas exchange.
(696, 701–702, 705)	2. Understand how the human respiratory system is related to the circulatory system, to cellular respiration, and to the nervous system.
(698–699)	3. Define *countercurrent flow* and explain how it works. State where such a mechanism is found.
(696, 698)	4. Describe how incoming oxygen is distributed to the tissues of insects, and contrast this process with the process that occurs in mammals.
(700–701, 704–705)	5. List all the principal parts of the human respiratory system and explain how each structure contributes to transporting oxygen from the external world to the bloodstream.
(701–702)	6. Describe the relationship of the human lung to the pleural sac and to the thoracic cavity.
(704–705)	7. Explain why oxygen diffuses from the bloodstream into the tissues far from the lungs. Then explain why carbon dioxide diffuses into the bloodstream from the same tissues.
(704–705)	8. Explain why oxygen diffuses from alveolar air spaces, through interstitial fluid, and across capillary epithelium. Then explain why carbon dioxide diffuses in the reverse direction.
(705)	9. Describe what happens to carbon dioxide when it dissolves in water under conditions normally present in the human body.
(705)	10. List the structures involved in detecting carbon dioxide levels in the blood and in regulating the rate of breathing. Name the location of each structure.
(706)	11. Distinguish bronchitis from emphysema. Then explain how lung cancer differs from emphysema.
(708–711)	12. List some of the ways that respiratory systems are adapted to unusual environments.

Integrating and Applying Key Concepts

Consider the amphibians—animals that generally have aquatic larval forms (tadpoles) and terrestrial adults. Outline the respiratory changes that you think might occur as an aquatic tadpole metamorphoses into a land-going juvenile.

Answers

Interactive Exercises

41-I. THE NATURE OF RESPIRATION (pp. 694–697) MODES OF RESPIRATION (pp. 698–699)

1. gill; 2. countercurrent flow; 3. tracheas; 4. 21; 5. aerobic metabolism; 6. oxygen (O_2); 7. carbon dioxide (CO_2); 8. respiration; 9. pressure gradient; 10. high; 11. lowest; 12. high; 13. lower; 14. surface area; 15. partial pressure; 16. Hemoglobin; 17. lung; 18. airways; 19. blood; 20. water out; 21. blood vessel (in gill filament); 22. oxygen-poor blood; 23. oxygen-rich blood; 24. blood; 25. water.

41-II. HUMAN RESPIRATORY SYSTEM (pp. 700–701) VENTILATION (pp. 702–703)

1. diaphragm; 2. rib cage; 3. increases; 4. drops; 5. ventilating; 6. pleural sac; 7. larynx; 8. glottis; 9. bronchi; 10. bronchioles; 11. alveoli; 12. intercostal muscles; 13. diaphragm; 14. pharynx; 15. epiglottis; 16. vocal cords; 17. trachea; 18. bronchus; 19. bronchioles; 20. pleural membrane; 21. abdominal cavity; 22. smooth muscle; 23. bronchiole; 24. alveolus, alveoli; 25. capillary; 26. a. Alveoli; b. Bronchial tree; c. Diaphragm; d. Larynx; e. Pharynx.

41-III. GAS EXCHANGE AND TRANSPORT (pp. 704–707) RESPIRATION IN UNUSUAL ENVIRONMENTS (pp. 708–711)

1. partial pressure; 2. Diffusion; 3. carbon dioxide; 4. Hypoxia; 5. bicarbonate; 6. oxyhemoglobin (hemoglobin); 7. systemic (low-pressure); 8. partial pressure; 9. reticular formation (medulla oblongata); 10. Emphysema; 11. lung cancer; 12. N_2 (nitrogen gas); 13. joints; 14. decompression.

Self-Quiz

1. c; 2. e; 3. a; 4. c; 5. a; 6. a; 7. a; 8. a; 9. a; 10. d; 11. I; 12. F; 13. K; 14. L; 15. H; 16. O; 17. G; 18. M; 19. B; 20. D; 21. A; 22. E; 23. C; 24. N; 25. J.

42

DIGESTION AND HUMAN NUTRITION

Interactive Exercises

42-I. DIGESTIVE SYSTEMS (pp. 714–717)
FOOD PREPARATION AND STORAGE (pp. 718–719)

Selected Italicized Words

caries, gingivitis, periodontal disease, heartburn, peptic ulcer

Boldfaced, Page-Referenced Terms

(715) ruminants _____

_____

(715) nutrition _____

_____

(716) digestive system _____

_____

(716) incomplete digestive system _____

_____

(716) complete digestive system _____

_____

(716) motility _____

_____

(716) secretion _____

_____

(716) digestion _____

_____

(716) absorption _____

_____

(716) elimination _____

_____

(718) oral cavity _____

_____

(718) tongue _____

_____

(718) tooth _____

_____

(718) saliva _____

_____

(718) pharynx _____

_____

(718) esophagus _____

_____

(718) sphincter _____

_____

(718) stomach _____

_____

(718) gastric fluid _____

_____

(718) chyme _____

_____

Fill-in-the-Blanks

(1) _____ is a large concept that encompasses processes by which food is ingested, digested, absorbed, and later converted to the body's own (2) _____, lipids, proteins, and nucleic acids. A digestive system is some form of body cavity or tube in which food is reduced first to (3) _____ and then to small (4) _____. Digested nutrients are then (5) _____ into the internal environment. A(n) (6) _____ digestive system has only one opening, two-way traffic, and a highly branched gut cavity that serves both digestive and (7) _____ functions. A(n) (8) _____ digestive system has a tube or cavity with regional specializations and a(n) (9) _____ at each end. (10) _____ involves the muscular movement of the gut wall, but (11) _____ is the release into the lumen of enzyme fluids and other substances required to carry out digestive functions.

The human digestive system is a tube, 21 feet to 30 feet long in an adult, that has regions specialized for different aspects of digestion and absorption; they are, in order, the mouth, pharynx, esophagus, (12) _____, (13) _____ _____, large intestine, rectum, and (14) _____. Various (15) _____ structures secrete enzymes and other substances that are also essential to the breakdown and absorption of nutrients; these include the salivary glands, liver, gallbladder, and (16) _____. The (17) _____ system distributes nutrients to cells throughout the body. The (18) _____ system supplies oxygen to the cells so that they can oxidize the carbon atoms of food molecules, thereby changing them to the waste product, (19) _____ _____, which is eliminated by the same system. And if excess water, salts, and wastes accumulate in the blood, the (20) _____ system and skin will maintain the volume and composition of blood and other body fluids.

Saliva contains an enzyme, (21) _____ _____, that breaks down starch. Contractions force the larynx against a cartilaginous flap called the (22) _____, which closes off the trachea. The (23) _____ is a muscular tube that propels food to the stomach. The stomach stores and mixes food; its secretions dissolve and (24) _____ food, and it helps control the rate that food passes into the small intestine. (25) _____ are enzymes that work in the stomach to begin to break down proteins.

Labeling

Identify each numbered structure in the accompanying illustration.

26. _____ _____

27. _____ _____

28. _____

29. _____

30. _____

31. _____ _____

32. _____ _____

33. _____

34. _____

35. _____

36. _____

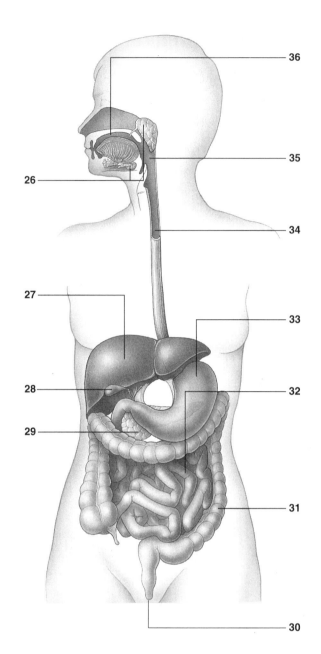

42-II. DIGESTION IN THE SMALL INTESTINE (pp. 720–721)
ABSORPTION FROM THE SMALL INTESTINE (p. 722)
FUNCTIONS OF THE LARGE INTESTINE (p. 723)
CONTROLS OVER THE SYSTEM (p. 723)

Selected Italicized Words

segmentation, peristalsis, appendicitis

Boldfaced, Page-Referenced Terms

(720) pancreas _____

_____

(720) liver _____

_____

(720) gallbladder _____

_____

(721) bile _____

_____

(721) emulsification _____

_____

(722) villus, villi _____

_____

(722) microvillus, microvilli _____

_____

(722) micelles _____

_____

(723) colon _____

_____

(723) appendix _____

_____

(723) bulk _____

_____

Complete the Table

1. Complete the following table by naming the organs described.

Organ	Main Functions
a.	mechanically breaks down food, mixes it with saliva
b.	moisten food; start polysaccharide breakdown; buffer acidic foods in mouth
c.	stores, mixes, dissolves food; kills many microorganisms; starts protein breakdown; empties in a controlled way
d.	digests and absorbs most nutrients
e.	produces enzymes that break down all major food molecules; produces buffers against hydrochloric acid from stomach
f.	secretes bile for fat emulsification; secretes bicarbonate, which buffers hydrochloric acid from stomach
g.	stores, concentrates bile from liver
h.	stores, concentrates undigested matter by absorbing water and salts
i.	controls elimination of undigested and unabsorbed residues

Fill-in-the-Blanks

Carbohydrates include sugars and (2) _____, the name commonly given to polysaccharides. Rice, cereal, pasta, bread, and white potatoes are composed of many polysaccharide molecules that are too large to be absorbed into the internal environment. If these foods are chewed thoroughly, (3) _____ _____ in the mouth digests them to the (4) _____ (double sugar) level. Because no carbohydrate digestion occurs in the (5) _____, if you gulped down your food, starch digestion would again begin in the (6) _____ _____ where (7) _____ produced by the pancreas would do what should have been done in the mouth. Digestion of (4) to monosaccharides (simple sugars) also occurs in the (8) _____ _____. The enzymes responsible are (9) _____ with names such as sucrase, lactase, and maltase.

Proteins are digested to protein fragments, beginning in the (10) _____ by (11) _____ secreted by the lining of the stomach. Protein fragments are subsequently digested to smaller protein fragments in the (12) _____ _____ by enzymes known as trypsin and chymotrypsin, produced by the (13) _____. Eventually the smaller protein fragments are digested to (14) _____ _____ by carboxypeptidase, produced by the pancreas, and aminopeptidase, produced by glands in the intestinal lining.

Another name for fat is triglycerides. (15) _____, produced by the pancreas but acting in the (16) _____ _____, breaks down one triglyceride molecule into three (17) _____ _____ molecules and one glycerol molecule. (18) _____, which is made by the liver, stored in the (19) _____, and does not contain digestive enzymes, emulsifies the fat droplet (converts it into small droplets coated with bile salts), thereby increasing the surface area of the substrate on which (20) _____ can act.

Nutrients are also mostly digested and absorbed in the (21) _____ _____. (18) is made by the liver, is stored in the gallbladder, and works in the (22) _____ _____. (23) _____-_____ is an example of an enzyme that is made by the pancreas but works in the small intestine to convert protein fragments to amino acids. (24) _____ _____ are made in the pancreas, but convert DNA and RNA into nucleotides in the small intestine. Any alternating progression of contracting and relaxing muscle movements along the length of a tube is known as (25) _____.

True-False

If the statement is true, write a T in the blank. If false, make it true by changing the underlined word.

_____ 26. Amylase digests starch, lipase digests lipids, and proteases break peptide bonds.

_____ 27. ATP is the end product of digestion.

_____ 28. The appendix has no known digestive functions.

_____ 29. Water and <u>sodium</u> <u>ions</u> are absorbed into the bloodstream from the lumen of the large intestine.

_____ 30. Fatty acids and monoglycerides recombine into fats inside epithelial cells lining the <u>colon</u>.

42-III. HUMAN NUTRITIONAL REQUIREMENTS (pp. 724–725)
VITAMINS AND MINERALS (pp. 726–727)
ENERGY AND BODY WEIGHT (pp. 728–729)

Selected Italicized Words

"food pyramid," "starvation," anorexia nervosa, bulimia, yo-yo dieting

Boldfaced, Page-Referenced Terms

(725) essential fatty acids _____

_____

(725) essential amino acids _____

_____

(725) net protein utilization, NPU _____

_____

(726) vitamins _____

_____

(726) minerals _____

_____

(728) kilocalories _____

_____

(728) obesity _____

_____

Complete the Table

1. Complete the following table by determining how many kilocalories the people described should take in daily, given the stated exercise level, in order to *maintain* their weight. Consult page 728 of the text.

Height	Age	Sex	Level of Physical Activity	Present Weight (lbs.)	Number of Kilocalories/Day
5'6"	25	Female	Moderately active	138	a.
5'10"	18	Male	Very active	145	b.
5'8"	53	Female	Not very active	143	c.

Fill-in-the-Blanks

People who are (2) _____ percent heavier than "ideal" are considered to be obese. (3) _____ _____ are the body's main source of energy; they should make up (4) _____ to _____ percent of the human daily caloric intake. (5) _____ and cholesterol are components of animal cell membranes.

Fat deposits are used primarily as (6) _____ _____, but they also cushion many organs and provide insulation. Lipids should constitute less than (7) _____ percent of the human diet. One tablespoon a day of polyunsaturated oil supplies all (8) _____ _____ _____ that the body cannot synthesize. (9) _____ are digested to twenty common amino acids, of which eight are (10) _____ (cannot be synthesized) and must be supplied by the diet. Animal proteins such as (11) _____ and (12) _____ contain high amounts of essential amino acids (that is, they are complete). (13) _____ are organic substances needed in small amounts to build enzymes or help them catalyze metabolic reactions. (14) _____ are inorganic substances needed for a variety of uses.

Related Problems

You are a 19-year-old male who is very sedentary (TV, sleep, and computers), with a 6'1", medium frame, and you weigh 195 pounds.

15. Use Fig. 42.14, page 728, to calculate the number of calories required to sustain your desired weight. Use the average of that range. Are you underweight, overweight, or just right? _____
16. How many calories are you allowed to ingest every day? _____

Use the new, improved food pyramid (page 724) to construct a one-day diet that would eventually allow you to reach that weight if you ate a similar diet every day. Place your choices in the table below as a diet for the person described above.

How many servings from each group below are you allowed to have daily?	What, specifically, could you choose to eat?
17. complex carbohydrates	17 b.
18. fruits	18 b.
19. vegetables	19 b.
20. dairy group	20 b.
21. assorted proteins	21 b.
22. the "sin" group at the top	22 b.

Fill-in-the-Blanks

Use the (23) _____ _____ diagram at the left, as revised in 1992, to devise a well-balanced diet for yourself. Group 1, the trapezoidal base, represents the group of complex (24) _____, which includes rice, pasta, cereal, and (25) _____. (26) (choose 1) ❐ 0 ❐ 2–3 ❐ 2–4 ❐ 3–5 ❐ 6–11 servings from this group are needed every day to supply energy and fiber. Group 2 represents the (27) _____ group. Use the choices in (26) to indicate the number of servings (28) _____ that are needed from this group each day. Group 3 is the (29) _____ group, from which (30) _____ servings are needed each day. Choices include mango, oranges, (31) _____, cantaloupe, pineapple, or 1 cup of fresh (32) _____. Group 4 includes foods that are a source of nitrogen: nuts, poultry, fish, legumes, and (33) _____. From this group, the body's (34) _____ and nucleic acids are constructed. (35) _____ servings are required every day because the human body cannot synthesize eight of the twenty essential (36) _____ _____ that are used to construct proteins and must get them in its food supply. The foods in Group 5, the (37) _____, yogurt, and cheese group, supply calcium and vitamins A, D, B$_2$, and B$_{12}$. You need (38) _____ servings every day. The foods in Group 6 provide extra calories but few vitamins and minerals; (39) _____ servings are needed every day.

42-IV. NUTRITION AND ORGANIC METABOLISM (pp. 730–731)

Short Answer

To answer the following questions, consult Table 42.1 and Fig. 42.15.

1. What is the pool of amino acids used for in the human body? _____

_____

2. Which breakdown products result from carbohydrate and fat digestion? _____

_____

3. Monosaccharides, free fatty acids, and monoglycerides all have three uses; identify them. _____

_____

Amino acid conversions in the liver form (4) _____, which is potentially toxic to cells; the liver imme-

diately converts this substance to (5) _____, a much less toxic waste product that is expelled by the

urinary system from the body. Beta cells of the pancreas secrete (6) _____, and alpha cells secrete

(7) _____, a hormone that commands liver cells to convert (8) _____ (a storage starch) into glu-

cose subunits. Under hypothalamic commands, the adrenal medulla begins secreting (9) _____ and

(10) _____, which stop (11) _____ synthesis in the liver, stop (12) _____ uptake in muscles,

and promote the shift from "burning" (13) _____ during cellular respiration to "burning" fats.

Self-Quiz

Multiple Choice

___ 1. The process that moves nutrients into the blood or lymph is _____.

    a. ingestion
    b. absorption
    c. assimilation
    d. digestion
    e. none of the above

___ 2. The enzymatic digestion of proteins begins in the _____.

    a. mouth
    b. stomach
    c. liver
    d. pancreas
    e. small intestine

___ 3. The enzymatic digestion of starches begins in the _____.

    a. mouth
    b. stomach
    c. liver
    d. pancreas
    e. small intestine

___ 4. The greatest amount of absorption of digested nutrients occurs in the

    _____.

    a. stomach
    b. pancreas
    c. liver
    d. colon
    e. small intestine

___ 5. Glucose moves through the membranes of the small intestine mainly by _____.

    a. peristalsis
    b. osmosis
    c. diffusion
    d. active transport
    e. bulk flow

___ 6. Which of the following is *not* found in bile?

    a. lecithin
    b. salts
    c. digestive enzymes
    d. cholesterol
    e. pigments

___ 7. The average American consumes approximately _____ pounds of sugar per year.

    a. 25
    b. 50
    c. 75
    d. 100
    e. 125

___ 8. Of the following, _____ has (have) the highest net protein utilization.

    a. milk
    b. eggs
    c. fish
    d. meat
    e. bread

___ 9. Obesity is defined as being _____ percent above the ideal weight, which has been defined by insurance companies.

a. 5
b. 10
c. 15
d. 20
e. 25

___ 10. One hour after a meal, the blood richest in nutrients would be in the _____.

a. abdominal aorta
b. hepatic portal vein
c. hepatic vein
d. pulmonary artery
e. vena cava

___ 11. The element needed by humans for blood clotting, nerve impulse transmission, and bone and tooth formation is _____.

a. magnesium
b. iron
c. calcium
d iodine
e. zinc

Matching

Match the best lettered item with its correct numbered item at the left.

12. __ anorexia nervosa

13. __ antioxidants

14. __ bulimia

15. __ complex carbohydrates

16. __ essential amino acids

17. __ essential fatty acids

18. __ mineral

19. __ rickets

20. __ scurvy

21. __ vitamin

A. Linoleic acid is one example
B. Phenylalanine, lysine, and methionine are three of eight
C. Combine with free radicals; counteract their destructive effects on DNA and cell membranes
D. Obsessive dieting and skewed perception of body weight
E. Vitamin C deficiency
F. Vitamin D deficiency in young children
G. Organic substances that help enzymes to do their jobs; required in small amounts for good health
H. Feasting followed by vomiting or taking laxatives
I. Inorganic substances required for good health
J. Long chains of simple sugars; in pasta and white potatoes

Chapter Objectives/Review Questions

This section lists general and detailed chapter objectives that can be used as review questions. You can make maximum use of these items by writing answers on a separate sheet of paper. To check for accuracy, compare your answers with information given in the chapter or glossary.

Page	Objectives/Questions
(716)	1. Distinguish between incomplete and complete digestive systems and tell which are characterized by (a) specialized regions, (b) two-way traffic, and (c) discontinuous feeding.
(716)	2. Define and distinguish among motility, secretion, digestion, and absorption.
(716–717)	3. List all parts (in order) of the human digestive system through which food actually passes. Then list the auxiliary organs that contribute one or more substances to the digestive process.
(718–721)	4. Explain how, during digestion, food is mechanically broken down. Then explain how it is chemically broken down.
(720, 722)	5. Describe how the digestion and absorption of fats differ from the digestion and absorption of carbohydrates and proteins.
(720)	6. Tell which foods undergo digestion in each of the following parts of the human digestive system and state what the food is broken into: oral cavity, stomach, small intestine, large intestine.

(720)   7. List the enzyme(s) that act in (a) the oral cavity, (b) the stomach, and (c) the small intestine. Then tell where each enzyme was originally made.

(720–722)   8. Describe the cross-sectional structure of the small intestine and explain how its structure is related to its function.

(722)   9. List the items that leave the digestive system and enter the circulatory system during the process of absorption.

(723)   10. State which processes occur in the colon (large intestine).

(724)   11. Reproduce from memory the food pyramid diagram as revised in 1992. Identify each of the six components, list the numerical range of servings permitted from each group, and also list some of the choices available.

(724)   12. Construct an ideal diet for yourself for one 24-hour period. Calculate the number of calories necessary to maintain your weight (see p. 728) and then use the food pyramid (p. 724) to choose exactly what to eat and how much.

(724–727)   13. Compare the contributions of carbohydrates, proteins, and fats with the contributions of vitamins and minerals to human nutrition.

(724–728)   14. Summarize current ideas for promoting health by eating properly.

(724–728)   15. Summarize the daily nutritional requirements of a 25-year-old woman who works at a desk job and exercises very little. State what she needs in energy, carbohydrates, proteins, and lipids, and name at least six vitamins and six minerals that she needs to include in her diet every day.

(726)   16. Name five minerals that are important in human nutrition and state the specific role of each.

(726–727)   17. Distinguish vitamins from minerals, and state what is meant by net protein utilization.

(730–731)   18. Explain how the human body manages to meet the energy and nutritional needs of the various body parts even though the person may be feasting sometimes and fasting at other times.

Integrating and Applying Key Concepts

Suppose you could not eat solid food for two weeks and you had only water to drink. List in correct sequential order the measures your body would take to try to preserve your life. Mention the command signals that are given as one after another critical point is reached, and tell which parts of the body are the first and the last to make up for the deficit.

Answers

Interactive Exercises

42-I. DIGESTIVE SYSTEMS (pp. 714–717)
FOOD PREPARATION AND STORAGE (pp. 718–719)

1. Nutrition; 2. carbohydrates; 3. particles; 4. molecules; 5. absorbed; 6. incomplete; 7. circulatory; 8. complete; 9. opening; 10. Motility; 11. secretion; 12. stomach; 13. small intestine; 14. anus; 15. accessory; 16. pancreas; 17. circulatory; 18. respiratory; 19. carbon dioxide (CO_2); 20. urinary; 21. salivary amylase; 22. epiglottis; 23. esophagus; 24. degrade (digest); 25. Pepsin; 26. salivary glands; 27. liver; 28. gallbladder; 29. pancreas; 30. anus; 31. large intestine; 32. small intestine; 33. stomach; 34. esophagus; 35. pharynx; 36. mouth.

42-II. DIGESTION IN THE SMALL INTESTINE (pp. 720–721)
ABSORPTION FROM THE SMALL INTESTINE (p. 722)
FUNCTIONS OF THE LARGE INTESTINE (p. 723)
CONTROLS OVER THE SYSTEM (p. 723)

1. a. Mouth; b. Salivary glands; c. Stomach; d. Small intestine; e. Pancreas; f. Liver; g. Gallbladder; h. Large intestine; i. Rectum; 2. starches; 3. salivary amylase; 4. disaccharide; 5. stomach; 6. small intestine; 7. amylase; 8. small intestine; 9. disaccharidases; 10. stomach; 11. pepsins; 12. small intestine; 13. pancreas; 14. amino acids; 15. Lipase; 16. small intestine; 17. fatty acid; 18. Bile; 19. gallbladder; 20. lipase; 21. small intestine; 22. small intestine; 23. Carboxypeptidase; 24. Pancreatic nucleases; 25. peristalsis; 26. T; 27. cellular respiration; 28. T; 29. T; 30. small intestine.

42-III. HUMAN NUTRITIONAL REQUIREMENTS (pp. 724–725)
VITAMINS AND MINERALS (pp. 726–727)
ENERGY AND BODY WEIGHT (pp. 728–729)

1. a. 2,070; b. 2,900; c. 1,230; 2. 25; 3. Complex carbohydrates; 4. 50 to 60; 5. Phospholipids; 6. energy reserves; 7. 30; 8. essential fatty acids; 9. Proteins; 10. essential; 11. milk (eggs); 12. eggs (milk); 13. Vitamins; 14. Minerals; 15. Consulting Fig. 42.14 (men's column, 6'1",

medium frame) yields 160–174 as the ideal weight range. The average is 160 + 174 = 334 ÷ 2 = **167** pounds, which is 195 – **167** = 28 pounds overweight; 16. Multiply 167 times 10 (see p. 728) to obtain 1670 kilocalories (the daily number of calories that *maintains* weight in the correct size range). The excess 28 pounds should be lost gradually by adopting an everyday exercise program that over many months would gradually eliminate the excess kilocalories that are stored mostly in the form of fat; The smallest range of serving sizes shown in Figure 42.12 will help keep the total caloric intake to about 1600 kcal:

17. a. 6 servings b. bread, cereal, rice , pasta
18. a. 2 servings b. fruits
19. a. 3 servings b. vegetables
20. a. 2 servings b. milk, yogurt, or cheese
21. a. 2 servings b. legumes, nuts, poultry, fish, or meats
22. a. Scarcely any; b. added fats and simple sugars
23.–38. Choose from Figure 42.12.
23. food pyramid; 24. carbohydrates; 25. bread; 26. 6–11; 27. vegetable; 28. 3–5; 29. fruit; 30. 2–4; 31. apples; 32. berries; 33. meat; 34. proteins; 35. 2–3; 36. amino acids; 37. milk; 38. 2–3; 39. 0.

42-IV. NUTRITION AND ORGANIC METABOLISM (pp. 730–731)

1. constructing hormones, nucleotides, proteins, and enzymes; 2. monosaccharides, free fatty acids, and glycerol; 3. The three uses are (a) to construct components of cells, storage forms (such as glycogen), and specialized derivatives, such as steroids and acetylcholine; (b) to convert to amino acids as needed; and (c) to serve as a source of energy; 4. ammonia; 5. urea; 6. insulin; 7. glucagon; 8. glycogen; 9. epinephrine (norepinephrine); 10. norepinephrine (epinephrine); 11. glycogen; 12. glucose; 13. glucose.

Self-Quiz

1. b; 2. b; 3. a; 4. e; 5. d; 6. c; 7. e; 8. b; 9. e; 10. b; 11. c; 12. D; 13. C; 14. H; 15. J; 16. B; 17. A; 18. I; 19. F; 20. E; 21. G.

43

WATER-SOLUTE BALANCE AND TEMPERATURE CONTROL

Interactive Exercises

43-I. MAINTAINING THE EXTRACELLULAR FLUID (pp. 734–737) KIDNEY STRUCTURE AND FUNCTION (pp. 738–739)

Selected Italicized Words

interstitial fluid, blood, glomerular capillaries, peritubular capillaries

Boldfaced, Page-Referenced Terms

(736) extracellular fluid _____

_____

(736) urinary system _____

_____

(736) urinary excretion _____

_____

(737) kidneys _____

_____

(737) urine _____

_____

(737) ureter _____

_____

(737) urinary bladder _____

_____

(737) urethra _____

_____

(738) nephrons _____

_____

(738) glomerulus _____

_____

(738) Bowman's capsule _____

_____

(738) proximal tubule _____

_____

(738) loop of Henle _____

_____

(738) distal tubule _____

_____

(739) filtration _____

_____

(739) reabsorption _____

_____

(739) secretion _____

_____

Fill-in-the-Blanks

The body gains water by absorbing it from the slurry in the lumen of the small intestine and from

(1) _____, during condensation reactions. The mammalian body loses water mostly by excretion of

(2) _____, evaporation through the skin and (3) _____ _____, elimination of feces from the

gut, and (4) _____ as the body is cooled. (5) _____ behavior, in which the brain compels the indi-

vidual to seek liquids, influences the gain of water.

The body gains solutes by absorption of substances from the gut, by the secretion of hormones and

other substances, and by (6) _____, which produces CO_2 and other waste products of degradative

reactions. Besides CO_2, there are several major metabolic wastes that must be eliminated: (7) _____, formed by deamination reactions; (8) _____, which is produced in the liver during reactions that link two ammonia molecules to CO_2 and release a molecule of water; and (9) _____ _____, which is formed in reactions that break down nucleic acids.

Labeling

Identify each indicated part of the accompanying illustrations.

10. _____

11. _____

12. _____ _____

13. _____

14. _____

15. _____

16. _____

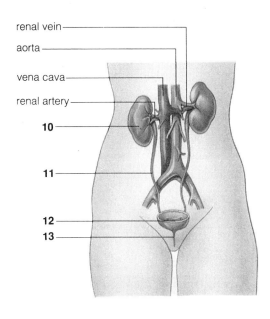

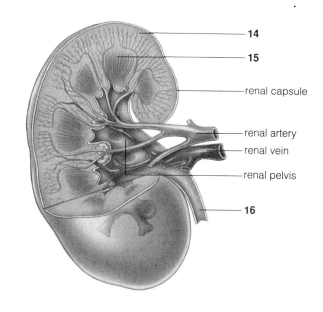

Fill-in-the-Blanks

In mammals, urine formation occurs in a pair of (17) _____. Each contains about a million tubelike blood-filtering units called (18) _____. The function of (17) depends on intimate links between the (18) and the (19) _____. In every (18), blood flows from a(n) (20) _____ into a set of capillaries inside the (21) _____ _____, then into a second set of capillaries that thread around the tubular parts of the nephron, then back to the bloodstream, leaving the kidney.

Urine composition and volume depend on three processes: *filtration* of blood at the (23) _____ of a nephron, with (24) _____ _____ providing the force for filtration; *reabsorption*, in which water and (25) _____ move out of tubular parts of the nephron and back into adjacent (26) _____ _____; and (27) _____, in which excess ions and a few foreign substances move out of those capillaries and back into the nephron so that they are disposed of in the urine. (28) _____ carry urine away from the kidney to the (29) _____ _____, where it is stored until it is released via a tube called the (30) _____, which carries urine to the outside.

By adjusting the blood's (22) _____ and composition, kidneys help maintain conditions in the extracellular fluid.

43-II. URINE FORMATION (pp. 740–741)
ACID-BASE BALANCE (p. 742)
KIDNEY MALFUNCTIONS (p. 742)
ON FISH, FROGS, AND KANGAROO RATS (p. 743)

Selected Italicized Words

"thirst," neutralize, eliminate, hypertension, kidney stones, hemodialysis, peritoneal dialysis

Boldfaced, Page-Referenced Terms

(741) ADH, antidiuretic hormone _____

_____

(741) aldosterone _____

_____

(742) acid-base balance _____

_____

(742) bicarbonate–carbon dioxide buffer system _____

_____

(742) kidney dialysis machine _____

_____

True-False

If false, explain why.

___ 1. When the body rids itself of excess water, urine becomes more dilute.

___ 2. Water reabsorption into capillaries is achieved by diffusion and active transport.

Fill-in-the-Blanks

Two hormones, ADH and (3) _____, adjust the reabsorption of water and (4) _____ along the distal tubules and collecting ducts. An increase in the secretion of aldosterone causes (5) (choose one) ❏ more ❏ less sodium to be excreted in the urine. When the body cannot rid itself of excess sodium, it inevitably retains excess water, and this leads to a rise in (6) _____ _____. Abnormally high blood pressure is called (7) _____; it can damage the kidneys, vascular system, and brain. One way to control hypertension is to restrict the intake of (8) _____ _____. Increased secretion of (9) _____ enhances water reabsorption at distal tubules and collecting ducts when the body must conserve water. When excess water must be excreted, ADH secretion is (10) (choose one) ❏ stimulated ❏ inhibited.

Labeling

Identify each indicated part of the accompanying illustration.

11. _____ _____

12. _____ _____

13. _____ _____

14. _____ _____

15. _____ _____

16. _____ _____

17. _____ _____

_____

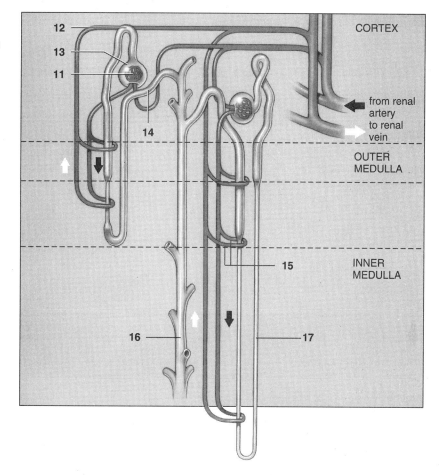

Fill-in-the-Blanks

The (18) _____ control the acid-base balance of body fluids by controlling the levels of dissolved ions, especially (19) _____ ions. The extracellular pH of humans must be maintained between 7.37 and (20) _____. (21) (choose one) ❑ Acids ❑ Bases lower the pH and (22) (choose one) ❑ acids ❑ bases raise it. If you were to drink a gallon of orange juice, the pH would be (23) (choose one) ❑ raised ❑ lowered, but the effect is minimized when excess (24) _____ ions are neutralized by (25) _____ ions in the bicarbonate–carbon dioxide buffer system. Only the (26) _____ system eliminates excess H^+ and restores buffers. Desert-dwelling kangaroo rats have very long (27) _____ _____ _____ so that nearly all (28) _____ that reaches their very long collecting ducts is reabsorbed. In freshwater, bony fishes and amphibians tend to gain (29) _____ and lose (30) _____; they produce (31) (choose one) ❑ very dilute ❑ very concentrated urine.

43-III. MAINTAINING THE BODY'S CORE TEMPERATURE (pp. 744–745)
CLASSIFICATION OF ANIMALS BASED ON TEMPERATURE (pp. 746–747)
TEMPERATURE REGULATION IN MAMMALS (pp. 748–749)

Selected Italicized Words

hypothermia, frostbite, "panting," hyperthermia, fever

Boldfaced, Page-Referenced Terms

(744) core temperature _____

_____

(744) radiation _____

_____

(744) conduction _____

_____

(744) convection _____

_____

(745) evaporation _____

_____

(746) ectotherms _____

_____

(746) behavioral temperature regulation _____

_____

(746) endotherms _____

_____

(747) heterotherms _____

_____

(748) peripheral vasoconstriction _____

_____

(748) pilomotor response _____

_____

(748) shivering _____

_____

(748) nonshivering heat production _____

_____

(749) peripheral vasodilation _____

_____

(749) evaporative heat loss _____

_____

Fill-in-the-Blanks

In the brain of mammals, the (1) _____ is the seat of temperature control. Thermoreceptors located deep in the body are called (2) _____ thermoreceptors. The (3) _____ _____ creates an insulating layer of still air when smooth muscles that erect hairs or feathers contract. (4) _____ _____

is a response to cold stress in which the bloodstream's convective delivery of heat to the body's surface is reduced. A drop in body temperature below tolerance levels is referred to as (5) _____.

True-False

If false, explain why.

___ 6. Jackrabbits are endotherms.

___ 7. When the core temperature of the human body is about 86°F, consciousness is lost and heart muscle action becomes irregular.

___ 8. When the core temperature of the human body reaches 77°F, ventricular fibrillation sets in and death soon follows.

Matching

Choose the most appropriate answer to match with the following terms.

9. ___ conduction
10. ___ convection
11. ___ ectotherm
12. ___ endotherm
13. ___ evaporation
14. ___ heterotherm
15. ___ radiation

A. body temperature determined more by heat exchange with the environment than by metabolic heat
B. heat transfer by heat-bearing air or water currents moving away from or toward a body
C. body temperature determined largely by metabolic activity and by precise controls over heat produced and heat lost
D. direct transfer of heat energy between two objects in direct contact with each other
E. the emission of energy in the form of infrared or other wavelengths that are converted to heat by the absorbing body
F. body temperature fluctuating at some times and heat balance controlled at other times
G. in changing from the liquid state to the gaseous state, the energy required is supplied by the heat content of the liquid

---

Self-Quiz

___ 1. The most toxic waste product of metabolism is _____.
   a. water
   b. uric acid
   c. urea
   d. ammonia
   e. carbon dioxide

___ 2. An entire subunit of a kidney that purifies blood and restores solute and water balance is called a _____.
   a. glomerulus
   b. loop of Henle
   c. nephron
   d. ureter
   e. none of the above

___ 3. In humans, the thirst center is located in the _____.
   a. adrenal cortex
   b. thymus
   c. heart
   d. adrenal medulla
   e. hypothalamus

___ 4. The longer the _____, the greater an animal's capacity to conserve water and to concentrate solutes to be excreted in the urine.
   a. loop of Henle
   b. proximal tubule
   c. ureter
   d. Bowman's capsule
   e. collecting tubule

___ 5. During reabsorption, sodium ions cross the proximal tubule walls into the interstitial fluid principally by means of _____.

a. phagocytosis
b. countercurrent multiplication
c. bulk flow
d. active transport
e. all of the above

___ 6. Filtration of the blood in the kidney takes place in the _____.

a. loop of Henle
b. proximal tubule
c. distal tubule
d. Bowman's capsule
e. all of the above

___ 7. _____ primarily controls the concentration of solutes in urine.

a. Insulin
b. Glucagon
c. Antidiuretic hormone
d. Aldosterone
e. Epinephrine

___ 8. Hormonal control over excretion primarily affects _____.

a. Bowman's capsules
b. distal tubules
c. proximal tubules
d. the urinary bladder
e. loops of Henle

___ 9. The last portion of the excretory system passed by urine before it is eliminated from the body is the _____.

a. renal pelvis
b. bladder
c. ureter
d. collecting ducts
e. urethra

___ 10. Normally, the extracellular pH of the human body must be maintained between _____ and _____; only the urinary system eliminates excess _____ and restores _____.

a. 6.45–7.30; NH_4^+, urea
b. 7.37–7.45; H^+, buffers
c. 7.50–7.85; H^+, glucose
d. 7.90–8.30; NH_4^+, urea
e. 8.15–8.35; OH^-, glucose

Chapter Objectives/Review Questions

This section lists general and detailed chapter objectives that can be used as review questions. You can make maximum use of these items by writing answers on a separate sheet of paper. To check for accuracy, compare your answers with information given in the chapter or glossary.

Page Objectives/Questions

(734–736) 1. List some of the factors that can change the composition and volume of body fluids.
(736) 2. List three soluble by-products of animal metabolism that are potentially toxic.
(737) 3. List successively the parts of the human urinary system that constitute the path of urine formation and excretion.
(739–740) 4. Locate the processes of filtration, reabsorption, and tubular secretion along a nephron and tell what makes each process happen.
(740–741) 5. State explicitly how the hypothalamus, posterior pituitary, and distal tubules of the nephrons are interrelated in regulating water and solute levels in body fluids.
(742) 6. Describe the role of the kidney in maintaining the pH of the extracellular fluids between 7.37 and 7.45.
(742) 7. List two kidney disorders and explain what can be done if kidneys become too diseased to work properly.
(746) 8. Distinguish between ectotherms and endotherms and give two examples of each group.
(746) 9. List the ways in which ectotherms are disadvantaged by not being able to maintain a particular body temperature. Describe the things ectotherms can do to lessen their vulnerability.
(748–749) 10. Explain how endotherms maintain their body temperature when environmental temperatures fall.

(748–749) 11. Define *hypothermia* and state the situations in which a human might experience the disorder.

(749) 12. Explain how endotherms maintain their body temperature when environmental temperatures rise 3 degrees to 4 degrees Fahrenheit above standard body temperature.

Crossword Puzzle—Urinary Systems

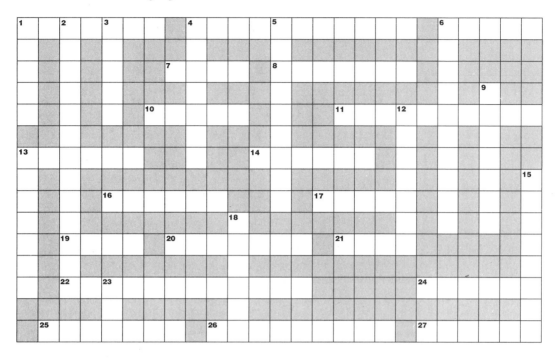

ACROSS

1. Connects urinary bladder to exterior.
4. _____ enhances sodium reabsorption and is produced by the adrenal cortex.
6. Extracellular _____ bathes the body's cells.
7. _____ acid is the least toxic nitrogenous waste and is constructed from nucleic acid breakdown.
8. Bowman's _____ encloses the glomerulus.
10. _____ is an enzyme secreted by kidney cells that detaches part of a protein circulating in the blood so that the new protein can be made into a hormone that acts on the adrenal cortex.
11. A cluster of capillaries enclosed by a Bowman's capsule that filters blood.
13. Connects a kidney to the urinary bladder.
14. An organ that adjusts the volume and composition of blood and helps maintain the composition of the extracellular fluid.
16. A small tube.
17. Relating to kidney function.
19. _____ of Henle.
20. Most toxic nitrogenous waste.
21. Nitrogenous waste formed in the liver and relatively harmless.
22. Water and solutes move out of the nephron tubule then into adjacent capillaries.
24. The urinary _____ includes three organs and three larger waste-containing tubes.

25. More than a million of these units are packed inside each fist-sized kidney.
26. The elimination of fluid wastes.
27. The _____ tubule is the part of the nephron that is farthest from the Bowman's capsule.

DOWN

1. Product of kidneys.
2. _____ fluid bathes the body's cells.
3. Loop of _____.
4. _____ glands are perched on top of the kidneys.
5. _____ moves excess H⁺ and a few other substances by active transport from the capillaries into the cells of the nephron wall and then into the urine.
6. Water and small-molecule solutes are forced from the blood into the Bowman's capsule.
9. Urinary _____ stores urine.
12. The middle region of the kidney.
13. _____ excretion is a process that dumps excess mineral ions and metabolic wastes.
15. The _____ tubule is nearest to its Bowman's capsule.
18. The outer region of kidney, adrenal gland, or brain.
23. The "water conservation" hormone produced by the posterior pituitary.

Integrating and Applying Key Concepts

The hemodialysis machine used in hospitals is expensive and time-consuming. So far, artificial kidneys capable of allowing people who have nonfunctional kidneys to purify their blood by themselves, without having to go to a hospital or clinic, have not been developed. Which aspects of the hemodialysis procedure do you think have presented the most problems in development of a method of home self-care? If you had an unlimited budget and were appointed head of a team to develop such a procedure and its instrumentation, what strategy would you pursue?

Answers

Interactive Exercises

43-I. MAINTAINING THE EXTRACELLULAR FLUID
(pp. 736–737)
KIDNEY STRUCTURE AND FUNCTION
(pp. 738–739)
1. metabolism; 2. urine; 3. respiratory surfaces; 4. sweating; 5. Thirst; 6. metabolism; 7. ammonia; 8. urea; 9. uric acid; 10. kidney; 11. ureter; 12. urinary bladder; 13. urethra; 14. cortex; 15. medulla; 16. ureter; 17. kidneys; 18. nephrons; 19. bloodstream; 20. arteriole; 21. Bowman's capsule; 22. glomerulus; 23. blood pressure; 24. solutes; 25. blood capillaries (peritubular capillaries); 26. secretion; 27. Ureters; 28. urinary bladder; 29. urethra; 30. volume.

43-II. URINE FORMATION (pp. 740–741)
ACID-BASE BALANCE (p. 742)
KIDNEY MALFUNCTIONS (p. 742)
ON FISH, FROGS, AND KANGAROO RATS
(p. 743)
1. T; 2. T; 3. aldosterone; 4. sodium; 5. less; 6. blood pressure; 7. hypertension; 8. table salt (sodium chloride); 9. ADH; 10. inhibited; 11. glomerular capillaries; 12. proximal tubule; 13. Bowman's capsule; 14. distal tubule; 15. peritubular capillaries; 16. collecting duct; 17. loop of Henle; 18. kidneys; 19. H^+; 20. 7.45; 21. Acids; 22. bases; 23. lowered; 24. H^+; 25. bicarbonate (HCO_3^-); 26. urinary; 27. loops of Henle; 28. water; 29. water; 30. solutes; 31. very dilute.

43-III. MAINTAINING THE BODY'S CORE TEMPERATURE (pp. 744–745)
CLASSIFICATION OF ANIMALS BASED ON TEMPERATURE (pp. 746–747)
TEMPERATURE REGULATION IN MAMMALS
(pp. 748–749)
1. hypothalamus; 2. core; 3. pilomotor response; 4. Peripheral vasoconstriction; 5. hypothermia; 6. T; 7. T; 8. T; 9. D; 10. B; 11. A; 12. C; 13. G; 14. F; 15. E.

Self-Quiz
1. d; 2. c; 3. e; 4. a; 5. d; 6. d; 7. d; 8. b; 9. e; 10. b.

Crossword Puzzle

44

PRINCIPLES OF REPRODUCTION AND DEVELOPMENT

Interactive Exercises

44-I. THE BEGINNING: REPRODUCTIVE MODES (pp. 752–755)
STAGES OF DEVELOPMENT (pp. 756–757)

Selected Italicized Words

reproductive timing, oviparous, viviparous, ovoviviparous

Boldfaced, Page-Referenced Terms

(754) sexual reproduction _____

_____

(754) asexual reproduction _____

_____

(755) yolk _____

_____

(756) embryo _____

_____

(756) gamete formation _____

_____

(756) fertilization _____

_____

(756) cleavage _____

_____

(756) gastrulation _____

_____

(756) ectoderm _____

_____

(756) endoderm _____

_____

(756) mesoderm _____

_____

(756) organ formation _____

_____

(756) tissue specialization _____

_____

Fill-in-the-Blanks

New sponges budding from parent sponges and a flatworm dividing into two flatworms represent examples of (1) _____ reproduction. This type of reproduction is useful when gene-encoded traits are strongly adapted to a limited set of (2) _____ conditions. Asexual processes of reproduction include (3) _____, which is common in animals such as *Hydra* and other cnidarians. Separation into male and female sexes requires special reproductive structures, control mechanisms, and behaviors; this cost is offset by a selective advantage: (4) _____ in traits among the offspring.

(5) _____ _____ is considered the first stage of animal development. Rich stores of substances become assembled in localized regions of the (6) _____ cytoplasm. When sperm and egg unite and their DNA mingles and is reorganized, the process is referred to as (7) _____. At the end of fertilization, a(n) (8) _____ is formed. (9) _____ includes the repeated mitotic divisions of a zygote that segregate the egg cytoplasm into a cluster of cells; the entire cluster is known as a (10) _____. (11) _____ is the process that arranges cells into three germ layers. Ectoderm eventually will give rise to skin epidermis and the (12) _____ system; endoderm forms the inner lining of the (13) _____ and associated digestive glands. Mesoderm forms the circulatory system, the (14) _____, and the muscles. (15) _____ _____ activated at fertilization direct the initial stages of development until gastrulation occurs.

Copperheads show (16) _____: Fertilization is internal, the fertilized eggs develop inside the mother's body without additional nourishment, and the young are born live. Birds show (17) _____: Eggs with large yolk reserves are released from and develop outside the mother's body.

Each stage of (18) _____ development builds on structures that were formed during the stage preceding it. (19) _____ cannot proceed properly unless each stage is successfully completed before the next begins.

True-False

If false, explain why.

__ 20. In the chick during the first few mitotic divisions of cleavage, cell membranes totally segregate the yolk and albumin into different daughter cells.

__ 21. Sperm penetration into the cytoplasm of the egg brings about specific structural changes and chemical reactions.

__ 22. Most animals reproduce sexually.

__ 23. Sexual reproduction is less advantageous in predictable environments; asexual reproduction is more advantageous in predictable environments.

__ 24. Nearly all land-dwelling animals depend on internal fertilization to help assure the survival of their offspring.

__ 25. Gastrulation precedes organ formation.

Sequence

Arrange the following events in correct chronological sequence. Write the letter of the first step next to 26, the letter of the second step next to 27, and so on.

26. ___ A. Gastrulation
27. ___ B. Fertilization
28. ___ C. Cleavage
29. ___ D. Growth, tissue specialization
30. ___ E. Organ formation
31. ___ F. Gamete formation

32. Complete the table below by entering the correct germ layer (ectoderm, mesoderm, or endoderm) that forms the tissues and organs listed.

Tissues/Organs	Germ Layer
Muscle, circulatory organs	a.
Nervous tissues	b.
Inner lining of the gut	c.
Circulatory organs (blood vessels, heart)	d.
Outer layer of the integument	e.
Reproductive and excretory organs	f.
Organs derived from the gut	g.
Most of the skeleton	h.
Connective tissues of the gut and integument	i.

44-II. A VISUAL TOUR OF FROG AND CHICK DEVELOPMENT (pp. 758–759)
EARLY MARCHING ORDERS (pp. 760–761)
EMERGENCE OF THE EARLY EMBRYO (pp. 762–763)

Selected Italicized Words

"*maternal instructions,*" *animal pole, vegetal pole,* "*inner cell mass,*" *gastrula*

Boldfaced, Page-Referenced Terms

(760) oocyte _____

_____

(760) gray crescent _____

_____

(762) cytoplasmic localization _____

_____

(762) blastula _____

_____

True-False

If false, explain why.

___ 1. During gastrulation, <u>maternal</u> controls over gene activity are activated and begin the process of differentiation in each cell's nucleus.

___ 2. In complex eukaryotes, development until gastrulation is governed by <u>DNA in the nucleus of the zygote</u>.

___ 3. In a developing chick embryo, the heart begins to beat at some time between <u>30 and 36 hours</u> after fertilization.

Fill-in-the-Blanks

The third stage of animal development, (4) _____, is characterized by the subdividing and compartmentalizing of the zygote; no growth occurs at this stage, and usually a hollow ball of cells, the (5) _____, is formed. The fourth stage, (6) _____, is concerned with the formation of ectoderm, mesoderm, and endoderm, the (7) _____ layers of the embryo; at the end of this stage, the (8) _____ is formed.

Much of the information that determines how structures will be spatially organized in the embryo begins with the distribution of (9) _____ _____ in the oocyte. As cleavage membranes divide up the cytoplasm, cytoplasmic determinants become localized in different daughter cells; this process, called (10) _____ _____, helps seal the developmental fate of the descendants of those cells.

In amphibian eggs, sperm penetration on one side of an egg causes pigment granules on the opposite side of the egg to flow toward the (11) _____ _____. A lightly pigmented area called the (12) _____ _____ results. It is a visible marker of the site where the (13) _____ _____ will be established and where gastrulation will begin.

Cleavage of a frog's zygote is total because there is so little yolk that cleavage membranes can subdivide the entire cytoplasmic mass; in the chick, however, there is so much yolk that cleavage membranes cannot subdivide the entire mass. Cleavage is therefore said to be incomplete, and the chick grows from a primitive streak on the surface of the (14) _____ mass into a chick embryo complete with wing and leg buds and beating heart during the first (15) _____ days.

44-III. FORMATION OF SPECIALIZED TISSUES AND ORGANS (pp. 764–765)
PATTERN FORMATION (pp. 766–767)
FROM THE EMBRYO ONWARD (pp. 768–769)

Selected Italicized Words

nonidentical twins, active cell migration, incomplete metamorphosis, complete metamorphosis

Boldfaced, Page-Referenced Terms

(764) cell differentiation _____

(764) identical twins _____

_____

(764) morphogenesis _____

_____

(764) neural plate _____

_____

(764) neural tube _____

_____

(765) localized growth _____

_____

(765) controlled cell death _____

_____

(766) pattern formation _____

_____

(766) embryonic induction _____

_____

(766) morphogens _____

_____

(767) homeobox genes _____

_____

(768) molting _____

_____

(768) metamorphosis _____

_____

(768) aging _____

_____

(768) limited division potential _____

_____

Fill-in-the-Blanks

Through (1) _____ _____, a single fertilized egg gives rise to an assortment of different types of specialized cells; these differentiated cells have the same number and same kinds of (2) _____ because they are all descended by mitosis from the same zygote. Through gene controls, however, (3) _____ are placed on which genes may be expressed (translated) in a given cell. (4) _____ involves the growth, shaping, and spatial coordination necessary to form functional body units.

In (5) _____ _____ _____, cells move about by means of (6) _____, which are temporary projections from the main cell body. They move in response to chemical gradients, a behavior called (7) _____. These gradients are created when different cells release specific substances. Cells also move in response to (8) _____ cues provided by recognition proteins on other cell surfaces.

Morphogenesis depends on (9) _____ _____, which contributes to changes in the sizes, shapes, and proportions of body parts. A kitten's eyes after birth are an example of (10) _____ _____ _____ in action.

As the embryo develops, one group of cells may produce a substance (say, a growth factor) that diffuses to another group of cells and turns on protein synthesis in those cells. Such interaction among embryonic cells is called (11) _____ _____. Sometimes entire organs (such as testes in human males) change position in the developing organism, but the inward or outward folding of (12) _____ _____ is seen more often. Spemann demonstrated that the process known as (13) _____ _____ occurs in salamander embryos, where one body part differentiates because of signals it receives from an adjacent body part. (14) _____ mutations affect regulatory genes that activate <u>sets</u> of genes concerned with development; they are generally disadvantageous and cause mistakes in the developmental program of an organism.

In many animals, the embryo develops into a motile, independent (15) _____, which extends the food supply and range of the population. A larva necessarily must undergo (16) _____ in order to become a juvenile. Fruitflies show (17) _____ metamorphosis. Normal cell types have a(n) (18) _____ _____ _____, and mitosis is scheduled to quit after so many cell divisions.

True-False

If false, explain why.

__ 19. The aging and death of a cell may be coded in large part in its DNA; <u>external</u> signals activate those DNA messages and tell the cell that it is time to die.

__ 20. A process of predictable cellular deterioration is built into the life cycle of all organisms that consist of <u>differentiated cells that show considerable specialization</u>.

__ 21. Body parts become folded, tubes become hollowed out, and eyelids, lips, noses, and ears all become slit or perforated by <u>controlled cell death</u>.

Self-Quiz

___ 1. Animals such as birds lay eggs with large amounts of yolk; embryonic development happens within the egg covering outside the mother's body. This developmental strategy is called _____.

   a. ovoviviparity
   b. viviparity
   c. oviparity
   d. parthenogenesis
   e. none of the above

___ 2. The process of cleavage most commonly produces a(n) _____.

   a. zygote
   b. blastula
   c. gastrula
   d. third germ layer
   e. organ

___ 3. Imaginal disks are characteristic of the embryonic development of _____.

   a. frogs
   b. fruit flies
   c. chickens
   d. sea urchins
   e. humans

___ 4. The formation of three germ (embryonic) tissue layers occurs during _____.

   a. gastrulation
   b. cleavage
   c. pattern formation
   d. morphogenesis
   e. neural plate formation

___ 5. The differentiation of a body part in response to signals from an adjacent body part is _____.

   a. contact inhibition
   b. ooplasmic localization
   c. embryonic induction
   d. pattern formation
   e. none of the above

___ 6. A homeotic mutation _____.

   a. may cause a leg to develop on the head where an antenna should grow
   b. affects the expression of imaginal disks
   c. affects morphogenesis
   d. may alter the path of development
   e. all of the above

___ 7. Shortly after fertilization, the zygote is subdivided into a multicelled embryo during a process known as _____.

   a. meiosis
   b. parthenogenesis
   c. embryonic induction
   d. cleavage
   e. invagination

___ 8. Muscles differentiate from _____ tissue.

   a. ectoderm
   b. mesoderm
   c. endoderm
   d. parthenogenetic
   e. yolky

___ 9. The gray crescent is _____.

   a. formed where the sperm penetrates the egg
   b. next to the dorsal lip of the blastopore
   c. the yolky region of the egg
   d. where the first mitotic division begins
   e. formed opposite from where the sperm enters the egg

___ 10. The nervous system differentiates from _____ tissue.

   a. ectoderm
   b. mesoderm
   c. endoderm
   d. yolky
   e. homeotic

Chapter Objectives/Review Questions

This section lists general and detailed chapter objectives that can be used as review questions. You can make maximum use of these items by writing answers on a separate sheet of paper. To check for accuracy, compare your answers with information given in the chapter or glossary.

Page   *Objectives/Questions*

(754)   1. Understand how asexual reproduction differs from sexual reproduction. Know the advantages and problems associated with having separate sexes.

(754)   2. Explain why evolutionary trends in many groups of organisms tend toward developing more complex, sexual strategies rather than retaining simpler, asexual strategies.

(755, 758, 760)   3. Explain how the amount of yolk in an ovum can influence an animal's cleavage pattern.

(755)   4. Define *oviparous, viviparous,* and *ovoviviparous.* For each of the three developmental strategies, cite an example of an animal that goes through it.

(756)   5. Name each of the three embryonic tissue layers and the organs formed from each.

(756–759)   6. Describe early embryonic development and distinguish among the following: oogenesis, fertilization, cleavage, gastrulation, and organ formation.

(757–761)   7. Compare the early stages of frog and chick development (see Figs. 44.5 and 44.6) with respect to egg size and type of cleavage pattern (incomplete or complete).

(760–761)   8. Explain what causes polarity to occur during oocyte maturation in the mother and state how polarity influences later development.

(763)   9. Define *gastrulation* and state what process begins at this stage that did not happen during cleavage.

(764)   10. Define *differentiation* and give two examples of cells in a multicellular organism that have undergone differentiation.

(764)   11. Explain why the differentiation of cells in a multicellular organism goes hand in hand with the division of labor and the integration of life processes.

(764–768)   12. Explain how a spherical zygote becomes a multicellular adult with arms and legs.

(768)   13. Distinguish complete from incomplete metamorphosis.

(768)   14. Define what is meant by *larva.* Distinguish metamorphosis from morphogenesis.

Integrating and Applying Key Concepts

If embryonic induction did not occur in a human embryo, how would the eye region appear? What would happen to the forebrain and epidermis? If controlled cell death did not happen in a human embryo, how would its hands appear? Its face?

Answers

44-I. THE BEGINNING: REPRODUCTIVE MODES
(pp. 754–755)
STAGES OF DEVELOPMENT (pp. 756–757)

1. asexual; 2. environmental; 3. budding; 4. variation; 5. Gamete formation; 6. egg; 7. fertilization; 8. zygote; 9. Cleavage; 10. blastula; 11. Gastrulation; 12. nervous; 13. gut; 14. skeleton; 15. RNA transcripts (maternal messages); 16. ovoviviparity; 17. oviparity; 18. embryonic; 19. Development; 20. F, only part of the yolk surface is contained within the cells created by the first few mitotic divisions; 21. T; 22. T; 23. T; 24. T; 25. T; 26. F; 27. B; 28. C; 29. A; 30. E; 31. D; 32. a. mesoderm; b. ectoderm; c. endoderm; d. mesoderm; e. ectoderm; f. mesoderm; g. endoderm; h. mesoderm; i. mesoderm.

44-II. A VISUAL TOUR OF FROG AND CHICK
DEVELOPMENT (pp. 758–759)
EARLY MARCHING ORDERS (pp. 760–761)
EMERGENCE OF THE EARLY EMBRYO
(pp. 762–763)

1. F, nuclear; 2. F, by "maternal instructions"; 3. T; 4. cleavage; 5. blastula; 6. gastrulation; 7. germ; 8. gastrula; 9. maternal instructions; 10. cytoplasmic localization; 11. animal pole; 12. gray crescent; 13. body axis; 14. yolk; 15. three.

44-III. FORMATION OF SPECIALIZED TISSUES
AND ORGANS (pp. 764–765)
PATTERN FORMATION (pp. 766–767)
FROM THE EMBRYO ONWARD (pp. 768–769)

1. cell differentiation; 2. genes; 3. restrictions; 4. Morphogenesis; 5. active cell migration; 6. pseudopods; 7. chemotaxis; 8. adhesive; 9. localized growth; 10. controlled cell death; 11. embryonic induction; 12. ectodermal sheets; 13. embryonic induction; 14. Homeotic; 15. larva; 16. metamorphosis; 17. complete; 18. limited division potential; 19. F, internal; 20. T; 21. T.

Self-Quiz
1. c; 2. b; 3. b; 4. a; 5. c; 6. e; 7. d; 8. b; 9. e; 10. a.

45

HUMAN REPRODUCTION AND DEVELOPMENT

Interactive Exercises

45-I. MALE REPRODUCTIVE SYSTEM (pp. 772–775)
MALE REPRODUCTIVE FUNCTION (pp. 776–777)

Selected Italicized Words

testicular cancer, prostate gland, seminal vesicles, vas deferens, bulbourethral glands, epididymis, urethra, ejaculatory ducts, scrotum

Boldfaced, Page-Referenced Terms

(772) blastocyst _____

---

(773) ovaries (ovary, sing.) _____

_____

(773) testes (testis, sing.) _____

_____

(773) secondary sexual traits _____

_____

(774) seminiferous tubules _____

_____

(774) semen _____

_____

(776) Sertoli cells _____

_____

(776) Leydig cells _____

_____

(776) testosterone _____

_____

(777) LH, luteinizing hormone _____

_____

(777) FSH, follicle stimulating hormone _____

_____

Fill-in-the-Blanks

The numbered items on the illustration that follows represent missing information; complete the numbered blanks in the narrative below to supply the missing information on the illustration. Some illustrated structures are numbered more than once to aid identification.

Within each testis and following repeated (1) _____ divisions of undifferentiated diploid cells just inside the (2) _____ tubule walls, (3) _____ occurs to form haploid, mature (4) _____. Males produce sperm continuously from puberty onward. Sperm leaving a testis enter a long coiled duct, the (5) _____; the sperm are stored in the last portion of this organ. When a male is sexually aroused, muscle contractions quickly propel the sperm through a thick-walled tube, the (6) _____ _____, then to ejaculatory ducts, and finally the (7) _____, which opens at the tip of the penis. During the trip to the urethra, glandular secretions become mixed with the sperm to form semen. (8) _____ _____ secrete fructose to nourish the sperm and prostaglandins to induce contractions in the female reproductive tract. (9) _____ _____ secretions help neutralize vaginal acids. (10) _____ glands secrete mucus to lubricate the penis, aid vaginal penetration, and improve sperm motility.

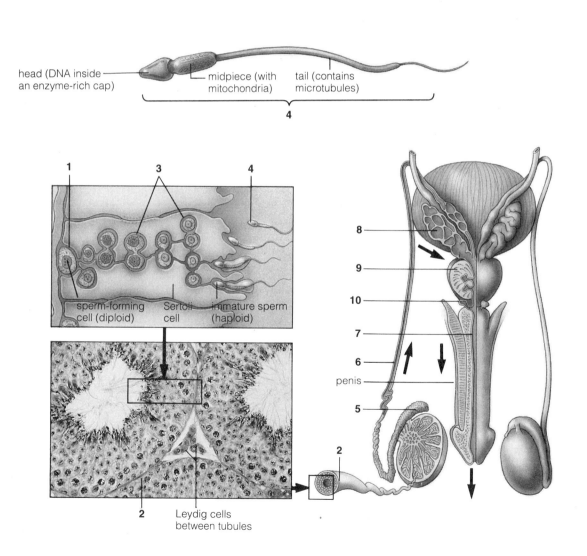

head (DNA inside an enzyme-rich cap) — midpiece (with mitochondria) — tail (contains microtubules) — 4

sperm-forming cell (diploid) — Sertoli cell — immature sperm (haploid)

Leydig cells between tubules

penis

Dichotomous Choice

Circle one of two possible answers given between parentheses in each statement.

11. Testosterone is secreted by (Leydig/hypothalamus) cells.
12. (Testosterone/FSH) governs the growth, form, and functions of the male reproductive tract.
13. Sexual behavior, aggressive behavior, and secondary sexual traits are associated with (LH/testosterone).
14. LH and FSH are secreted by the (anterior/posterior) lobe of the pituitary gland.
15. The (testes/hypothalamus) govern(s) sperm production by controlling interactions among testosterone, LH, and FSH.
16. When blood levels of testosterone (increase/decrease), the hypothalamus stimulates the pituitary to release LH and FSH, which travel the bloodstream to the testes.
17. Within the testes, (LH/FSH) acts on Leydig cells; they secrete testosterone, which enters the sperm-forming tubes.
18. FSH enters the sperm-forming tubes and diffuses into (Sertoli/Leydig) cells to improve testosterone uptake.
19. When blood testosterone levels (increase/decrease) past a set point, negative feedback loops to the hypothalamus slow down testosterone secretion.

Labeling

20. _____

21. _____ _____

22. _____ _____

23. _____ _____

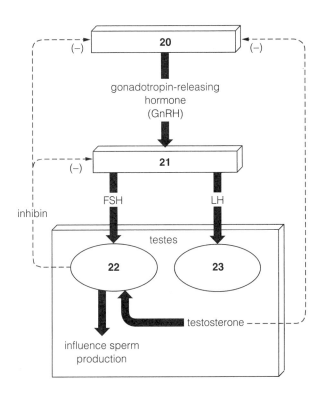

45-II. FEMALE REPRODUCTIVE SYSTEM (pp. 778–779)
FEMALE REPRODUCTIVE FUNCTION (pp. 780–781)
SUMMING UP—KEY EVENTS OF THE MENSTRUAL CYCLE (p. 782)

Selected Italicized Words

estrous, endometriosis, primary oocyte, secondary oocyte, vagina, clitoris, labium major, labium minor, cervix, myometrium

Boldfaced, Page-Referenced Terms

(778) oocyte _____

_____

(778) oviduct _____

_____

(778) uterus _____

_____

(778) endometrium _____

_____

(778) menstrual cycle _____

_____

(778) follicular phase _____

_____

(778) ovulation _____

_____

(778) luteal phase _____

_____

(778) estrogens _____

_____

(778) progesterone _____

_____

(780) follicle _____

_____

(780) zona pellucida _____

_____

(781) corpus luteum _____

_____

Fill-in-the-Blanks

The numbered items on the illustration that follows represent missing information; complete the numbered blanks in the narrative below to supply the missing information on the illustration. Some illustrated structures are numbered more than once to aid identification.

An immature egg (oocyte) is released from one (1) _____ of a pair. From each ovary, a(n) (2) _____ forms a channel for transport of the immature egg to the (3) _____, a hollow, pear-shaped organ where the embryo grows and develops. The lower, narrowed part of the uterus is the (4) _____. The uterus has a thick layer of smooth muscle, the (5) _____, lined inside with connective tissue, glands, and blood vessels; this lining is called the (6) _____. The (7) _____, a muscular tube, extends from the cervix to the body surface; this tube receives sperm and functions as part of the birth canal. At the body surface are external genitals (vulva) that include organs for sexual stimulation. Outermost is a pair of fat-padded skin folds, the (8) _____ _____. Those folds enclose a smaller pair of skin folds, the (9) _____ _____. The smaller folds partly enclose the (10) _____, an organ sensitive to stimulation. The location of the (11) _____ is about midway between the clitoris and the vaginal opening. (12) _____ occurs in the ovaries so that a normal female infant has about 2 million primary oocytes with the division process halted in the (13) ❏ I ❏ II stage. By age seven, only about (14) _____ remain. A primary oocyte surrounded by a nourishing layer of granulosa cells in called a(n) (15) _____.

When a femal enter puberty, the (16) _____ secretes a hormone (GnRH) that makes the (17) _____ _____ secrete follicle stimulating hormone (FSH) and luteinizing hormone (LH). These hormones are carried by the blood to all parts of the body, but each month, one (or more) follicles respond by growing and secreting the steroid hormones called (18)_____.

The primary oocyte completes the (12) (13) division 8–10 hours before (19) _____ (the time when the

secondary oocyte is released from the ovary). (19) _____ occurs in response to a surge of

(20) _____ being produced by (17) on day 12 or 13 of a 28-day cycle.

The first 5 days of the cycle are occupied by (21) _____ , the deterioration and expulsion of

(22) _____ tissues that line the uterus. During the next week, (18) _____ stimulate new

(22) _____ tissues to be constructed.

45-III. PREGNANCY HAPPENS (p. 783)
FORMATION OF THE EARLY EMBRYO (pp. 784–785)
EMERGENCE OF THE VERTEBRATE BODY PLAN (p. 786)
ON THE IMPORTANCE OF THE PLACENTA (p. 787)
EMERGENCE OF DISTINCTLY HUMAN FEATURES (pp. 788–791)

Selected Italicized Words

teratogens, thalidomide, fetal alcohol syndrome

Boldfaced, Page-Referenced Terms

(783) ovum, ova _____

_____

(784) embryonic period _____

_____

(784) fetal period _____

_____

(785) blastocyst _____

_____

(785) implantation _____

_____

(785) embryonic disk _____

_____

(785) amnion _____

_____

(785) yolk sac _____

_____

(785) chorion _____

_____

(785) allantois _____

_____

(785) HCG _____

_____

(786) neural tube _____

_____

(786) somites _____

_____

(787) placenta _____

_____

Fill-in-the-Blanks

Fertilization generally takes place in the (1) _____; five or six days after conception, (2) _____ begins as the blastocyst sinks into the endometrium. Extensions from the chorion fuse with the endometrium of the uterus to form a(n) (3) _____, the organ of interchange between mother and fetus. By the beginning of the (4) _____ trimester, all major organs have formed; the offspring is now referred to as a(n) (5) _____.

Labeling

Identify each indicated part of the accompanying illustrations.

6. _____ _____

7. _____ _____

8. _____ _____

9. _____ _____

10. _____ _____

11. _____

12. _____

13. _____

14. _____ _____

15. _____ _____

16. _____

17. _____ _____

18. _____

19. _____

20. _____

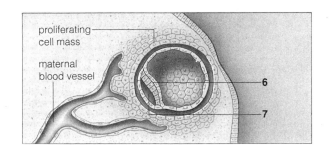

proliferating cell mass

maternal blood vessel

6

7

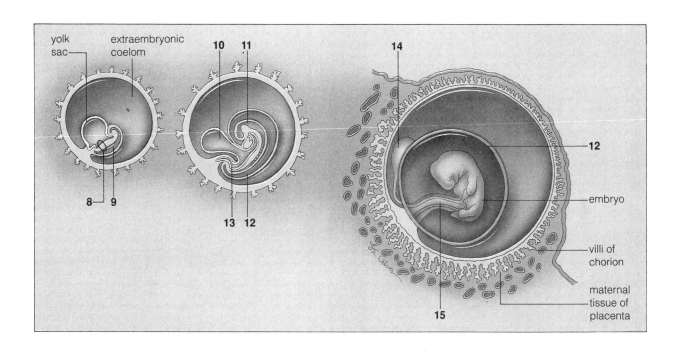

yolk sac

extraembryonic coelom

10   11

14

12

embryo

villi of chorion

maternal tissue of placenta

8   9

13   12

15

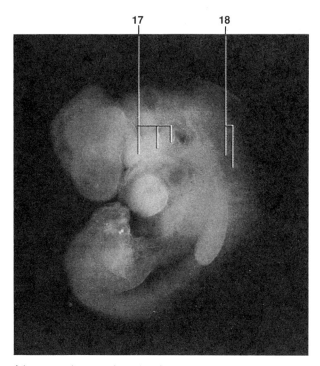

17   18

A human embryo at **16** weeks after conception.

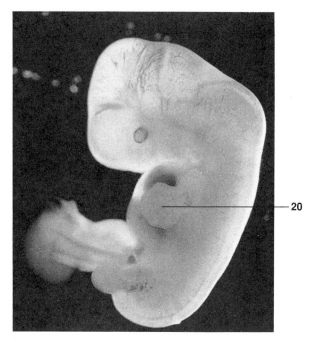

20

A human embryo at **19** weeks after conception.

Human Reproduction and Development   **525**

45-IV. FROM BIRTH ONWARD (pp. 792–793)
CONTROL OF HUMAN FERTILITY (pp. 794–797)
TWO ISSUES OF CLINICAL AND ETHICAL IMPORTANCE (p. 798)

Selected Italicized Words

identical twins, fraternal twins, breast cancer, prenatal, postnatal, abstinence, rhythm method, withdrawal, douching, vasectomy, tubal ligation, spermicidal foam, spermicidal jelly, diaphragm, condoms, the Pill, RU-486, AIDS, genital warts, gonorrhea, syphilis, chancre, pelvic inflammatory disease, genital herpes, chlamydial infection, abortion, in vitro fertilization

Boldfaced, Page-Referenced Terms

(792) labor _____

_____

(792) lactation _____

_____

Fill-in-the-Blanks

On Earth each week, (1) _____ more babies are born than people die. Each year in the United States, we still have about (2) _____ unwed teenage mothers and (3) _____ abortions. The most effective method of preventing conception is complete (4) _____. (5) _____ are about 85 to 93 percent reliable and help prevent venereal disease. A(n) (6) _____ is a flexible, dome-shaped disk, used with a spermicidal foam or jelly, that is placed over the cervix. In the United States, the most widely used contraceptive is the Pill—an oral contraceptive of synthetic (7) _____ and (8) _____ that suppress the release of (9) _____ from the pituitary and thereby prevent the cyclic maturation and release of eggs. Two forms of surgical sterilization are vasectomy and (10) _____ _____.

Matching

Match each of the following with *all* applicable diseases.

11. ___ Can damage the brain and spinal cord in ways leading to various forms of insanity and paralysis

12. ___ Has no cure

13. ___ Can cause violent cramps, fever, vomiting, and sterility due to scarring and blocking of the oviducts

14. ___ Caused by a motile, corkscrew-shaped bacterium, *Treponema pallidum*

15. ___ Affects about 14 million women a year, causing scarred oviducts, abnormal pregnancies, and sterility

16. ___ Infected women typically have miscarriages, stillbirths, or sickly infants

17. ___ Caused by a bacterium with pili, *Neisseria gonorrhoeae*

18. ___ Caused by direct contact with the viral agent; about 5 million to 20 million people in the United States infected by it

19. ___ Produces a chancre (localized ulcer) one to eight weeks following infection

20. ___ Chronic infections by this can lead to cervical cancer

21. ___ Caused by an intracellular parasite that infects 3 million to 10 million Americans per year, especially college students

22. ___ Can lead to lesions in the eyes that cause blindness in babies born to mothers with this

23. ___ Acyclovir decreases the healing time and may also decrease the pain and viral shedding from the blisters

24. ___ Following infection, the parasites migrate to lymph nodes, which become enlarged and tender; may lead to pronounced tissue swelling

25. ___ May be cured by antibiotics but can infect again

26. ___ Can be treated with tetracycline and sulfonamides

27. ___ Generally preventable by correct condom usage

A. AIDS
B. Chlamydial infection
C. Genital herpes
D. Gonorrhea
E. Pelvic inflammatory disease
F. Syphilis

Self-Quiz

For questions 1–5, choose from the following answers:

        a. AIDS
        b. Chlamydial infection
        c. Genital herpes
        d. Gonorrhea
        e. Syphilis

___ 1. _____ is a disease caused by a spherical bacterium (*Neisseria*) with pili; it is curable by prompt diagnosis and treatment.

___ 2. _____ is a disease caused by a spiral bacterium (*Treponema*) that produces a localized ulcer (a chancre).

___ 3. _____ is an incurable disease caused by a retrovirus (an RNA-based virus).

___ 4. _____ is a disease caused by an obligate, intracellular parasite that migrates to regional lymph nodes, which swell and become tender.

___ 5. _____ is an extremely contagious viral infection (DNA-based) that causes sores on the facial area and reproductive tract; it is also incurable.

For questions 6–8, choose from the following answers:

        a. blastocyst
        b. allantois
        c. yolk sac
        d. oviduct
        e. cervix

___ 6. The _____ lies between the uterus and the vagina.

___ 7. The _____ is a pathway from the ovary to the uterus.

___ 8. The _____ results from the process known as cleavage.

For questions 9–12, choose from the following answers:

        a. interstitial cells
        b. seminiferous tubules
        c. vas deferens
        d. epididymis
        e. prostate

___ 9. The _____ connects a structure on the surface of the testis with the ejaculatory duct.

___ 10. Testosterone is produced by the _____.

___ 11. Meiosis occurs in the _____.

___ 12. Sperm mature and become motile in the _____.

Chapter Objectives/Review Questions

This section lists general and detailed chapter objectives that can be used as review questions. You can make maximum use of these items by writing answers on a separate sheet of paper. To check for accuracy, compare your answers with information given in the chapter or glossary.

Page		Objectives/Questions
(773)	1.	Distinguish between primary and secondary sexual traits and between gonads and accessory reproductive organs.
(774)	2.	Follow the path of a mature sperm from the seminiferous tubules to the urethral exit. List every structure encountered along the path and state the contribution to the nurture of the sperm.
(776–777)	3.	Diagram the structure of a sperm, label its components, and state the function of each.
(776; 780–781)	4.	Compare the function of the Leydig cells of the testis with the function of the ovarian follicle and the corpus luteum.
(776)	5.	List in order the stages that compose spermatogenesis.
(776–777)	6.	Name the four hormones that directly or indirectly control male reproductive function. Diagram the negative feedback mechanisms that link the hypothalamus, anterior pituitary, and testes in controlling gonadal function.
(778)	7.	Distinguish the follicular phase of the menstrual cycle from the luteal phase and explain how the two cycles are synchronized by hormones from the anterior pituitary, hypothalamus, and ovaries.
(780–781)	8.	State which hormonal event brings about ovulation and which other hormonal events bring about the onset and finish of menstruation.
(783)	9.	List the physiological factors that bring about erection of the penis during sexual stimulation and the factors that bring about ejaculation.
(783)	10.	List the similar events that occur in both male and female orgasm.
(778; 783)	11.	Trace the path of a sperm from the urethral exit to the place where fertilization normally occurs. Mention in correct sequence all major structures of the female reproductive tract that are passed along the way and state the principal function of each structure.
(772–773; 784–789)	12.	Describe the events that occur during the first month of human development. State how much time cleavage and gastrulation require, when organogenesis begins, and what is involved in implantation and placenta formation.
(790–791)	13.	Explain why the mother must be particularly careful of her diet, health habits, and lifestyle during the first trimester after fertilization (especially during the first six weeks).
(789)	14.	State when the embryo begins to be referred to as a fetus and at what point at least 10 percent of births result in survival.
(775; 793)	15.	Describe how a woman examines herself for breast cancer and how a man examines himself for testicular cancer.
(794–795)	16.	Identify the factors that encourage and discourage methods of human birth control.
(795)	17.	Identify the three most effective birth control methods used in the United States and the four least effective birth control methods.
(796)	18.	State which birth control methods help prevent venereal disease.
(794)	19.	Describe two different types of sterilization.
(798)	20.	State the physiological circumstances that would prompt a couple to try in vitro fertilization.
(796–797)	21.	For each STD described in the Commentary, know the causative organism and the symptoms of the disease.

Integrating and Applying Key Concepts

What rewards do you think a society should give a woman who has at most two children during her lifetime? In the absence of rewards or punishments, how can a society encourage women not to have abortions and yet ensure that the human birth rate does not continue to increase?

Answers

Interactive Exercises

45-I. MALE REPRODUCTIVE SYSTEM (pp. 774–775)
MALE REPRODUCTIVE FUNCTION (pp. 776–777)
1. mitotic (mitosis); 2. seminiferous; 3. meiosis (spermatogenesis); 4. sperm; 5. epididymis; 6. vas deferens; 7. urethra; 8. Seminal vesicles; 9. Prostate gland; 10. Bulbourethral; 11. Leydig; 12. Testosterone; 13. testosterone; 14. anterior; 15. hypothalamus; 16. decrease; 17. LH; 18. Sertoli; 19. increase; 20. hypothalamus; 21. anterior pituitary; 22. Sertoli cells; 23. Leydig cells.

45-II. FEMALE REPRODUCTIVE SYSTEM
(pp. 778–779)
FEMALE REPRODUCTIVE FUNCTION
(pp. 780–781)
SUMMING UP—KEY EVENTS OF THE
MENSTRUAL CYCLE (p. 782)
1. ovary; 2. oviduct; 3. uterus; 4. cervix; 5. myometrium; 6. endometrium; 7. vagina; 8. labia majora; 9. labia minora; 10. clitoris; 11. urethra; 12. Meiosis; 13. I; 14. 300,000; 15. follicle; 16. hypothalamus; 17. anterior pituitary; 18. estrogens; 19. ovulation; 20. LH; 21. menstruation; 22. endometrial.

45-III. PREGNANCY HAPPENS (p. 783)
FORMATION OF THE EARLY EMBRYO
(pp. 784–785)
EMERGENCE OF THE VERTEBRATE BODY PLAN
(p. 786)
ON THE IMPORTANCE OF THE PLACENTA
(p. 787)
EMERGENCE OF DISTINCTLY HUMAN
FEATURES (pp. 788–791)
1. oviduct; 2. implantation; 3. placenta; 4. second; 5. fetus; 6. embryonic disk; 7. amniotic cavity; 8. embryonic disk; 9. amniotic cavity; 10. yolk sac; 11. embryo; 12. amnion; 13. allantois; 14. yolk sac; 15. umbilical cord; 16. four; 17. gill arches; 18. somites; 19. five; 20. forelimb.

45-IV. FROM BIRTH ONWARD (pp. 792–793)
CONTROL OF HUMAN FERTILITY (pp. 794–797)
TWO ISSUES OF CLINICAL AND ETHICAL
IMPORTANCE (p. 798)
1. 1,700,000; 2. 200,000; 3. 1,500,000; 4. abstinence; 5. Condoms; 6. diaphragm; 7. estrogens (progesterones); 8. progesterones (estrogens); 9. gonadotropins; 10. tubal ligation; 11. A, F; 12. A, C; 13. D, E; 14. F; 15. E; 16. F; 17. D; 18. C; 19. F; 20. C; 21. B; 22. C, F; 23. C; 24. B, F; 25. D, F; 26. B; 27. A, B, C, D, E, F.

Self-Quiz
1. d; 2. e; 3. a; 4. b; 5. c; 6. e; 7. d; 8. a; 9. c; 10. a; 11. b; 12. d.

46

POPULATION ECOLOGY

Interactive Exercises

46-I. CHARACTERISTICS OF POPULATIONS (pp. 802–805)
POPULATION SIZE AND EXPONENTIAL GROWTH (pp. 806–807)

Selected Italicized Words

biotic, abiotic, clumped dispersion, uniform dispersion, random dispersion, births, immigration, deaths, emigration

Boldfaced, Page-Referenced Terms

(803) ecology _____

_____

(803) population _____

_____

(803) habitat _____

_____

(803) community _____

_____

(803) ecosystem _____

(803) biosphere _____

(804) population size _____

(804) population density _____

(804) population distribution _____

(804) age structure _____

(804) reproductive base _____

(806) zero population growth _____

(806) net reproduction per individual per unit time, *r* _____

(806) exponential growth _____

(807) doubling time _____

(807) biotic potential _____

Fill-in-the-Blanks

(1) _____ is the study of the ways in which organisms interact with one another and with their physical and chemical environment. Populations of all species occupying a habitat is defined as the (2) _____. A(n) (3) _____ is the ecological level that includes a biotic community plus its nonliving environment. The (4) _____ includes the entire Earth realm in which organisms live.

A group of individuals of the same species occupying a given area at a specific time is a(n) (5) _____. The place where a population lives is its (6) _____. (7) _____ _____ is the number of individuals per unit of area or volume. Examples of nonrandom distribution are uniform dispersion and (8) _____ dispersion. Population size, N, increases by birth and (9) _____ , and decreases by death and (10) _____ . Those members of a population that are in their offspring-producing period represent the (11) _____ _____ of the population.

Any population that is not restricted in some way will show a pattern of (12) _____ growth, because any increase in population size enlarges the (13) _____ base. When the course of such growth is plotted on a graph, a(n) (14) _____-shaped curve is obtained.

The ability to produce the maximum possible number of new individuals is called the population's (15) _____ _____.

For exercises 16–18, consider the equation $G=rN$, where $G=$ the population growth rate, $r=$ the N reproduction per individual per unit of time, and $N=$ the number of individuals in the population.

16. Assume that r remains constant at 0.2.
    a.   As the value of G increases, what happens to the value of N?_____
    b.   If the value of G decreases, what happens to the value of N?_____
    c.   If the net reproduction per individual stays the same and the population grows faster, then what must happen to the number of individuals in the population?_____
17. If a society decides it is necessary to lower its value of N through reproductive means because supportive resources are dwindling, it must lower either its net reproduction per individual per unit of time or its

    _____

    _____
18. The equation $G=rN$ expresses a direct relationship between G and $r \times N$. If G remains constant and N increases, what must the value of r do? (In this situation, r varies *inversely* with N.) _____

    _____

    _____
19. Look at line (a) in the graph at the right. After seven hours have elapsed, approximately how many individuals are in the population? _____
20. Look at line (b) in the same graph.
    a.   After 24 hours have elapsed, approximately how many individuals are in the population? _____
    b.   After 28 hours have elapsed, approximately how many individuals are in the population? _____

46-II. LIMITS ON THE GROWTH OF POPULATIONS (pp. 808–809)
LIFE HISTORY PATTERNS (pp. 810–813)

Selected Italicized Words

density-dependent controls; bubonic plague; "survivorship;" type I, type II, and type III survivorship curves

Boldfaced, Page-Referenced Terms

(808) limiting factor _____

_____

(808) carrying capacity, K _____

_____

(808) logistic growth _____

_____

(810) cohort _____

_____

(811) survivorship curves _____

_____

Fill-in-the-Blanks

If (1) _____ factors (essential resources in short supply) act on a population, population growth tapers off.

(2) _____ _____ refers to the maximum number of individuals of a population that can be sustained indefinitely by the environment. S-shaped growth curves are characteristic of (3) _____ population growth. The plot of (3) growth levels off once the (4) _____ _____ is reached. In the equation $G = r_{max}N [(K - N)/K]$, as the value of N approaches the value of K, and K and r_{max} remain constant, the value of G (5) (choose one) ❒ increases ❒ decreases ❒ cannot be determined by humans, even if they know algebra. As the value of r_{max} increases and G and N remain constant, the value of K (the carrying capacity) (6) (choose one) ❒ increases ❒ decreases ❒ cannot be determined by humans, even if they know algebra. In an overcrowded population, predators, parasites, and disease agents serve as (7) _____-_____ controls. When an event such as a freak summer snowstorm in the Colorado Rockies causes more deaths or fewer births in a butterfly population (with no regard to crowding or dispersion patterns), the controls are said to be (8) _____ _____.

(9) _____ _____ use information summarized from life tables, which show trends in mortality and life expectancy. Type (10) _____ populations have low survivorship early in life. Food availability is a density-(11) _____ factor that works to cut back population size when it approaches the environment's (12) _____ _____. Environmental disruptions such as forest fires and floods are density-(13) _____ factors that may push a population above or below its tolerance range for a given variable.

Matching

Choose the most appropriate answer for each.

___ 14. cohort

___ 15. type III survivorship curves

___ 16. life tables

___ 17. type I survivorship curves

___ 18. type II survivorship curves

A. survivorship curves that reflect a fairly constant death rate at all ages; typical of some song birds, lizards, and small mammals

B. survivorship curves that reflect high survivorship until fairly late in life; produce a few large offspring provided with extended parental care; examples are elephants and humans

C. a group tracked by researchers from birth until the last survivor dies

D. survivorship curves that reflect a high death rate early in life; typical of sea stars and other invertebrate animals, insects, many fishes, plants, and fungi

E. summaries of age-specific patterns of birth and death

Matching

Choose the most appropriate answer(s) for each. A letter may be used more than once, and a blank may contain more than one letter.

_____ 19. cohort

_____ 20. density-independent factors

_____ 21. density-dependent factors

_____ 22. Reznick and Endler

A. drought, floods, earthquakes
B. reindeer in the Pribilof Islands in 1935
C. food availability
D. adrenal enlargement in wild rabbits in response to crowding
E. all of the 1987 human babies of New York City
F. life history patterns of Trinidadian guppies

46-III. HUMAN POPULATION GROWTH (pp. 814–815)
CONTROL THROUGH FAMILY PLANNING (pp. 816–817)
POPULATION GROWTH AND ECONOMIC DEVELOPMENT (pp. 818–819)
QUESTIONS CONCERNING ZERO POPULATION GROWTH (p. 820)

Selected Italicized Words

"replacement rate;" preindustrial, transitional, industrial, and postindustrial stages

Boldfaced, Page-Referenced Terms

(816) family planning programs _____

_____

(818) demographic transition model _____

_____

1. Graph the following data in the space provided on page 542.

Year	Estimated World Population
1650	500,000,000
1850	1,000,000,000
1930	2,000,000,000
1975	4,000,000,000
1986	5,000,000,000
1991	5,400,000,000

a. Estimate the year that the world contained 3 billion humans. _____

b. Estimate the year that Earth will accommodate 8 billion humans. _____

c. Do you expect Earth to shelter 8 billion humans within your lifetime? _____

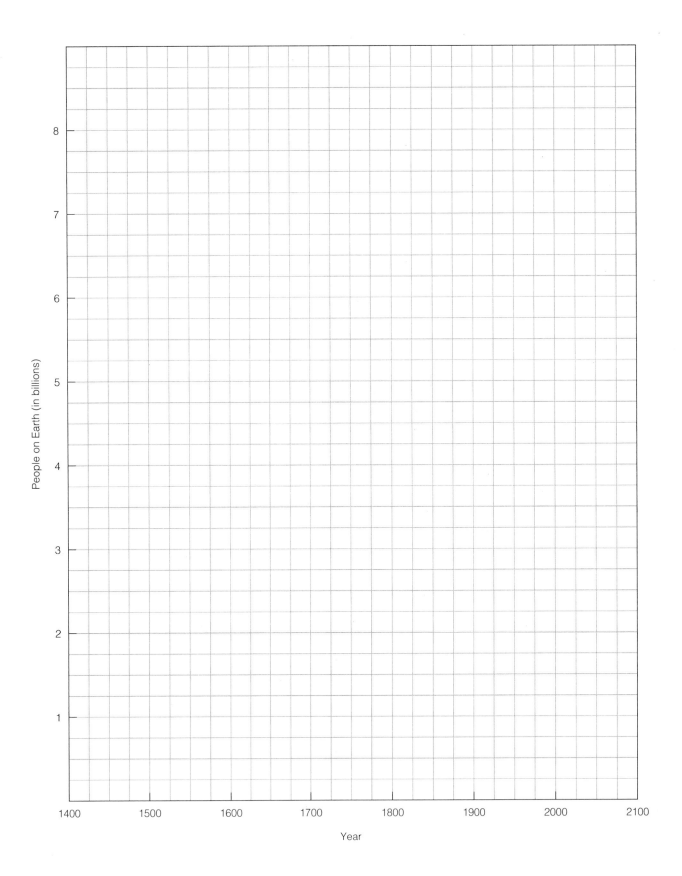

Fill-in-the-Blanks

The (2) _____ _____ of a population shows how individuals are distributed into age groups. In 1994, the most populous country was China with 1.17 billion people; next most populous was (3) _____ with 812 million. By comparison, the USA had 261 million people.

If the present rate of human population growth continues, (4) _____ (number) billion people will live on the planet in just thirty years. These numbers will have adverse effects on resource supplies, and crowding will also invite the return of severe density-(5) _____ controls.

Many governments attempt to control population growth rates by restricting (6) _____, but others attempt to reduce population pressures by encouraging (7) _____. However, most countries focus on reducing (8) _____ rates through economic security and family planning. The key point of the (9) _____ _____ model is that changes in population growth can be correlated with changes that occur during four stages of economic development. The United States, Canada, Australia, Japan, the former Soviet Union, and most countries of Western Europe are in the (10) _____ stage of the demographic transition model; Mexico and other less developed countries are in the (11) _____ stage. To achieve global zero population growth, the average "replacement rate" is about (12) _____ (number) children per woman in less developed countries and (13) _____ (number) in more developed countries. Age structure diagrams of actively growing human populations show that more than a third of the world population falls in the broad (14) _____ base category (high numbers of men and women of reproductive age).

A simple way to slow things down would be to encourage (15) _____ _____. In 1994, the human population on Earth reached (16) _____ (number) billion. In that year, almost (17) _____ (number) million more individuals were added.

True/False

If the statement is true, write a T in the blank. If the statement is false, make it correct by changing the underlined word(s) and writing the correct word(s) in the answer blank.

_____ 18. During 1994, <u>one in two</u> humans on the planet was malnourished or starving, without clean drinking water and without adequate shelter.

_____ 19. Even if we could double our present food supply, death from starvation could still reach <u>20 million</u> to <u>40 million</u> people a year.

_____ 20. Compared with the geographic spread of other organisms, it has taken the human population a relatively <u>long</u> period of time to expand into new environments.

_____ 21. Managing food supplies through agriculture has had the effect of increasing the <u>carrying capacity</u> for human populations.

_____ 22. Humans have sidestepped the <u>biotic potential</u> of their environment by bringing many disease agents under control and by tapping into concentrated, existing stores of energy.

Sequence

Arrange the following stages of the demographic transition model in correct chronological sequence. Write the letter of the first step next to 23, the letter of the second step next to 24, and so on.

23. ___   A. Industrial stage: population growth slows and industrialization is in full swing
24. ___   B. Preindustrial stage: harsh living conditions, high birth rates and low death rates, slow population growth
25. ___   C. Postindustrial stage: zero population growth is reached; birth rate falls below death rate, and population size slowly decreases
26. ___   D. Transitional stage: industrialization begins, food production rises, and health care improves; death rates drop, and birth rates remain high, causing rapid population growth

---

Self-Quiz

Multiple Choice

___ 1. The total number of individuals of the same species that occupy a given area at a given time is _____.
   a. the population density
   b. the population growth
   c. the population birth rate
   d. the population size

___ 2. The average number of individuals of the same species per unit area at a given time is _____.
   a. the population density
   b. the population growth
   c. the population birth rate
   d. the population size

___ 3. A population that is growing exponentially in the absence of limiting factors can be illustrated accurately by a(n) _____.
   a. S-shaped curve
   b. J-shaped curve
   c. curve that terminates in a plateau phase
   d. tolerance curve

___ 4. If reproduction occurs early in the life cycle, _____.
   a. higher population levels tend to result
   b. it represents an extrinsic factor that limits population size
   c. it represents a density-dependent factor that limits population size
   d. all of the above

___ 5. A situation in which the birth rate plus immigration over the long term equal the death rate plus emigration is called

_____.
   a. an intrinsic limiting factor
   b. exponential growth
   c. saturation
   d. zero population growth

___ 6. The rate of increase for a population (r) refers to the _____ the birth rate and death rate plus immigration and minus emigration.
   a. sum of
   b. product of
   c. doubling time between
   d. difference between

___ 7. _____ is a way to express the growth rate of a given population.
   a. Doubling time
   b. Population density
   c. Population size
   d. Carrying capacity

___ 8. Any population that is not restricted in some way will grow exponentially

_____.
   a. except in the case of bacteria
   b. irrespective of doubling time
   c. if the death rate is even slightly greater than the birth rate
   d. all of the above

___ 9. Interaction between resource availability and a population's tolerance to prevailing environmental conditions defines _____.

   a. the carrying capacity of the environment
   b. exponential growth
   c. the doubling time of a population
   d. density-independent factors

___ 10. In natural communities, some feedback mechanisms operate whenever populations change in size; they are _____.

   a. density-dependent factors
   b. density-independent factors
   c. always intrinsic to the individuals of the community
   d. always extrinsic to the individuals of the community

___ 11. Which of the following is *not* an intrinsic factor that can influence population size?

   a. behavior
   b. metabolism
   c. predation
   d. fertility

___ 12. Which of the following is *not* characteristic of logistic growth?

   a. S-shaped curve
   b. leveling off of growth as carrying capacity is reached
   c. unrestricted growth
   d. slow growth of a low-density population followed by rapid growth

___ 13. The population growth rate (*G*) is equal to the _____ net reproduction per individual (*r*) and number of individuals (*N*).

   a. sum of
   b. product of
   c. doubling of
   d. difference between

___ 14. $G = r_{max} N [(K - N)/K]$ represents _____.

   a. exponential growth
   b. population density
   c. population size
   d. logistic growth

___ 15. The beginning of industrialization, a rise in food production, improvement of health care, rising birth rates, and declining death rates describe the _____ stage of the demographic transition model.

   a. preindustrial
   b. transitional
   c. industrial
   d. postindustrial

___ 16. The maximum number of individuals of a population (or species) that can be sustained by a given environment defines _____.

   a. the carrying capacity
   b. exponential growth
   c. logistic growth
   d. density-independent factors

___ 17. The survivorship curve typical of industrialized human populations is type _____.

   a. I
   b. II
   c. III
   d. none of the above types

Dichotomous Choice

Circle one of two possible answers given between parentheses in each statement.

18. Population (size/density) is the number of individuals per unit of area or volume.
19. Population (size/density) refers to the number of members that make up the gene pool.
20. The general pattern of dispersal of members of a population through the habitat, for example, clumped or random, is its (density/distribution).
21. Dividing a population into pre-reproductive, reproductive, and post-reproductive categories characterizes its (age/distribution) structure.
22. The "reproductive base" of a population refers to the number of individuals in the (pre-reproductive/reproductive) age structure category.
23. When the number of population members stabilizes due to a balance in births + immigrations and deaths + emigrations, a (biotic potential/zero population growth) is demonstrated.
24. Any population whose growth is not restricted in some way will show a pattern of (exponential/logistic) growth.
25. Any increase in population size enlarges the population (distribution/reproductive) base.
26. When the course of exponential growth is plotted on a graph, a (J-shaped/S-shaped) curve is obtained.
27. When limiting factors (essential resources in short supply) act on a population, population growth (increases/decreases).

Chapter Objectives/Review Questions

This section lists general and detailed chapter objectives that can be used as review questions. You can make maximum use of these items by writing answers on a separate sheet of paper. Fill in answers where blanks are provided. To check for accuracy, compare your answers with information given in the chapter or glossary.

Page	Objectives/Questions
(804–805)	1. List the three ways that individuals can be distributed in space and provide an example of each.
(806)	2. Calculate a population growth rate (G); use values for birth, death, and number of individuals (N) that seem appropriate.
(807)	3. In the equation $G = rN$, as long as r holds constant, any population will show _____ growth.
(806)	4. Define *zero population growth* and describe how achieving it would affect the human population of the United States.
(807)	5. State how increasing the death rate of a population affects its doubling time.
(807–808)	6. Contrast the conditions that promote J-shaped curves with those that promote S-shaped curves in populations.

(808–809) 7. Understand the meaning of the logistic growth equation and know how to calculate values of G by using the logistic growth equation. Understand the meaning of r_{max} and K.

(808–809) 8. Define *limiting factors* and tell how they influence population curves.

(809) 9. Define *density-dependent controls*, give two examples, and indicate how density-dependent factors act on populations.

(809) 10. Define *density-independent controls*, give two examples, and indicate how such controls affect populations.

(810–811) 11. Understand the significance and use of life tables; be able to interpret survivorship curves.

(810–811) 12. Explain how the construction of life tables and survivorship curves can be useful to humans in managing the distribution of scarce resources.

(818–819) 13. Describe the four stages of the demographic transition model.

(804, 817) 14. Define *age structure* and explain why this is the principal reason it would be 70 to 100 years before the world population would stabilize even if the world average became 2.5 children per family.

(817) 15. Explain how timing of reproduction can affect the degree of intraspecific competition for available resources.

Integrating and Applying Key Concepts

Assume that the world has reached zero population growth. The year is 2110, and there are 10.5 billion individuals of *Homo sapiens pollutans* on Earth. You have seen stories on the community television screen about how people used to live 120 years ago. List the ways that life has changed and comment on the events that no longer happen because of the enormous human population.

Answers

Interactive Exercises

46-I. CHARACTERISTICS OF POPULATIONS
(pp. 802–805)
POPULATION SIZE AND EXPONENTIAL GROWTH (pp. 806–807)

1. Ecology; 2. community; 3. ecosystem; 4. biosphere; 5. population; 6. habitat; 7. Population density; 8. clumped; 9. immigration; 10. emigration; 11. reproductive base; 12. exponential; 13. reproductive; 14. J; 15. biotic potential; 16. a. It increases; b. It decreases; c. It must increase; 17. population growth rate; 18. It must decrease; 19. 100,000; 20. a. 100,000; b. 300,000.

46-II. LIMITS ON THE GROWTH OF POPULATIONS
(pp. 808–809)
LIFE HISTORY PATTERNS (pp. 810–813)

1. limiting; 2. Carrying capacity; 3. logistic; 4. carrying capacity; 5. decreases; 6. decreases; 7. density-dependent; 8. density independent; 9. Insurance companies; 10. III; 11. dependent; 12. carrying capacity; 13. independent; 14. C; 15. D; 16. E; 17. B; 18. A; 19. [B], E; 20. A; 21. B, C, D; 22. F.

46-III. HUMAN POPULATION GROWTH
(pp. 814–815)
CONTROL THROUGH FAMILY PLANNING (pp. 816–817)
POPULATION GROWTH AND ECONOMIC DEVELOPMENT (pp. 818–819)
QUESTIONS CONCERNING ZERO POPULATION GROWTH (p. 820)

1. a. 1962–1963; b. 2025 or sooner; c. depends on the age and optimism of the reader; 2. age structure; 3. India; 4. 4.85; 5. dependent; 6. immigration; 7. emigration; 8. birth; 9. demographic transition; 10. industrial; 11. transition; 12. 2.5; 13. 2.1; 14. reproductive; 15. delayed reproduction; 16. 5.6; 17. 90; 18. one in five; 19. T; 20. short (brief); 21. T; 22. limiting factors; 23. B; 24. D; 25. A; 26. C.

Self-Quiz

1. d; 2. a; 3. b; 4. a; 5. d; 6. d; 7. a; 8. b; 9. a; 10. a; 11. c; 12. c; 13. b; 14. d; 15. b; 16. a; 17. a; 18. density; 19. size; 20. distribution; 21. age; 22. reproductive; 23. zero population growth; 24. exponential; 25. reproductive; 26. J-shaped; 27. decreases.

47

COMMUNITY INTERACTIONS

Interactive Exercises

47-I. COMMUNITY CHARACTERISTICS (pp. 822–824)
MUTUALISM (p. 825)

Selected Italicized Words

"feeding levels"

Boldfaced, Page-Referenced Terms

(824) habitat _____

_____

(824) community _____

_____

(824) niche _____

_____

(824) neutral interaction _____

_____

(824) commensalism _____

_____

(824) mutualism _____

_____

(824) interspecific competition _____

_____

(824) predation _____

_____

(824) parasitism _____

_____

(825) symbiosis _____

_____

Fill-in-the-Blanks

A community is characterized by the kinds and diversity of species, as well as by the numbers and
(1) _____ of their individuals throughout the habitat. A(n) (2) _____ is limited to all the different
populations that occupy and are adapted to a defined area. The (3) _____ of a population is the sort of
place where it is typically located; in contrast, the (4) _____ of a population is defined by its role in a
community, including all the ecological requirements and interactions that influence that population in its
community. Robins and fruit flies are (5) _____ with human populations. The flowering plants and
their pollinators form (6) _____ relationships from which both participating populations benefit.
(7) _____ _____ links two species that have no direct effect on each other, but are related indirectly through their relationships with other species.

Matching

Choose the most appropriate answer to match with each term.

8. ___ habitat

9. ___ community

10. ___ commensalism

11. ___ mutualism

12. ___ interspecific competition

13. ___ predation

14. ___ parasitism

A. An interaction that directly benefits one species but does not harm or help the other
B. Has adverse effects on both of the interacting species
C. The parasite benefits, the host is harmed
D. The type of place one finds a particular organism; characterized by physical and chemical features as well as other species
E. The predator benefits, the prey is harmed
F. An interaction from which both species benefit
G. The associations of all populations of species in any given habitat

Complete the Table

15. Complete the following table to describe how each of the organisms listed is intimately dependent on the other for survival and reproduction in a mutualistic symbiotic interaction.

Organism	Dependency
Yucca moth	a.
Yucca plant	b.

47-II. COMPETITIVE INTERACTIONS (pp. 826–827)
PREDATION (pp. 828–829)
PREY DEFENSES (pp. 830–831)
ADAPTIVE RESPONSES TO PREY (p. 832)
PARASITIC INTERACTIONS (p. 833)

Selected Italicized Words

intraspecific competition, interspecific competition, host, speed mimicry, social parasite, Paramecium caudatum, Paramecium aurelia

Boldfaced, Page-Referenced Terms

(826) competitive exclusion _____

_____

(827) resource partitioning _____

_____

(828) predator _____

_____

(828) parasite _____

_____

(830) coevolution _____

_____

(830) camouflage _____

_____

(831) mimicry _____

_____

(833) parasitoids _____

_____

Fill-in-the-Blanks

In (1) _____ _____, one population or individual exploits the same limited resources as another, different population or interferes with another sufficiently to keep it from gaining access to the resources. According to the concept of (2) _____ _____, when two species are competing for the same resource, one tends to exclude the other from the area of niche overlap. To a greater or lesser extent, one would have the advantage; the other would be forced to modify its (3) _____. (4) _____ minimizes the intensity of competition among prey species. When two or more populations share resources in different ways, in different areas, or at different times, coexistence is possible through (5) _____ _____.

Populations of the Canadian lynx and snowshoe hare undergo (6) _____ _____, providing support for a popular model of predator-prey interactions. When predation prevents a prey population from overshooting its (7) _____ _____, the predator and prey populations tend to coexist at relatively stable levels.

(8) _____ often occurs through reciprocal selection pressures operating on two ecologically interacting populations. Prey populations attempt to avoid predation by adaptations for flight, hiding, fighting, or (9) _____. (10) _____ refers to adaptations in form, patterning, color, or behavior that enable an organism to blend with its background and escape detection. Sometimes, weaponless prey species strikingly resemble unpalatable or dangerous species; this type of (11) _____ helps protect prey.

Matching

Match each of the following with the most appropriate answer. The same letter may be used more than once. Use only one letter per blank.

12. ___ polar bears against snow

13. ___ cornered beetles, skunks, and stink beetles producing awful
     odors

14. ___ tapeworms and humans

15. ___ yucca moth and yucca plant

16. ___ Canadian lynx and snowshoe hare

17. ___ tigers against tall-stalked and golden grasses

18. ___ striped skunk, yellow-banded wasp, and bright-orange monarch
     butterfly

19. ___ fungal mycelia and plant root hairs

20. ___ insect larvae that always kill larvae or pupae of other insect
     species and eat them

21. ___ baboon on the run turns to give canine tooth display to a pursuing leopard

22. ___ an aggressive yellow jacket is the probable model for similar-looking but edible flies

23. ___ least bittern with coloration similar to surrounding withered reeds

A. parasitism
B. mutualism
C. predator-prey relationship
D. camouflage
E. warning coloration
F. mimicry
G. moment-of-truth defense
H. parasitoid

Dichotomous Choice

Circle one of two possible answers given between parentheses in each statement.

24. Intraspecific competition is (more/less) fierce than interspecific competition.
25. When all individuals have equal access to a required resource and some are better at exploiting it, the interaction tends to (reduce/increase) the common supply of the shared resource unless it is abundant.
26. Two species are (less/more) likely to coexist in the same habitat when they are very similar in their use of scarce resources.
27. Gause used two species of *Paramecium* competing for (the same/different) bacterial cells to illustrate competitive exclusion.
28. Gause also found that two species of *Paramecium* that did not overlap as much in their requirements were (more/less) likely to coexist.

47-III. SUCCESSION—DIRECTIONAL CHANGE IN COMMUNITIES (pp. 834–835)
REGARDING COMMUNITY STABILITY (pp. 836–837)
EFFECTS OF DISPERSAL (pp. 838–839)
PATTERNS OF BIODIVERSITY (pp. 840–841)

Selected Italicized Words

primary succession, secondary succession, facilitation model, inhibition model, jump dispersal, cichlid

Boldfaced, Page-Referenced Terms

(834) succession _____

_____

(834) pioneer species _____

_____

(834) climax community _____

_____

(835) climax-pattern model _____

_____

(836) keystone species _____

_____

(838) geographic dispersal _____

_____

(841) distance effect _____

_____

(841) area effect _____

_____

Problems

1. After consideration of the distance effect, the area effect, and species diversity patterns as related to the equator, answer the following question. There are two islands (B and C) of the same size and topography that are equidistant from the African coast (A), as shown in the illustration. Which will have the higher species diversity values?

   _____

2. Name two factors responsible for creating the higher species diversity values as related to the distance of land and sea from the equator.

   a. _____

   b. _____

Fill-in-the-Blanks

As summarized by the ecologist Charles Krebs, (3) _____ _____ is a dynamic process, resulting from a balance between the colonizing ability of some species and the (4) _____ capacity of other species; (3) might not always proceed from simple networks of interactions to complex.

(5) _____ _____ occurs after a disturbance wipes out one or more communities; in the years that follow, earlier community stages become reestablished. (6) _____ _____ is the ratio of the total number of different species in an area to the total number of all individuals in the same area; its value (7) (choose one) ❑ increases ❑ decreases ❑ stays the same as one moves from Earth's poles toward the equator.

Sequence

Arrange the following in correct sequence of their appearance in the primary succession of glaciated regions of Alaska, from first to last.

8. ____	A. alders
9. ____	B. cottonwoods and willows
10. ____	C. mountain avens (Dryas)
11. ____	D. Sitka spruce and western hemlock

True-False

If false, explain why.

___ 12. If two islands are equivalent distances from source areas, larger islands tend to support more species than smaller islands.

___ 13. The closer an island is to a source of potential colonists, the more different species it supports.

Short Answer

14. Explain how modest fire disturbances among groves of dominant giant sequoia trees of the Sierra Nevada in California benefit that climax community.

_____

_____

_____

_____

_____

True/False

If the statement is true, write a T in the blank. If the statement is false, make it correct by changing the underlined word(s) and writing the correct word(s) in the answer blank.

_____ 15. Nine species of fruit-eating pigeons living in the same forest are a good example of resource <u>disturbance</u>.

_____ 16. Water hyacinths and killer bees are examples of <u>predator</u> introductions.

_____ 17. In resource partitioning, <u>similar</u> species generally share the same kind of resource in different ways, in different areas, or at different times.

_____ 18. The number of algal species is greatest in tidepools with <u>high</u> densities of algae-eating periwinkles.

_____ 19. Predation reduces the density of prey populations but can also <u>increase</u> competition between prey species and promote their coexistence.

_____ 20. In the rocky intertidal zone, mussels are the main prey of sea stars; the reduction in numbers of mussels maintains a <u>greater</u> diversity of other invertebrate prey species.

Choice

For questions 21–30, choose from the following:

> a. primary succession   b. secondary succession

___ 21. Following a disturbance, a patch of habitat or a community moves once again toward the climax state.

___ 22. Successional changes begin when a pioneer population colonizes a barren habitat.

___ 23. Involves populations that are adapted to growing in habitats that cannot support most other populations.

___ 24. Many plants in this succession arise from seeds or seedlings that are already present when the process begins.

___ 25. Early successional populations inhibit the growth of later ones, which become dominant only when some disturbance removes the established competitors.

___ 26. The first plants are typically small with short life cycles; each year they produce an abundance of quickly dispersed small seeds.

___ 27. A successional pattern that occurs in ponds, shallow lakes, abandoned fields, and in parts of established forests.

___ 28. On land, early and late species often are able to grow together under prevailing conditions.

___ 29. Might occur on a new volcanic island or on land exposed by the retreat of a glacier.

___ 30. Includes populations adapted to growing in areas exposed to intense sunlight, wide temperature swings, and nutrient-deficient soil.

Self-Quiz

___ 1. All the populations of different species that occupy and are adapted to a given habitat are referred to as a(n) _____.

    a. biosphere
    b. community
    c. ecosystem
    d. niche

___ 2. The range of all factors that influence whether a species can obtain resources essential for survival and reproduction is called the _____ of a species.

    a. habitat
    b. niche
    c. carrying capacity
    d. ecosystem

___ 3. A one-way relationship in which one species benefits and the other is directly harmed is called _____.

    a. commensalism
    b. competitive exclusion
    c. parasitism
    d. mutualism

___ 4. A lopsided interaction that directly benefits one species but does not harm or help the other much, if at all, is _____.

    a. commensalism
    b. competitive exclusion
    c. predation
    d. mutualism

___ 5. An interaction in which both species benefit is best described as _____.

    a. commensalism
    b. mutualism
    c. predation
    d. parasitism

___ 6. The brown-headed cowbird removes an egg from the nest of another kind of bird and lays one as a "replacement." This is an example of _____.

    a. commensalism
    b. competitive exclusion
    c. mutualism
    d. social parasitism

___ 7. When an inexperienced predator attacks a yellow-banded wasp, the predator receives the pain of a stinger and will not attack again. This is an example of _____.

    a. mimicry
    b. camouflage
    c. a prey defense
    d. warning coloration
    e. both c and d

___ 8. _____ is represented by foxtail grass, mallow plants, and smartweed because their root systems exploit different areas of the soil in a field.

    a. Succession
    b. Resource partitioning
    c. A climax community
    d. A disturbance

___ 9. During the process of community succession, _____.

    a. pioneer populations adapt to growing in habitats that cannot support most species
    b. pioneers set the stage for their own replacement
    c. later successional populations crowd out the pioneers
    d. species composition eventually is stable in the form of the climax community
    e. all of the above

___ 10. G. Gause utilized two species of *Paramecium* in a study that described _____.

   a. interspecific competition and competitive exclusion
   b. resource partitioning
   c. the establishment of territories
   d. coevolved mutualism

___ 11. The most striking patterns of species diversity on land and in the seas relate to _____.

   a. distance effect
   b. area effect
   c. immigration rate for new species
   d. distance from the equator

___ 12. In a community, friction between two competing populations might be minimized by _____.

   a. partitioning the niches in time
   b. partitioning the niches spatially
   c. fights to the death
   d. both a and b

___ 13. Robins probably have a(n) _____ relationship with humans.

   a. parasitic
   b. mutualistic
   c. obligate
   d. commensal

___ 14. The relationship between an insect and the plants it pollinates (for example, apple blossoms, dandelions, and honeysuckle) is best described as _____.

   a. mutualism
   b. competitive exclusion
   c. parasitism
   d. commensalism

___ 15. The relationship between the yucca plant and the yucca moth that pollinates it is best described as _____.

   a. camouflage
   b. commensalism
   c. competitive exclusion
   d. coevolution

Chapter Objectives/Review Questions

This section lists general and detailed chapter objectives that can be used as review questions. You can make maximum use of these items by writing answers on a separate sheet of paper. Fill in answers where blanks are provided. To check for accuracy, compare your answers with information given in the chapter or glossary.

Page	Objectives/Questions
(824)	1. The type of place where you normally find a maple is its _____.
(824)	2. List five factors that shape the structure of a biological community.
(824)	3. The full range of environmental and biological conditions under which its members can live, grow, and reproduce is called the _____ of that species.
(824)	4. The interaction of a bird's nest and a tree is known as _____.
(825)	5. Define *symbiosis*.
(825)	6. The interdependence of the yucca plant and yucca moth is an example of _____.
(826)	7. Describe a study that demonstrates laboratory evidence in support of the competitive exclusion concept.
(827)	8. Cite one example of resource partitioning.
(828)	9. A predator gets food from other living organisms, its _____.
(828–829)	10. List three factors that influence the outcome of predator-prey interactions.
(830)	11. Suggest why coevolving might serve the interests of the populations concerned.
(830–831)	12. Be able to completely define and give examples of the following prey defenses: warning coloration and mimicry, moment-of-truth defenses, and camouflage.
(833)	13. Parasites tend to _____ with their hosts in ways that produce less-than-fatal effects.
(833)	14. _____ are insect larvae that always kill larvae or pupae of other insect species for food.
(834)	15. Distinguish between primary and secondary succession.
(836)	16. Describe how fire disturbances positively affect sequoia communities.
(836)	17. Explain how the presence of a keystone species can increase biodiversity in a community.

(838–840) 18. Explain how the introduction of nonnative species can be disastrous. List five specific examples of species introductions into the United States that have had adverse results (see Table 47.1).

(839–940) 19. Tell the story of the cichlids in Lake Victoria and relate it to the United States law that mandates an environmental impact statement be prepared before any major potentially disruptive project is begun.

(840) 20. Estimate qualitatively the differences in species diversity and abundance of organisms likely to exist on two islands with the following characteristics: Island A has an area of 6,000 square miles, and Island B has an area of 60 square miles; both islands lie at $10°$ N latitude and are equidistant from the same source area of colonizers.

Integrating and Applying Key Concepts

1. If you were Ruler of All People on Earth, how would you organize industry and human populations in an effort to solve our most pressing pollution problems?

2. Is there a *fundamental niche* that is occupied by humans? If you think so, describe the minimal abiotic and biotic conditions required by populations of humans in order to live and reproduce. (Note that "thrive and be happy" are not criteria.) If you do not think so, state why.

3. These minimal niche conditions can be viewed as resource categories that must be protected by populations if they are to survive. Do you believe that the cold war between the United States and the Soviet Union primarily involved protection of minimal niche conditions, or do you believe that the cold war was based on other, more (or less) important factors?

    a. If the former, how do you think *minimal* niche conditions might have been guaranteed for all humans willing and able to accept certain responsibilities as their contribution toward enabling this guarantee to be met?

    b. If the latter, identify what you think those factors are and explain why you consider them more (or less) important than minimal niche conditions.

Answers

Interactive Exercises

47-I. COMMUNITY CHARACTERISTICS (p. 824)
  MUTUALISM (p. 825)

1. dispersion; 2. community; 3. habitat; 4. niche; 5. commensal (symbiotic); 6. mutualistic; 7. neutral interaction; 8. D; 9. G; 10. A; 11. F; 12. B; 13. E; 14. C; 15. a. Cannot complete life cycle in any other plant and its larvae eat only yucca seeds; b. The yucca moth is the plant's only pollinator.

47-II. COMPETITIVE INTERACTIONS (pp. 826–827)
  PREDATION (pp. 828–829)
  PREY DEFENSES (pp. 830–831)
  ADAPTIVE RESPONSES TO PREY (p. 832)
  PARASITIC INTERACTIONS (p. 833)

1. interspecific competition; 2. competitive exclusion; 3. niche; 4. interference competition (Predation); 5. resource partitioning; 6. irregular fluctuations; 7. carrying capacity; 8. Coevolution; 9. camouflage; 10. Camouflage; 11. mimicry; 12. D; 13. G; 14. A; 15. B; 16. C; 17. D; 18. E; 19. B; 20. H; 21. G (also predator-prey); 22. F; 23. D; 24. more; 25. reduce; 26. less; 27. the same; 28. more.

47-III. SUCCESSION—DIRECTIONAL CHANGE IN COMMUNITIES (pp. 834–835)
  REGARDING COMMUNITY STABILITY (pp. 836–837)
  EFFECTS OF DISPERSAL (pp. 838–839)
  PATTERNS OF BIODIVERSITY (pp. 840–841)

1. Island C; 2. a. Greater annual amount of sunlight promotes greater resource availability; b. Species diversity is self-reinforcing. When more plant species coexist, more herbivore species emerge. More predators and parasites evolve in response to the diversity of prey and hosts; 3. primary succession; 4. competitive; 5. Secondary succession; 6. Species diversity; 7. increases; 8. C; 9. A; 10. B; 11. D; 12. T; 13. T; 14. Sequoia seeds germinate only in the absence of smaller, shade-tolerant plant species. Modest fires eliminate trees and shrubs that compete with young sequoias but do not damage mature sequoias; 15. partitioning; 16. species; 17. T; 18. intermediate or moderate; 19. reduce or decrease; 20. T; 21. b; 22. a; 23. a; 24. b; 25. b; 26. a; 27. b; 28. b; 29. a; 30. a.

Self-Quiz

1. b; 2. b; 3. c; 4. a; 5. b; 6. d; 7. e; 8. b; 9. e; 10. a; 11. d; 12. d; 13. d; 14. a; 15. d.

48

ECOSYSTEMS

Interactive Exercises

48-I. THE NATURE OF ECOSYSTEMS (pp. 844–847)
ENERGY FLOW THROUGH ECOSYSTEMS (pp. 848–849)
CASE STUDY: ENERGY FLOW AT SILVER SPRINGS, FLORIDA (p. 850)

Selected Italicized Words

herbivores, carnivores, parasites, omnivores, energy input, nutrient inputs, nutrient outputs, gross and net primary productivity

Boldfaced, Page-Referenced Terms

(846) primary producers _____

_____

(846) consumers _____

_____

(846) decomposers _____

_____

(846) detritivores _____

_____

(846) ecosystem _____

_____

(846) trophic levels _____

_____

(847) food chain _____

_____

(847) food webs _____

_____

(848) primary productivity _____

_____

(848) grazing food webs _____

_____

(848) detrital food webs _____

_____

(848) ecological pyramid _____

_____

Fill-in-the-Blanks

Photosynthetic (1) _____ capture (2) _____ and concentrate (3) _____ for ecosystems. All
organisms in a community that are the same number of energy transfers away from the initial energy input
into their ecosystem constitute a(n) (4) _____ _____. All (5) _____ are primary consumers.
(6) _____ obtain their energy from the organic remains and products of other organisms.
(7) _____ are invertebrates that feed on partially decomposed bits of organic matter. Energy flows
(8) _____-_____ through an ecosystem; materials are (9) _____ through an ecosystem.
During every transfer of energy, (10) _____ is released as a by-product.

The energy remaining after respiration and contained in plant biomass is its (11) _____ _____
_____. (12) _____ typically is expressed as grams (dry weight) of organic matter per unit of area.
In a(n) (13) _____ _____ _____, decomposers feed on organic wastes, dead tissues, and
decomposed organic matter. Earthworms, millipedes, and fly and beetle larvae are examples of
(14) _____ feeders. Only about (15) _____ to (16) _____ percent of the energy entering one
trophic level becomes available to organisms at the next level.

Energy flows into ecosystems from an outside source, which in most cases is the (17) _____.
Energy flows through the food (18) _____ of ecosystems. Living tissues of photosynthesizers are the

basis of (19) _____ food webs. The remains of photosynthesizers and consumers are the basis of (20) _____ food webs. Energy leaves ecosystems mainly by loss of metabolic (21) _____, which each organism generates. The trophic structure of an ecosystem is often represented by an "ecological (22) _____." In such schemes, (23) _____ form the base for successive tiers of consumers above them. Some pyramids are based on (24) "_____" as determined by the weight of all the members at each trophic level. Some pyramids of (25) _____ are "upside down" with the smallest tier at the bottom; they represent an ecosystem in which reproductive rates and body sizes differ at each trophic level. A(n) (26) _____ pyramid reflects the energy losses at each transfer to a different trophic level and are always "right-side up" with the largest energy tier at the base.

Complete the Table

27. Complete the following table of productivity definitions.

Productivity Type	Definition
Primary productivity	a.
Gross primary productivity	b.
Net primary productivity	c.

Dichotomous Choice

Circle one of two possible answers given between parentheses in each statement.

28. The amount of energy actually stored in an ecosystem depends on how many (plants/animals) are present and the balance between photosynthesis and aerobic respiration in the plants.
29. In a harsh ecosystem environment, productivity would be expected to be (lower/higher).
30. Heat losses represent a one-way flow of (energy/materials) out of the ecosystem.
31. In (grazing/detrital) food webs, energy flows from plants to herbivores, then through some array of carnivores.
32. In (grazing/detrital) food webs, energy flows mainly from plants through decomposers and detritivores.

True-False

___ 33. Energy pyramids are always "right-side up" with a large energy base at the bottom.

___ 34. The amount of useful energy flowing through the consumer trophic levels of an ecosystem increases at each energy transfer as metabolically generated heat is lost and as energy tied up in the bonds of food molecules is shunted into organic wastes.

Choice

For questions 35–49, choose from the following:

a. primary producer   b. consumer   c. detritivore   d. decomposer

___ 35. mule deer

___ 36. earthworm

___ 37. parasites

___ 38. the only category lacking heterotrophs

___ 39. omnivores

___ 40. tapeworm

___ 41. herbivores

___ 42. wolf

___ 43. fungi and bacteria

___ 44. carnivores

___ 45. green plants

___ 46. *Homo sapiens*

___ 47. crabs

___ 48. autotrophs

___ 49. grasshopper

Dichotomous Choice

Circle one of two possible answers given between parentheses in each statement.

50. Ecosystems are (open/closed) systems and so are not self-sustaining.
51. Minerals carried by erosion into a lake represent nutrient (input/output).
52. Energy (can/cannot) be recycled.
53. Nutrients typically (can/cannot) be cycled.
54. Ecosystems have energy and nutrient inputs as well as nutrient output and (have/lack) energy output.

Matching

Match each organism in the Antarctic food chain given below with the principal trophic level it occupies. Some letters may not be needed. (See Fig. 48.3 in the text.)

55. ___ emperor penguins
56. ___ krill
57. ___ blue whale
58. ___ diatoms
59. ___ leopard seal
60. ___ fishes, small squids
61. ___ killer whale

A. Chemosynthetic autotrophs
B. Herbivores
C. Photosynthetic autotrophs
D. Secondary consumers
E. Tertiary consumers

48-II. BIOGEOCHEMICAL CYCLES—AN OVERVIEW (p. 851)
HYDROLOGIC CYCLE (pp. 852–853)
CARBON CYCLE (pp. 854–857)
NITROGEN CYCLE (pp. 858–859)
PHOSPHORUS CYCLE (p. 860)
ECOSYSTEM MODELING (p. 861)

Selected Italicized Words

nutrients, hydrologic cycle, atmospheric cycles, sedimentary cycles, "holding stations," "greenhouse gases," Rhizobium

Boldfaced, Page-Referenced Terms

(851) biogeochemical cycles _____

_____

(852) hydrologic cycle _____

_____

(852) watersheds _____

_____

(854) carbon cycle _____

_____

(856) greenhouse effect _____

_____

(858) nitrogen cycle _____

_____

(858) nitrogen fixation _____

_____

(858) decomposition _____

_____

(858) ammonification _____

_____

(858) nitrification _____

_____

(858) denitrification _____

_____

(860) phosphorus cycle _____

_____

(861) ecosystem modeling _____

_____

(861) biological magnification _____

_____

Short Answer

1. In what form are elements used as nutrients usually available to producers? _____

_____

2. How is the ecosystem's reserve of nutrients maintained? _____

_____

3. How does the amount of a nutrient being cycled through most major ecosystems compare with the amount entering or leaving in a given year? _____

_____

4. What are the input sources for an ecosystem's nutrient reserves? _____

_____

5. What are the output sources of nutrient loss for land ecosystems? _____

_____

Complete the Table

6. Complete the following table to summarize the functions of the three types of biogeochemical cycles.

Biogeochemical Cycle Type	General Function(s)
Hydrologic cycle	a.
Atmospheric cycles	b.
Sedimentary cycles	c.

Matching

Choose the most appropriate answer for questions 7–14; 11–14 may have more than one answer.

7. _____ solar energy

8. _____ source of most water

9. _____ watershed

10. _____ water and plants taking up water

11. _____ forms of precipitation falling to land

12. _____ deforestation

13. _____ have important roles in the global hydrologic cycle

14. _____ forms of atmospheric water

A. mostly rain and snow
B. mechanisms by which nutrients move into and out of ecosystems
C. water vapor, clouds, and ice crystals
D. where precipitation of a specified region becomes funneled into a single stream or river
E. may have long-term disruptive effects on nutrient availability for an entire ecosystem
F. slowly drives water through the atmosphere, on or through land mass surface layers, to oceans, and back again
G. ocean currents and wind patterns
H. evaporation from the oceans

Matching

Choose the one most appropriate answer for each.

15. ___ greenhouse gases
16. ___ carbon dioxide fixation
17. ___ carbon cycle
18. ___ ways carbon enters the atmosphere
19. ___ greenhouse effect
20. ___ carbon dioxide (CO_2)
21. ___ oceans and plant biomass accumulation

A. form of most of the atmospheric carbon
B. aerobic respiration, fossil fuel burning, and volcanic eruptions
C. CO_2, CFCs, CH_4, and N_2O
D. photosynthesizers incorporate carbon atoms into organic compounds
E. annual "holding stations" for about half of all atmospheric carbon
F. carbon reservoirs → atmosphere and oceans → through organisms → carbon reservoirs
G. warming of Earth's lower atmosphere due to accumulation of greenhouse gases

Choice

For questions 22–26, choose from the following:

    a. nitrogen cycle   b. nitrogen fixation   c. ammonification   d. nitrification   e. denitrification

___ 22. Ammonia or ammonium ions in soil are stripped of electrons, and nitrite (NO_2^-) is the result; other bacteria convert nitrite to nitrate (NO_3^-).

___ 23. Bacteria convert nitrate or nitrite to N_2 and a bit of nitrous oxide (N_2O).

___ 24. Bacteria and fungi break down nitrogen-containing wastes and plant and animal remains; released amino acids and proteins are used for growth with the excess given up as ammonia or ammonium ions that plants can use.

___ 25. Occurs in the atmosphere (largest reservoir); only certain bacteria, volcanic action, and lightning can convert N_2 into forms that can enter food webs.

___ 26. A few kinds of bacteria convert N_2 to ammonia (NH_3), which dissolves quickly in water to form ammonium (NH_4^+).

Short Answer

27. List reasons that an insufficient soil nitrogen supply is a problem for land plants.

_____

_____

_____

_____

_____

Fill-in-the-Blanks

In the (28) _____ _____, Earth's surface is warmed by sunlight and radiates (29) _____ (infrared wavelengths) to the atmosphere and space. Greenhouse gases such as (30) _____, which are used as refrigerants, solvents, and plastic foams, and (31) _____ _____, which is released from burning fossil fuels, deforestation, car exhaust, and factory emissions, will together be responsible for about 75 percent of the global warming trend by 2020.

The (32) _____ _____ studies demonstrated that stripping the land of vegetation disrupts nutrient retention by an entire ecosystem for a long time; in the watershed, (33) _____ _____ efficiently mined the soil for calcium, moving it up into the growing plant parts. (34) _____ and weathering of rocks brought calcium replacements back into the watershed. In (35) _____ cycles, the element does not have a gaseous phase. In most major ecosystems, the amount of a nutrient that is (36) _____ within the ecosystem is greater than the amount entering or leaving in a given year. Water molecules stay in the atmosphere for approximately (37) _____ days.

(38) _____ can assimilate nitrogen from the air in the process known as (39) _____ _____. In (40) _____, either ammonia or ammonium ions are stripped of electrons, and (41) _____ (NO_2^-) is released as a product of the reactions. Under some conditions, nitrate is converted into (42) _____ _____ by denitrifying bacteria.

During World War II, DDT was sprayed in the tropical Pacific to control (43) _____ responsible for transmitting the organisms that cause a dangerous disease, (44) _____. Because of its stability, DDT is a prime candidate for (45) _____ _____—the increasing concentration of a nondegradable substance as it moves up through trophic levels. (46) _____ _____ is a method of combining crucial bits of information about an ecosystem through computer programs and models in order to predict the outcome of the next disturbance.

Self-Quiz

___ 1. A network of interactions that involve the cycling of materials and the flow of energy between a community and its physical environment is a(n) _____.

    a. population
    b. community
    c. ecosystem
    d. biosphere

___ 2. _____ consume dead or decomposing particles of organic matter.

    a. Herbivores
    b. Parasites
    c. Detritivores
    d. Carnivores

___ 3. In the Antarctic, blue whales feed mainly on _____.

    a. petrels
    b. krill
    c. seals
    d. fish and small squids

___ 4. Which of the following is a primary consumer?

    a. cow
    b. dog
    c. hawk
    d. all of the above

___ 5. In a natural community, the primary consumers are _____.

    a. herbivores
    b. carnivores
    c. scavengers
    d. decomposers

___ 6. A straight-line sequence of who eats whom in an ecosystem is sometimes called a(n) _____.

    a. trophic level
    b. food chain
    c. ecological pyramid
    d. food web

___ 7. A growing animal must take in _____ of photosynthetically derived food energy to produce every 1 kilocalorie of energy stored in its body.

    a. 1 kilocalorie
    b. 10 kilocalories
    c. 100 kilocalories
    d. 2,000 kilocalories

___ 8. Chemosynthesizers utilize as their direct energy source _____.

    a. energy released from degrading the organic remains and products of all other organisms
    b. energy from sunlight
    c. energy released during dehydration synthesis (condensation) reactions
    d. energy released from the oxidation of inorganic substances

___ 9. Of the 1,700,000 kilocalories of solar energy that entered an aquatic ecosystem in Silver Springs, Florida, investigators determined that about _____ percent of incoming solar energy was trapped by photosynthetic autotrophs.

    a. 1
    b. 10
    c. 25
    d. 74

___ 10. One square meter of an ecosystem receives 1,600 kilocalories of light energy each day. How much energy is likely to be passed on to the ecosystem's herbivores from that square meter?

    a. 160 to 480 kilocalories
    b. 16 to 48 kilocalories
    c. 1.6 to 4.8 kilocalories
    d. 0.16 to 0.48 kilocalories

___ 11. _____ is a process in which nitrogenous waste products or organic remains of organisms are decomposed by soil bacteria and fungi that use the amino acids being released for their own growth and release the excess as ammonia or ammonium, which plants take up.

    a. Nitrification
    b. Ammonification
    c. Denitrification
    d. Nitrogen fixation

___ 12. Water that passes from a plant to its surrounding air arrives there by _____.

    a. evaporation
    b. transpiration
    c. detention
    d. precipitation

___ 13. In the carbon cycle, carbon enters the atmosphere through _____.

    a. carbon dioxide fixation
    b. respiration, burning, and volcanic eruptions
    c. oceans and accumulation of plant biomass
    d. release of greenhouse gases

___ 14. _____ refers to an increase in concentration of a nondegradable (or slowly degradable) substance in organisms as it is passed along food chains.

    a. Ecosystem modeling
    b. Nutrient input
    c. Biogeochemical cycle
    d. Biological magnification

Chapter Objectives/Review Questions

This section lists general and detailed chapter objectives that can be used as review questions. You can make maximum use of these items by writing answers on a separate sheet of paper. To check for accuracy, compare your answers with information given in the chapter or glossary.

Page	Objectives/Questions
(846)	1. List the principal trophic levels in an ecosystem of your choice; state the source of energy for each trophic level and give one or two examples of organisms associated with each trophic level.
(846)	2. Explain why nutrients can be completely recycled but energy cannot.
(846)	3. A(n) _____ is a complex of organisms and their physical environment, all interacting through a flow of energy and a cycling of materials.
(846)	4. Members of an ecosystem fit somewhere in a hierarchy of energy transfers (feeding relationships) called _____ levels.
(847)	5. Distinguish between food chains and food webs.
(846–848)	6. Understand how materials and energy enter, pass through, and exit an ecosystem.
(848)	7. Distinguish between net and gross primary productivity.
(848–849)	8. Compare grazing food webs with detrital food webs. Present an example of each.
(848)	9. Ecological pyramids are based on _____ as determined by the weight of all the members of each trophic level; _____ pyramids reflect the energy losses at each transfer to a different trophic level.
(851)	10. In _____ cycles, the nutrient is transferred from the environment to organisms, then back to the environment—which serves as a large reservoir for it.
(852–853)	11. Be able to discuss water movements through the hydrologic cycle.
(852–853)	12. Explain what studies in the Hubbard Brook watershed have taught us about the movement of substances (water, for example) through a forest ecosystem.

(854)	13.	The carbon cycle traces carbon movement from reservoirs in the _____ and oceans, through organisms, then back to reservoirs.
(856)	14.	Certain gases cause heat to build up in the lower atmosphere, a warming action known as the _____ effect.
(858)	15.	A major element found in all proteins and nucleic acids moves in an atmospheric cycle called the _____ cycle.
(858–859)	16.	Define the chemical events that occur during nitrogen fixation, nitrification, ammonification, and denitrification.
(858–859)	17.	Explain why agricultural methods in the United States tend to put more energy into mechanized agriculture in the form of fertilizers, pesticides, food processing, storage, and transport than is obtained from the soil in the form of energy stored in foods.
(861)	18.	Describe how DDT damages ecosystems; discuss biological magnification.
(861)	19.	Through _____ modeling, crucial bits of information about different ecosystem components are identified and used to build computer models for predicting outcomes of ecosystem disturbances.

Integrating and Applying Key Concepts

In 1971, *Diet for a Small Planet* was published. Frances Moore Lappé, the author, felt that people in the United States wasted protein and ate too much meat. She said, "We have created a national consumption pattern in which the majority, who can pay, overconsume the most inefficient livestock products [cattle] well beyond their biological needs (even to the point of jeopardizing their health), while the minority, who cannot pay, are inadequately fed, even to the point of malnutrition." Cases of *marasmus* (a nutritional disease caused by prolonged lack of food calories) and *kwashiorkor* (caused by severe, long-term protein deficiency) have been found in Nashville, Tennessee, and on an Indian reservation in Arizona, respectively. Lappé's partial solution to the problem was to encourage people to get as much of their protein as possible directly from plants and to supplement that with less meat from the more efficient converters of grain to protein (chickens, turkeys, and hogs) and with seafood and dairy products. Most of us realize that feeding the hungry people of the world is not just a matter of distributing the abundance that exists—that it is being prevented in part by political, economic, and cultural factors. Devise two full days of breakfasts, lunches, and dinners that would enable you to exploit the lowest acceptable trophic levels to sustain yourself healthfully.

Answers

Interactive Exercises

48-I. THE NATURE OF ECOSYSTEMS (pp. 846–847)
ENERGY FLOW THROUGH ECOSYSTEMS (pp. 848–849)
CASE STUDY: ENERGY FLOW AT SILVER SPRINGS, FLORIDA (p. 850)

1. autotrophs; 2. sunlight; 3. energy; 4. trophic level; 5. herbivores; 6. Decomposers; 7. Detritivores (crabs, nematodes, or earthworms); 8. one-way; 9. cycled; 10. heat; 11. net primary production; 12. Biomass; 13. detrital food web; 14. detrital; 15. 6; 16. 16; 17. Sun; 18. webs; 19. grazing; 20. detrital; 21. heat; 22. pyramid; 23. producers; 24. biomass; 25. biomass; 26. energy; 27. a. The rate at which the ecosystem's producers capture and store a given amount of energy in a given length of time; b. The total rate of photosynthesis for an ecosystem during a specified period; c. The rate of energy storage in plant tissues in excess of the rate of aerobic respiration by the plants themselves; 28. plants; 29. lower; 30. energy; 31. grazing; 32. detrital; 33. T; 34. F; 35. b; 36. c; 37. b; 38. a; 39. b; 40. b; 41. b; 42. b; 43. d; 44. b; 45. a; 46. b; 47. b, c; 48. a; 49. b; 50. open; 51. input; 52. cannot; 53. can; 54. have; 55. E; 56. B; 57. D; 58. C; 59. E; 60. D; 61. E.

48-II. BIOGEOCHEMICAL CYCLES—AN OVERVIEW (p. 851)
HYDROLOGIC CYCLE (pp. 852–853)
CARBON CYCLE (pp. 854–857)
NITROGEN CYCLE (pp. 858–859)
PHOSPHORUS CYCLE (p. 860)
ECOSYSTEM MODELING (p. 861)

1. As mineral ions such as ammonium (NH_4^+); 2. Inputs from the physical environment and the cycling activities of decomposers and detritivores; 3. The amount of a nutrient being cycled through the ecosystem is greater; 4. Rainfall or snowfall, metabolism (such as nitrogen fixation), and weathering of rocks; 5. Losses by runoff and evaporation; 6. a. Oxygen and hydrogen move in the form of water molecules; b. A large portion of the nutrient is in the form of atmospheric gas such as carbon and nitrogen; c. The nutrient is not in gaseous forms; it moves from land to the seafloor and only "returns" to land through geological uplifting of long duration; phosphorus is an example; 7. F; 8. H; 9. D; 10. B; 11. A (C); 12. E (B); 13. G (H); 14. C (A); 15. C; 16. D; 17. F; 18. B; 19. G; 20. A; 21. E; 22. d; 23. e; 24. c; 25. a (b); 26. b; 27. Soil nitrogen compounds are vulnerable to being leached and lost from the soil; some fixed nitrogen is lost to air by denitrification; nitrogen fixation comes at high metabolic cost to plants that are symbionts of nitrogen-fixing bacteria; losses of nitrogen are enormous in agricultural regions through the tissues of harvested plants, soil erosion, and leaching processes; 28. greenhouse effect; 29. heat; 30. CFCs; 31. carbon dioxide; 32. Hubbard Brook; 33. tree roots; 34. Erosion (Rainfall); 35. sedimentary; 36. cycled; 37. ten; 38. Bacteria (*Rhizobium*); 39. nitrogen fixation; 40. nitrification; 41. nitrite; 42. nitrous oxide; 43. mosquitoes; 44. malaria; 45. biological magnification; 46. Ecosystem modeling.

Self-Quiz

1. c; 2. c; 3. b; 4. a; 5. a; 6. b; 7. b; 8. d; 9. a; 10. c; 11. b; 12. b; 13. b; 14. d.

49

THE BIOSPHERE

Interactive Exercises

49-I. AIR CIRCULATION PATTERNS AND REGIONAL CLIMATES (pp. 864–867)
THE OCEAN'S EFFECT ON REGIONAL CLIMATES (p. 868)
EFFECTS OF TOPOGRAPHY ON REGIONAL CLIMATES (p. 869)

Selected Emphasized Words

"slope," "tug," "topography"

Boldfaced, Page-Referenced Terms

(865) biosphere _____

_____

(865) hydrosphere _____

_____

(865) lithosphere _____

_____

(865) atmosphere _____

_____

(865) climate _____

_____

(866) ozone layer _____

_____

(866) temperature zones _____

_____

(869) monsoons _____

_____

(869) rain shadow _____

_____

Matching

Choose the one most appropriate answer for each.

1. ___ biosphere
2. ___ climate
3. ___ Earth's regional climates
4. ___ Earth's seasonal climate variations
5. ___ global patterns of air circulation
6. ___ greenhouse effect
7. ___ influence the distribution and dominant features of ecosystems
8. ___ lithosphere
9. ___ monsoon
10. ___ ozone layer
11. ___ rain shadow
12. ___ surface currents and drifts in the ocean
13. ___ westerly

A. due to tilt of Earth's axis, which yields annual incoming solar radiation variation; provides daylength and temperature differences
B. caused by molecules of the lower atmosphere absorbing some heat and then reradiating it back to Earth; this heat drives Earth's weather systems
C. influenced by mountains, valleys, and other land formations
D. the entire Earth realm where organisms live
E. due to warm equatorial air rising, then spreading southward and northward; Earth's rotation creates worldwide belts of prevailing east and west winds
F. reduction in rainfall on the leeward side of high mountains
G. created by Earth's rotation, prevailing surface winds, and water temperature variations
H. refers to prevailing weather conditions such as temperature, humidity, wind speed, cloud cover, and rainfall
I. contains atmospheric molecules that absorb potentially lethal ultraviolet wavelengths
J. the interaction of atmospheric circulation patterns, ocean currents, and landforms
K. a wind that blows in an eastward direction
L. rocks, soils, sediments, and outer portions of crust
M. very wet summers alternate with very dry winters

Identification/ Matching

Notice the numbered currents shown in the map. Match their numbers with the lettered choices below.

14. ___	18. ___	A. Benguela current	E. Humboldt current
15. ___	19. ___	B. California current	F. Japan current
16. ___	20. ___	C. Canary current	G. Labrador current
17. ___		D. Gulf Stream	H. North Atlantic drift

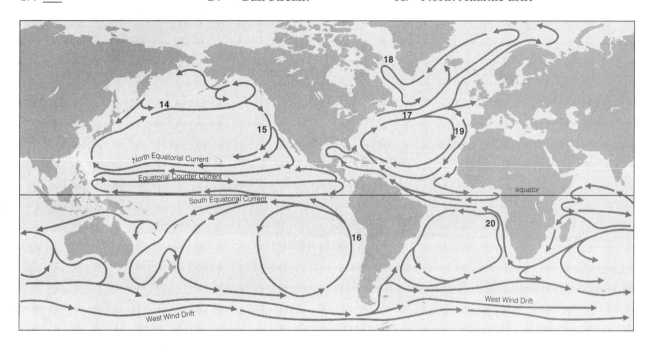

Fill-in-the-Blanks

Besides being the primary energy source for ecosystems, (21) _____ _____ influences their distri-bution on Earth's surface. (22) _____ energy derived from the sun warms the atmosphere and drives Earth's weather systems. Global air circulation patterns, ocean currents, and topographic features interact to produce regional variations in temperature and (23) _____, which influence the composition of (24) _____ and sediments, which, in turn, influence the growth and distribution of the primary (25) _____ and, through them, the distribution of (26) _____ and ecosystems. [(26) are large land regions with characteristic climax vegetation.] When descending dry air warms, it picks up moisture from the land and leaves it very dry so that (27) _____ are created; most of these occur at about 30° latitudes.

49-II. THE WORLD'S BIOMES (pp. 870–871)
SOILS OF MAJOR BIOMES (p. 872)
DESERTS (p. 873)
DRY SHRUBLANDS AND WOODLANDS (p. 874)
GRASSLANDS AND SAVANNAS (pp. 874–875)

Selected Italicized Words

ungulates

Boldfaced, Page-Referenced Terms

(870) biogeographic realms _____

_____

(870) biome _____

_____

(872) soil _____

_____

(873) deserts _____

_____

(873) desertification _____

_____

(874) dry shrublands _____

_____

(874) dry woodlands _____

_____

(874) shortgrass prairie _____

_____

(874) tallgrass prairie _____

_____

(874) savannas _____

_____

(874) monsoon grasslands _____

_____

Fill-in-the-Blanks

(1) _____ is formed as the products of living and dead organisms (humus) are mixed with weathered and eroded loose rock. As rainwater percolates down through the topsoil, minerals dissolve in it [that is, are (2) _____] and are carried along with particles to the (3) _____. Tree roots penetrate far below the soil surface, breaking up rocks as they go and absorbing minerals in solution.

Arid regions that support little primary production are known as (4) _____. Scrubby plants are dominant primary producers in (5) _____ _____, and cacti, yuccas, and other drought-resistant plants dominate (6) _____ _____. At one time, (7) _____ extended westward from the Mississippi River region to the Rocky Mountains, from Canada through Texas. (8) _____ _____ were converted into vast fields of corn and wheat, and (9) _____ _____ were converted into ranges for imported cattle. Overgrazing and ill-advised attempts to farm the region led to massive erosion, which crippled the primary productivity of much of this land.

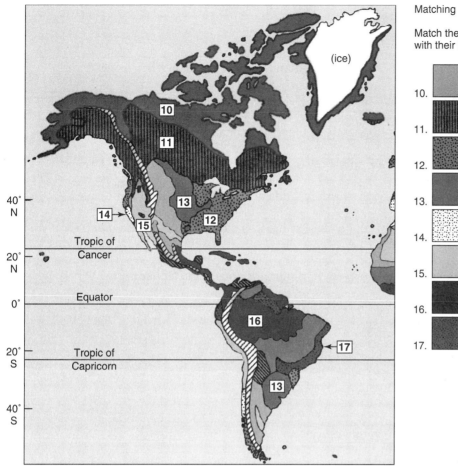

Match the indicated regions of the world map with their identifiers at right.

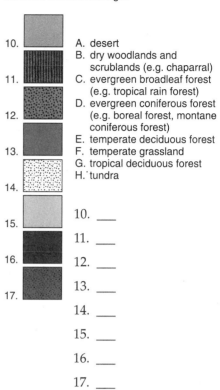

10.

11.

12.

13.

14.

15.

16.

17.

A. desert
B. dry woodlands and scrublands (e.g. chaparral)
C. evergreen broadleaf forest (e.g. tropical rain forest)
D. evergreen coniferous forest (e.g. boreal forest, montane coniferous forest)
E. temperate deciduous forest
F. temperate grassland
G. tropical deciduous forest
H. tundra

10. ___

11. ___

12. ___

13. ___

14. ___

15. ___

16. ___

17. ___

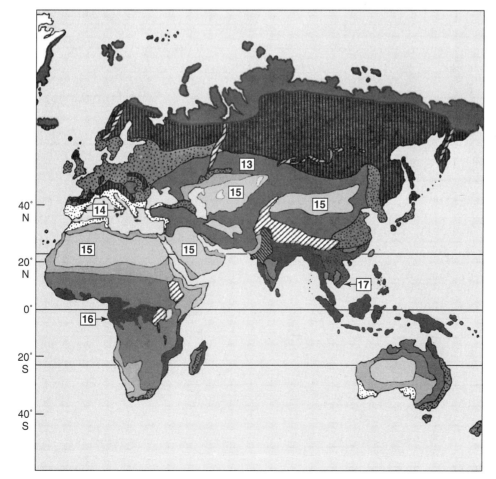

Choice

For questions 18–23, choose from the following:

a. deserts   b. dry shrublands   c. dry woodlands   d. grasslands   e. savannas

___ 18. Biome where the potential for evaporation greatly exceeds rainfall

___ 19. Steinbeck's *Grapes of Wrath* speaks eloquently of the disruption of this biome

___ 20. The plants are woody, multibranched, and a few meters tall; fynbos and chaparral

___ 21. Eucalyptus woodlands of southwestern Australia and oak woodlands of the Pacific states

___ 22. Home to deep-rooted evergreen shrubs, fleshy-stemmed, shallow-rooted cacti, saguaro, short prickly pear, and ocotillo

___ 23. Home of the baobab and acacia trees, African hornbills, baboons, and elephants

49-III. BROADLEAF FORESTS (pp. 876–877)
CONIFEROUS FORESTS (p. 878)
TUNDRA (p. 879)

Selected Italicized Words

evergreen broadleafs, deciduous broadleafs, evergreen conifers

Boldfaced, Page-Referenced Terms

(876) tropical rain forests _____

_____

(877) tropical deciduous forests _____

_____

(877) monsoon forests _____

_____

(877) temperate deciduous forests _____

_____

(878) boreal forests _____

_____

(878) montane coniferous forests _____

_____

(878) temperate rain forests _____

_____

(878) pine barrens _____

_____

(879) arctic tundra _____

_____

(879) permafrost _____

_____

(879) alpine tundra _____

_____

Matching

For the next nine questions, choose from these choices:

a. deciduous broadleaf forests   b. evergreen coniferous forests
c. evergreen broadleaf forests   d. tundra

___ 1. Nearly continuous sunlight in summer

___ 2. Near the equator, highly productive, rainfall regular and heavy

___ 3. Decomposition and mineral cycling are rapid in the hot, humid climate

___ 4. Maple and beech forests

___ 5. Boreal forests or taiga

___ 6. Spruce and balsam fir dominate this North American biome

___ 7. Biome of the temperate zone; cold winter temperatures; many trees drop all their leaves in winter

___ 8. Lies between the polar ice cap and belts of boreal forests in North America, Europe, and Asia

___ 9. Soils are highly weathered, humus-poor, and not good nutrient reservoirs

Fill-in-the-Blanks

Warm temperatures combined with uniform, abundant rainfall encourage the growth of (10) _____

_____ _____, which contain (11) _____ communities dominated by trees spreading their

(12) _____ far above the forest floor. This type of biome exists where the annual mean temperature is

about (13) _____°C and where heavy, regular rainfall coincides with (14) _____ of 80 percent or

more and produces more litter than any other forest biome; however, (15) _____ and mineral cycling

are rapid in the hot, humid climate, and its soils are not a good reservoir of nutrients. In (16) _____

_____ forests, most plants lose their leaves during part of the year as an adaptation to the dry seasons

that alternate with the wet monsoons. In (17) _____ _____ forests, most of the dominant trees

drop their leaves—not in response to dry seasons, but as protection against (18) _____ _____. At

lower elevations of the montane coniferous forest, (19) _____ and Douglas fir predominate.

(20) _____ is synonymous with boreal forest. Just beneath the surface of the (21) _____ is the

permafrost. The main limiting factor in the tundra is the low level of (22) _____ _____.

49-IV. REGARDING THE WATER PROVINCES (p. 880)
LAKE ECOSYSTEMS (pp. 880–881)
STREAM ECOSYSTEMS (p. 882)
COASTAL ECOSYSTEMS (pp. 882–883)

Selected Italicized Words

phytoplankton, zooplankton, oligotrophic, eutrophic, rocky shore, sandy shore, muddy shore, Spartina

Boldfaced, Page-Referenced Terms

(880) lake _____

_____

(880) plankton _____

_____

(880) spring overturn _____

_____

(881) fall overturn _____

_____

(881) eutrophication _____

_____

(882) stream _____

_____

(882) estuary _____

_____

(883) intertidal zone _____

_____

Fill-in-the-Blanks

Water is densest at 4°C; at this temperature, it sinks to the bottom of its basin, displacing the nutrient-rich bottom water upward and giving rise to spring and fall (1) _____. Portions of a lake where light penetrates to the bottom compose the (2) _____ zone, where rooted vegetation and decomposers are abundant. The (3) _____ zone includes areas of a lake *below* the depth where sufficient light penetration balances the respiration and photosynthesis of the primary producers; anaerobic bacterial decomposers are the principal organisms here. (4) _____ lakes are nutrient-poor. In bodies of water that are sufficiently deep, there is a depth at which the temperature of the water decreases rapidly with a small increase in depth; this region is known as the (5) _____. A region where freshwater mixes with saltwater is a(n) (6)_____.

Choice

For questions 7–18, choose from the following:

    a. stream ecosystem   b. estuary   c. intertidal zone of open coastline with no freshwater input
                      d. lake ecosystem   e. none of the preceding

___ 7. Riffles, pools, and runs.

___ 8. A partly enclosed coastal ecosystem where seawater mixes with nutrient-rich freshwater from rivers, streams, and runoff from land.

___ 9. Begin as freshwater springs or seeps.

___ 10. Chesapeake Bay and San Francisco Bay are broad, shallow examples; narrow and deep examples are found in Alaska, British Columbia, and Norway.

___ 11. Photosynthetic activity is restricted to its upper waters.

___ 12. Average flow volume and temperature depend on rainfall, snowmelt, geography, altitude, and even shade cast by plants.

___ 13. They grow and merge as they flow downslope, then often combine to form a river.

___ 14. The most productive aquatic ecosystems on earth.

___ 15. Upper, middle, and lower littoral zones of rocky shores.

___ 16. Chemosynthetic bacteria are the primary producers of hydrothermal vents in the abyssal zone.

___ 17. Home to a great diversity of invertebrates and a rest stop for migratory birds; subject to much human pollution.

___ 18. Alternately submerged and exposed by tides; land animals feed here at low tides, and high tides bring predatory fishes.

Complete the Table

19. Complete the following table, which describes the three zones of a lake.

Lake Zone	Description
Littoral	a.
Limnetic	b.
Profundal	c.

Dichotomous Choice

Circle one of two possible answers given between parentheses in each statement.

20. Water near the freezing point is the (least/most) dense and accumulates beneath the ice.
21. Water at 4°C is densest; it accumulates in deeper layers, which are a bit (warmer/cooler) than the surface layer in midwinter.
22. Daylength increases, air warms, lake ice melts, and surface water slowly warms to 4°C so that temperatures become uniform throughout; winds blowing across a lake cause a (spring/fall) overturn.
23. In (spring/fall) overturns, strong vertical movements carry dissolved oxygen from the lake's warming surface layer to its depths, and nutrients released by decomposition are brought from the bottom sediments to the surface layer.
24. By midsummer, the surface layer is well above 4°C and the lake may have a thermocline, a(n) (upper/middle) layer that changes abruptly in temperature and prevents vertical mixing.
25. At midsummer, (warmer/cooler) and less dense surface water floats on the thermocline.
26. In summer, decomposers may deplete the (upper/lower) water of dissolved oxygen.
27. The lake's upper layer cools, becomes denser, and sinks, and the thermocline vanishes. This describes the (spring/fall) overturn when the dissolved oxygen again moves down and nutrients move up.
28. Primary productivity of a lake increases following spring overturn, but as the season progresses, the thermocline prevents mixing of nutrients accumulating in (surface/deeper) waters and primary productivity declines.
29. By late summer, nutrient shortages limit photosynthesis. Following fall overturn, cycled nutrients drive a (short/long) burst of primary productivity.

Complete the Table

30. Complete the following table, which summarizes the trophic nature of lakes.

Trophic Nature	Description
Oligotrophic lake	a.
Eutrophic lake	b.

49-V. THE OPEN OCEAN (pp. 884–885)
CORAL REEFS AND BANKS (pp. 886–887)
UPWELLING (pp. 888–889)

Selected Italicized Words

El Niño Southern Oscillation (ENSO), fringing reefs, barrier reefs, atolls

Boldfaced, Page-Referenced Terms

(884) benthic province _____

_____

(884) pelagic province _____

_____

(884) hydrothermal vents _____

_____

(886) coral reefs _____

_____

(886) coral banks _____

_____

(888) upwelling _____

_____

(889) El Niño _____

_____

Labeling

Label each numbered part of the accompanying illustration (see text, Fig. 49.27).

1. _____ zone

2. _____ zone

3. _____ zone

4. _____ province

5. _____ _____

6. _____ zone

7. _____ zone

8. _____ zone

9. _____ province

10. _____ water

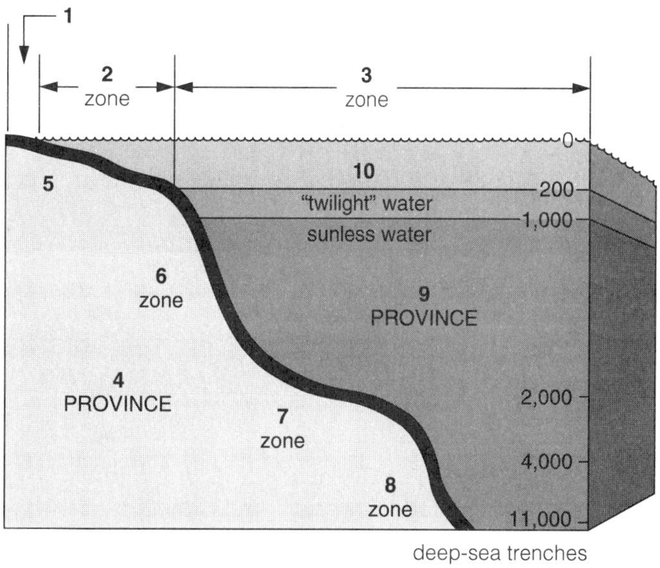

deep-sea trenches

Fill-in-the-Blanks

Organisms that depend on currents and drifts to distribute them are collectively referred to as
(11) _____. Most marine ecosystems fall within the shallow (12) _____ _____ (between the
high and low tide marks) and the (13) _____ _____ (between 2,000 and 4,000 meters beneath the
ocean's surface). The (14) _____ _____ is a geothermal ecosystem 2,500 meters beneath the
ocean's surface, where sulfur-oxidizing bacteria are the primary producers. (15) _____ is a marsh
grass that is the dominant primary producer in New England salt marshes.

(16) _____ is an upward movement of deep, nutrient-rich ocean water along the margins of continents. This occurs when (17) _____ force surface waters to move away from a coastline and deep
water moves in vertically to replace it. Every two to seven years, surface waters of the western equatorial
Pacific move eastward. This is a massive displacement of (18) _____ water that influences the prevailing winds to accelerate eastward flow. This flow displaces the cooler waters of the Humboldt Current and
prevents upwelling. Meteorologists call the phenomenon ENSO, or the (19) _____ _____
Southern Oscillation. ENSO has a catastrophic effect on (20) _____, particularly on anchoveta-eating
birds, and the anchoveta industry.

Self-Quiz

___ 1. The distribution of different types of ecosystems is influenced by _____.

　　a. air currents
　　b. variation in the amount of solar radiation reaching the Earth through the year
　　c. ocean currents
　　d. all of the above

___ 2. In a(n)_____, water draining from the land mixes with seawater carried in on tides.

　　a. pelagic province
　　b. rift zone
　　c. upwelling
　　d. estuary

___ 3. A biome with grasses as primary producers and scattered trees adapted to prolonged dry spells is known as a _____.

　　a. warm desert
　　b. savanna
　　c. tundra
　　d. taiga

___ 4. Located at latitudes of about 30° north and south, limited vegetation, and rapid surface cooling at night describe a _____ biome.

　　a. shrubland
　　b. savanna
　　c. taiga
　　d. desert

___ 5. In tropical rain forests, _____.

　　a. competition for available sunlight is intense
　　b. diversity is limited because the tall forest canopy shuts out most of the incoming light
　　c. conditions are extremely favorable for growing luxuriant crops
　　d. all of the above

___ 6. In a lake, the open sunlit water with its suspended phytoplankton is referred to as its _____ zone.

　　a. epipelagic
　　b. limnetic
　　c. littoral
　　d. profundal

___ 7. The lake's upper layer cools, the thermocline vanishes, lake water mixes vertically, and once again dissolved oxygen moves down and nutrients move up. This describes the _____.

　　a. spring overturn
　　b. summer overturn
　　c. fall overturn
　　d. winter overturn

___ 8. The _____ is a permanently frozen, water-impermeable layer just beneath the surface of the _____ biome.

　　a. permafrost; alpine tundra
　　b. hydrosphere; alpine tundra
　　c. permafrost; arctic tundra
　　d. taiga; arctic tundra

___ 9. _____ refers to a season of heavy rain that corresponds to a shift in prevailing winds over the Indian Ocean.

　　a. Geothermal ecosystem
　　b. Upwelling
　　c. Taiga
　　d. Monsoon

___ 10. All of the water above the continental shelves is in the _____.

　　a. neritic zone of the benthic province
　　b. oceanic zone of the pelagic province
　　c. neritic zone of the pelagic province
　　d. oceanic zone of the benthic province

___ 11. The Galápagos Rift, a geothermal ecosystem 2,500 meters beneath the ocean's surface, has _____ as its primary producers.

　　a. blue-green algae
　　b. protistans
　　c. nitrogen-fixing organisms
　　d. chemosynthetic organisms

___ 12. Monocrops of fast-growing softwood trees _____.

　　a. are replacing climax forests in parts of the national forest system
　　b. are less vulnerable to insect predators than constituents of a mixed deciduous climax forest
　　c. use fewer nutrients than climax forests
　　d. are desirable for their nitrogen-fixing ability

Chapter Objectives/Review Questions

This section lists general and detailed chapter objectives that can be used as review questions. You can make maximum use of these items by writing answers on a separate sheet of paper. To check for accuracy, compare your answers with information given in the chapter or glossary.

Page   *Objectives/Questions*

(865)   1. The _____ is the entire realm in which organisms live; the portion containing oceans, polar ice caps, and all other forms of water constitutes the _____; the envelope of gases and airborne particles is the _____.

(865)   2. _____ refers to prevailing weather conditions, such as temperature, humidity, wind speed, cloud cover, and rainfall.

(866)   3. State the reason that most forms of life depend on the ozone layer.

(866)   4. _____ energy drives Earth's weather systems.

(866–867) 5. Be able to describe the causes of global air circulation patterns.

(867)   6. Describe how the tilt of the Earth's axis affects annual variation in the amount of incoming solar radiation.

(868)   7. Atmospheric and oceanic _____ patterns influence the distribution of different types of ecosystems.

(869)   8. Mountains, valleys, and other land formations influence _____ climates.

(869)   9. Describe the cause of the rain shadow effect.

(870)   10. Broadly, there are six distinct land realms, the _____ realms that were named by W. Sclater and Alfred Wallace.

(870)   11. Each major type of ecosystem is a(n) _____.

(872)   12. _____ is a mixture of rock, mineral ions, and organic matter in some state of physical and chemical breakdown.

(873–879) 13. Be able to list the major biomes and briefly characterize them in terms of climate, topography, and organisms.

(880)   14. Define *plankton*, *phytoplankton*, and *zooplankton*.

(880–881) 15. Describe the spring and fall overturn in a lake in terms of causal conditions and physical outcomes.

(881)   16. A _____ is a body of freshwater with littoral, limnetic, and profundal zones.

(881)   17. _____ refers to nutrient enrichment of a lake or some other body of water.

(881)   18. _____ lakes are often deep, poor in nutrients, and low in primary productivity; _____ lakes are often shallow, rich in nutrients, and high in primary productivity.

(882)   19. Describe a stream ecosystem.

(882–884) 20. Be able to fully describe estuary, intertidal zone, and benthic and pelagic provinces of the ocean.

(884)   21. In the abyssal zone, remarkable communities thrive at _____ vents.

(888)   22. State the significance of ocean upwelling.

(888–889) 23. Describe conditions of ENSO occurrence and its significance to the ocean.

Integrating and Applying Key Concepts

One species, *Homo sapiens*, uses about 40 percent of all of Earth's productivity, and its representatives have invaded every biome, either by living there or by dumping waste products there. Many of Earth's residents are being denied the minimal resources they need to survive, while human populations continue to increase exponentially. Can you suggest a better way of keeping Earth's biomes healthy while providing at least the minimal needs of all Earth's residents (not just humans)? If so, outline the requirements of such a system and devise a way in which it could be established.

Answers

Interactive Exercises

49-I. AIR CIRCULATION PATTERNS AND REGIONAL CLIMATES (pp. 864–867)
  THE OCEAN'S EFFECT ON REGIONAL CLIMATES (p. 868)
  EFFECTS OF TOPOGRAPHY ON REGIONAL CLIMATES (p. 869)
1. D; 2. H; 3. C (J); 4. A; 5. E; 6. B; 7. J (C); 8. L; 9. M; 10. I; 11. F; 12. G; 13. K; 14. F; 15. B; 16. E; 17. D; 18. G; 19. C; 20. A; 21. solar radiation; 22. Heat; 23. rainfall (climate); 24. soils; 25. producers; 26. biomes; 27. deserts (rain shadows).

49-II. THE WORLD'S BIOMES (pp. 870–871)
  SOILS OF MAJOR BIOMES (p. 872)
  DESERTS (p. 873)
  DRY SHRUBLANDS AND WOODLANDS (p. 874)
  GRASSLANDS AND SAVANNAS (pp. 874–875)
1. Soil; 2. leached; 3. subsoil; 4. deserts; 5. dry shrublands (cold deserts); 6. warm deserts; 7. prairies (grasslands); 8. Tallgrass prairies; 9. shortgrass prairies; 10. H; 11. D; 12. E; 13. F; 14. B; 15. A; 16. C; 17. G; 18. a; 19. d; 20. b; 21. c; 22. a; 23. e.

49-III. BROADLEAF FORESTS (pp. 876–877)
  CONIFEROUS FORESTS (p. 878)
  TUNDRA (p. 879)
1. d; 2. c; 3. c; 4. a; 5. b; 6. b; 7. a; 8. d; 9. c (d); 10. tropical rain forests; 11. vertical (stratified); 12. canopies; 13. 25; 14. humidity; 15. decomposition; 16. tropical deciduous; 17. temperate deciduous; 18. freezing weather; 19. pine; 20. Taiga; 21. tundra; 22. soil nutrients (solar radiation).

49-IV. REGARDING THE WATER PROVINCES (p. 880)
  LAKE ECOSYSTEMS (pp. 880–881)
  STREAM ECOSYSTEMS (p. 882)
  COASTAL ECOSYSTEMS (pp. 882–883)
1. overturns; 2. littoral; 3. profundal; 4. Oligotrophic; 5. thermocline; 6. estuary; 7. a; 8. b; 9. a; 10. b; 11. d

(b); 12. a; 13. a; 14. b; 15. c; 16. e; 17. b; 18. c; 19. a. Shallow, usually well-lit zone extending all around the shore to the depth at which rooted aquatic plants stop growing; diversity of organisms is greatest in the littoral; b. Open, sunlit water beyond the littoral, down to a depth where photosynthesis is insignificant; has plankton; c. All the open water below the depth at which wavelengths suitable for photosynthesis can penetrate; detritus sinks through the profundal to the bacterial decomposers of bottom sediments; decomposition releases nutrients into the water; 20. least; 21. warmer; 22. spring; 23. spring; 24. middle; 25. warmer; 26. lower; 27. fall; 28. deeper; 29. short; 30. a. Often deep, poor in nutrients, and low in primary productivity; b. Often shallow, rich in nutrients, and high in primary productivity.

49-V. THE OPEN OCEAN (pp. 884–885)
  CORAL REEFS AND BANKS (pp. 886–887)
  UPWELLING (pp. 888–889)
1. intertidal; 2. neritic; 3. oceanic; 4. benthic; 5. continental shelf; 6. bathyal; 7. abyssal; 8. hadal; 9. pelagic; 10. sunlit; 11. plankton; 12. intertidal zone; 13. abyssal zone (bathypelagic); 14. hydrothermal vent; 15. *Spartina*; 16. Upwelling; 17. winds; 18. warm; 19. El Niño; 20. productivity.

Self-Quiz

1. d; 2. d; 3. b; 4. d; 5. a; 6. b; 7. c; 8. c; 9. d; 10. c; 11. d; 12. a.

50

HUMAN IMPACT ON THE BIOSPHERE

Interactive Exercises

50-I. HUMAN POPULATION GROWTH AND THE ENVIRONMENT (pp. 892–894)
LOCAL EFFECTS OF AIR POLLUTION (p. 895)
GLOBAL EFFECTS OF AIR POLLUTION (pp. 896–897)

Boldfaced, Page-Referenced Terms

(895) pollutants _____

_____

(895) thermal inversion _____

_____

(895) industrial smog _____

_____

(895) photochemical smog _____

_____

(895) PANs (peroxyacyl nitrates) _____

_____

(896) dry acid deposition _____

_____

(896) acid rain _____

_____

(897) ozone hole _____

_____

(897) chlorofluorocarbons (CFCs) _____

_____

Fill-in-the-Blanks

(1) _____ _____ _____ attack marble, metals, mortar, rubber, and plastic; they also form droplets of (2) _____ _____ that create holes in nylon stockings among other things. When a layer of dense, cool air gets trapped beneath a layer of warm air, the situation is known as a(n) (3) _____ _____. When (4) _____ and nitrogen dioxide are exposed to sunlight, they are converted into poisonous substances collectively called (5) _____ smog. (6) _____ _____ kills fish in northeastern U.S. lakes and streams, while greenhouse gases such as (7) _____ destroy the ozone layer and raise the (8) _____ _____ _____ on Earth, melting the ice caps and flooding coastal lands.

The human population has undergone rapid exponential growth since the mid- (9) _____ century. Ecosystems have had no prior evolutionary experience with (10) _____ and therefore have no mechanisms for dealing with them. The precipitation in much of eastern North America is (11) (choose one) ❏ 30 ❏ 50 ❏ 75 ❏ 100 times more acidic than it was several decades ago, and croplands and (12) _____ are suffering. The (13) _____ layer in the lower stratosphere has been thinning, due mostly to odorless, invisible (14) _____.

Short Answer

15. Briefly list the major environmental effects of human population growth. _____

_____

_____

_____

Choice

For questions 16–37, choose from the following aspects of atmospheric pollution:

a. thermal inversion   b. industrial smog   c. photochemical smog
d. acid deposition   e. chlorofluorocarbons   f. ozone layer

___ 16. Develops as a brown, smelly haze over large cities.

___ 17. Includes the "dry" and "wet" types.

___ 18. Contributes to ozone reduction more than any other factor.

___ 19. Where winters are cold and wet, this develops as a gray haze over industrialized cities that burn coal and other fossil fuels.

___ 20. Weather conditions trap a layer of cool, dense air under a layer of warm air.

___ 21. The cause of London's 1952 air pollution disaster, in which 4,000 died.

___ 22. Each year, from September through mid-October, it thins down by as much as half above Antarctica.

___ 23. By some estimates, nearly all of this released between 1955 and 1990 is still making its way up to the stratosphere.

___ 24. Intensifies a phenomenon called smog.

___ 25. Today most of this forms in cities of China, India, and other developing countries, as well as in Hungary, Poland, and other countries of Eastern Europe.

___ 26. Contains airborne pollutants, including dust, smoke, soot, ashes, asbestos, oil, bits of lead and other heavy metals, and sulfur oxides.

___ 27. Depending on soils and vegetation, some regions are more sensitive than others to this.

___ 28. Has contributed to some of the worst local air pollution disasters.

___ 29. Chemically attack marble buildings, metals, mortar, rubber, plastic, and even nylon stockings.

___ 30. Its reduction allows more ultraviolet radiation to reach the Earth's surface.

___ 31. Reaches harmful levels where the surrounding land forms a natural basin, as it does around Los Angeles and Mexico City.

___ 32. Tall smokestacks were added to power plants and smelters in an unsuccessful attempt to solve this problem.

___ 33. Those already in the air will be there for over a century before they are neutralized by natural processes.

___ 34. A dramatic rise in skin cancers, eye cataracts, immune system weakening, and harm to photosynthesizers is related to its reduction.

___ 35. Widely used as propellants in aerosol spray cans, refrigerator coolants, air conditioners, industrial solvents, and plastic foams; enter the atmosphere slowly and resist breakdown.

___ 36. The main culprit is nitric oxide, produced mainly by cars and other vehicles with internal combustion engines.

___ 37. Oxides of sulfur and nitrogen dissolve in water to form weak solutions of sulfuric acid and nitric acid that may fall with rain or snow.

50-II. WATER SCARCITY AND WATER POLLUTION (pp. 898–899)
CHANGES IN THE LAND (p. 900)
DESERTIFICATION (p. 901)
DEFORESTATION (pp. 902–903)

Selected Italicized Words

primary, secondary, tertiary treatment; subsistence agriculture, animal-assisted agriculture, mechanized agriculture

Boldfaced, Page-Referenced Terms

(898) salination _____

_____

(898) water table _____

_____

(898) wastewater treatment _____

_____

(900) green revolution _____

_____

(901) desertification _____

_____

(902) deforestation _____

_____

(902) shifting cultivation _____

_____

Fill-in-the-Blanks

In (1) _____ wastewater treatment, mechanical screens and sedimentation tanks are used to force coarse suspended solids out of the water to become a sludge. Chlorination does not kill all of the pathogens that remain, and some of the resulting chlorinated organic compounds are (2) _____. In (3) _____ treatment, microbial populations break down organic matter after primary treatment. (4) _____ treatment involves advanced methods of precipitation of suspended solids and phosphate compounds.

(5) _____ on steep slopes leads to soil erosion and watershed disruption. Farmers take so much water from the (6) _____ _____ that the overdraft nearly equals the annual flow of the Colorado River. (7) _____ accounts for almost two-thirds of the human population's annual use of freshwater; (8) _____ may not ever be cost-effective enough to supply the needs of agriculture. Mechanized agriculture is based on massive inputs of fertilization, pesticides, and ample irrigation to sustain (9) _____-_____ crops. (10) _____ _____ energy instead of animal energy is used to drive farm machines, and crop yields are (11) (choose one) ❏ 4 ❏ 8 ❏ 20 times as high as in green revolution countries, but the modern practices use up (12) (choose one) ❏ 5 ❏ 25 ❏ 50 ❏ 100 times more energy and mineral resources than does subsistence agriculture. Today, (13) _____ on marginal lands is the main cause of large-scale desertification.

There is a tremendous amount of water in the world, but most is too (14) _____ for human consumption. Today, we produce about one-third of our food supply on (15) _____ land; this often changes the land's suitability for (16) _____. Irrigation water is often loaded with (17) _____ salts. In regions with poorly draining soils, evaporation may cause (18) _____; such soils can stunt growth, decrease yields, and in time kill crop plants. Improperly drained irrigated lands can become (19) _____. Water accumulating underground can gradually raise the water (20) _____; when this is close to the ground's surface, soil becomes (21) _____ with saline water that can damage plant roots. Salinity and waterlogging can be corrected at great cost. (22) _____ is often the principal use of groundwater.

Complete the Table

23. Complete the following table, which summarizes three levels of treatment methods for maintaining the water quality of polluted wastewater.

Treatment	Description
Primary treatment	a.
Secondary treatment	b.
Tertiary treatment	c.

Matching

Choose the most appropriate answer to match with each term.

24. ___ throwaway mentality
25. ___ green revolution
26. ___ shifting cultivation
27. ___ animal-assisted agriculture
28. ___ deforestation
29. ___ recycling
30. ___ forested watersheds
31. ___ subsistence agriculture
32. ___ desertification
33. ___ mechanized agriculture

A. runs on energy inputs from sunlight and human labor
B. an affordable, technologically feasible alternative to "throw-away technology"
C. act like giant sponges that absorb, hold, and gradually release water
D. conversion of large tracts of grasslands, rain-fed cropland, or irrigated cropland to a more desertlike state with a 10 percent or more productivity decrease
E. runs on energy inputs from oxen and other draft animals
F. an attitude prevailing in the United States and other developed countries that greatly adds to solid waste accumulation
G. requires massive inputs of fertilizers, pesticides, fossil fuel energy, and ample irrigation to sustain high-yield crops
H. research directed toward improving crop plants for higher yields and exporting modern agricultural practices and equipment to developing countries
I. trees are cut and burned, then ashes tilled into the soil; crops are grown for one to several seasons on quickly leached soils
J. removal of all trees from large land tracts; leads to loss of fragile soils and disrupts watersheds

50-III. A QUESTION OF ENERGY INPUTS (pp. 904–905)
ALTERNATIVE SOURCES OF ENERGY (pp. 906–907)

Selected Italicized Words

net energy

Boldfaced, Page-Referenced Terms

(904) fossil fuels _____

_____

(904) meltdown _____

_____

(906) breeder reactors _____

_____

(906) fusion power _____

_____

(906) solar-hydrogen energy _____

_____

Fill-in-the-Blanks

The Chernobyl power station experienced a(n) (1) _____, and (2) _____ was released into the atmosphere as the plant's containment structures were breached. Nuclear-powered electricity-generating plants produce (3) _____ (number) percent of world energy consumption. Only (4) _____ has good long-term, intermediate, and short-term availability as an energy option for the United States. Unlike conventional nuclear-fission reactors, (5) _____ reactors could potentially explode like atomic bombs. (6) _____ _____ is the energy left over after subtracting the energy used to locate, extract, transport, store, and deliver energy to consumers. Energy supplies in the form of (7) _____ _____ are nonrenewable and dwindling, and their extraction and use come at high environmental cost. (8) _____ _____ normally does not pollute the environment as much as fossil fuels do, but the costs and risks associated with fuel (9) _____ and with storage of (10) _____ wastes are enormous.

Dichotomous Choice

Circle one of two possible answers given between parentheses in each statement.

11. Paralleling the (S-shaped/J-shaped) curve of human population growth is a steep rise in total and per capita energy consumption.
12. The increase in per capita energy consumption is due to (increased number of energy users and extravagant consumption and waste/energy used to locate, extract, transport, store, and deliver energy to consumers).
13. (Total energy/Net energy) is that left over after subtracting the energy used to locate, extract, transport, store, and deliver energy to consumers.
14. Fossil fuels are the carbon-containing remains of (plants/plants and animals) that lived hundreds of millions of years ago.
15. Even with strict conservation efforts, known petroleum and natural gas reserves may be used up during the (current/next) century.
16. The net energy (decreases/increases) as costs of extraction and transportation to and from remote areas increase.
17. World coal reserves can meet human energy needs for several centuries, but burning releases sulfur dioxides into the atmosphere and adds to the global problem of (photochemical smog/acid deposition).
18. By 1990 in the United States, it cost slightly (less/more) to generate electricity by nuclear energy than by using coal.
19. The danger in the use of radioactivity as an energy supply during normal operation is with potential (radioactivity escape/meltdown).

Sequence-Classify

Arrange the consumption of world resources in correct hierarchical order. Enter the letter of the energy source of highest consumption next to 20, the letter of the next highest next to 21, and so on. Enter an (n) in the parentheses following the letter of the resource if it is nonrenewable and an (r) if the resource is renewable.

20. ____ ( )   A. Hydropower
21. ____ ( )   B. Natural gas
22. ____ ( )   C. Oil
23. ____ ( )   D. Nuclear power
24. ____ ( )   E. Biomass
25. ____ ( )   F. Coal

Self-Quiz

___ 1. Which of the following processes is not generally considered a component of secondary wastewater treatment?

  a. Screens and settling tanks remove sludge.
  b. Microbial populations are used to break down organic matter.
  c. Removal of all nitrogen, phosphorus, and toxic substances.
  d. Chlorine is often used to kill pathogens in the water.

___ 2. When fossil-fuel burning gives dust, smoke, soot, ashes, asbestos, oil, bits of lead, other heavy metals, and sulfur oxides, we have _____.

  a. photochemical smog
  b. industrial smog
  c. a thermal inversion
  d. both (a) and (c)

___ 3. _____ result(s) when nitrogen dioxide and hydrocarbons react in the presence of sunlight.

  a. Photochemical smog
  b. Industrial smog
  c. A thermal inversion
  d. Both (a) and (c)

___ 4. Between _____ of urban wastes consists of paper products.

  a. 10 and 20 percent
  b. 20 and 40 percent
  c. 50 and 65 percent
  d. 75 and 90 percent

___ 5. About _____ of the wastewater in the United States is not even receiving primary treatment.

  a. 20 percent
  b. 30 percent
  c. 50 percent
  d. 65 percent

___ 6. The wastes from a nuclear power reactor must be kept out of the environment for _____ years before they are safe.

  a. 25
  b. 250
  c. 1 million
  d. 250,000

___ 7. What proportion of the world's human population does *not* have enough pure water?

  a. one-fourth
  b. one-half
  c. three-fourths
  d. two-thirds

___ 8. For every million liters of water in the world, only about _____ liters are in a form that can be used for human consumption or agriculture.

  a. 6
  b. 60
  c. 600
  d. 6,000

___ 9. Each day in the United States, _____ metric tons of pollutants are discharged into the atmosphere.

  a. 1,000
  b. 100,000
  c. 700,000
  d. 5,000,000

___ 10. The most abundant fossil fuel in the United States is _____.

  a. carbon monoxide
  b. oil
  c. natural gas
  d. coal

___ 11. When weather conditions trap a layer of cool, dense air under a layer of warm air, _____ occurs.

  a. photochemical smog
  b. a thermal inversion
  c. industrial smog
  d. acid deposition

___ 12. Sulfur and nitrogen dioxides dissolve in atmospheric water to form a weak solution of sulfuric acid and nitric acid; this describes _____.

a. photochemical smog
b. industrial smog
c. ozone and PANs
d. acid rain

___ 13. Which of the following statements is false?

a. Ozone reduction allows more ultraviolet radiation to reach the Earth's surface.
b. CFCs enter the atmosphere and resist breakdown.
c. Salination of soils aids plant growth and increases yields.
d. CFCs already in the air will be there for over a century.

Chapter Objectives/Review Questions

This section lists general and detailed chapter objectives that can be used as review questions. You can make maximum use of these items by writing answers on a separate sheet of paper. To check for accuracy, compare your answers with information given in the chapter or glossary.

Page	*Objectives/Questions*
(895)	1. Identify the principal air pollutants, their sources, their effects, and the possible methods for controlling each pollutant.
(895–896)	2. Define *thermal inversion* and indicate its cause.
(895–896)	3. Distinguish photochemical smog from industrial smog.
(896–897)	4. Explain what acid rain does to an ecosystem. Contrast those effects with the action of CFCs.
(898)	5. Examine the effects that modern agriculture has wrought on desert and grassland ecosystems.
(899)	6. Define *primary*, *secondary*, and *tertiary wastewater treatment* and list some of the methods used in each of the three types of treatment.
(900)	7. _____ agriculture runs on energy inputs from sunlight and human labor; _____ _____ agriculture has energy inputs from oxen and other draft animals; and _____ agriculture requires massive inputs of _____, pesticides, and ample irrigation to sustain high-yield crops.
(902–903)	8. Explain the repercussions of deforestation that are evident in soils, water quality, and genetic diversity in general.
(904)	9. State when the U.S. natural gas reserves are expected to run out.
(904–907)	10. Describe how our use of fossil fuels, solar energy, and nuclear energy affects ecosystems.
(904–906)	11. Compare the functions of a conventional nuclear power plant reactor with those of a breeder reactor and assess the principal benefits and risks of each.
(904–905)	12. Explain what a meltdown is.
(894–907)	13. Understand the magnitude of pollution problems in the United States.
(906–907)	14. List five ways in which you could become personally involved in ensuring that institutions serve the public interest in a long-term, ecologically sound way.

Integrating and Applying Key Concepts

If you were Ruler of All People on Earth, how would you encourage people to depopulate the cities and adopt a way of life by which they could supply their own resources from the land and dispose of their own waste products safely on their own land?

Explain why some biologists believe that the endangered species list now includes all species.

Answers

Interactive Exercises

50-I. HUMAN POPULATION GROWTH AND THE ENVIRONMENT (pp. 892–894)
LOCAL EFFECTS OF AIR POLLUTION (p. 895)
GLOBAL EFFECTS OF AIR POLLUTION (pp. 896–897)

1. Wet acid depositions; 2. sulfuric acid (nitric acid); 3. thermal inversion; 4. hydrocarbons; 5. photochemical; 6. Acid rain; 7. CFCs (chlorofluorocarbons); 8. average annual temperature; 9. eighteenth; 10. pollutants; 11. 30; 12. forests; 13. ozone; 14. CFCs; 15. Human population growth became exponential and may be demanding more than the biosphere can sustain; as we take energy and resources, we give back monumental amounts of waste; we are destroying the stability of ecosystems, tainting the hydrosphere, and changing the composition of the atmosphere; millions in developing countries are already starving to death and hundreds of millions more suffer from malnutrition and inadequate health care; 16. c; 17. d; 18. e; 19. b; 20. a; 21. b; 22. f; 23. e; 24. a; 25. b; 26. b; 27. d; 28. a (b); 29. d; 30. f; 31. c; 32. d; 33. e; 34. f; 35. e; 36. c; 37. d.

50-II. WATER SCARCITY AND WATER POLLUTION (pp. 898–899)
CHANGES IN THE LAND (p. 900)
DESERTIFICATION (p. 901)
DEFORESTATION (pp. 902–903)

1. primary; 2. carcinogenic; 3. secondary; 4. Tertiary; 5. Deforestation; 6. Ogallala aquifer; 7. Agriculture; 8. desalination; 9. high-yield; 10. Fossil fuel; 11. 4; 12. 100; 13. overgrazing; 14. salty; 15. irrigated; 16. agriculture; 17. mineral; 18. salination; 19. waterlogged; 20. table; 21. saturated; 22. Irrigation; 23. a. Screens and settling tanks remove sludge, which is dried, burned, dumped in landfills, or treated further; chlorine is often used to kill pathogens in water, but does not kill them all; b. Microbial populations are used to break down organic matter after primary treatment but before chlorination; c. Removes nitrogen, phosphorus, and toxic substances, including heavy metals, pesticides, and industrial chemicals; it is largely experimental and expensive; 24. F; 25. H (G); 26. I; 27. E; 28. J; 29. B; 30. C; 31. A; 32. D; 33. G.

50-III. A QUESTION OF ENERGY INPUTS (pp. 904–905)
ALTERNATIVE SOURCES OF ENERGY (pp. 906–907)

1. meltdown; 2. radiation; 3. 5; 4. coal; 5. breeder; 6. Net energy; 7. fossil fuels; 8. Nuclear energy; 9. containment; 10. radioactive; 11. J-shaped; 12. increased number of energy users and extravagant consumption and waste; 13. Net energy; 14. plants; 15. next; 16. decreases; 17. acid deposition; 18. more; 19. meltdown; 20. C(n); 21. F(n); 22. B(n); 23. E(r); 24. A(r); 25. D(n).

Self-Quiz

1. c; 2. b; 3. a; 4. c; 5. b; 6. d; 7. c; 8. a; 9. c; 10. d; 11. b; 12. d; 13. c.

51

AN EVOLUTIONARY
VIEW OF BEHAVIOR

THE HERITABLE BASIS OF BEHAVIOR
    Genes and Behavior
    Hormones and Behavior

INSTINCTIVE AND LEARNED BEHAVIOR

THE ADAPTIVE VALUE OF BEHAVIOR
    Selection Theory and Feeding Behavior
    Selection Theory and Mating Behavior

Interactive Exercises

51-I. THE HERITABLE BASIS OF BEHAVIOR (pp. 910–913)

Selected Italicized Words

intermediate response, organizing hormones, activating hormones

Boldfaced, Page-Referenced Terms

(912) animal behavior _____

_____

(913) sound system _____

_____

Dichotomous Choice

Circle one of two possible answers given between parentheses in each statement.

1. For snake populations living along the California coast, the food of choice is (the banana slug/tadpoles and small fishes).
2. In Stevan Arnold's experiments, newborn garter snakes that were offspring of coastal parents usually (ate/ignored) a chunk of slug as the first meal.
3. Newborn garter snake offspring of (coastal/inland) parents ignored cotton swabs drenched in essence of slug and only rarely ate the slug meat.
4. The differences in the behavioral eating responses of coastal and inland snakes (were/were not) learned.
5. Hybrid garter snakes with coastal and inland parents exhibited a feeding response that indicated a(n) (environmental/genetic) basis for this behavior.
6. In white-throated sparrows and some other songbirds, singing behavior begins with melatonin, a hormone secreted by the (gonads/pineal gland).
7. In spring, melatonin secretion is suppressed and gonads grow; their secretion of estrogen and testosterone is increased to (indirectly/directly) influence singing behavior.

8. In very young male songbirds, a high (estrogen/testosterone) level triggers development of a masculinized brain.

9. Later, at the start of the breeding season, a male's enlarged gonads secrete increasing amounts of (estrogen/testosterone) that acts on cells in the sound system to prepare the bird to sing when properly stimulated.

10. (Hormones/Genes) influence the organization and activation of mechanisms required for particular forms of behavior.

51-II. INSTINCTIVE AND LEARNED BEHAVIOR (pp. 914–915)

Selected Italicized Words

classical conditioning, operant conditioning, habituation, spatial or latent learning, insight learning

Boldfaced, Page-Referenced Terms

(914) instinctive behavior _____

_____

(914) fixed action pattern _____

_____

(914) learned behavior _____

_____

(914) imprinting _____

_____

Complete the Table

1. Complete the following table to examine examples of instinctive and learned behavior.

Category	Example(s)
a. Instinctive behavior	
b. Fixed action pattern	
c. Learned behavior	

Matching

Choose the one most appropriate answer for each.

2. ___ imprinting

3. ___ classical conditioning

4. ___ operant conditioning

5. ___ habituation

6. ___ spatial or latent learning

7. ___ insight learning

A. Birds living in cities learn not to flee from humans or cars, which pose no threat to them.
B. Chimpanzees abruptly stack several boxes and use a stick to reach suspended bananas out of reach.
C. In response to a bell, dogs salivate even in the absence of food.
D. Bluejays storing information about dozens or hundreds of places where they have stashed food.
E. Adult roosters courting ducks instead of hens of their own species.
F. A toad learns to avoid stinging or bad-tasting insects after voluntary attempts to eat them.

51-III. THE ADAPTIVE VALUE OF BEHAVIOR (pp. 916–917)

Selected Italicized Words

Harpobittacus apicalis

Boldfaced, Page-Referenced Terms

(916) reproductive success _____

_____

(916) adaptive behavior _____

_____

(916) selfish behavior _____

_____

(916) altruistic behavior _____

_____

(916) natural selection _____

_____

(916) territory _____

_____

(916) sexual selection _____

_____

(917) lek _____

_____

Matching

Match the following organisms with the appropriate description relating to feeding or mating selection theory.

1. ___ bighorn sheep
2. ___ wandering ravens
3. ___ hangingflies
4. ___ sage grouse
5. ___ elk and bison

A. Females widely dispersed; during mating season, males congregate in a lek where a display ground is staked out; females select and mate with one male only and nest by themselves.

B. Males compete intensely for access to clusters of sexually receptive females; competition for ready-made harems favors combative males.

C. Males fight only to control areas where receptive females gather during winter rutting season; winners mate often with many females; rather than challenging stronger, larger males, gangs of losers invade guarded areas and overwhelm male winners; losers attempt to mate on the run.

D. Selfishly advertise carcasses; this gives them a chance at otherwise off-limits food (within a territory defended by powerful adults).

E. Male captures and kills an insect, then releases pheromones to attract females to the nuptial gift; females choose choicest offering and permit mating following eating.

Self-Quiz

___ 1. The observable responses that animals make to stimuli are what we call _____.

    a. imprinting
    b. instinct
    c. behavior
    d. learning

___ 2. Newly hatched goslings follow any large moving objects to which they are exposed shortly after hatching; this is an example of _____.

    a. homing behavior
    b. imprinting
    c. piloting
    d. migration

___ 3. In _____, components of the nervous system allow an animal to carry out complex, stereotyped responses to certain environmental cues, which are often simple.

    a. natural selection
    b. altruistic behavior
    c. sexual selection
    d. instinctive behavior

___ 4. Newly hatched, blind cuckoos contact an egg, maneuver it onto their back, and then push it out of the nest. This instinctive response is called _____.

    a. resource defense
    b. fixed action pattern
    c. learned behavior
    d. altruism

___ 5. A young toad flips its sticky-tipped tongue and captures a bumblebee that stings its tongue; in the future, the toad leaves bumblebees alone. This is _____.

    a. instinctive behavior
    b. a fixed action pattern

    c. altruistic
    d. learned behavior

___ 6. Brain regions that govern vocal organ muscles and that formed under the influence of estrogen secretion represent _____.

    a. an imprinting system
    b. habituation
    c. a sound system
    d. a mating adaptation

___ 7. Pavlov's dog experiments represent an example of _____.

    a. classical conditioning
    b. latent learning
    c. operant conditioning
    d. habituation

___ 8. Self-sacrificing behavior is also known as _____.

    a. selfish behavior
    b. adaptive behavior
    c. altruistic behavior
    d. mating behavior

___ 9. A _____ is an area that one or more individuals defend against competitors.

    a. lek
    b. territory
    c. communal display
    d. battleground

___ 10. _____ behavior is a behavior by which an individual protects or increases its own chance of producing offspring, regardless of the consequences for the group to which it belongs.

    a. Altruistic
    b. Adaptive
    c. Selfish
    d. Feeding

Chapter Objectives/Review Questions

This section lists general and detailed chapter objectives that can be used as review questions. You can make maximum use of these items by writing answers on a separate sheet of paper. To check for accuracy, compare your answers with information given in the chapter or glossary.

Page	Objectives/Questions
(912)	1. Animal _____ refers to the observable responses an animal makes to stimuli.
(912)	2. By influencing the development of the nervous system, _____ influence the observable, coordinated responses that animals make to stimuli.
(912)	3. What explains the fact that coastal and inland garter snakes of the same species have different food preferences?
(913)	4. _____ influence the organization and activation of mechanisms required for particular forms of behavior.
(913)	5. Describe the origin and formation of a sound system.
(914)	6. In _____ behavior, components of the nervous stem allow an animal to carry out complex, stereotyped responses to certain environmental cues, which are often simple.
(914)	7. Describe and cite an example of a fixed action pattern.
(914)	8. When animals incorporate and process information gained from specific experiences and then use the information to vary or change responses to stimuli, it is _____ behavior.
(914)	9. Genes and _____ contribute to mechanisms underlying both instinctive and learned behavior.
(914)	10. A rooster was exposed to a mallard duck during a critical period in the rooster's life and is attracted to the mallard; this time-dependent form of learning is called _____.
(915)	11. Summarize Pavlov's experiments with dogs and classical conditioning.
(915)	12. When a toad begins to avoid bad-tasting insects following voluntary attempts to eat them, this represents _____ behavior.
(915)	13. Give an example of the learned behavior known as habituation.
(915)	14. An example of _____ or latent learning is the storage of information about hundreds of places in which food is stashed by bluejays.
(915)	15. An abrupt solution of a problem by an animal (without trial and error attempts) is known as _____ learning.
(916)	16. What is meant by reproductive success?
(916)	17. _____ behavior is a behavior by which an individual protects or increases its own chance of producing offspring, regardless of the consequences for the group to which it belongs.
(916)	18. _____ behavior is self-sacrificing behavior.
(916)	19. _____ selection is the outcome of individuals of one sex competing for matings with individuals of the opposite sex.
(916)	20. _____ behavior promotes the propagation of an individual's genes and so tends to occur at increased frequency in successive generations.
(917)	21. During the breeding season, sage grouse congregate in a large territory, the _____.

Integrating and Applying Key Concepts

Explain whether or not you think humans have any critical periods for establishing the ability to learn certain kinds of knowledge. State whether or not you think humans undergo imprinting.

Answers

Interactive Exercises

51-I. THE HERITABLE BASIS OF BEHAVIOR
(pp. 912–913)

1. the banana slug; 2. ate; 3. inland; 4. were not; 5. genetic; 6. pineal gland; 7. directly; 8. estrogen; 9. testosterone; 10. Hormones.

51-II. INSTINCTIVE AND LEARNED BEHAVIOR
(pp. 914–915)

1. a. Cuckoo birds are social parasites in that adult females lay eggs in the nests of other bird species; young cuckoos instinctively eliminate the natural-born offspring and then receive the undivided attention of their unsuspecting foster parents; b. The reaction pattern whereby newly hatched cuckoos contact an egg, maneuver it onto their back, then push it out of the nest; this is an instinctive response, triggered by a well-defined, simple stimulus—once set in motion, it is performed in its entirety; c. Young toads instinctively capture edible insects with sticky tongues; if a bumblebee is captured and then stings the tongue, the toad learns to leave bumblebees alone; 2. E; 3. C; 4. F; 5. A; 6. D; 7. B.

51-III. THE ADAPTIVE VALUE OF BEHAVIOR
(pp. 916–917)

1. C; 2. D; 3. E; 4. A; 5. B.

Self-Quiz

1. c; 2. b; 3. d; 4. b; 5. d; 6. c; 7. a; 8. c; 9. b; 10. c.

52

ADAPTIVE VALUE
OF SOCIAL BEHAVIOR

SIGNALING MECHANISMS
    Functions of Communication Signals
    Types of Communication Signals

COSTS AND BENEFITS OF SOCIAL LIFE
    Disadvantages to Sociality
    Advantages of Sociality

SOCIAL LIFE AND SELF-SACRIFICE
    Self-Sacrificing Behavior and Dominance
      Hierarchies
    Costs and Benefits of Parenting

THE EVOLUTION OF ALTRUISM
    Focus on Science: About Those Self-Sacrificing
      Naked Mole-Rats

AN EVOLUTIONARY VIEW OF HUMAN SOCIAL
BEHAVIOR

Interactive Exercises

52-I. SIGNALING MECHANISMS (pp. 920–923)

Selected Italicized Words

cue, signaling pheromones, priming pheromones, threat display, courtship displays

Boldfaced, Page-Referenced Terms

(922) social behavior _____

_____

(922) communication signals _____

_____

(922) signalers _____

_____

(922) signal receivers _____

_____

(922) illegitimate receiver _____

_____

(922) illegitimate signalers _____

_____

(922) pheromones _____

_____

(923) visual signals _____

_____

(923) acoustical signals _____

_____

(923) tactile signals _____

_____

Choice

For questions 1–13, choose from the following; some answers may require two letters.

a. chemical signal   b. visual signal   c. acoustical signal
d. tactile signal   e. illegitimate receiver   f. illegitimate signaler

____ 1. Worker termites bang their heads to alert brown soldier termites.

____ 2. A handshake, hug, caress, or shove.

____ 3. The bioluminescent signals of male and female fireflies of the same species.

____ 4. A soldier termite detects ant scent that serves as a cue; the termite then kills the ant.

____ 5. Sex pheromones of hangingfly males.

____ 6. Signals of some predatory female fireflies seeking a meal of males belonging to other firefly species.

____ 7. Information-seeking bees stay in physical contact with a dancing bee who has returned to the bee-hive after foraging.

____ 8. Fringe-lipped bat as a receiver of the tungara frog's call.

____ 9. Soldier termite shooting thin jets of silvery goo from its nose; the goo emits volatile odors that attract more soldiers to battle danger.

____ 10. Assassin bugs hook dead termite bodies on their dorsal surfaces and acquire termite scent; this deception allows assassin bugs to hunt termite victims more easily.

____ 11. Distinctive song of the male white-throated sparrow.

____ 12. Exposing formidable canine teeth, male baboons "yawn" at each other when they compete for a receptive female; this is a threat display that may precede an attack.

____ 13. Courtship displays of lekking sage grouse and albatross.

52-II. COSTS AND BENEFITS OF SOCIAL LIFE (pp. 924–925)
SOCIAL LIFE AND SELF-SACRIFICE (pp. 926–927)

Selected Italicized Words

reproductive success of the individual, Parus major

Boldfaced, Page-Referenced Terms

(924) cost-benefit approach _____

_____

(925) selfish herd _____

_____

(926) self-sacrificing behavior _____

_____

(926) dominance hierarchy _____

_____

Choice

For questions 1–8, choose from the following:

      a. disadvantages to sociality   b. cooperative predator avoidance   c. the selfish herd
    d. dominance hierarchy or self-sacrificing behavior, or both   e. costs and benefits of parenting

___ 1. Disturbed clumps of Australian sawfly caterpillars collectively rear up, writhe about, and regurgitate partially digested eucalyptus leaves; the leaves contain chemical compounds toxic to most other animals, including predatory birds.

___ 2. Huge colonies of cliff swallows, prairie dogs, or penguins; an individual or its offspring is more likely to be weakened by parasites, which are more likely to be transmitted from host to host in large groups.

___ 3. Subordinate wolves and baboons move up the social ladder when dominant members slip down a rung or fall off.

___ 4. Male baboons relinquish safe sleeping sites, choice bits of food, even receptive females to others upon receiving a threat signal from another male.

___ 5. Members of a wolf pack show helpful behavior, as when they fend off predators or share food; only one male and one female wolf produce pups.

___ 6. Building of adjacent male bluegill nests on lake bottoms; the largest, most powerful males claim the central locations where eggs laid in the nests are less subject to attack by snails and largemouth bass.

___ 7. Breeding pairs of herring gulls in a nesting colony, if given opportunity, will cannibalize the eggs or young chicks of their neighbors in an instant.

___ 8. A pair of Caspian terns incubate their eggs, then shelter and feed the nestlings, defend them from predators, and accompany them when they begin to fly and forage; in this case, the benefit of immediate reproductive success may outweigh the cost of reduced reproductive success at some later time.

52-III. THE EVOLUTION OF ALTRUISM (pp. 928–930)

Selected Italicized Words

DNA fingerprinting

Boldfaced, Page-Referenced Terms

(928) altruistic behavior _____

_____

(928) theory of indirect selection _____

_____

Dichotomous Choice

Circle one of two possible answers given between parentheses in each statement.

1. By one theory, individuals (as with sterile termite soldiers or honeybee workers in insect societies) can (directly/indirectly) pass on their genes by helping relatives survive and reproduce.
2. When a sexually reproducing parent helps offspring, it (is/is not) helping exact genetic copies of itself.
3. According to Hamilton's theory of (direct/indirect) selection, caring for nondescendant relatives favors the genes associated with helpful behavior.
4. An uncle helps a niece or nephew survive long enough to reproduce; if the cost of his altruistic action is (more/less) than the benefit, genes the uncle has in common with his relatives will be propagated.
5. DNA fingerprints made for naked mole-rat colonies indicate that individuals from a given colony are very close relatives due to high inbreeding; such lineages have extremely (increased/reduced) genetic variability.
6. An altruistic naked mole-rat helps to perpetuate a very (low/high) proportion of the genes it carries.

52-IV. AN EVOLUTIONARY VIEW OF HUMAN SOCIAL BEHAVIOR (p. 931)

Selected Italicized Words

adaptive behavior, socially desirable behavior

Dichotomous Choice

Circle one of two possible answers given between parentheses in each statement.

1. Many people seem to believe that attempts to identify the adaptive value of a particular (animal/human) trait are attempts to define its moral or social advantage.
2. "Adaptive" refers to (a trait with moral value/a trait valuable in gene transmission).
3. In many species, adults that have lost their offspring will, if presented with a substitute, (adopt/reject) it.
4. John Alcock suggests that "husbands and wives who have lost an only child or who fail to produce children themselves should be especially prone to adopt (strangers/relatives)."
5. The human adoption process (can/cannot) be considered adaptive when indirect selection favors adults who direct parenting assistance to relatives.
6. Joan Silk showed that in some traditional societies, children (are not/are) adopted overwhelmingly by relatives.
7. In modern societies in which agencies and other means of adoption exist, adoption of relatives (is/is not) predominant.

8. It (is/is not) possible to test evolutionary hypotheses about the adaptive value of human behaviors.
9. *Adaptive* behavior and *socially desirable* behavior (are not/are) separate issues.
10. Strong parenting mechanisms evolved in the past, and it may be that their redirection toward a (relative/nonrelative) says more about human evolutionary history than it does about the transmission of one's genes.

Self-Quiz

___ 1. _____ provides an example of an illegitimate signaler.
   a. A soldier termite killing an ant on cue
   b. An assassin bug with acquired termite odor
   c. A termite pheromone alarm signal
   d. The "yawn" of a dominant male baboon

___ 2. The claiming of the more protected central locations of the bluegill colony by the largest, most powerful males suggests _____.
   a. cooperative predator avoidance
   b. the selfish herd
   c. a huge parent cost
   d. self-sacrificing behavior

___ 3. A chemical odor in the urine of male mice triggers and enhances estrus in female mice. The source of stimulus for this response is a _____.
   a. generic mouse pheromone
   b. signaling pheromone
   c. priming pheromone
   d. cue from male mice

___ 4. Female insects often attract mates by releasing sex pheromones. This is an example of a(n) _____ signal.
   a. chemical
   b. visual
   c. acoustical
   d. tactile

___ 5. Worker termites bang their heads specifically to attract soldiers; male birds sing to stake out territories, attract females, and discourage males. These are examples of _____ signals.
   a. chemical
   b. visual
   c. acoustical
   d. tactile

___ 6. An example of dominance hierarchy and self-sacrificing behavior is _____.
   a. cannibalistic behavior of a breeding pair of herring gulls in a huge nesting colony
   b. members of wolf packs helping others by sharing food or fending off predators even though they do not breed
   c. clumps of regurgitating Australian sawfly caterpillars
   d. a huge colony of prairie dogs being ravaged by a parasite

___ 7. When musk oxen form a "ring of horns" against predators, it is _____.
   a. a selfish herd
   b. cooperative predator avoidance
   c. self-sacrificing behavior
   d. dominance hierarchy

___ 8. Caring for nondescendant relatives favors the genes associated with helpful behavior and is classified as _____.
   a. dominance hierarchy
   b. indirect selection
   c. altruism
   d. both b and c

___ 9. "_____" means only that a given trait has proved beneficial in the transmission of an individual's genes.
   a. Dominance hierarchy
   b. Indirect selection
   c. Adaptive
   d. Altruism

___ 10. _____ also favors adults who direct parenting behavior toward relatives and so indirectly perpetuate their shared genes.
   a. Indirect selection
   b. Moral selection
   c. Redirected selection
   d. Perpetuated selection

Chapter Objectives/Review Questions

This section lists general and detailed chapter objectives that can be used as review questions. You can make maximum use of these items by writing answers on a separate sheet of paper. To check for accuracy, compare your answers with information given in the chapter or glossary.

Page   *Objectives/Questions*

(922)   1. At the heart of social behavior is the ability of animals to _____ with one another, using a complex array of signals to exchange information.

(922)   2. Examples of _____ signals are chemical, visual, acoustical, and tactile.

(922)   3. Define the roles of signalers and signal receivers.

(922)   4. When soldier termites kill ants on cue, the termites are said to be _____ receivers of a communication signal.

(922)   5. Assassin bugs covered with termite scent are able to use deception to hunt termite victims more easily and as such are acting as _____ signalers.

(922)   6. Distinguish between *signaling* pheromones and *priming* pheromones; cite an example of each.

(923)   7. A baboon threat display, lekking sage grouse, and the bioluminescent flashes of a male firefly are all examples of _____ signals.

(923)   8. The male white-throated sparrow's song and the "whine" and "chuck" of a male tungara frog are examples of _____ signals.

(923)   9. An example of a(n) _____ signal is the physical contact of bees in a hive during the dance of a returning foraging bee.

(923)   10. What criteria determine what constitutes a communication signal between individuals of the same species?

(923)   11. Natural selection tends to favor communication signals that promote _____ success.

(924)   12. Explain the "cost-benefit approach" that evolutionary biologists take to find answers to the questions about social life.

(924)   13. List disadvantages to sociality.

(924)   14. Studies of Australian sawfly caterpillars indicate _____ predator avoidance.

(925)   15. Define *the selfish herd*; cite an example.

(926)   16. List two examples of self-sacrifice in a dominance hierarchy.

(927)   17. List the costs and benefits of parenting in the example of adult Caspian terns.

(928)   18. With _____ behavior, an individual behaves in a self-sacrificing way that helps others but decreases its own chance of reproductive success.

(928)   19. Hamilton's theory of _____ selection states that caring for nondescendant relatives favors genes associated with helpful behavior.

(931)   20. It is possible to test evolutionary hypotheses about the _____ value of human behaviors; discuss human adoption practices in this context.

(931)   21. "_____" means only that a given trait has proved beneficial in the transmission of an individual's genes.

Integrating and Applying Key Concepts

Think about communication signals that humans use and list them. Do you believe a dominance hierarchy exists in human society? Think of examples.

Answers

Interactive Exercises

52-I. SIGNALING MECHANISMS (pp. 922–923)
1. c; 2. d; 3. b; 4. a, e; 5. a; 6. b, f; 7. d; 8. c, e; 9. a; 10. a, f; 11. c; 12. b; 13. b.

52-II. COSTS AND BENEFITS OF SOCIAL LIFE (pp. 924–925)
SOCIAL LIFE AND SELF-SACRIFICE (pp. 926–927)
1. b; 2. a; 3. d; 4. d; 5. d; 6. c; 7. a; 8. e.

52-III. THE EVOLUTION OF ALTRUISM (pp. 928–930)
1. indirectly; 2. is not; 3. indirect; 4. less; 5. reduced; 6. high.

52-IV. AN EVOLUTIONARY VIEW OF HUMAN SOCIAL BEHAVIOR (p. 931)
1. human; 2. a trait valuable in gene transmission; 3. adopt; 4. strangers; 5. can; 6. are; 7. is not; 8. is; 9. are; 10. nonrelative.

Self-Quiz
1. b; 2. b; 3. c; 4. a; 5. c; 6. b; 7. b; 8. d; 9. c; 10. a.